Java Methods for Financial Engineering

Philip Barker

Java Methods for Financial Engineering

Applications in Finance and Investment

Philip Barker, BSc (HONS), MBCS, MCMI
BWA Technologies LTD, Roslin
Midlothian, Scotland

British Library Cataloguing in Publication Data
A catalogue record for this book is available from the British Library

ISBN-13: 978-1-84996-932-1 e-ISBN-13: 978-1-84628-741-1

Printed on acid-free paper

9 8 7 6 5 4 3 2 1

Springer Science + Business Media
springer.com

To my wife Avril
Whose support, encouragement and patience
made this book possible

Contents

Preface

The early chapters of this book are based on many years of teaching the material to my final year honours students in computer science and accountancy at Heriot-Watt University, Edinburgh. My approach has been guided by their response. The latter material has benefited greatly from my postgraduate student's, many of whom contributed to addressing issues of practical, efficient implementation of derivative models. Although, in those days, the work was largely C++ based the principles (and problems) of object based deployment remain the same. In making use of the appropriate built-in Java data structures and the general development methodology I have relied heavily on my experience over the last decade in directing technical and operations teams deploying internet and intranet distributed financial tools for a wide range of financial organisations.

Many applications in finance and investment are readily solved using analytical methods. The use of analytical techniques such as the calculus cannot be used directly on a standard computer. The analytical methods require numerical approximation techniques to be applied, which allow a standard computer to be programmed. The resultant programs give an approximate solution (approximate to the solution which would be found by direct use of analytical techniques). The approximation methods provide solutions that are only ever partially correct (to a given degree of accuracy).

A number of applications in finance and investment require the use of methods which involve time-consuming and laborious iterative calculations. The direct application of analytical techniques would not be of any help, so there is little option but to use trial and error or 'best guess' techniques. In other situations many of the valuation methods used in financial engineering have no closed form solutions and require analytical methods; these need to be approximated for solution on a standard computing platform.

The issues mentioned above are tackled within this book by providing a series of fundamental or core classes which will allow the implementation of analytical techniques. The core classes also provide methods for the solution of problems involving the tedious repetition or best guess route. There are fundamental methods available for the provision of 'better' approximations. However there is a point at which the continuous adjustment to an approximation exhibits diminishing returns. The decision taken here is to include the most widely used and robust methods which are used as the basis of a large number of financial engineering tools.

Statistical methods are widely used in investment and finance applications. Many statistical methods rely heavily on analytical techniques to solve problems, thus for computer implementation of these statistical methods one needs to make use of numerical methods to approximate the analytical components. The java classes developed in this book will provide a series of statistical methods which allow the direct application of the statistical techniques. The statistical classes, in some cases will make use of underlying core classes to provide the needed numerical methods. Other statistical classes will not inherit any of the core classes but will themselves be the fundamental class.

Application classes are the end product of the building process. An application class implements a solution to a given problem in financial engineering or finance and investment. The application classes embody the techniques that are used throughout financial engineering practice. Those techniques will invariably use the underlying statistical classes (which in turn may use the methods of core classes), the core classes or a combination of both.The categorisation of classes into core(CoreMath), statistical(BaseStats) and application(FinApps) allows the independent development of a library system which can be added to over time, without affecting the operation of applications already built.

The methodology employed here is to make the core classes static, where the function is unchanging in the application or as abstract as possible, where the function is largely affected by the application context. The core classes are used (extended or implemented) by calling classes, which become increasingly more concrete as application classes. The core, statistical and application classes are organised as packages. CoreMath is the package containing all of the classes dealing with numerical algorithms, BaseStats contains the statistical classes and FinApps contains the application classes.

The chapters follow a largely linear progression from an investigation of fundamental concepts of finance and investment tools through to implementation of the techniques which underpin a wide range of the option products being used in Financial Engineering in *Chapter 1*. There is a brief discussion of number representation and accuracy which sets the scene for much of the termination criteria and levels of acceptable accuracy used in algorithm development. The first two chapters cover the implementation of financial tools and portfolio management techniques and introduce some Java data structures. *Chapters 3 and 4* develop the technical issues in Bond markets and provide Java implementations of Bond valuation methods. *Chapters 5 and 6* provide an introduction to the basis of option markets and aspects of practical techniques. *Chapters 7–9* give the theoretical basis for much of the work shown in later chapters. For those who are starting in financial engineering, the three chapters will provide the necessary background to understand the methods and limitations of the standard tools. For the experienced practitioner, these chapters will guide an understanding of the Java class implementations that follow. *Chapters 10–18* provide the models and implementation of a wide range of Financial Engineering methods. These chapters are accessible directly by the practitioner who wishes to implement a specific type of model or specific methods within a model.

The Java classes and methods are designed to be used as modular 'object-based' tools that can be used as-is to implement the many techniques covered. However it is expected that the imaginative practitioner will want to combine the many methods to develop their own products; particular to their unique application context. The class structures developed here will encourage this approach.

1 Introduction

1.1. Numerical Accuracy & Errors

Since we are largely dealing with numeric approximations or iterative convergence to a desired solution the discussion of accuracy and error are important.

In the decimal system irrational (e.g. $\sqrt{2}$) and transcendental (e.g. π) numbers cannot have a precise representation, most rational numbers are also not represented precisely, in decimal notation. Representing $\frac{1}{3}$ as a decimal can be approximated by 0.3 or 0.333333333 or some other arbitrarily large representation. Providing a decimal value for $\frac{1}{3}$, means representing the division as a floating point number.

In Java floating point numbers can be stored up to 15 digits in length (as type double with 64 bits). The Java BigDecimal is capable of storing an arbitrarily large decimal number with no loss of accuracy. However the transition from rational to binary representation provides the opportunity for potential loss of accuracy. Floating point numbers are represented in the form $M \times R^e$, where the Mantissa is the integer part and Radix is the base of a particular computer's numbering system, this is usually 2, but can be 10 (in calculator processors) or 16. The exponent e can be up to 38 in type float and 308 in type double. The Mantissa provides us with the available precision of a number and the exponent provides the range.

Since representing numbers in floating point arithmetic can have varying results dependent on the underlying machine architecture and the data type (single or double precision floating point) it is important to know the target machine limitations and also use the appropriate number representations in the executing code.

One of the more common errors encountered with floating point representation is rounding error. This results from having more digits in a number than can be accommodated by the system. As an example consider a computer with a particularly small representation of real numbers. In this machine we can store four integers in the Mantissa and have a single exponent.

This simple machine has a largest value of 0.9999E9; the next lowest value is 0.9998E9. The value in decimal of 0.9999E9 is 999900000; the value of 0.9998E9 is 999800000. If we do the subtraction $999900000 - 999800000 = 100000$, we see that there is no way of representing 100,000 intermediate values. So, 999855000 is represented as 999900000, as is 999895000 and so on.

Rounding errors are machine number related and are an artefact of using fixed lengths (machine word lengths) of bits to represent an infinite variety of numbers. Because rounding errors are related to machine architecture, it is useful to have some knowledge of the target platform.

One of the benefits in programming with Java is that the code is portable in the sense that it will run on any platform with a JVM. Unfortunately code written on one machine architecture with a different number representation to the target machine is not guaranteed. Floating point calculations that have a large dependency on accuracy should be configurable at run time with knowledge of the runtime architecture. We will return to the issue of accuracy and error at points in following chapters as individual algorithms introduce their own particular representational characteristics.

1.2. Core Math's Classes

All of the Core classes are contained within the package CoreMath. This package covers functions, interpolation & extrapolation, roots of functions, series, linear algebra, Wiener, Brownian and Ito processes. The Java code for each of these classes is given in Appendix 1.

Core classes are designed as static or abstract classes, which in many cases require extending in other implementing classes (usually application classes). Some of the core classes are designed as standard Java classes, where it can be reasonably expected that the interface will be modified in the application context. The examples used throughout the text are working and tested 'off the shelf' Java code but are not developed as user ready applications. The intention is to show and explain Java methods that will run and perform a given function without adding the overhead of error trapping and exception handling.

We will often make use of core classes that provide roots (or zeros) of functions. The general methodology adopted for the book is best explained with the aid of an example that deals with providing the roots of a function. Our first example will be the development of a class that makes use of bracketing and bisection techniques to converge on a root with a given precision. The class is called **IntervalBisection** and is in the package CoreMath.

1.2.1. Root Finding - Interval Bisection

Figure 1.1 shows the Interval Bisection technique being applied to the function $y = 2 - e^x$ Interval bisection solves for a root of the equation by starting with two outlying values (the end points X_0 and X_1) that bracket the root. This is shown by arrow (1). The assumption is that the initial range of these end points contains the root. By evaluating the function at these points $f(x_0)$, $f(x_1)$ and checking that the function changes sign we know the root is within the range. The assumption is also made that the function is continuous at the root and thus

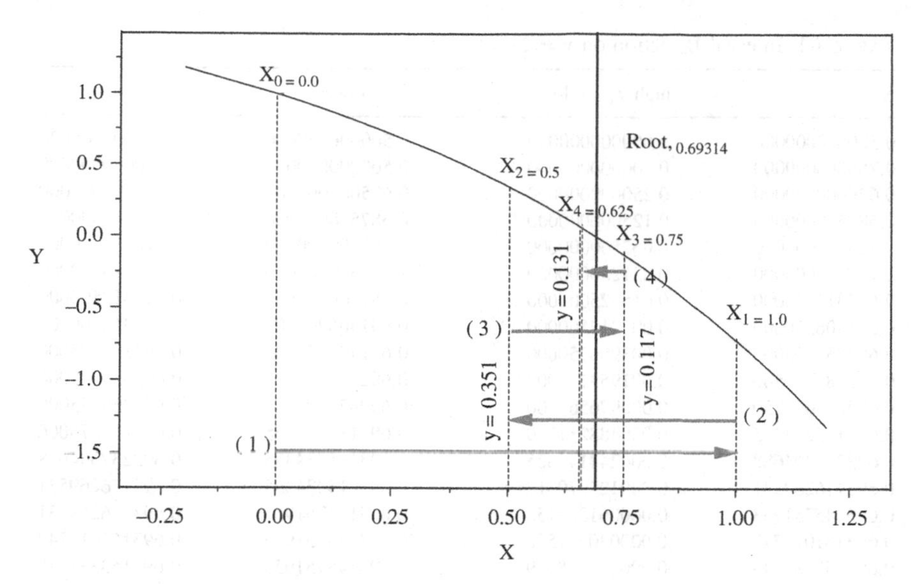

FIGURE 1.1. Bisection on $y = 2 - e^x$.

has at least one zero. The method takes the first approximation x_2 to the root by halving the initial range. So

$$x_2 = \frac{1}{2}(X_0 + X_1) \tag{1.2.1}$$

The function $f(x_2)$ is evaluated; there are three possible outcomes. First, in the interval, $[x_2, x_0]$. $f(x_0)f(x_2) < 0$. Means there is at least one root between these endpoints. Second, $f(x_0)f(x_2) = 0$ (we assume that $f(x_0) \neq 0$). This indicates that we have found the root $f(x_2)$. Third, $f(x_0)f(x) > 0$. This means the root is in the other interval half, $[x_2, x_1]$. Given that the second outcome is not initially achieved we continue with the process of halving the uncertainty until a root is found within the desired precision. From Figure 1.1 we see that the initial range, shown by arrow (1) is halved at $x = 0.5$. The function evaluates to 0.351, the function evaluates to -0.7183 at $x = 1.0$. The root therefore lies within the range shown by arrow (2). The halved value is at $x = 0.75$. The function evaluates to -0.117. The root therefore lies within the range now in the direction shown by arrow (3). The halved value is evaluated to be 0.131. The range is now in the direction shown by arrow (4) This process is continued until the desired precision is reached. The data in Table 1.1 shows the convergence for the function of Figure 1.1.

Listing 1.1 shows the method *evaluateRoot* in the abstract class **IntervalBisection**. The abstract method *ComputeFunction* is implemented in the extending class. The method *evaluateRoot* provides functionality for the bisection algorithm. This is outlined below in Listing 1.1.

TABLE 1.1. Interval Bisection on $y = 2 - e^x$

X_n	$higher_{n-1} - lower_{n-1}$	$lower_n$	$higher_n$
0.500000000000	1.000000000000	0.500000000000	1.000000000000
0.750000000000	0.500000000000	0.500000000000	0.750000000000
0.625000000000	0.250000000000	0.625000000000	0.750000000000
0.687500000000	0.125000000000	0.687500000000	0.750000000000
0.718750000000	0.062500000000	0.687500000000	0.718750000000
0.703125000000	0.031250000000	0.687500000000	0.703125000000
0.695312500000	0.015625000000	0.687500000000	0.695312500000
0.691406250000	0.007812500000	0.691406250000	0.695312500000
0.693359375000	0.003906250000	0.691406250000	0.693359375000
0.692382812500	0.001953125000	0.692382812500	0.693359375000
0.692871093750	0.000976562500	0.692871093750	0.693359375000
0.693115234375	0.000488281250	0.693115234375	0.693359375000
0.693237304688	0.000244140625	0.693115234375	0.693237304688
0.693176269531	0.000122070312	0.693115234375	0.693176269531
0.693145751953	0.000061035156	0.693145751953	0.693176269531
0.693161010742	0.000030517578	0.693145751953	0.693161010742
0.693153381348	0.000015258789	0.693145751953	0.693153381348
0.693149566650	0.000007629395	0.693145751953	0.693149566650
0.693147659302	0.000003814697	0.693145751953	0.693147659302
0.693146705627	0.000001907349	0.693146705627	0.693147659302

```
public double evaluateRoot(double lower, double higher)
        //lower and higher are the initial estimates//
{
    double fa;          //fa and fb are the initial 'guess' values.//
    double fb;
    double fc;          //fc is the function evaluation , f(x)//
    double midvalue=0;
    double precvalue=0;
    fa=computeFunction(lower);  //ComputeFunction is implemented
                                //by the caller//
    fb=computeFunction(higher);

    //Check to see if we have the root within the range bounds//
    if (fa*fb>0)
    {                   //If fa*fb>0 then both are either positive//
                            //or negative and don't bracket zero.//
        midvalue=0;//Terminate program//
    }
    else
    do
    {
        precvalue=midvalue;//preceding value for testing
                                        //relative precision//
        midvalue=lower+0.5*(higher-lower);
        fc=computeFunction(midvalue) //Computes the f(x)//
                                        //for the mid value//
```

```
            if(fa*fc<0)
            {
                        higher=midvalue;
            }
            else
            if(fa*fc>0)
            {
                        lower=midvalue;
            }
} while((abs(fc)>precisionvalue&i<iterations));
//loops until desired number of iterations or precision is reached//
return midvalue;
}
```

LISTING 1.1. Method evaluateRoot from class **IntervalBisection** in package CoreMath

The return value, midvalue in this case is output when the converging solution is < 0.001. Note we might have used different precision criteria that would rely on the relative change in precision from one evaluation to the other. For example using the loop: while((abs(midvalue-precvalue)>precisionvalue&i<iterations)); would terminate when successive values of the intermediate evaluations are < 0.001.

Table 1.1 shows the output from **IntervalBisection** when evaluating the equation $y = 2 - e^x$.

Column one shows the approximation output from the computation. Column two shows previous higher estimate minus the previous lower estimate. Columns three and four show the high and low estimates. For our example the initial 'guesses' were higher = 1.0 and lower = 0.5.

The approximation after 19 iterations reaches the desired precision to 1E-06, which is accurate to the 'real' solution by around -1E-07.

Listing 1.2 shows the complete class for **IntervalBisection**. *ComputeFunction* is an abstract method which has to be implemented in the calling class (which provides the actual function, in our example this is $y = 2 - e^x$). The constructor defaults to 20 iterations of the algorithm and the precision is set to 1e-3. The using class can pass other values through the alternate constructor (int iterations, double precisionvalue). Access to the internal values is via the get methods.

```
public abstract class IntervalBisection
{
//computeFunction is implemented to evaluate successive root estimates//
    public abstract double computeFunction(double rootvalue);
    protected double precisionvalue;
    protected int iterations;
    protected double lowerBound;
    protected double upperBound;

    //default constructor//
    protected IntervalBisection()
    {
        iterations=20;
```

```
            precisionvalue= 1e-3;
        }
        //Constructor with user defined repetitions and precision//
        protected IntervalBisection(int iterations, double precisionvalue)
        {
        this.iterations=iterations;
        this.precisionvalue=precisionvalue;
    }

    public int getiterations()
    {
        return iterations;
    }
    public double getprecisionvalue()
    {
        return precisionvalue;
    }
    public double evaluateRoot(double lower, double higher)
    {
        double fa;
        double fb;
        double fc;
        double midvalue=0;
        double precvalue=0;
        fa=computeFunction(lower);
        fb=computeFunction(higher);
        //Check to see if we have the root within the range bounds//
        if (fa*fb>0)
        {
        midvalue=0;//Terminate program//
        }
        else
        do
        {
            precvalue=midvalue;//preceding value for testing//
                                //relative precision//
            midvalue=lower+0.5*(higher-lower);
            fc=computeFunction(midvalue);
            if(fa*fc<0)
            {
                higher=midvalue;
            }
            else
            if(fa*fc>0)
            {
                lower=midvalue;
            }
            }
    while ((abs(fc)>precisionvalue<iterations));
        //loops until desired number of iterations or precision is reached//
    return midvalue;
    }
    }
```

LISTING 1.2. **IntervalBisection** in package CoreMath

A class such as **IntervalBisection** has its core functionality controlled by the using class. To compute the function $y = 2 - e^x$ we had to use an application class which extended the abstract method *ComputeFunction*. The using class provided the controlling logic to provide the equation into **IntervalBisection**. The abstract class is there to provide a core technique (interval bisection) and not to perform other functionality. The strategy of keeping core functionality within static or abstract classes allows us to re-use the class in a variety of applications without the need to re-design or add to the core.

In this example we have used the class **IntervalBisection** to implement interval bisection on the function $y = 2 - e^x$. Later we will use this same class to implement interval bisection on a yield equation. It will perform exactly the same functionality on a completely different equation; the controlling class (an application class) will implement the abstract method *computeFunction* with the various input equations.

We will see later that more than one class is often required before we can implement an application. The interval bisection algorithm although generally robust is slower to converge than other root finding algorithms. The Newton Raphson algorithm (abbreviated to Newton's method) is a method for a root finding algorithm which converges to a root much more quickly than interval bisection. Although the Newton method is quicker to converge, it requires the derivative of the function to be used in the solution. This is a good example of a series of classes being used to implement an application.

1.2.2. Newton's Method

To use Newton's method we will need to use the class **Derivative** from the CoreMath package. This abstract class provides the method derivation to provide functionality for providing the derivative of a single function. The class has its abstract method *deriveFunction* extended by the using class which provides the controlling logic to provide the single functions for evaluation. Listing 1.3 provides the complete abstract class for **Derivative**.

The method derivation uses the technique of difference quotients to arrive at an approximation of a function. The method being implemented is based on the general definition of the derivative.

$$f'(x) = \lim_{h \to 0} \frac{f(x+h) - f(x)}{h} \tag{1.2.2}$$

The implementation used is based on the approximation which gives best accuracy with lower computational cost:

$$f'(x) = \frac{f(x+h) - f(x-h)}{2h} \tag{1.2.3}$$

The method which implements algorithm 1.1.3 is given below in Listing 1.3.

```
public double derivation (double InputFunc)
        {
                double value;
                double X2=deriveFunction(InputFunc-h);
                double X1=deriveFunction(InputFunc+h);
                value=((X1-X2)/(2*h));
                return value;        }
```

LISTING 1.3. Method derivation in class **Derivative** package CoreMath

X1 and X2 take the value from the abstract method *deriveFunction* and implement the arithmetic from equation 1.1.3. The value of h is chosen to provide optimum accuracy. The smaller we can make h, the greater the accuracy we achieve (from theory). The analytic answer to the derivative of e^x for $x = 1$, is e itself. Column three in Table 1.2 shows the error in the derived approximation from the actual value of e. Column four shows the ratio of previous to present

TABLE 1.2. Output from derivation for InputFunc = e^x

$\frac{1}{h}$	f'	$\varepsilon = e - f'(x)$	Ratio$(\varepsilon_{n-1}/\varepsilon_n)$
0.5	2.8329678	−0.114685971	
0.25	2.746685882	−0.028404053	4.03766
0.125	2.72536622	−0.007084391	4.00939
0.0625	2.720051889	−0.00177006	4.00234
0.03125	2.718724279	−4.42*E*-04	4.00059
0.015625	2.718392437	−1.11*E*-04	4.00015
0.0078125	2.71830948	−2.77*E*-05	4.00004
0.00390625	2.718288741	−6.91*E*-06	4.00001
0.001953125	2.718283557	−1.73*E*-06	4
9.77E-04	2.718282261	−4.32*E*-07	4
4.88E-04	2.718281936	−1.08*E*-07	4.00001
2.44E-04	2.718281855	−2.70*E*-08	3.99996
1.22E-04	2.718281835	−6.75*E*-09	3.99984
6.10E-05	2.71828183	−1.69*E*-09	3.99719
3.05E-05	2.718281829	−4.19*E*-10	4.02772
1.53E-05	2.718281829	−9.19*E*-11	4.56177
7.63E-06	2.718281828	−1.92*E*-11	4.79628
3.81E-06	2.718281829	−4.83*E*-11	0.39706
1.91E-06	2.718281828	9.94*E*-12	−4.85718
9.54E-07	2.718281828	1.26*E*-10	0.07865
4.77E-07	2.718281828	1.26*E*-10	1
2.38E-07	2.718281829	−3.39*E*-10	−0.37238
1.19E-07	2.718281828	5.92*E*-10	−0.57314
5.96E-08	2.718281828	5.92*E*-10	1
2.98E-08	2.718281835	−6.86*E*-09	−0.08632
1.49E-08	2.718281835	−6.86*E*-09	1
7.45E-09	2.718281835	−6.86*E*-09	1
3.73E-09	2.718281806	2.29*E*-08	−0.29893
1.86E-09	2.718281865	−3.67*E*-08	−0.62584
9.31E-10	2.718281984	−1.56*E*-07	0.2352

error. The ratio of improvement is about 4. For each halving of h. This is true until 1/h is at 4.88E-04, thereafter the improvement oscillates widely.

This illustrates a phenomenon mentioned earlier in the introduction, namely machine (rather than theoretical) error. The errors being introduced are largely the result of rounding. The effects of repeated divisions of f(x+h) and f(x-h), together with machine representation (we are using type double for all floating calculations) are introducing practical implementation errors. If we used type float (32 bit) in the calculation things would be worse and we could expect significant error to be shown at around an h of -6.9E-06. We can achieve accuracy of -1.92E-11 before things deteriorate. For most applications this is good enough, but for some it could pose problems. You can use Table 1.1 to assess the size of h that might be suitable for your particular application.

Table 1.2 below shows the output from derivation with input = e^x. The values of h are decreasing from 0.5 down to 9.31E-10. The computed function f', is gradually converging on the 'correct' (high precision) answer.

Listing 1.4 gives the complete abstract class for **Derivative**.

```
package CoreMath;
public abstract class Derivative//
{
        public abstract double deriveFunction(double fx);
        //returns a double..... the function//
                public double h;// degree of accuracy in the calculation//
                public double derivation(double InputFunc)
                {
                        double value;
                        double X2=deriveFunction(InputFunc-h);
                        double X1=deriveFunction(InputFunc+h);
                        value=((X1-X2)/(2*h));
                                return value;
                        }
}
```

LISTING 1.4. **Derivative**

Now we know something about the characteristics of our core class **Derivative** let's examine the use of it in Newton's method.

Newton's method is based on linear approximations to the function. The approximation is based on the tangent line to the function curve.

$$\tan\theta = f'(x_0) = \frac{f(x_0)}{(x_0 - x_1)} \text{Thus, } x_1 = x_0 - \frac{f(x_0)}{f'(x_0)} \text{ and } x_2 = x_1 - \frac{f(x_1)}{f'(x_1)}$$

In general, $x_{n+1} = x_n - \frac{f(x_n)}{f'(x_n)}$ for n = 1, 2, 3 ... (0.2.4)

To use Newton's method we only require a single approximation for the root and the derivative of the function f(x). There can be problems with Newton's method; one is where the derivative is zero near the root. In this case; $f(x_n)/f'(x_n) \to \infty$.

Small values of $f'(x_n)$ can cause large differences between iterations and slow convergence; also the calculation of $f'(x_n)$ itself can be complicated.

Package CoreMath contains the class **NewtonRaphson**. This is an abstract class which implements the Newton Raphson algorithm and extends the abstract method *deriveFunction*.

```
public void newtraph(double lowerbound)
{
    double fx=newtonroot(lowerbound);// y=2-e^x in our example//
    double Fx=derivation(lowerbound);
    double x=(lowerbound-(fx/Fx));  //=x^{n+1}=x_n - f(x_n)/f'(x_n)//
    while((abs(x-lowerbound)>precisionvalue&counter< iterate))
                  {
                  lowerbound=x;
                  fx=newtonroot(lowerbound);
                  Fx=derivation(lowerbound);
                  x=(lowerbound-(fx/Fx));
                          counter++;
 }
 }
```

LISTING 1.5. Shows the method *newtraph*. This implements the algorithm of 1.1.4

The method *newtraph* takes the approximation as lowerbound. The abstract method *newtonroot* is extended in the calling class which provides the function for evaluation. The method *derivation* is used to calculate the derivative. The method iterates through calls until the desired precision or predefined number of iterations is reached. This is controlled through the while loop which implements: $|x_{n+1} - x_n| > \varepsilon \& <$ Iterations. The precision value ε is defined in the method accuracy as is the value of the desired maximum number of iterations.

Table 1.3 shows the output from **NewtonRaphson** for $y = 2 - e^x$.

Column one shows the approximation (guesses) input. The initial approximation was 1.0. The second column shows the actual (analytical) solution to the function minus the approximation. The third column shows log base 10 of the differences. From this column it can be intuitively appreciated that the error in the successive approximations is halving each time.

It is instructive to compare the number of iterations and the convergence characteristics shown in Table 1.1 for the bisection algorithm and Tables 1.3 and

TABLE 1.3. Newton Raphson method on $y = 2 - e^x$

N	x_n	*(actual-x_n)*	*log_{10}(actual-x_n)*
1	1.000000000000000	−0.3068528194400547	−0.5130698819559578
2	0.735759896178339	−0.04261271561839375	−1.3704607882472601
3	0.6940422724627328	−8.950919027874704E-4	−3.048132371584211
4	0.6931475844952419	−4.039352966556109E-7	−6.3936881956713165
5	0.6931471805598971	4.8183679268731794E-14	−13.317100040678492

TABLE 1.4. Output from method *newtraph*

N	f(x)	f'(x)	(x)
1	−0.718281828459	−2.718292257953	0.735759896178
2	−0.087067344576	−2.087063855072	0.694042272463
3	−0.001790985234	−2.001798726781	0.693147584495
4	−0.000000807871	−2.000000165481	0.693147180560
5	0.000000000000	−2.000000165481	0.693147180560

1.4 below. Clearly Newton's method converges within four iterations whereas the bisection method takes 19 iterations for the same degree of precision.

Table 1.4 shows the output from the method *newtraph*. Column one is the function evaluation with the 'guess' value as input. Column two is the derivative of the function with the variable set to the 'guess' value. Column three shows the successive approximations for x.

Listing 1.6 gives the complete class for **NewtonRaphson**. Since the Newton Raphson method requires the use of the derivative, this class extends the abstract class **Derivative**. It was mentioned earlier that we often use several classes to provide an application with the needed methods. In this case we have **NewtonRaphson** making use of **Derivative** (and extending the abstract method). However the class **NewtonRaphson** is itself only designed to provide the means for carrying out Newton's algorithm. To do the computation on an actual function, **NewtonRaphson** needs to have its abstract method *newtonroot* extended by an application which provides the function to be evaluated. **Newton-Raphson** also needs to pass this function to the **Derivative** method which requires it.

```
package CoreMath ;
public abstract class NewtonRaphson extends Derivative
{
public abstract double newtonroot(double rootvalue);
        //the requesting function implements the calculation fx//
public double precisionvalue;
public int iterate;
public void accuracy(double precision,int iterations)
         //method gets the desired accuracy//
{
super.h=precision;//sets the superclass derivative//
         //to the desired precision//
this.precisionvalue=precision;
this.iterate=iterations;
}
    public double newtraph(double lowerbound)
    { int counter=0;
         double fx=newtonroot(lowerbound);
         double Fx=derivation(lowerbound);
         double x=(lowerbound-(fx/Fx));
```

```
        while(abs(abs(x)-abs(lowerbound))
              > precisionvalue||counter<iterate)
              {
                     lowerbound=x;
                     newtraph(lowerbound);//recursive call//
                                               // to newtraph//
        counter++;
    }
        return x;
    }
    public double deriveFunction(double inputa)
    {
        double x1=newtonroot(inputa);
                     return x1;
              }
}
```

LISTING 1.6. **NewtonRaphson**

1.3. Statistical Classes

The statistical classes implement methods for the manipulation and analysis of data. Statistical classes provide standard re-usable techniques as methods for use in application classes where the specific functionality of the methods are needed to create a sophisticated technique (from possibly many methods). The class structures are minimal in the sense that a particular technique will usually be applied through the use of a series of classes that implement a particular part of the technique.

For example, the data in Table 1.5 is to be used to provide the standard deviation for the sample. The standard deviation will not be directly computed, rather the mean, followed by the variance then standard deviation will be used to provide the desired result Table 1.5 shows data with associated probability. Table 1.6 contains data only.

TABLE 1.5. Input dataset

Data Item : 1 Data 12.000000 Probability 0.100000
Data Item : 2 Data 7.000000 Probability 0.200000
Data Item : 3 Data 11.000000 Probability 0.100000
Data Item : 4 Data 23.000000 Probability 0.100000
Data Item : 5 Data 44.000000 Probability 0.075000
Data Item : 6 Data 58.000000 Probability 0.025000
Data Item : 7 Data 22.000000 Probability 0.200000
Data Item : 8 Data 33.000000 Probability 0.100000
Data Item : 9 Data 56.000000 Probability 0.050000
Data Item : 10 Data 76.000000 Probability 0.050000

TABLE 1.6. Input dataset equal probability

Data Item : 1 Data 12.000000
Data Item : 2 Data 7.000000
Data Item : 3 Data 11.000000
Data Item : 4 Data 23.000000
Data Item : 5 Data 44.000000
Data Item : 6 Data 58.000000
Data Item : 7 Data 22.000000
Data Item : 8 Data 33.000000
Data Item : 9 Data 56.000000
Data Item :10 Data 76.000000

When we make use of the statistical methods in BaseStats to provide us with the mean and variance of this data it's reasonable to expect that we can use the same methods. This is achieved by making extensive use of method overloading. The majority of the statistical classes are implemented as static methods. Statistical classes are placed in the package BaseStats.

1.3.1. Measures of Dispersion

The class **DataDispersion** in the package BaseStats is a general purpose class with static methods for direct use. A range of methods that deal with aspects of data dispersion are supplied to enable an application class to make direct use of specific techniques, or to combine methods into a more global technique.

The data in Tables 1.5 and 1.6 are input from a controlling class which makes use of the class **DataDispersion** to evaluate the standard deviation of the data. The methods used for this operation are shown in Listing 1.7

// uses the algorithm $\frac{1}{n}\sum_{i=1}^{n} X_i$ //

```
public static double mean(double[] x)
        //arithmetic mean for a single list//
    {
        double total=0.0;
        for(int i=0;i<x.length;i++)
        total+=x[i];
                  return total/x.length;
    }
```

// uses the algorithm $\sum_{i=1}^{n}(X_i P_i)$ //

```
public static double mean(double[][] x) //returns expected value//
                                   //for variable * probability//
    {

        double total=0.0;
        double probability=0.0;


        for(int i=0;i<x.length;i++)//the number of rows//
```

```
        {
        total+=(x[i][0]*x[i][1]);
        probability+=x[i][1];
        }
    if(probability!=1.0)
    System.out.println("WARNING ! The probabilities do not sum to 1.0");
        return total;

    }
```

// uses the algorithm $\frac{\sum_{i=1}^{n} X_i^2 - \sum_{i=1}^{n} X_i}{n-1}$ //

```
        public static double variance(double[] v1)
             //variance of a single variable with equal likeliehood//
        {
             double sumd=0.0;
         double total=0.0;
             for(int i=0;i<v1.length;i++)
             {
                 total+=v1[i];
                 sumd+= pow(v1[i],2);//sum of x sqrd
             }
        return (sumd-total)/((v1.length)-1);
             //true value of convergence as length is large//
        }
```

// uses $\sum_{i=1}^{n}(X_i^2 P_i) - \sum_{i=1}^{n}(X_i P_i)^2$ //

```
        public static double variance(double[][] v1)
             //variance of a variable with different
             //probability of otcome//
        {
         double sumd=0.0;
         double total=0.0;
         double totalpow=0.0;
         double probability=0.0;
             for(int i=0;i<v1.length;i++)
             {
                 total+=(v1[i][0]*v1[i][1]);//mean or expected value//
                  totalpow+=(pow(v1[i][0],2)*v1[i][1]);//E[X2]//
                  probability+=v1[i][1];
                 }
                 if(probability!=1.0)
         System.out.println("WARNING !The probabilities
                                   do not approximate to sum to 1.0");
             total=pow(total,2);
        return (totalpow-total);
        }
        public static double standardDeviation(double s1)
             // computes standard deviation for variance s1//
        {
             double sdev;
                         return sdev=sqrt(s1);
        }
```

LISTING 1.7. Methods used to provide Standard Deviation

In each case the controlling class passes the data to the method. The JVM determines which particular implementation of mean and variance to use. In the case of Table 1.5 mean and variance for the single list take the data. For Table 1.6 data, the methods mean and variance for the double list take the data.

The method *standardDeviation* takes the variance and produces the measure.

This simple example shows that a combination of overloading methods and constructing a global technique (or algorithm) from simpler ones, is a very efficient technique. The output from Tables 1.5 and 1.6 are:

Table 1.5 : Mean : 25.050000 Variance : 342.297500Standard Deviation : 18.501284
Table 1.6: Mean : 34.200000 Variance : 547.955556Standard Deviation : 23.408451

Class **DataDispersion** is used in a later example when we make use of covariance and standard deviation. The remaining methods to complete the listing for DataDispersion is shown in Listing 1.8.

```
package BaseStats;
import java.util.ArrayList;
import java.io.*;
import static.java.lang.Math.*;
public class DataDispersion
{
//$\frac{1}{n}\sum_{i=1}^{n} X_i$ for both entries //
    public static double[] dumean(double[][] x)//arithmetic mean//
                                                //for a double list//
    {
        double x1=0.0;
        double y=0.0;
        double[] total=new double[2];
        for(int i=0;i<x.length;i++)
        {
            x1+=x[i][0];
            y+=x[i][1];
    }
    total[0]=x1/x.length;
    total[1]=y/x.length;
    return total;
    }
//use algorithm $\frac{1}{n-1}\sum_{i=1}^{n} X_i$//
    public static double convmean(double[] x)// for large length//
    {
        double total=0.0;
        for(int i=0;i<x.length;i++)
        total+=x[i];
            return total/(x.length-1);
    }
```

```
// uses the algorithm
```
$\frac{\sum_{i=1}^{n} X_i^2 - \sum_{i=1}^{n} X_i}{n-1}$
```
 for each input //
     public static double[] variances(double[][]v1)
          //variance of a single variable with equal likeliehood//
          //for double inputs//
{
     double[] output=new double[2];
     double sumd=0.0;
     double sumd1=0.0;
 double total=0.0;
 double total1=0.0;
     for(int i=0;i<v1.length;i++)
     {
          total+=v1[i][0];
          total1+=v1[i][1];

          sumd+= pow(v1[i][0],2);//sum of x sqrd
          sumd1+= pow(v1[i][1],2);//sum of x sqrd
     }
     total=(pow(total,2)/v1.length);//sum of [x]sqrd/n
     total1=(pow(total1,2)/v1.length);//sum of [x]sqrd/n
     output[0]=((sumd-total)/((v1.length)-1));
     output[1]=((sumd1-total1)/((v1.length)-1));
return output;
}
//uses algorithm
```
$\frac{\sum_{i=1}^{n} (X_i - \bar{X}_i)(Y_i - \bar{Y}_i)}{n}$
```
 //
public static double covar(double[][] outcomes)
                    //equally likely outcomes//
{
     double sa=0.0;
     double sb=0.0;
     double product=0.0;
     int size=outcomes.length;
     for(int i=0;i<size;i++)
     {
          sa+=outcomes[i][0];//x values or proprtions//
          sb+=outcomes[i][1];//y values or proportions//
          }
          double samn=sa/size;//expected value of x//
          double sbmn=sb/size;//expected value of y//
          for(int i=0;i<size;i++)
          {
               product+=((outcomes[i][0]-samn) *
                         (outcomes[i][1]-sbmn));
                         //sum of the products ofdeviations//
          }
          return product/size;//covariance//
}
//use algorithm
```
$\sum_{i=1}^{n} (X_i - \bar{X})(Y_i - \bar{Y})^* P_i$
```
 //
     public static double covar2(double[][] outcomexyp)
                         //inputs of non equal joint outcomes//
     {
```

```
        // data in the form A value, B value . Probability(P)
        // of B and A the same//
        double productx=0.0;
        double producty=0.0;
        int size=outcomexyp.length;
        double covariance=0.0;
        for(int i=0;i<size;i++)
        {
            //A[n][0],B[n][1],P[n][2].........//
            productx+= outcomexyp[i][0]*outcomexyp[i][2];//
probability * observed value//
            producty+=outcomexyp[i][1]*outcomexyp[i][2];
        }
            for(int j=0;j<size;j++)
            {
                double xdevs=outcomexyp[j][0]-productx;
                double ydevs=outcomexyp[j][1]-producty;
                double devproduct=xdevs*ydevs;
                double covprobs=devproduct*outcomexyp[j][2];
                covariance+=covprobs;
            }
            return covariance;
    }
    // ρij = σij/(σiσj) //
    public static double correlation(double cov,double sd1,double sd2)
    {
        double cor=cov/sd1*sd2;
        return cor;
    }
}
```

LISTING 1.8. Class **DataDispersion**

1.4. Application Classes

The application classes comprise those classes which provide the functionality for solving application problems in Financial Engineering computation. Application class methods provide the controlling logic for calling other classes from the CoreMath and BaseStats packages. The majority of application classes are self contained within the package FinApps, a limited number of the classes involve combinations of methods from others within FinApps. Application classes comprise the focus of this book. An application class which makes use of the root finding algorithms discussed earlier is the Yield evaluator and another which makes use of a range of methods from the CoreMath and BaseStats packages is the Portfolio evaluator. See Elton & Gruber (1995) for background theory.

Yield evaluation can be accomplished with a range of numerical methods. For our examples we will use the bisection algorithm implemented in **IntervalBisection** and the Newton Raphson algorithm implemented in the **NewtonRaphson** class. This example gives us the opportunity to see how the concrete classes that make up the application set make use of increasingly abstract classes. When using application classes the emphasis is on the use of controlling structures to access and manipulate data and choosing the appropriate methods to solve an application problem.

We can use application classes to implement business logic without having any real concern about the implementation lower down the scale. The solution of an internal rate of return problem can be accomplished by using one of the yield classes that have been developed for this book. That yield class might use the bisection algorithm or could make use of the Newton Raphson algorithm, which in turn makes use of the derivative algorithms etc. If the main imperative is to design an application that solves a problem in finance, that application can be solved using the classes as is. Alternatively if the application requires a new construct, the application package can be added to by developing a class which uses say, some methods from BaseStats and some from CoreMath in a novel way. This is the essence of the approach taken throughout the rest of this book.

The application classes provide the controlling logic and implement business rules to solve problems in finance and investment domains. The core math and base statistics classes are concerned with providing solutions which are numerical methods and statistical methods per se. We will now examine our example class for Yield evaluation. Yield calculations are core to valuing bonds. See Martelleni et al (2003).

This application class is designed to solve for the internal rate of return (IRR). The equation of value for IRR connects the amounts paid into an investment with the amounts going out of that investment. The specific class we will examine takes the discounting to the present time.

1.4.1. Internal Rate of Return

The specific formula is:

$$MP = \frac{C_{a_{\overline{2n}}}}{2} + \frac{100}{(1+i)^{2n}} \tag{1.4.1}$$

Mp is the market price, C the coupon rate, $a_{\overline{2n}}$ is the present value of a series of payments for n periods paid twice per n.

This specific equation is used to find the market price, given the yield, based on the simplifying assumption that settlement occurs on an interest paying date. Alternatively, to find the yield given the published price involves the use of one of the root finding algorithms.

$$\text{Thus, } 0 = -MP + \frac{C_{a_{\overline{2n}}}}{2} + \frac{100}{(1+i)^{2n}} \tag{1.4.2}$$

1.4.2. Deriving yield approximations – Bisection method

We will initially use **IntervalBisection** from package CoreMath to compute the function derived from 1.1.6. The application class is therefore called **YieldBisect.**

TABLE 1.7. Interval Bisection on: $f(x) = \frac{5}{i}\left\{1 - \frac{1}{(1+i)^{2n}}\right\} + \frac{100}{(1+i)^{2n}} - 104.5$ (0.4.3)

N	Higher	x	Lower	|f(x)|
1	0.0500000000	0.0500000000	0.0300000000	4.5000000000
2	0.0500000000	0.0400000000	0.0400000000	0.7421368567
3	0.0450000000	0.0450000000	0.0400000000	1.9210637586
4	0.0425000000	0.0425000000	0.0400000000	0.6001950030
5	0.0425000000	0.0412500000	0.0412500000	0.0682627610
6	0.0418750000	0.0418750000	0.0412500000	0.2666399816
7	0.0415625000	0.0415625000	0.0412500000	0.0993574717
8	0.0414062500	0.0414062500	0.0412500000	0.0155896204
9	0.0414062500	0.0413281250	0.0413281250	0.0263259979
10	0.0414062500	0.0413671875	0.0413671875	0.0053655464
11	0.0413867187	0.0413867187	0.0413671875	0.0051126975
12	0.0413867187	0.0413769531	0.0413769531	0.0001262593
13	0.0413818359	0.0413818359	0.0413769531	0.0024932603
13	0.0413793945	0.0413793945	0.0413769531	0.0011835108
15	0.0413781738	0.0413781738	0.0413769531	0.0005286283
16	0.0413775635	0.0413775635	0.0413769531	0.0002011851

Within **YieldBisect** the method *computeFunction* provides the functionality to implement equation 1.1.6. *computeFunction* is listed below in Listing 1.9. The complete class is shown in Listing 1.9. The run-time code for the example is appended.

```
public double computeFunction(double rootinput)
     //implements the abstract method from interval bisection
{
   double poscashflow,solution;
   poscashflow=rateperTerm;//cashflow out per term//
                                   //as monthly amount * termperiod//
   solution=(poscashflow/rootinput*
            (1.0-1.0/(pow(1.0+rootinput,rateindex))))
            +(nominalstockprice/(pow(1.0+rootinput, rateindex)))
            -marketpricevalue;
   return solution;
}
```

LISTING 1.9. Method *computeFunction* in class **YieldBisect**

Table 1.7 shows the output for the data of Equation 1.1.7. This represents the yield on an investment with the following characteristics: Nominal Price 100.0. Market Price 104.5. Interest Paid at 5 every six months (twice yearly). Term to redemption 3 years(n)Coupon rate 10% Per annum. The initial estimates for yield are Low = 3%. High = 7%.

The solution is 4.1377%. This takes 15 iterations to complete with an accuracy of 1e-6.

```
package FinApps;
import java.text.*;
import java.lang.*;
import static.java.lang.Math.*;
import CoreMath.IntervalBisection;

public class YieldBisect extends IntervalBisection {
   public YieldBisect() //default constructor//
   {
   }
   public YieldBisect(int Nofiterations, double Precision,
                        double high, double low) {
      super(Nofiterations,Precision);
          //alternate constructor with changed values for precision
          // and number of iterations Interval Bisection//

      inputevaluelow=low;
      inputvaluehigh=high;
   }

   protected double nominalstockprice;
   protected double termperiod;
   protected double couponrate;
   protected double marketpricevalue;
   protected double inputevaluelow;
   protected double inputvaluehigh;
   protected double rateperTerm;
   protected double maturityperiod;
   protected double rateindex;

   public double computeFunction(double rootinput)
      //implements the abstract method from interval bisection
   {
      double poscashflow,solution;
      poscashflow=rateperTerm;//cashflow out per term
                                 //as monthly amount * termperiod//
      solution=(poscashflow/rootinput*(1.0-
1.0/(pow(1.0+rootinput,rateindex))))+(nominalstockprice/
(pow(1.0+rootinput, rateindex)))-marketpricevalue;
   return solution;
   }


   public double yieLd(double noms, double term, double coupon,
                        double mktp, double period) {
      nominalstockprice=noms;
      termperiod=term;
      couponrate=coupon;
      marketpricevalue=mktp;
      rateperTerm=((coupon/term));
      maturityperiod=period;
      rateindex=(maturityperiod*term);
      return evaluateRoot(inputevaluelow,inputvaluehigh);
          //evaluateRoot is in the class: CoreMath IntervalBisection
   }
```

```
    public static void main(String[] args) {
        YieldBisect CalcBond= new YieldBisect(20,1e-6,0.07,0.03);
        double yieldvalue=CalcBond.yieLd(100.0,2.0,10.0,104.5,3.0);
        System.out.println("The required yield is =="+yieldvalue);
    }

}
```

LISTING 1.10. Class **YieldBisect**

1.4.3. Deriving Yield Approximations -the Newton Raphson Method

Consider now that our class Yield evaluator is to be implanted with the Newton Raphson method. This time our application class is called **NewtonYield**. It implements the same equation as before. The results are shown in Table 1.8.

Initial estimate for yield is 5%.

This has achieved the desired solution in three iterations. Compare this with the number of iterations required for the interval bisection technique.

The application class **NewtonYield** shares a significant amount of code with **YieldBisect**. The difference is in the implementation of the abstract method *newtonroot* from **NewtonRaphson**, rather than the implementation of the abstract method *computeFunction* from **IntervalBisection**. Also the class extends **NewtonRaphson** in the header . The listing for **NewtonYield** is shown in Listing 1.11.

We have examined how different configurations of Yield have been implemented through application classes. The two examples have used methods from another package (CoreMath). We will now look at two application classes that make use of two packages (CoreMath and BaseStats) and also use methods from the FinApps package.

The two examples we will examine are the classes **Portfolio** and **SelectPortfolio** in the package FinApps.

1.4.4. Portfolio Management

Class **SelectPortfolio** is concerned with supplying methods to handle the diverse type of input data that can be used to perform portfolio analysis. For our examples

TABLE 1.8. Newton's method on $f(x) = \frac{5}{i}\left\{1 - \frac{1}{(1+i)^{2n}}\right\} + \frac{100}{(1+i)^{2n}} - 104.5$

N	*Estimate(x)*	*f(x)*	*f'(x)*	*Estimate(x+1)*
1	0.0500000000	−4.5000000000	−507.5692067322	0.0411342139
2	0.0411342139	0.1304545467	−537.3274875922	0.0413769980
3	0.0413769980	0.0001021894	−536.4859306667	0.0413771885

we are using methods to handle raw data and methods to handle pre-processed data based on expected returns and expected end price. The size of SelectPortfolio is obviously arbitrary and will vary dependant on the variety of data formats that require handling; the more types required can simply be added. When dealing with large datasets from external sources, we will need to make use of Java file handling and data stream handling. Constructing a general purpose data handler can therefore be a fairly large job. Methods dealing specifically with streaming and file management for use in SelectPortfolio are fortunately provided by Java.

```
package FinApps;
import CoreMath.NewtonRaphson;
import BaseStats.inputmod;
import static.java.lang.Math.*;
public class NewtonYield extends NewtonRaphson
{

  public NewtonYield(double initialval, double precision,
                        int iterations ) {
     inputvalue=initialval;
    iteration=iterations;
    prec=precision;
  }
  protected double nominalstockprice;
  protected double termperiod;
  protected double couponrate;
  protected double marketpricevalue;
  protected double inputvalue;
  protected double rateperTerm;
  protected double maturityperiod;
  protected double rateindex;
  int iteration;
  double prec;
     public double newtonroot(double rootinput)
          //implements the abstract method from interval bisection
  {
    double poscashflow,solution;
    poscashflow=rateperTerm;//cashflow out per term as monthly
                                //amount * termperiod//
    solution=(poscashflow/rootinput*(1.0-
1.0/(pow(1.0+rootinput,rateindex))))+(nominalstockprice/
(pow(1.0+rootinput, rateindex)))-marketpricevalue;
    return solution;
  }
     public double yieLd(double noms, double term, double coupon,
                           double mktp, double period) {
    nominalstockprice=noms;
    termperiod=term;
    couponrate=coupon;
    marketpricevalue=mktp;
```

```
        rateperTerm=((coupon/term));
        maturityperiod=period;
        rateindex=(maturityperiod*term);
        accuracy(prec,iteration);
        return newtraph(inputvalue);
    }

        public static void main(String[] args)
        {

            NewtonYield CalcBond= new NewtonYield(0.05,1e-6,20);
            System.out.println("RESULT "+CalcBond.yieLd(100.0,2.0,
                                    11.0,108.120,3.0));

            }
}
```

LISTING 1.11. Class **NewtonYield**

We will use a 'console' based approach to handling input/output for the setting up of data handling parameters. A console application is largely independent of the java version being used. We would normally expect to use a graphical interface to provide user choice, but the interfaces to the various components are JVM sensitive.

The Markowitz approach to investment appraisal is based on the end of period value of an asset or portfolio of assets. Emphasis is thus on the expected returns and deviation (as a measure of risk) of a portfolio. The data in Table 1.9 shows the pre-processed data for ten securities. This data is input to SelectPortfolio (shown in Listing 1.12) and used to provide the expected return for the portfolio of these assets. The output from the class Portfolio is shown in Table 1.10.

SelectPortfolio uses data handling techniques (case switch) to categorise the input type and organise appropriate data structures (ArrayList). Based on the input style selection, the appropriate method from class Portfolio is used to implement the algorithm for calculating the expected return for a portfolio.

TABLE 1.9. Input data for SelectPortfolio

Asset	No Shares	Initial Price	Expected Return %
A	100	40	16.2
B	200	35	24.6
C	100	62	22.8
D	150	30	21.3
E	100	31	22.1
F	300	17	16.6
G	180	22	15.0
H	200	10	13.7
I	120	40	12.5
J	100	54	11.3

TABLE 1.10. Output from method retInitprice in class Portfolio

Expected return for a portfolio:			
A investment per share	4000.0	proportion	0.08684324793747286
B investment per share	7000.0	proportion	0.1519756838905775
C investment per share	6200.0	proportion	0.13460703430308293
D investment per share	4500.0	proportion	0.09769865392965697
E investment per share	3100.0	proportion	0.06730351715154147
F investment per share	5100.0	proportion	0.1107251411202779
G investment per share	3960.0	proportion	0.08597481545809814
H investment per share	2000.0	proportion	0.04342162396873643
I investment per share	4800.0	proportion	0.10421189752496743
J investment per share	5400.0	proportion	0.11723838471558837

Starting valuation: 46060.0
Expected portfolio return 18.132870169344333

ArrayList, which is available from Java 1.2 and above, can store heterogeneous data as a single list structure. This is a powerful structure which will dynamically grow to accommodate the input. The ArrayList class is a utility in the Java.util package. The ArrayList object stores other objects not primitive data types. We will see in later chapters how this datatype can be fully exploited in dealing with large data volumes from external sources.

In this first example we will look at the general structure being used to implement the basic algorithms. This simple example uses the same broad methodology that more complicated algorithms will use. Much of the code is supporting the data manipulation required for data sets of unknown size and content mix. We are using case selection “2” where the input is the pre processed data of Table 1.9. SelectPortfolio uses the flexibility of the ArrayList datatype to store input data. The supporting code is used to do basic housekeeping on the data array. The case switch calls the methods folioreturns and retInitprice from the class Portfolio. The remaining case switches can select different input data styles and choose the appropriate processing methods from the Portfolio class.

```
package FinApps;
import BaseStats.inputmod;
import static.java.lang.Math.*

public class SelectPortfolio extends Portfolio
{

    public static void main(String[] args)
    {
          int vals=0;
          double data=0.0;
          int numshares=0;
          int numshare=0;
          double price=0.0;
          double initialvalue=0.0;
```

```
        double expectret=0.0;
        double endprice=0.0;
        String name=" ";
        SelectPortfolio port=new SelectPortfolio();

        System.out.println("CHOOSE TYPE of DATA INPUT");
        System.out.print("ENTER 1 For Raw Data(monthly %): 2 For
                         % Expected Returns: 3 For
Expected End Price:");
        int inputtype=inputmod.readInt();
        switch(inputtype)
        {
              case 1: System.out.println("Enter the NUMBER of
                                    securities to be processed");
        int numelements=inputmod.readInt();
        port.insertnumsec(numelements);
        System.out.println("Enter the NUMBER of monthly returns to
                            be processed for all ");
              vals=inputmod.readInt();
              port.datasize(vals);
        for (int i=0;i<numelements; i++)
        {
              System.out.println("Enter the NAME of the securiy to
                                  be processed");
              name=inputmod.readString();
              port.insertstring(name);
                      //adds new securities to the list//
              port.offsetsize(1);//default value.. can accommodate a//
                                //series of other non-data headers//

              for(int j=0;j<vals;j++)
              {
              System.out.println("Enter the EXPECTED Monthly % Return
                                 for the security "+name);
              data=inputmod.readDouble();
              port.insertdata(data);//adds data to the list//
              }

        }


        port.propanalysis();
        break;

        case 2: System.out.println("Enter the number of securities
                                    to be processed");
        int nums=inputmod.readInt();
        for (int i=0;i<nums; i++)
        {
              System.out.println("Enter the NAME of securiy
                                  to be processed");
              name=inputmod.readString();
        System.out.println("Enter the NUMBER of issues purchased
                           for "+name);
              numshares=inputmod.readInt();
        System.out.print("Enter the INITIAL PRICE of securities
```

```
                            for "+name);
              initialvalue=inputmod.readDouble();
              System.out.println("Enter the EXPECTED % RETURN
                                for the security "+name);
              expectret=inputmod.readDouble();
              port.folioreturns(name,numshares,initialvalue,
                               expectret);//Adds data to the list//
       }

              port.retInitprice();
              break;
              case 3: System.out.println("Enter the NUMBER of
                                        securities to be processed");
        int numelem=inputmod.readInt();
              for (int i=0;i<numelem; i++)
       {
              System.out.println("Enter the NAME of securiy to be
                                processed");
              name=inputmod.readString();
       System.out.println("Enter the NUMBER of issues purchased for
                            "+name);
        numshare=inputmod.readInt();
       System.out.print("Enter the INITIAL PRICE of securities for
                             "+name);
               initialvalue=inputmod.readDouble();
              System.out.print("Enter the EXPECTED END PRICE
                                of securities for "+name);
               endprice=inputmod.readDouble();
                     port.folioendvals(name,numshare,initialvalue,
                                        endprice);
                                         //Adds data to the list//

       }

    port.retendvals();

       break;
       default: System.out.println("Enter the type of
                                    securities !!!!!!!!! ");
    }

    }

}
```

Listing 1.12. Class **SelectPortfolio**

Examining 'case 2', in Listing 1.12. The input loop is set by the number of securities to be processed. The variables name, numshares, initialvalue and expectret are passed to the method *port.folioreturns* (method in class **Portfolio**). Since we are using the ArrayList to store data the method *folioreturns* manipulates the data into Object wrappers. Listing 1.13 shows the method *folioreturns*. The method adds primitive data from *SelectPortfolio* to the ArrayList structure by wrapping each primitive datatype in its Object wrapper. The wrapper classes are in the package Java. Lang.

```
public void folioreturns(String sname,int numshares,
                          double initialprice,double
expectedrets)// for
expected end period share price//
        {
                ArrayList folioentry=getFolio();
                folioentry.add(sname);//0 entry....index number//
                folioentry.add(new Integer(numshares)); // WRAPPERS //
                folioentry.add(new Double(initialprice));
                folioentry.add(new Double(expectedrets));//3 entry//

        }
```

LISTING 1.13. Method *folioreturns* from class **Portfolio**

Class Portfolio implements the following equation in the method retInitprice: The method is shown in Listing 1.14.

$\overline{R_P} = \sum_{i=1}^{n} X_i \overline{R}_i$ (0.4.4), where the $\overline{R}_P$ = the expected return for the portfolio X_i = the proportion invested for security I,$\overline{R}_i$ = the expected return of security I and n = the number of securities.

```
public void retInitprice()
{
   double tots=0.0;
   double proportion=0.0;
   double initialportval=0.0;
   double totalinvest=0.0;
   double portfolioreturn=0.0;
   ArrayList folioentry=getFolio();
   final int collectionsize=folioentry.size();
                                    //get the size of the array//

   for (int i=3;i<collectionsize;)
   {                                // LOOP 1//
      Double totals=(Double)folioentry.get(i-1);
                                    //Initial market price//
      Integer totalnums=(Integer)folioentry.get(i-2);
                                       //number of shares//
      initialportval+=(totals.doubleValue()*totalnums.intValue());

      i=i+4;

   }
      for (int j=3;j<collectionsize;)
   {                                    // LOOP 2 //
      Double sums=(Double)folioentry.get(j);//Expected returns//
      Double totalsinitial=(Double)folioentry.get(j-1);
                                       //Initial market price//
```

```
        Integer totalnumsinitial=(Integer)folioentry.get(j-2);
                                            //number of shares//
        String security=(String)folioentry.get(j-3);
                                            //name of security//
        tots=(totalsinitial.doubleValue()*
              totalnumsinitial.intValue());//
total investment per share//
        proportion=tots/initialportval;
                  //as a proportion of the initial porfolio valuation//
        totalinvest=sums.doubleValue()*proportion;
                              //expected return (%) * proportion//
        portfolioreturn+=totalinvest;
                              // Gross portfolio expected return//
        j=j+4;

        }
        System.out.print (" Start Period Valuation of Portfolio
                             :"+initialportval);
        System.out.print ("Expected Valuation of Portfolio Return
                             :"+portfolioreturn);

        folioentry.clear ();

}
```

LISTING 1.14. Method retInitprice in class **Portfolio**

From listing 1.13. Loop 1 retrieves Objects from folioentry. This is done within the for loop by taking 'blocks' of data. We know the position of the data as String 0, Integer 1, Double 2, Double 3. The loop also does some arithmetic on the data and produces the total initial value of the portfolio. Loop 2 returns the primitive data types from the Object wrappers and performs the remaining arithmetic for the portfolio.

1.4.5. Portfolio Risk Measurement

Our last example for the application classes makes use of methods from FinApps and BaseStats. This application is to produce a risk measure for a portfolio. The risk associated with a portfolio is related to the dispersion of individual assets around the expected value and the degree of correlation between assets. We can analyse a portfolio for dispersion in terms of the standard deviation of the portfolio, where the variances and covariance's are combined (Variance-Covariance matrix) or we can separate the variances from the covariances.

The equation for the variance of a portfolio where variance and covariance terms are separated is:

$$\sigma_P^2 = \sum_{j=1}^{n} X_j^2 \sigma_j^2 + \sum_{j=1}^{n} \sum_{\substack{k=1 \\ k \neq j}}^{n} X_j X_k \sigma_{jk} \tag{1.4.5}$$

TABLE 1.11. Monthly Stock Returns for three assets.

A	B	C
13.05	14.0	5.0
14.2	3.0	8.88
3.76	7.2	5.56
1.76	26.1	30.0
2.98	0.07	4.0
−2.8	7.0	0.33
−7.0	−5.9	6.0
−2.0	2.0	0.01
2.0	15.0	1.2
14.0	−6.7	12.0
6.0	−1.1	2.99
−0.99	10.0	1.5

The equation for the combined variance of a portfolio is:

$$\sigma_P^2 = \left[\sum_{i=1}^{n}\sum_{j=1}^{n} X_i X_j \sigma_{ij}\right] \tag{1.4.6}$$

In this example we will use Equation 1.1.10 to compute the standard deviation of the portfolio. This is simply:

$$\sigma_P = \left[\sum_{i=1}^{n}\sum_{j=1}^{n} X_i X_j \sigma_{ij}\right]^{\frac{1}{2}} \tag{1.4.7}$$

The class SelectPortfolio uses the case statement to select the appropriate data structures and methods in class Portfolio to implement Equation 1.1.11. The input data is from Table 1.11. SelectPortfolio uses case '1' to select the option for processing raw data. Referring back to Listing 1.11. we see that the various data structures are set up and data entered to the ArrayList. On completion the method *proPanalysis* in class **Portfolio** is called, which then implements Equation 1.1.11. *proPanalysis* is shown in listing 1.15

```
public void proPanalysis()
      {
            ArrayList riskdata=getRisk();
            ArrayList folioentry=getFolio();
            ArrayList rawdata=getRaw();
            int blocksize=getDatalength();
            int nos=getEntrynums();
            double[] compare=new double[blocksize];
            double[] codata=new double[blocksize];
            double[][] covalues=new double[blocksize][2];
            int size=rawdata.size();
            double riskvalue=0.0;
```

```
        Double er;
        Double covars;
        int a=0;
        int comp=0;
        int end=0;
        int gets=0;
        while ( end<nos)
                                {
        for(a=0;a<blocksize;a++)
        {
                Double value=(Double)rawdata.get(comp);
                compare[a]= value.doubleValue();     // LOOP 1 //
                covalues[a][0]=value.doubleValue();
                comp++;
        }
    for(int counter=0;counter<size;)  // LOOP 2 //
    {
                        for(int b=0;b<blocksize;b++)
                {
                Double covalue=(Double)rawdata.get(counter);
                codata[b]=covalue.doubleValue();
                                                // LOOP 3 //
                covalues[b][1]=covalue.doubleValue();
                                counter++;
                }
        double cors=DataDispersion.covar(covalues);
        double[] answer2=DataDispersion.variances(covalues);
        double[] meanvals=DataDispersion.dumean(covalues);
                        riskdata.add(new Double(meanvals[0]*
meanvals[1]*cors));
                        }
    end++;
    }
    for(int d=0;d<riskdata.size();d++)
    {
        er=(Double)riskdata.get(d);
        double tempout=er.doubleValue();            // LOOP 4 //
        riskvalue+=tempout;
                        riskvalue=sqrt(riskvalue);
    }
    }
```

LISTING 1.15. Method *proPanalysis* in class Portfolio

Loop 1 deals with basic data handling and retrieves the raw data into intermediate array structures. Loop 2 performs the basis of the double summation calculation for Equation 1.1.11. Calls are made to three methods from **DataDispersion** these are *cover, variances and dumean.* Loop 2 also computes the factor; mean x1 * mean x2 * covariance for entry to variance-covariance matrix. Loop 2 provides the data which 'virtually' produces the matrix as shown in Table 1.12.

Loop 4 uses the *riskdata* ArrayList to directly compute the standard deviation.

The values within the matrix are then multiplied by the expected returns for each security and entered into the *riskdata* ArrayList. Finally Loop 4 provides

TABLE 1.12. Variance Covariance Matrix

	A	B	C
A	43.952338888888896	1.961638888888887	30.936955555555556
B	1.961638888888887	104.927290972222	21.83175555555555
C	30.936955555555556	21.83175555555555	61.94372222222221

Portfolio Standard Deviation: 83.10626977168779

the completion of Equation 1.1.11 by summing all of the entries and returning the square root.

References

Elton, E.J. and E., Gruber. *Modern Portfolio Theory and Investment Analysis* 4th edition. Wiley.

Martelleni, L., P. Priaulet and S. Priaulet. *Fixed-Income Securities Valuation, Risk Management and Portfolio Strategies*. Wiley Finance.

2
Interest Rate Calculations

2.1. Compound Interest

2.1.1. Nominal and Effective Interest

Spot rates are interest rates based on the yield from pure discount bonds (The one year Treasury bond is usual for the one year rate). We will cover bonds and bond yields later on in this chapter. For now we will take it that the spot rate is calculated from the yield on a bond contract. Spot rates are quoted on a period (time related) basis as the 1 year, 2 year or n-year spot rate.

The n-year spot rate is given by:

$$P_n = \frac{V_n}{(1+i_n)} \tag{2.1.1}$$

Where P_n is the current market valuation of a pure discount bond with n years and has a face value of V_n. The spot rate is i_n. In the absence of pure longer term Treasury bonds, the n-year spot rate is calculated by taking the one year calculation and using coupon bearing bond prices. Taking the market price as P_2 and the face value of V_2, with the coupon payment equal to C_1 the 2 year spot rate can be calculated by solving:

$$P_2 = \frac{C_1}{(1+i_1)} + \frac{V_2}{(1+i_2)^2} \tag{2.1.2}$$

This process can continue in an iterative fashion to provide n year spot rates, based on n-year coupons. When spot rates are known the future values of investments or debt can be discounted. For an overview see Adams et al (1993).

Interest rates are time dependent. A rate of 10% per annum is the effective rate over 12 months, whereas the rate of 10% over 3 months is the effective rate for a quarter year and not the effective rate per annum. If a quarterly rate of 10% is extrapolated to an annual rate, the simple value is 10%*4 = 40%, convertible quarterly. Interest offered on deposits with consumer banks is often expressed as the nominal rate with a lesser period conversion. Legislative restrictions for the rate being offered on deposits are sometimes circumvented by offering the nominal rate, but having conversion at frequent periods. The converse is of course true for lending rates.

Effective rates refer to the time periods being quoted and have to adjust to any other period which differs from the quoted basis. As an example an effective annual rate of 4% is adjusted to an effective monthly rate by the relationship; $(1+m)^{12} = 1+i$, which gives us the value $1+m = (1.04)^{\frac{1}{12}}$, which is an effective monthly interest rate of 0.327%.

Nominal interest can be viewed as the quoted rate and the real rate is the quoted rate adjusted for inflation changes as reflected in the cost of living index. If we take the cost of living index as C_0 being the year start and C_1 the year end, the relationship is given as:

$$i_r = \frac{C_0(1+i_n)}{C_1} - 1 \tag{2.1.3}$$

Where i_r is the real interest rate and i_n is the nominal interest rate.

Example 2.0

During a base year with a nominal interest rate of 8% the cost of living index started at 155 and ended at 159. What is the real interest rate ?

$$i_r = \frac{155(1+0.08)}{159} - 1 = 0.0528 = 5.28\%.$$

Equation 2.1.4 shows the general formula for conversion of an annual nominal rate back to the annual effective rate.

$$i = (1+\frac{i_{(n)}}{n})^n - 1 \tag{2.1.4}$$

Where i is the effective annual interest, $i_{(n)}$ the nominal interest and n the number of conversions per annum . Where n>1, the effective interest is greater than the nominal.

Example 2.1

What is the effective annual interest rate for 8.9% per annum convertible 3 monthly?

$i = (1+\frac{0.089}{4})^4 - 1 = 0.09201$. The annual effective rate is therefore 9.2%.

As $n \to \infty$, the limit is reached and continuous compounding of interest takes place. The effective rate is then referred to as the force of interest, e^f. The new annual rate is given by $i = e^f - 1$.

Example 2.2

Given an effective annual rate of 9.2%, what is the force of interest?

$i = e^{0.092} - 1 = 0.088$, which gives us 8.8%.

An annuity certain is the term given to a series of payments (in arrears or advance) for n periods. The total (accumulated) value for such a series of payments is $S_{\overline{n}|}{}^{r\%}$ and is arrived at as follows:

Given a rate of interest r%, $S_{n\rceil}^{r\%} = 1 + (1+r) + \ldots + (1+r)^{n-1}$, a geometric series with common difference $(1+r)$ and n terms. The sum becomes $s_{n\rceil}^{r\%} = \frac{(1+r)^n - 1}{r}$. If the payments are in advance, this value is $S_{n\rceil}^{r\%} = \frac{(1+r)^n - 1}{r}(1+r)$.

Example 2.3

1. What is the accumulated value of $321.89, paid in arrears each period for 6 periods at an interest rate of 6%?
2. What is the accumulated value, if paid in advance?

1. $\frac{(1+0.06)^6 - 1}{0.06} = 6.975. = 6.975 * 321.89 = 2245.18$. This gives us an accumulated value of $2245.18.
2. $\frac{(1+0.06)^6 - 1}{0.06} * 1.06 = 7.393 = 7.393 * 321.89 = 2379.73$. This gives us an accumulated value of $2379.73.

These formulae are outlined in Listing 2.1 for the class **Intr.**

```
public final class Intr
{
  private static double cl0=0;
  private static double cl1=0;
   public Intr ()
   {
     this.cl0=100; // Sensible default values are put here for the index
     this.cl1=104;
   }
   public Intr (double a, double b)
   {
     this.cl0=a; // Calling with proper defaults is the preferred way
     this.cl1=b;
   }
   public static double realintr(double nintr)     //  implements
   {
     return 100*((cl0*(1+nintr)/cl1)-1.0);
   }

   public static double erate(double intr,double convertp

   {                                                    // Implements

     return pow((1+(intr/convertp)), convertp)-1;

   }

   public static double fint(double intr)

   {
     return log(1+intr);                      // implements
   }
```

$i_r \frac{C_0(1+i_n)}{C_1} - 1$

$i = (1 + \frac{i_{(n)}}{n})^n - 1$

$i = e^f - 1$

```
public static double ancertain(double intr, double n)
{                                          // Implements S_n]^r% = ((1+r)^n - 1)/r
   return ((pow((1+intr), n)-1)/intr;
}

public static double ancertainAd(double intr, double n)
{
   return (((pow((1+intr), n)-1)/intr)*(1+intr));
                              //Implements S_n]^r% = ((1+r)^n - 1)/r (1+r)
}

public static double pvancert(double intr, double n)
{
   return (1.0-(1/pow((1+intr),n)))/intr;
                                  // Implements ä_n] = (1 - 1/(1+r)^n)/r
}

public static double pvancertAd(double intr, double n)
{
            return((1+intr)*(1.0-(1/pow((1+intr),n)))/intr);

                          //Implemets ä_n] = (1 - 1/(1+r)^n)/r *(1+r)
}

   public static double pvainfprog(double intr, double growth,
                                   double value)
{
   return value/intr-growth;
                           //if growth =0, this is a perpetuity
                           // Implements A/(r-v)
}
   public static double pvanmult(double intr, double n)
{
   double value=1/(1+intr);
   return ((pvancertAd(intr, n))-(n*pow(value, n)))/intr;

                              // Implements (Ia)_n] = (ä_n] - na^n)/r
}
  public static double effectintp(double annualintr,double p)
  {
  return pow((1+annualintr), (1/p))-1;  //given the effective
                                          // annual int rate
                                          // returns the nominal rate
  }
```

```
    public static double effectann(double nomnualintr,double p )
     {
       return (pow((1+nomnualintr/p), p)-1);//given a nominal rate
                                               //returns the effective rate
     }
}
```

LISTING 2.1. Calculation of compound interest

2.2. Present Value (PV)

Present values relate the value at today's prices of an amount sometime in the future. Present value calculations are used to value the worth of future payments (the discount). Future value calculations can be based on a so-called *riskless* basis with known interest rates, or can be based on varying interest and changing risk environments.

A simple form of the PV formula for a single amount is given as:

$$PV = \sum_{i=1}^{n} \frac{1}{(1+r)^i} \tag{2.2.1}$$

This formula can be generalised for a series of varying amounts (A_t) as: $PV = \sum_{i=1}^{n} \frac{A_t}{(1+r)^t}$

This formula calculates the series of equal payments (in arrears) over a given time frame (n). Where, i is the time period and r is the interest rate.

2.2.1. *Compounding Cashflows*

The class ***PresentValue*** contains methods to calculate present values of a series of cashflows. There are three basic methods in the class. As shown in Listing 2.2 below.

```
package FinApps;
import static.java.lang.Math.*;
public final class PresentValue {

   /** creates a new instance of PresentValue */
   public PresentValue() {
   }
   public double pV(double[] discounts,double[]cashflows)
   {
      int n=cashflows.length;
      double presval=0;
```

```
    for(int i=0;i<n;i++)

    {

       presval+=discounts[i]*cashflows[i]; // returns sum of
                                           // discounted values..  1/(1+r)^i
                                           //for each period cashflow

    }
    return presval;
  }


  public double pV(double r,double[] cashflows)
  {
    int indx=1;
    double sum=0;
    for(int i=0;i<cashflows.length;i++)
    {

       sum+=(cashflows[i]/(pow((1+r),(indx))));//Implements
                                   //PV = \sum_{i=1}^{n} A_1/(1+r)^i

         indx++;
    }
    return sum;
  }

  public double pV(double r,double cash,int period)
  {
    double sum=0;
    int indx=1;
     for(int i=0;i<period;i++)
    {
       sum+=(cash/(pow((1+r),(indx))));// Implements PV = \sum_{i=1}^{n} 1/(1+r)^i
         indx++;
    }
    return sum;
  }
}
```

LISTING 2.2. **PresentValue**

A series of equal payments can also be represented more conveniently as:

$$a_{n\rceil} = \frac{1 - \dfrac{1}{(1+r)^n}}{r} \tag{2.2.2}$$

Formula 2.2.2 is dealt with in Listing 2.1 for the class Intr.

The formulas in 2.2.1 and 2.2.2 apply for cash flows with annual compounding and fixed interest rates for each period.

Example 2.4

What is the present value of \$10, paid annually, for 3 years at 5% interest.

Using 2.2.1 the PV of \$10 per annum over 3 years at 5% annual interest is:

$$PV = \sum_{i=1}^{3} \frac{1}{(1+r)^n} = \frac{10}{(1+0.05)^1} + \frac{10}{(1+0.05)^2} \frac{10}{(1+0.05)^3}$$
$$= 9.523 + 9.070 + 8.638$$

This gives the sum of \$27.31.

Using 2.2.2 we get, $a_{\overline{n}|} = \frac{1-\frac{1}{(1+r)^n}}{r} = \frac{1-\frac{1}{(1+0.05)^3}}{0.05} = 2.731$, multiply this by the payment of \$10, = \$27.31. The latter formula is used to construct standard tables of compound interest.

A series of equal advance payments is the same as 2.2.2, with each discount being a single period fewer. The formula is:

$$\ddot{a}_{\overline{n}|} = \frac{1 - \frac{1}{(1+r)^n}}{r} {}^*(1+r) \qquad (2.2.3)$$

Example 2.5

What is the present value of \$10, paid annually in advance, for 3 years at 5% interest?

$$\ddot{a}_{\overline{n}|} = \frac{1 - \frac{1}{(1+0.05)^3}}{0.05} {}^*(1+0.05) = 2.859^*\$10 = \$28.59$$

2.2.2. *Perpetuity and Annuity*

A *perpetuity* is a series of indefinite payments of fixed amounts and is related to 2.2.2 by:

$$perpetuity = \frac{1 - \frac{1}{(1+r)^\infty}}{r} * (amount\,\$) \qquad (2.2.4)$$

A growing *perpetuity* of indefinite payments is one where each payment is some multiple of the previous (growth). If the payments (A) grow by some amount (1+v) then the PV is given by:

$\frac{A}{(1+r)} + \frac{A(1+v)}{(1+r)^2} + \frac{A(1+v)^2}{(1+r)^3} + \ldots$ if growth remains below the interest rate (r>v) this series reduces to:

$$\frac{A}{(r-v)} \tag{2.2.5}$$

An increasing annuity is one where the first payment is one unit, the second is 2 units... the nth payment is n units. If each payment is represented as $a = \frac{1}{(1+r)}$, then the PV is given by the sequence $a + 2a^2 + \ldots + na^n$. Multiplying by $(1+\text{r})$. We get $1 + 2a + 3a^2 + \ldots + na^{n-1}$, subtracting $a + 2a^2 + \ldots + na^n$ gives us $\text{PV*r} = 1 + a + a^2 + \ldots + a^{n-1} - na^n$. The PV of an increasing annuity (usually written $(Ia)_{n]}$) can therefore be represented by:

$$(Ia)_{n]} = \frac{\ddot{a}_{n]} - na^n}{r} \tag{2.2.6}$$

Example 2.6

What is the value of an increasing annuity with a first payment of € 107 being paid for 6 periods with an interest rate of 6%?

$$(Ia)_{n]} = \frac{\ddot{a}_{n]} - na^n}{r} \text{ Firstly, calculate 2.2.3 , } \ddot{a}_{n]} = \frac{1 - \frac{1}{(1+0.06)^6}}{0.06} {}^*(1+0.06)$$

$$= 6.9753$$

$$\text{Thus, } (Ia)_{n]} = \frac{6.9753 - 6^* \left(\frac{1}{1+(0.06)}\right)^6}{0.06} = 16.3766. \text{ Multiplying by € 107}$$

$$= \text{€ } 1752.29$$

2.3. Internal Rate of Return

The internal rate of return can be defined as that interest rate which balances the total output of cashflow with the total input of cashflow for an investment over a given time frame. The basic equation for this is the *equation of value*, which equates the amount put in with the amount received. Satisfying this equation gives the yield for a particular investment. As an example consider an investment opportunity, which offers $8,000 after 7 years, for an initial outlay of $3,000.

The yield is r per annum. So, $3,000(1+r)^7 = 8,000 = 14.9\% = 14.9\%$.The equation of value discounts all amounts to the same point in time. The equation of value is summarised:

$$r = n\sqrt{\frac{return}{investment}} \tag{2.3.1}$$

The internal rate of return is therefore the rate of interest for which the amounts paid out equal the amount paid in. Although the yield can be calculated for any period, it is common to construct the calculation from *present time*. Yield equations with multiple cash flows produce polynomials of degree n, which means there are n possible roots to the equation. When solving these problems this is something we should check for.

The equation of value for Internal Rate of Return (IRR) connects the amounts paid into an investment with the amounts going out of that investment. The general equation for the polynomial representing an IRR is constructed to have the sum of positive cash flows minus the negative cash flows on one side equal to zero on the other side. If we have a series of cash flows (positive) from C_1 to C_n with an initial (negative) cash flow of C_0, the formula of 2.3.2 shows that solving for r, the rate of return will give us the value required to reach zero.

$$\sum_{n=1}^{n} \frac{C_n}{(1+r)^n} - C_0 = 0 \tag{2.3.2}$$

We have covered the IRR as an example in Chapter 1. The Listing shown in 1.10 for **YieldBisect** and 1.11 for **NewtonYield** provide details of these two classes being used to provide IRR calculations.

2.4. Term Structures

The term structure of interest rates defines the various rates applicable throughout the life of a particular financial instrument. Methods employed in defining the term structure allow us to provide a discount factor for PV calculations that incorporate varying spot rates and approximations of forward interest rates.

2.4.1. Rate Interchanges

The various rates; discount, forward and spot are all interchangeable. If we define the PV of a set of abstract cash flows (abstract, as they relate to a *default-free* security) as:

$$PV = \sum_{t=1}^{n} C_t d_t \tag{2.4.1}$$

The d_t-discount factors are equivalent to $(1+r_r)^{-t}, t = 1, 2, \ldots n$

The sum of the individual cash flows times the discount factor at a given period (t) gives the price (at the discounted prices). This is effective for discrete compounding periods. An analogous situation is the PV of a continuous compounding of interest:

$$PV = \sum_{t=1}^{n} C_t e^{-r_t} \tag{2.4.2}$$

Both of these market discount functions relate to the prevailing spot rates at times $t_1, ..t_2..t_n$

Thus, the spot rate from 2.4.2 can be evaluated from the discount rate in 2.4.1 as:

$$r_t = \frac{-l_n(d_t)}{t} \tag{2.4.3}$$

The discount rate is related to the spot rate by:

$$d_t - e^{r_t t} \tag{2.4.4}$$

The forward rate can be determined by the yield from a future borrowing time frame. If a future frame is defined as times $t_1, ..t_2$. The forward rate can be constructed as that rate which incorporates the spot rate from present to the start of the frame and the (estimated) forward (spot) for the time frame periods. From 2.4.2 and 2.4.3 this can be derived as: $R_f = e^{r_t} e^{f_{t_1 ..t_2}}$. , which can be expressed as :

$$R_f = \frac{l_n(\frac{d_{t_1}}{d_{t_2}})}{t_2 - t_1} \tag{2.4.5}$$

Where R_f is the forward rate and d_{t_1}, d_{t_2} are the discount rates at times t_1 and t_2.

Effectively a forward rate can be viewed as a series of discounts. As an example.

£10 due in two years with a two year spot rate of 10% and a one year spot rate of 8% has two equivalent series of values.

The PV = £8.264 for two years and the equivalent one year value discounted by the forward rate(the rate between one and two) thus, $\frac{\frac{10}{(1+R_f)}}{(1+0.8)}$ must equal £8.264, which is 12.03%.

In general, the relationship between the one year spot rates and forward rates can be expressed as:

$$(1 + R_{f_{1,2}}) = \frac{(1 + r_2)^2}{(1 + r_1)} \tag{2.4.6}$$

Forward rates can also be calculated from yields as:

$$R_{f_{1,2}} = r_2 \frac{t_2}{t_2 - t_1} - r_1 \frac{t_1}{t_2 - t_1} \tag{2.4.7}$$

The general formulae for interest terms is shown in class **Interms** shown in Listing 2.5.

Example 2.7

Given the one year spot rate of 7% and the two year spot rate of $7\frac{1}{4}$%. Provide the one and two year discount rate factors, together with the forward rate. Also, given a one year discount factor of 0.95 and a two year discount factor of 0.825. What are the spot and forward rates?

For the first part:

For time t=1. $d_{t=1} = e^{-0.07t=1} = 0.9323$. For time t = 2 $d_{t=2} = e^{-0.0725t=2} = 0.8650$. The forward rate is given by $(R_{f_{1,2}}) = \frac{(1+0.0725)^2}{(1+0.07)} - 1 = 0.0750$.

For the second part:

For time t = 1. $r_{t=1} = \frac{-l_n(0.95)}{1} = 0.0512$. For time t = 2 $r_{t=2} = \frac{-l_n(0.825)}{2} = 0.0961$. The forward rate is given by $R_f = \frac{l_n(\frac{0.95}{0.825})}{2-1} = 0.141$.

2.4.2. *Spot Rates*

Spot rates are related to the treasury yield curves . The yield curves are derived from prices of treasury securities. A simple 'bootstrapping' technique to derive spot rates from yield values can be used to derive a theoretical spot rate curve, which reflects the changing (daily) interest rate climate over the life of a security. The requirement for good data sources is paramount in constructing spot rates. This is true whether we use bootstrapping or more sophisticated curve fitting techniques.

The techniques involved with interest rate movements are based on government bond market prices, the 'on- the- run' prices of government (riskless) bonds are thought by some to be the best indicator of market rates, others prefer to include the whole bond market, including 'off- the- run' bonds. On- the- run securities such as T bills that pay no coupon and trade near to par are the preferred securities to start off the bootstrapping process, since coupon paying securities with a series of cash flows can be viewed as a series of zero coupon securities with maturity payments equal to the cash flows.

Since the spot rate is defined by a zero coupon government bond and subsequent spot rates are extracted from coupon bearing securities, using their price and maturity, the spot rate follows the yield curve of all the securities used in the spot rate construction process. The US treasury daily yield curve covers so called CMT's (Constant Maturity Treasury rates) for fixed periods of 1,3 and

6 months, together with 1,2,5,7,10 and 20 years. These rates can be used directly as the current spot rate. Alternatively we can compute our own spots based on direct market data of choice.

Example 2.8

A 1 year T bill with a yield of 6% is issued; this defines the present spot rate, at the same time a 2 year T note is issued with a 10% coupon and price of £95. The 2 year spot rate then becomes;

$$95 = \frac{10}{1.06} + \frac{110}{(1+s_2)^2}, \text{ so } 95 = 9.433 + \frac{110}{(1+s_2)^2}$$
$$\text{therefore } (1+s_2)^2 = 1.2855 \text{ and } s_2 = 13.37\%$$

For a semi annual coupon bond the spot rate for the *nth* period is given by the solution to:

$$P_n = C\sum_{j=1}^{i-1} \frac{1}{(1+s_j)^j} + \frac{C+100}{(1+s_i)^i} \quad (2.4.8)$$

Where P_n the price per 100 units of par, C is the semi annual coupon per 100 units of par and s is the spot rate.

The spot rate is then given by:

$$s_i = \left[\frac{C_n + 100}{P_n - C_n \sum_{j=1}^{i-1} \frac{1}{(1+s_j)^j}} \right]^{\frac{1}{i}} - 1 \quad (2.4.9)$$

When we have a list of the spot rates for a given period, it is possible to calculate the required coupon to price the security at par. This is given by:

$$C_n \frac{1 - \frac{1}{(1+s_n)^n}}{\sum_{i=1}^{n} \frac{1}{(1+s_i)^i}} \quad (2.4.10)$$

Equation 2.4.10 has a computationally more efficient form as;

$$C_n = \frac{1.0 - e^{-nl_n(l+s_n)}}{\sum_{i=1}^{n} e^{-il_n(1+s_i)}} \quad (2.4.11)$$

Exercise 2.3

The following prices and coupons are available from on the run issues,

Price	Coupon	Maturity	Yield
99.908	1.78	2	1.83
99.735	2.26	3	2.36
99.908	3.16	5	3.18
99.822	3.67	7	3.70
99.675	4.14	10	4.18
98.759	4.92	20	5.02

The T-bill yield for the 6 months is 1.03 and for the 12 months is 1.28. What is the 3 year spot rate?

From 2.4.8 :

99.735 = sum of cash flows discounted by the spot rates. The first period spot rate is 1.03 and the second period is 1.28. The coupon is 2.26, so the first two terms are;

$$99.735 = \frac{2.26}{(1+0.0128)} + \frac{2.26}{(1+0.0183)^2} \text{ so, } 99.735 - 4.410$$

$$= \frac{100+2.26}{(1+s_3)^3} \text{ therefore,}$$

$$95.3245 = \frac{102.26}{(1+s_3)^3} \text{ and } s_3 = ((\sqrt[3]{1.0727}) - 1) = 0.02368$$

s_3 is solved by using Equation 2.4.9. This is implemented in the class **Spots** shown in Listing 2.3

Spots provides the overloaded method *spotFcoupon* for calculating the spot rates, the first method provides spot rates for annual coupons the second invocation provides the addition of period adjustments to the algorithm. **Spots** also has a method *parCoupon* that provides functionality to compute the coupon for a par yield as outlined in Equations 2.4.10 and 2.4.11.The application code for Exercise 2.4 is shown in Listing Exercise 2.3.

```
package FinApps;
import java.util.*;
import java.lang.*;
import static java.lang.Math.*;

public class Spots {

    public Spots() {
    }
```

```
public double[] spotFcoupon(double[][]pcdata)
 {
   int n= pcdata.length;
   double[]spots=new double[n];
   double price;
   int s=0;
   double indx=1.0;
   spots[0]=((100.0/pcdata[0][0])-1);
   price=(pcdata[0][0]/100.0);
   for(s=1;s<n;s++){
      indx++;
      spots[s]=(exp(1/indx*log((pcdata[s][1]+100.0)
               /(pcdata[s][0]-(pcdata[s][1]*price))))-1);
      price+=(exp(-indx*log(1+spots[s])));

   }
   return spots;
 }

public double[] spotFcoupon(double[][]pcdata,int periods)
               // for period frequency of annual coupons
 {
   int n=pcdata.length;
   double[]spots=new double[n];
   double price;
   double temp=0;
   int s=0;
   double indx=1.0;
   spots[0]=((100.0/pcdata[0][0])-1);
   price=(pcdata[0][0]/100.0);/* first entry */
      for(s=1;s<n;s++){
      indx++;
      spots[s]=(exp(1/indx*log(((pcdata[s][1]/periods)+100.0)/
               (pcdata[s][0]-((pcdata[s][1]/
               periods)*price))))-1);
      price+=(exp(-indx*log(1+spots[s])));
 }
   return spots;
}
/* returns the n period coupon for par price given the spot rate */
public double parCoupon(double[]spots,int nperiod)
{
   int i=spots.length;
   int j=0;
   int counter=0;
   double flowdisc=0.0;
   double finaldisc=0.0;
   if(nperiod>i){

      return -1.0;

   }

   finaldisc=(1.0-(exp(-nperiod*log(1.0+spots[(nperiod-1)]))));
```

```
    for(double d:spots)
    {
       if (j<nperiod)
       {
    j++;
    flowdisc+=((exp(-j*log(1.0+d))));

       }
  }
       return(finaldisc/flowdisc);
}

}
```

LISTING 2.3. Implementation of class **spots**

Listing 2.3 shows the abstract class **Interms**. This class provides methods for calculating the various transformations of interest rates. The class is not implemented directly but is accessed through its extending class.

```
public class Exercise_2_3 {

  /** Creates a new instance of Exercise_2_3 */
  public Exercise_2_3() {
  }
  public static void main(String[] args) {
    //Assumes par value of 1,000
    NumberFormat formatter=NumberFormat.getNumberInstance();
    formatter.setMaximumFractionDigits(2);
    formatter.setMinimumFractionDigits(2);
    Spots s=new Spots();
    double[][]testdata=
    {

       {98.736,0.0},
       {99.908,1.78},
       {99.735,2.26},
       {99.908,3.16},
       {99.822,3.67},
       {99.873,4.14},
       {98.759,4.92}

    };

    double[]ansx=s.spotFcoupon(testdata);
    for(double i:ansx){
       System.out.println("THE SPOT RATE IS == "+i);
    }
  }
}
```

LISTING 2.4. Application code for Exercise 2.3

```
package FinApps;
public abstract class Interms {
public abstract void intermstimes();

 public Interms() {
 }

 public double disFromyld(double spotrate,double time)
 {

   return exp(-spotrate*time);          //Implements d_t = e^{-r_t t}

 }

 public double yldFromdisc(double discount,double time)
 {

   return -log(discount)/time;          //Implements r_t = -l_n(d_t)/t

 }

 public double forateFromspts(double spot1,double spot2)
 {

   return (pow((1+spot2),2)/(1+spot1)-1);
                             // Implements (1+R_{f1,2}) = (1+r_2)^2/(1+r_1)

 }

 public double forateFromdisc(double discount1,double discount2,
                              double time1,double      time2)
 {
   return (log(discount1/discount2)/(time2-time1));

                             //Implements R_f = l_n(d_{t1}/d_{t2})/(t_2 - t_1)

 }

 public double forateFromyld(double r1,double r2,double t1,double t2)
 {

   return (r2*(t2/(t2-t1)))-(r1*(t1/(t2-t1)));
                             //Implements R_{f1,2} = r_2 t_2/(t_2-t_1) - r_1 t_1/(t_2-t_1)

 }

}
```

LISTING 2.5. Implementation of class **Interms**

The interest term structures are divided into flat rate structures and the varying rate structures. Both sets of structures will make use of the underlying methods in the abstract class **Interms.**

2.4.3. Deriving the Spot Curve

Flat term structures take a single interest rate and provide the corresponding discount, spot or forward rate for the given time period. The varying rate structures take a range of input interest rates and compute the appropriate output rates. The input rates can be raw data taken from the markets (e.g. zero coupons) or more usually, rates from published yield curve constructions. Published curve data for the UK and US Debt Management Office (DMO) provides government estimates of the government bond market prices. DMO tables are constructed from the yield curve, which in turn, is constructed from coupon paying bonds that are adjusted through a cubic spline model. The U.S DMO model provides an estimate of the zero coupon yield, by reading off the daily curve at fixed points (maturities) that are 1 month, 3 months, 6 months, 1 year, 2 years, 3 years, 5 years, 7 years, 10 years and 20 years.

There are two basic issues in constructing current interest rate data structures. Firstly, the construction of an adequate model based on market data (this is an issue addressed by the DMO type techniques and methodology). Secondly, having data structures based on the adequate model which can provide valid estimates of periods between maturity periods. We will not examine the first issue, although it is worth noting that there are a range of possible construction models that could be used to derive actual market metrics for the yield rates. Anderson et al (2001).

The published daily DMO type data points offer discrete points from one month to 20 years, these points (which are themselves the result of interpolation from the 'yield curve') offer a direct value, or for intermediate values of time, say 4 years, data for further interpolation. An excerpt from the US daily yield data is shown in Table 2.1.

TABLE 2.1. US Treasury daily yield data

Date	1 mo	3 mo	6 mo	1 yr	2 yr	3 yr	5 yr	7 yr	10 yr	20 yr
08/02/04	1.28	1.50	1.78	2.12	2.66	3.06	3.68	4.10	4.48	5.22
08/03/04	1.37	1.48	1.77	2.11	2.66	3.05	3.67	4.08	4.45	5.20
08/04/04	1.34	1.49	1.76	2.11	2.66	3.05	3.66	4.08	4.45	5.20
08/05/04	1.34	1.48	1.75	2.09	2.64	3.04	3.64	4.05	4.43	5.18
08/06/04	1.36	1.44	1.67	1.91	2.40	2.79	3.40	3.84	4.24	5.04
08/09/04	1.40	1.51	1.73	1.97	2.45	2.86	3.45	3.88	4.28	5.06
08/10/04	1.42	1.50	1.75	2.01	2.55	2.94	3.52	3.94	4.32	5.08
08/11/04	1.40	1.44	1.73	2.00	2.54	2.91	3.51	3.92	4.30	5.07
08/12/04	1.31	1.43	1.73	1.99	2.52	2.89	3.47	3.89	4.27	5.05
08/13/04	1.32	1.44	1.72	1.97	2.47	2.85	3.42	3.85	4.22	5.02

For times within the raw data range that have no corresponding observed data points, interpolation is used to provide a yield. The interpolation method we will use is Lagrange interpolation, which is sufficiently accurate for up to around seven points. If we look at the lower end of DMO data, the Lagrange method would be applied for the periods between 1 month and 5 years. For points at the upper end we would use 3 years to 20 years. Interpolation between 10 and 20 years would not be of a high accuracy. Times selected outside of the data table are regarded as errors.

Listing 2.6 shows the class **Intermstructure** that provides methods for computing the basic and current rate structures. The class extends methods in the abstract class **Interms**. This gives access to basic transformations between the various discount, spot and forward rate structures. In addition this class introduces the facility for computing the current interest rate from table data, using a Lagrange interpolator from the class **Interpolate**.

Current yield data is enabled by the method *setCurrentRateData* that sets a *current_flag* variable and allows the current interest based methods to compute forward, spot and discount rates. The method *Errorcheck* provides the class with simple error checking for existence of yield data and validity of requests for time periods, ensuring that interpolation is only done within the data time periods.

```
package FinApps;
import CoreMath.*;

public class Intermstructure extends Interms {
 Interpolate It=new Interpolate();
                       //'It' is an object which allows the computation of
 private int current_flag=0;        // the Lagrange interpolation
 private double[][]currentdata;    // This is the yield data
                                    // in the order ..time/rates
 public Intermstructure()
 {
 }

 public double DiscpOne(double interate,double time_1) {

   return disFromyld(interate,time_1);// As in Listing 2.6

 }

 public double SpotpOne(double interate,double time_1) {

   return yldFromdisc(interate,time_1);// as in Listing 2.6
 }
 public double Forwdisc(double interate_1,double interate_2,
                         double time_1,double time_2) {
   return forateFromdisc(interate_1,interate_2,time_1,time_2);
                  // As in Listing 2.6
 }
 public double Forwyld(double interate_1,double interate_2,
                         double time_1,double time_2) {
```

```
    return forateFromyld(interate_1,interate_2, time_1,time_2);
                  // As in Listing 2.6
  }
  public void setCurrentRateData(double[][]yielddata) { // Provides the
                                                         // yield data
    currentdata=yielddata;
     current_flag=1; // sets a flag to register that current data is
                  // available
  }
  public double getCurrentDiscOne(double timepoint_1) {
                  //computes the current discount factor
      return Errorcheck(timepoint_1)==1? disFromyld(It.lagrange
                       (currentdata, timepoint_1),timepoint_1):0.0;
                  //Does error checking to see if request and data are
                  // valid if not this returns 0.0

  }

  public double getCurrentSpotOne(double timepoint_1) {// computes the
                  // current spot rate for the input timepoint
    return Errorcheck(timepoint_1)==1? It.lagrange(currentdata,
                                               timepoint_1):0.0;
  }
    public double getCurrentForwardrateYlds(double timepoint_1,
                                          double timepoint_2) {
                                     // computes forward spot rates
      return(Errorcheck(timepoint_1)==1&Errorcheck(timepoint_2)==1)?
      (forateFromyld(getCurrentSpotOne(timepoint_1),
      getCurrentSpotOne(timepoint_2), timepoint_1,timepoint_2)):0.0;
  }
  public void Intermstimes() {
                  // This implements the abstract method from Interms
  }
  private int Errorcheck(double timepoint){
                  // Method provides basic error checking
    if(current_flag==0)
                  // checks to see if there is data from current set method
    {
      System.out.println("Error:no data array found for yield");
      return 0;
    }
     int n=currentdata.length;
           if((timepoint<currentdata[0][0])||
                                        (timepoint>currentdata[n-1][0]))
                  // checks for bounds of calling method
{
      System.out.println("Error:time variable out of data range");
      return 0;
    }
    return 1;// if successful
  }
}
```

LISTING 2.6. Implementation of class Intermstructure

TABLE 2.2. Derived spot rates from yield curve

N	Spot	Yield
0.50	1.030	1.030
1.00	1.280	1.280
1.50	1.553	1.551
2.00	1.835	1.830
2.50	2.114	2.104
3.00	2.375	2.359

Exercise 2.4.

Extending Exercise 2.3. Determine the semi annual spot rates for a 1.5 and 2.5 year maturity.

Firstly we need to use interpolation and find the intermediate half year yield values then compute the price of each security given a coupon rate. Finally we compute the spot rates from the interpolated price data and the quoted coupon and coupon frequency. The solution is outlined in Table 2.2.

Thus, the spot rate for 1.5 year maturity is 1.553 and the spot for 2.5 years maturity is 2.114.

The application code is shown in Listing Exercise 2.7

```
package FinApps;
import java.text.*;
import java.io.*;
public class Exercise_2_4 {

  public Exercise_2_4() {
  }

  public static void main(String[] args) {

      NumberFormat formatter=NumberFormat.getNumberInstance();
    formatter.setMaximumFractionDigits(3);
    formatter.setMinimumFractionDigits(2);
    Intermstructure i=new Intermstructure();
    Spots s=new Spots();
    Volatility vol=new Volatility(100.0,2.0);
     double[][] xydat=
    {
       {0.5,1.03},
       {1.0,1.28},
       {2.0,1.83},
       {3.0,2.36}
    };
    int j=0;
    double mat=0.5;
    double firstpoint=i.lagraninterp(xydat,1.5);
    double secondpoint=i.lagraninterp(xydat,2.5);
    double[] yields={1.03,1.28,firstpoint,1.83,secondpoint,2.36};
```

```
    double[] coupons={0.0,0.0,(firstpoint-0.4),1.53,
                        (secondpoint-0.3),2.0};
    double[][]pcdata=new double[6][2];
    for(double yld:yields){
    pcdata[j][0] =vol.Bpricing(yld,mat,coupons[j]);
    pcdata[j][1]=coupons[j];
    mat+=0.5;
    j++;
    }

    int n=0;
     double[]ansx=s.spotFcoupon(pcdata,2);
       for(double x:ansx)
       {
       System.out.println("THE SPOT RATE IS ==
  "+formatter.format((200.0*x))+" For PRICE ==
  "+formatter.format(pcdata[n][0])+" and COUPON ==
  "+formatter.format(pcdata[n][1])+" YIELD ==
  "+formatter.format(yields[n]));
       n++;
       }
       }
}
```

LISTING 2.7. Application code for Example 2.4

Exercise 2.5

Using the daily yield data for years 1 to 7 from Table 2.1, for each security trading at par show the required par yield, in terms of the coupon level and compare to the yield curve. Calculate the discount rate and implied forward rates. Compare these rates with data from 2nd and 13th of August.

Since the table provides spot rates directly we only need to do simple interpolation of the data to get the 4 and 6 year values. Both data sets are shown in Table 2.3.

TABLE 2.3. Interpolated spot rates from Treasury yield estimates

Daily 02/08		Daily 13/08	
MATURITY	SPOT%	MATURITY	SPOT%
1.000	2.120	1.000	1.970
2.000	2.660	2.000	2.470
3.000	3.060	3.000	2.850
4.000	3.389	4.000	3.157
5.000	3.680	5.000	3.420
6.000	3.930	6.000	3.653
7.000	4.100	7.000	3.850

Using the data of Table 2.3, the required par yield for each maturity is calculated using Equation 2.4.10. This provides us with required coupon level for a security trading at par.

The required yield (both periods) for an annual coupon payment is shown in Table 2.4.

The comparisons of par yield to spot rates are shown in Figures 2.1A for the period 02/08 and 2.1B for period 13/08.

In both periods the par yield follows the yield curve shape, with the par rates being less than the yield.

The discount rate and forward rate are computed from the derived yield curve using Equations 2.4.4 and 2.4.7. The data is shown in Table 2.5.

The relationship between the various rates is shown in Figure 2.2 for the period 02/08.

TABLE 2.4. Par yield and Spot rates for varying maturities, based on par trading securities with an annual coupon.

MATURITY	PAR YIELD(02/08)	SPOT	PAR YIELD(13/08)	SPOT
1.0000	0.0212	0.0212	0.0197	0.0197
2.0000	0.0265	0.0266	0.0246	0.0247
3.0000	0.0304	0.0306	0.0283	0.0285
4.0000	0.0336	0.0339	0.0313	0.0316
5.0000	0.0363	0.0368	0.0338	0.0342
6.0000	0.0386	0.0393	0.0359	0.0365
7.0000	0.0402	0.0410	0.0378	0.0385

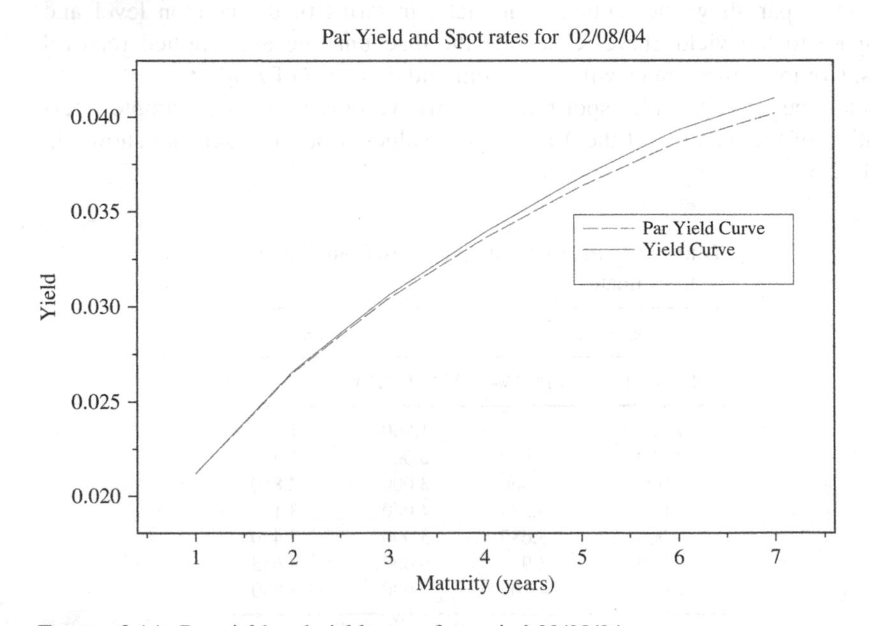

FIGURE 2.1A. Par yield and yield curve for period 02/08/04.

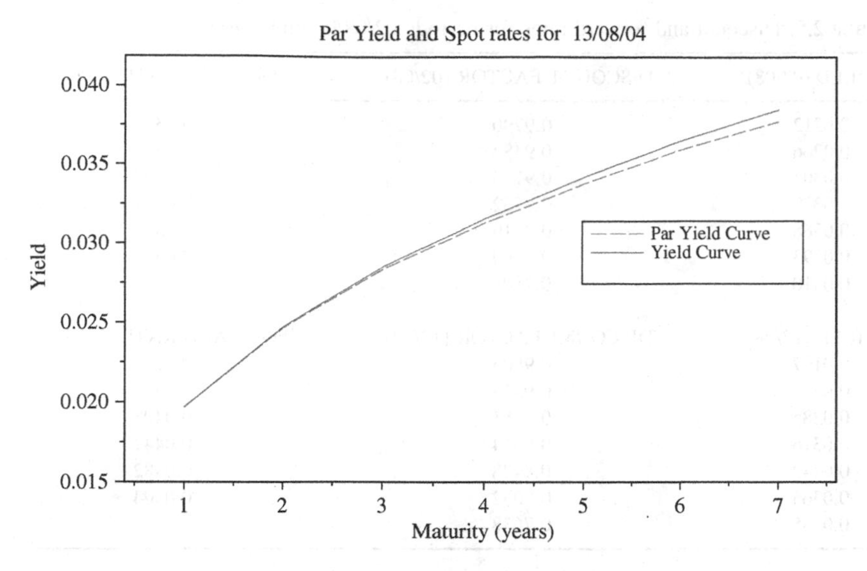

FIGURE 2.1B. Par yield and yield curve for period 13/08/04.

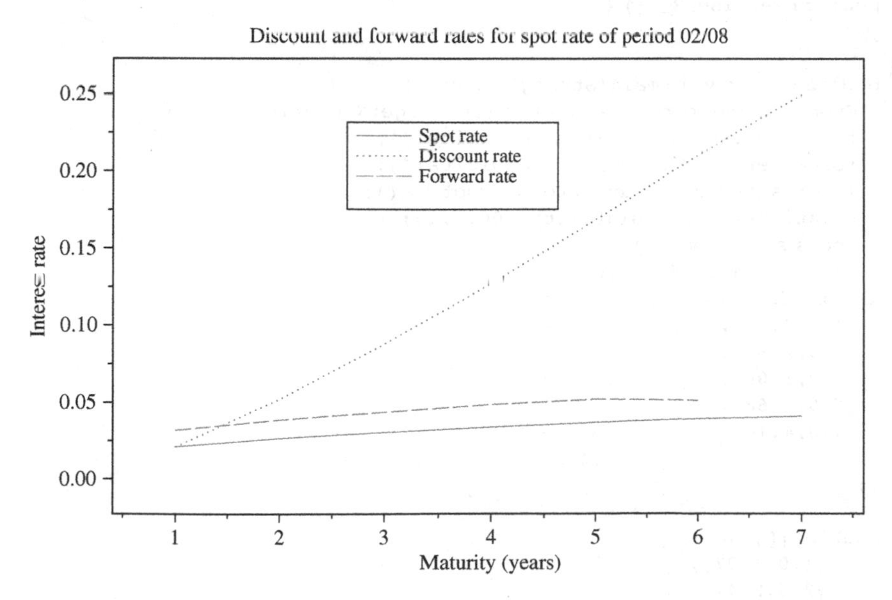

FIGURE 2.2. Relationship between Spot, Discount and Forward rates.

TABLE 2.5. Discount and forward rates for periods 02/08/04 and 13/08/04

YIELD (02/08)	DISCOUNT FACTOR (02/08)	FORWARD RATE (02/08)
0.0212	0.9790	0.0320
0.0266	0.9482	0.0386
0.0306	0.9123	0.0438
0.0339	0.8732	0.0484
0.0368	0.8319	0.0518
0.0393	0.7899	0.0512
0.0410	0.7505	
YIELD (13/08)	DISCOUNT FACTOR (13/08)	FORWARD RATE (13/08)
0.0197	0.9805	0.0297
0.0247	0.9518	0.0361
0.0285	0.9181	0.0408
0.0316	0.8814	0.0447
0.0342	0.8428	0.0482
0.0365	0.8032	0.0504
0.0385	0.7638	

The application code is shown in Listing Exercise 2.8

```
package FinApps;
import java.text.*;
import java.io.*;
public class Exercise_2_5 {

 public Exercise_2_5() {
 }

 public static void main(String[] args) {
   NumberFormat formatter=NumberFormat.getNumberInstance();
   formatter.setMaximumFractionDigits(4);
   formatter.setMinimumFractionDigits(4);
   Intermstructure i=new Intermstructure();
   Volatility v=new Volatility(100.0,1);
   Spots s=new Spots();
   double[]vals={4.0,6.0};
  double dat02[][]= {
   {1.0,2.12},
   {2.0,2.66},
   {3.0,3.06},
   {5.0,3.68},
   {7.0,4.10}

 };

 double[][]dat13= {
      {1.0,1.97},
      {2.0,2.47},
      {3.0,2.85},
```

```
    {5.0,3.42},
    {7.0,3.85}
};
double[] reqdyield_02;
double[] reqdyield_13;
 double[][]newdata ;
 double[][] dat_02;
 double[][] dat_13;
  int frwd;
frwd=2;// conter variables
  dat_02=i.lagraninterp(dat02,vals);
  dat_13=i.lagraninterp(dat13,vals);// get the interpolated data

  try{

    PrintWriter w=new PrintWriter(new FileWriter
                ("c:\\data_for_13_5B.txt"),true);
    PrintWriter pw=new PrintWriter(new FileWriter
                 ("c:\\data_for_02_5B.txt"),true);

  System.out.println(" Daily 02/08 Daily 13/08 ");
  System.out.println(" MATURITY SPOT MATURITY SPOT");


   for(int j=0;j<dat_02.length;j++)
  {
  System.out.println(" "+formatter.format(dat_02[j][0])+
                     " "+formatter.format(dat_02[j][1])+
                     " "+formatter.format(dat_13[j][0])+
                     " "+formatter.format(dat_13[j][1]));
  dat_02[j][1]=(dat_02[j][1]/100.0);
  dat_13[j][1]=(dat_13[j][1]/100.0);
       // put data into decimal from percentages
  }
 System.out.println(" MATURITY PAR YIELD(02/08)
                      SPOT PAR YIELD(13/08) SPOT");
 reqdyield_02=s.parCoupon(dat_02);
 reqdyield_13=s.parCoupon(dat_13);
       // get the required yield in terms of the coupon level required

  for(int j=0;j<dat_13.length;j++)
    {
      System.out.println(" "+formatter.format((j+1))+
                        " "+formatter.format(reqdyield_02[j])+
                        " "+formatter.format(dat_02[j][1])+
                        " "+formatter.format(reqdyield_13[j])+
                        " "+formatter.format(dat_13[j][1]));
  }
i.setCurrentRateData(dat_02);
    System.out.println( "YIELD (02/08) DISCOUNT RATE (02/08)
                         FORWARD RATE (02/08) ");
     for(int j=0;j<dat_02.length;j++)
  {
     System.out.print(formatter.format(dat_02[j][1])+" "
                     +formatter.format(i.getCurrentDiscOne((j+1))));
```

```
    pw.print((j+1)+","+dat_02[j][1]+","+
           ((1.0-(i.getCurrentDiscOne(j+1))))+",");
   if(frwd<(dat_02.length+1))
   {
      System.out.println(" "+formatter.format
               (i.getCurrentForwardrateYlds((frwd-1),(frwd))));
      pw.println(i.getCurrentForwardrateYlds((frwd-1),(frwd)));
frwd++;
}
}
System.out.println();
frwd=2;
   i.setCurrentRateData(dat_13);
    System.out.println( "YIELD (13/08) DISCOUNT RATE (13/08)
                            FORWARD RATE (13/08) ");
    for(int j=0;j<dat_13.length;j++)
  {
    System.out.print(formatter.format(dat_13[j][1])+" "
                +formatter.format(i.getCurrentDiscOne((j+1))));
    w.print((j+1)+","+dat_13[j][1]+","
            +((1.0-(i.getCurrentDiscOne(j+1))))+",");
   if(frwd<8)
   {
      System.out.println(" "+formatter.format
               (i.getCurrentForwardrateYlds((frwd-1),(frwd))));
      w.println(i.getCurrentForwardrateYlds((frwd-1),(frwd)));
   frwd++;
   }
   }
w.println(" ");
pw.println(" ");
      pw.close();
               w.close();
 }
 catch(IOException foe)
               {
               System.out.println(foe);
}
 }
}
```

Listing 2.8. Application code for Example 2.5

References

Adams, A., D. Bloomfield, P. Booth and P. England. *Investment Mathematics and Statistics.* Graham & Trotman.

Anderson, N. and J. Sleath. *New Estimates of the UK Real and Nominal Yield Curves.* Pub Group Bank of England 2001 (ISSN 1368-5562).

3 Bonds

3.1. Bonds – Fixed Interest

Bonds are loan products which can be traded before their repayment date. Bonds carry an interest rate (coupon) on the loan together with a face value (par). The market price is the current trading value of a bond. There are three basic types of bond: For an overview see the book by Fabozi.

1. Domestic, carrying local currency. These include products such as Treasury notes, bonds and local government notes/bonds
2. Foreign bonds carry local currency and are traded in the local markets. The borrower (Issuer) is however a foreign government or corporation
3. Eurobonds carry a variety of currencies and are issued by banking syndicates. It is a means of governments and corporations to borrow in the denominated currency from non domestic lenders. The various bond markets pay interest according to the rules applied for the market and type of bond. Bonds can be purchased at any time between interest payment (settlement) dates.

The accrued interest is that amount which is due between the last settlement date and the present transaction date. The prices quoted in the market does not generally include the accrued interest (which nevertheless is paid) and is referred to as the 'clean' price. Prices which do reflect accrued interest are referred to as 'dirty' prices.

The UK and Japanese markets calculate interest on a 365 day basis. In common with the US treasury bonds, these bonds pay interest twice yearly. Eurobonds pay interest annually. Markets vary in the way accrued interest is calculated, with Japan and UK being based on a 365 day year, Eurobonds a 360 day year and US treasury issues based on elapsed time between coupons. For any market the general points to consider for accrued interest are the coupon dates, the settlement date and for some bonds (including UK Gilts) whether the issues are *cum* or *ex* dividend (with or without the next coupon payment). UK Gilts go ex-dividend 5 weeks 2 days prior to the next payment date. With the consolidation of financial instruments throughout the Euro zone. Many European countries have harmonised to an Actual/actual day count convention. The table below refers to existing and pre-harmonising ('legacy') bonds in the market. Some countries have 'reconventioned' the day counts for existing bonds and others are in a ‘phasing’ stage of moving to a fully harmonised convention. Many European bonds have a mixture of conventions for the day count conventions; the specific issues provide the relevant details.

As an example Irish government Euro denominated bonds outstanding from June 2004 have benchmark bonds on an annual basis using the actual/actual convention with annual bonds using 30E/360 and semi annual bonds paying actual/365.

US Treasury Bills are issued once a week for the 13 and 26 week bills and once every fourth week for the 52 week bill. The selling method is by multiple price bidding. It is possible to buy these issues on a competitive or non competitive basis. Treasury bills are sold at auction with a discount and pay no coupon. The pricing consists of a bid near to the face value, e.g. a number of 13 week bills might be bid at 98.754. If this bid wins the investor buys a number of $1000 bills at $987.54 that can be redeemed for this face value in 13 weeks. The non competitive bidder agrees to accept the issues at the end of auction average price for the accepted competitive bids. The average price is calculated from the range of highest competitive bids for the issues which remain (after taking out the allocation for non competitive bidders). The weekly and four weekly current issues are referred to as on-the-run. Issues which are older than 1 week (or 4 weeks for the 52 week issue) are referred to as off the run; there is a secondary market for both on-the-run and off- the-run bills.

Treasury bills can be purchased by an investor through the US federal banks or on-line through the internet. The market is also served by domestic banks and brokerages. The pricing mechanisms offered by the secondary markets are published widely and conform to standard practices. The quoted prices are in terms of the ‘bank discount method’ on the bid (what the broker will pay) and ask (what the broker will sell at) price. The calculation year for bills has 360 days, so a bill quoted as 128 days to maturity with a bid of 4.95% means that this discount is effectively 360/128 of the actual discount. To find the actual discount multiply by 128/360. So 4.95% becomes an actual discount of 1.760% of the face value. For an ask price of 4.93, the actual discount is 1.752%. The difference between the bid and ask price is the so-called dealer spread $(1.7600 - 1.7528 = 0.0072)$. Published data usually includes the equivalent yield which is calculated by annualising the ask cash discount divided by the purchase price. For this example (a nominal $100) the ask discount is $(100\% - 1.7528\%) = \$98.247$ giving a yield of $1.752/98.247 = 0.01784. = 1.784\%$ (rate of return). Annualising this, $1.784 * 365/128 = 5.087\%$.

Treasury notes issued by the US government have redemption (maturities) periods from 1 to 10 years with coupons usually being twice annually. Notes are issued with nominal values of $1000 and above. Monthly auctions are held for the two and five year notes, every three months three and ten year notes are auctioned. The secondary market is also quite liquid for notes. The quoted prices are in terms of $\frac{1}{32}\%$ for bid and ask, with the annual rate being quoted as a percentage. In the published data a bid or ask shown in the form: $101:01$ which equates to $101\frac{1}{32}$. The published data also usually includes the gross yield to maturity based on the currently quoted ask. The ask yield is quoted as the annual effective value. The dealer's spread is calculated as the difference between the bid and ask prices.

A treasury note (from settlement on 8th October 2004) with the following quotes:

Rate	Maturity Mo/Yr	Bid	Asked	Chg	Ask Yld
$5\frac{7}{8}$	Nov 06	100:10	100:12	+2	5.68

Has dealer's spread of $100\frac{10}{32}\% - 100\frac{12}{32}\% = -\frac{2}{32}\%$. Thus for a nominal $1,000 note the spread is $0.625. The coupon is $5\frac{7}{8}\%$ paid semi annually at $2.9375\% =$ $29.37. The bid changed from the previous day by $+\frac{2}{32}\%$. The effective annual yield is 5.68%. This is based on the ask price of 100:12.

The annual yield can be calculated (if not available) as the internal rate of return. By using Equation 2.3.4 In the form:

$$\frac{5.875}{i}\left\{1 - \frac{1}{(1+i)^{\frac{2.0849}{0.5}}}\right\} + \frac{1000}{(1+i)^{\frac{2.0849}{0.5}}} - 1003.75$$

The yield quote for treasury bonds is based on the ask price. However the **actual** price paid, which would include the accrued interest will alter the ask yield to an actual yield based on the total market price. In our example (and assuming the coupon payment date is also the 8th) there is accrued interest of 153 days divide this by $183 = 0.836$. Multiplying by the annual rate applied twice yearly (5.875/2) gives us $24.55 per $1,000 note. The actual cost for the note is therefore $\$1003.75 + \$24.55 = \$1028.30$. The yield is now 4.438%.

The fraction 2.0849 is given by the time to maturity of 2 years 31 days. The 31 days are made up from 24 days in the settlement month (inclusive of settlement day) and 7 days in the maturity month (exclusive of the maturity day) giving, $2 + \left(\frac{31}{365}\right)$. The fraction 0.5 is the term period for coupon payments (6 months) as a ratio of 12 months. This particular equation will of course give the semi annual yield (the proportion of the term period).

Listing 3.1 Shows the class **Accruedconvention** which computes the accrued time between settlement and the next coupon date for the various day count conventions of the major markets as outlined in Table 3.1. The class has two public methods *daycounts* and *getPreviousCouponDays*. The *daycounts* method uses a case switch to select the six major day count conventions. The method *daycounts* is called with a variable *flagvalue* which determines the market selection implemented via the case switch. The two remaining variables are *Calendar* objects for the settlement date and next coupon. The accrued time in this class is based on the total number of days in the convention for the market. The number of days between the settlement date and the next coupon is taken, according to market convention. The number of days is then divided by the total to produce a weighting factor. The conventional expression of accrued interest takes into account the coupon basis (number of days in the coupon period).

TABLE 3.1. Bond day count conventions

Some Day Count Conventions		
Market	Coupon Frequency	Count
US Gov't	Semi annual	Actual/Actual
UK Gov't	Semi annual	Actual/Actual
Australian Gov't	Semi annual	Actual/Actual
Austrian Gov't*1	Annual	30E/360
Belgian Gov't	Annual	30E/360
Canadian Gov't	Semi annual	Actual/Actual
Danish Gov't	Annual	30E/360
Dutch Gov't	Annual	30E/360
Eurobond*1	Annual	30E/360
French Gov't*1	Annual	Actual/Actual
German Gov't*1	Annual	30E/360
Irish Gov't*1	Annual	Actual/365
Italian Gov't*1	Annual	30E/360
Japanese Gov't	Semi annual	Actual/365
New Zealand Gov't	Semi annual	Actual/Actual
Norwegian Gov't	Annual	Actual/365
Spanish Gov't	Semi annual	Actual/Actual
Swedish Gov't	Annual	30E/360
Swiss Gov't	Annual	30E/360

*1 denotes actual/actual reconventioned from 2000

The application program using this class would make the appropriate market basis adjustment by a multiple of the coupon frequency. The *daycounts* method returns the accrued value proportion for the period settlement to next coupon date. The method *getPreviousCouponDays* returns the proportionate value for the period last coupon to settlement, which is the portion of accrued time due to a seller.

```
package FinApps;
import java.util.*;
import static java.util.Calendar.*;
public class Accruedconvention {
   /** Creates a new instance of Daycounts according to the convention */
      //Default constructor assumes 2*6 monthly coupon payments...//
      //semi annual//
   public Accruedconvention() {
     this.coupons=2.0;
   }
   public Accruedconvention(double couponperiod)
      // non default constructor to set the coupon period monthly times
per annum
   {
     this.coupons=12.0/couponperiod;
   }
   public double getPreviousCoupondays() {
     return previouscoupondays;
```

```
}
private void setPreviousCoupondays(double prevcoupdate) {
  this.previouscoupondays=prevcoupdate;
}
Calendar previouscoupon=Calendar.getInstance();
private double coupons;
public double previouscoupondays;
public double daycounts(int flagvalue,Calendar settlementdate,
    Calendar nextcoupondate) {
  Calendar temp=Calendar.getInstance();
  previouscoupon.set(YEAR,(nextcoupondate.get(YEAR)));
  previouscoupon.set(MONTH,(nextcoupondate.get(MONTH)-6));
                                   // assumes default 6 monthly period
  previouscoupon.set(DATE,(nextcoupondate.get(DATE)));
  int actualday=0;
  int actualdays=0;
  int samedays=0;
  switch(flagvalue) {

    case 1:
        if(settlementdate.get(MONTH)==nextcoupondate.get(MONTH))
        //Actual/actual in period (eg, US gov)
    {
        samedays=(nextcoupondate.get(DATE)
              -settlementdate.get(DATE));
        for(int n=(previouscoupon.get(MONTH)+1);
              n<nextcoupondate.get(MONTH);n++) {
          temp.set(MONTH,n);
          actualdays+=temp.getActualMaximum(DAY_OF_MONTH);
        }
        actualdays+=(previouscoupon.getActualMaximum(DAY_OF_MONTH)
                -previouscoupon.get(DATE));
        setPreviousCoupondays((double)(actualdays-samedays)/
                           (actualdays));
        return (double)samedays/actualdays;
    }

    int setdays= (settlementdate.getActualMaximum(DAY_OF_MONTH)-
              settlementdate.get(DATE));
    actualday=setdays;

    for(int i=(settlementdate.get(MONTH)+1);
          i<nextcoupondate.get(MONTH);i++){
        temp.set(MONTH,i);
        actualday+=temp.getActualMaximum(DAY_OF_MONTH);


    }
    actualday+=nextcoupondate.get(DATE);
    actualdays=nextcoupondate.get(DATE);
    temp.clear();
    for(int n=(previouscoupon.get(MONTH)+1);
            n<nextcoupondate.get(MONTH);n++){
        temp.set(MONTH,n);
        actualdays+=temp.getActualMaximum(DAY_OF_MONTH);
    }
```

```
actualdays+=(previouscoupon.getActualMaximum(DAY_OF_MONTH)
            -previouscoupon.get(DATE));
setPreviousCoupondays((double)(actualdays-actualday)
                        /(actualdays));
return (double)actualday/actualdays; //returns fraction of the
                                      // coupon period


case 2:
    for(int n=(previouscoupon.get(MONTH)+1);
            n<settlementdate.get(MONTH);n++) {
      temp.set(MONTH,n);
      actualdays+=temp.getActualMaximum(DAY_OF_MONTH);
    }
    actualdays+=(previouscoupon.getActualMaximum
                (DAY_OF_MONTH)-previouscoupon.get(DATE));
    actualdays+=settlementdate.get(DATE);

    if(settlementdate.get(MONTH)==nextcoupondate.get(MONTH))
                      //Actual/365 (eg,UK gov)
    {


      samedays=(nextcoupondate.get(DATE)
                -settlementdate.get(DATE));
      setPreviousCoupondays((double)(((365.0/coupons)
                            -samedays)/(365.0/coupons)));
                  //requires annual multiple of coupon rate
      return (double)(samedays/(365.0/coupons));
                  //returns 1/365 ths of the annual coupon rate
    }
    actualday= (settlementdate.getActualMaximum(DAY_OF_MONTH)
              -settlementdate.get(DATE));

    for(int i=(settlementdate.get(MONTH)+1);
            i<nextcoupondate.get(MONTH);i++) {
      temp.set(MONTH,i);
      actualday+=temp.getActualMaximum(DAY_OF_MONTH);

    }
    actualday+=nextcoupondate.get(DATE);
   System.out.println("Actual days between coupon
                      and settlement =="+actualdays);
    setPreviousCoupondays((double)(actualdays/
                                    (365.0/coupons)));
    return (double)(((365.0/coupons)-actualdays)/
                    (365.0/coupons));


case 3:
    for(int n=(previouscoupon.get(MONTH)+1);
            n<settlementdate.get(MONTH);n++) {
      temp.set(MONTH,n);
      actualdays+=temp.getActualMaximum(DAY_OF_MONTH);
    }
    actualdays+=(previouscoupon.getActualMaximum
                (DAY_OF_MONTH)-previouscoupon.get(DATE));
```

```
        actualdays+=settlementdate.get(DATE);

        if(settlementdate.get(MONTH)==nextcoupondate.get(MONTH))
                                     //Actual/365 or 366 in leap year (eg,
        {
          samedays=(nextcoupondate.get(DATE)
                  -settlementdate.get(DATE));
          int total;
total=(previouscoupon.getActualMaximum(DAY_OF_YEAR)|nextcoupondate
                .getActualMaximum(DAY_OF_YEAR))==366?366:365;
          setPreviousCoupondays((double)(((total/coupons)
                                -samedays)/(total/coupons)));
          return (double)samedays/(total/coupons);
        }
        actualday= (settlementdate.getActualMaximum(DAY_OF_MONTH)
                  -settlementdate.get(DATE));
        for(int i=(settlementdate.get(MONTH)+1);
              i<nextcoupondate.get(MONTH);i++) {
          temp.set(MONTH,i);
          actualday+=temp.getActualMaximum(DAY_OF_MONTH);
        }

        actualday+=nextcoupondate.get(DATE);
        int totaldays;
totaldays=(previouscoupon.getActualMaximum(DAY_OF_YEAR)|
            nextcoupondate.getActualMaximum(DAY_OF_YEAR))
            ==366?366:365;
        System.out.println("Actual days between coupon
                      and settlement =="+actualdays);
        setPreviousCoupondays((double)(actualdays/
                                    (totaldays/coupons)));
        return (double)(((totaldays/coupons)-actualdays)/
                      (totaldays/coupons));
      case 4:
        if(settlementdate.get(MONTH)==nextcoupondate.get(MONTH))
                                     // Coupon annual eg US Gov't agency..
        {
          samedays=(nextcoupondate.get(DATE)
                  -settlementdate.get(DATE));
          setPreviousCoupondays((double)((360.0)-samedays)/
                                (360.0));
          return (double)samedays/(360.0);
        }
        actualday= (settlementdate.getActualMaximum(DAY_OF_MONTH)
                  - settlementdate.get(DATE));
        for(int i=(settlementdate.get(MONTH)+1);
              i<nextcoupondate.get(MONTH);i++) {
          temp.set(MONTH,i);
          actualday+=temp.getActualMaximum(DAY_OF_MONTH);
        }
        actualday+=nextcoupondate.get(DATE);
        setPreviousCoupondays((double)((360.0)-actualday)/
                                    (360.0));
```

```
        return (double)actualday/(360.0);
    case 5:
        if(settlementdate.get(MONTH)==nextcoupondate.get(MONTH))
                          //30/360..one day only..(eg,US corporate)
        {
          int numsetd=settlementdate.get(DATE);
          numsetd=numsetd==31?30:numsetd;
          int numd=nextcoupondate.get(DATE);
          numd=((numd==31)&(numsetd==30))?30:numd;
          samedays=numd-numsetd;
          samedays=(nextcoupondate.get(DATE)
                          -settlementdate.get(DATE));
          samedays=samedays==31?30:samedays;
          setPreviousCoupondays
            ((double)((360.0/coupons)-samedays)/(360.0/coupons));
          return (double)samedays/(360.0/coupons);
        }
        int couponset;
        int dayset=settlementdate.getActualMaximum(DAY_OF_MONTH);
        int dateset=settlementdate.get(DATE);
        dayset=dayset==31?30:dayset;
        dateset=dateset==31?30:dateset;
        actualday=dayset-dateset;
        for(int i=(settlementdate.get(MONTH)+1);
                i<nextcoupondate.get(MONTH);i++) {
          temp.set(MONTH,i);
          couponset=temp.getActualMaximum(DAY_OF_MONTH);
          couponset=((couponset==31)&(dayset==30))?30:couponset;
          actualday+=couponset;
        }
        int coupdate=nextcoupondate.get(DATE);
        coupdate=((coupdate==31)&(dateset==30))?30:coupdate;
        actualday+=coupdate;
        setPreviousCoupondays((double)((360.0/coupons)
                            -actualday)/(360.0/coupons));
        return (double)actualday/(360.0/coupons);
    case 6:
        if(settlementdate.get(MONTH)==nextcoupondate.get(MONTH))
                                                      //30e/360
        {
          int numdays=nextcoupondate.get(DATE);
          numdays=numdays==31?30:numdays;
          int numsetdays=settlementdate.get(DATE);
          numsetdays=numsetdays==31?30:numsetdays;
          samedays=numdays-numsetdays;
          setPreviousCoupondays((double)((360.0/coupons)
                            -samedays)/(360.0/coupons));
          return (double)samedays/(360.0/coupons);
        }
        int couponsettle;
        int daysettle=settlementdate.getActualMaximum
                            (DAY_OF_MONTH);
        int datesettle=settlementdate.get(DATE);
```

```
            daysettle=daysettle==31?30:daysettle;
            datesettle=datesettle==31?30:datesettle;
            actualday=daysettle-datesettle;
            for(int i=(settlementdate.get(MONTH)+1);
                  i<nextcoupondate.get(MONTH);i++){
               temp.set(MONTH,i);
               couponsettle=temp.getActualMaximum(DAY_OF_MONTH);
               couponsettle=couponsettle==31?30:couponsettle;
               actualday+=couponsettle;
            }
            int coupondate=nextcoupondate.get(DATE);
            coupondate=coupondate==31?30:coupondate;
            actualday+=coupondate;
            System.out.println("actualdays "+actualday);
            setPreviousCoupondays((double)((360.0/coupons)-actualday)
                                         /(360.0/coupons));
            return (double)actualday/(360.0/coupons);
            default: throw new AssertionError
                              ("Unknown market :"+flagvalue);
        }
      }
   }
```

LISTING 3.1. Computation of accrued time for major markets

Exercise 3.1

A US corporate issue has an annual coupon of 12% with payment dates of May 17th and November 17th. The settlement date is June 3rd. What % of the coupon is accrued interest for this issue? For 100 units of par value what is the accrued value due to the seller? What value would be due to the seller if this issue was US treasury or an Irish government semi annual issue?

For the first part: Total days between payments = 180. Total time from first payment to settlement = 16 days. The semi annual coupon is 6%. Accrued rate is therefore 16/180 = 8.889%. The value due to the seller is 8.889*0.12/2 = $0.533. For the second part: US treasury has 184 days between coupons. Total time between first payment and settlement =17 days. The accrued value is therefore 9.239*0.12/2 = $0.554. For the Irish Government Issue the value is 17 days for time between coupon payment and settlement. For the Irish issues there are 365 days in the year, which is 365/2 between coupons. The accrued value is therefore 17/(365.0/2.0) = 9.315%. The value is € 0.558904.

The application code for Exercise 3.1 is shown in Listing 3.2.

```
package FinApps;
import java.util.*;
import static java.util.Calendar.*;
import java.text.*;
public class Exercise_3_0 {

   public Exercise_3_0() {
```

```
    }

    public static void main(String[] args) {
        NumberFormat formatter=NumberFormat.getNumberInstance();
        formatter.setMaximumFractionDigits(6);
        formatter.setMinimumFractionDigits(3);
        Calendar settlement=Calendar.getInstance();
        Calendar nextcoupon=Calendar.getInstance();
        settlement.set(2004,5,3);//
        nextcoupon.set(2004,10,17);//next coupon
        double coupon=0.12;
        double couponfrequency=2.0;
        Accruedconvention days=new Accruedconvention();
        enum Markets{
            UStreasury(1),Irishsemi(2),UScorporate(5);
            private final int value;
            Markets(int value){ this.value= value;}
            public int mkt(){return value;}
        };

        for(Markets m:Markets.values())
        {

            days.daycounts(m.mkt(),settlement,nextcoupon);
            double daytimes=days.getPreviousCoupondays();
        System.out.println("For The "+m+ "The accrued % of the coupon ==
                "+formatter.format((daytimes*100.0))+"%"+" accrued value
                per $100 par == $"+formatter.format((daytimes*(coupon
                /couponfrequency)*100.0)));
        }

        }
}
```

LISTING 3.2. Application code for Example 3.1

3.2. Bond Prices

Bond pricing can be based on so-called 'risk-free' (e.g., government issues) or on 'risk-carrying' (public corporations and the like) issues. The simplest bond is risk free and pricing can be based on the value of coupons, market price, face value and environmental factors such as the tax regime, inflation and changing interest rates over the bond's lifetime. The simple bond price is calculated as:

$M_p = \sum_{t=1}^{T} \frac{C_t}{(l+r)^t}$ Where M_P is the market price and C_t is the cashflow at time t, r is the interest rate. Thus the simple fixed interest rate risk free bond is priced by the sum of the discounted cash flows. It is assumed in this equation that the final cashflow payment includes the face value at time $t = T$.

3.2.1. *Interest Yields*

Interest rates for bond yields can be calculated in several ways. The choice depends on the market and the product together with considerations of environmental factors.

Gross and net yields are of interest to investors who are assessing potential income from a particular set of issues. Gross and net yields are calculated by:

Gross Yield (Y_g) = coupon/clean price *100%. Net yield takes into account the investors tax liabilities. Net Yield $(Y_n) = (1-t)^*(Y_g)$. Where t is the tax rate. Expenses can be added (for purchase) or deducted (for sale) as required.

Example 3.1

Bond A is quoted as $218 with a nominal $200 and a coupon of 10%. Bond B is quoted as $327 with a nominal $300 and a coupon of 10%. Calculate the Gross yield. If the investor has a tax liability of 22% calculate the Net yield.

$$\text{For A}: \ (Y_g) = 20/218^*100 = 9.17\%. \ (Y_n) = (1-0.22)^*9.17 = 7.15\%.$$

$$\text{For B}: \ (Y_g) = 30/327^*100 = 9.17\%. \ (Y_g) = (1-0.22)^*9.17 = 7.15\%.$$

3.2.2. *Yield to Maturity*

Bond prices are sensitive to interest rates and the internal rate of return for a bond gives an indication of the required yield to maintain the cashflows. The yield to maturity of a bond is the internal rate of return. Equation 2.3.2 has already been applied to a general IRR calculation. This equation is also applicable to calculating the yield to maturity of a bond.

The Gross Redemption Yield is that yield which is returned annually to a bondholder. The gross yield takes no account of tax or transaction costs. For the US and UK markets with bond's paying a coupon twice annually. Equation 2.3.3 is applicable.

Example 3.2

A US bond has a term of 10 years. The gross yield is 6% and the coupon rate is 7%, paid twice annually. What is the current price?

$$\text{Using Equation 2.3.3., } Mp = \frac{C}{2}a_{2n\rceil} + \frac{100}{(1+r)^{2n}}.$$

We get $\frac{C}{2}a_{2n\rceil} = 3.5^*14.877$. And $\frac{100}{(1+r)^{2n}} = 55.367$. This gives a price of 107.438.

For European bonds the true effective rate would be used, rather than the nominal figure quoted for the US market. The answer in this case would be slightly lower at an initial true effective Gross rate of 6.09%, giving an answer of 106.741. If we have the market price and need to calculate the yield (the usual state of affairs). Equation 2.3.4 and the methods for root finding are applicable.

Japanese bonds use a simple yield calculation rather than the US/European method of compounding. The simple yield is given as:

$$Y = \frac{C}{M_p} + \left\{ \frac{100 - M_p}{M_p} * \frac{1}{n} \right\} \tag{3.1.1}$$

The market price is the dirty price and coupon payments are normally twice annually. Since market prices are usually available, the yield is easily calculated. Japanese government bond's have the coupon rate determined by the auction price and are based on a face value of ¥100.00. The basic dealing unit for purchase is ¥50,000.00.

3.3. Static Spread

The price of a corporate bond is affected by a range of factors which can be benchmarked against the rates and returns of treasury bonds. For a given maturity, coupon and spot rate, a corporate bond may be expected to be priced lower than a treasury issue as a reflection of the inherently higher risk of default.

The static spread of a bond is a reflection of the spread from the spot curve for the entire maturity period. The static spread is therefore some measure of the risk reflected in the bond price. Static spread differs from yield spread which measures the difference in yields to maturity.

The price of a bond which incorporates static spread is given as:

$$P = \sum_{i=1}^{n} \frac{C_i}{(1 + r_s + r(i))^i} + \frac{100}{(1 + r_s + r(n))^n} \tag{3.2.1}$$

Where r is the i period spot rate and r_s is the constant spread rate.

Example 3.3

A series of 11 period spot rates ;{2.5,2.8,3.0,3.5,3.8,4.0,4.9,5.5,6.5,7.9,8.9} are used to provide the riskless bond price of 103.795 for an 11 period, 8% semi annual coupon, with 100 unit par. A corporate bond with the same coupon and maturity is offered at par. Determine the yield and static spread. Also show the static spread for an 8% coupon based on the same spot rates and maturity, for a range of corporate bonds selling from 2.5% below to 2.5% above par. Verify that the static spreads do produce the correct cash flows.

Static spread for P = 100.0: 0.5405

TABLE 3.2. Spread ranges

Price (% of PAR)	Static Spread	Yield Spread
97.500	0.9119	0.8643
98.000	0.8366	0.793
98.500	0.7618	0.7221
99.000	0.6875	0.6518
99.500	0.6138	0.5819
100.000	0.5405	0.5125
100.500	0.4579	0.4435
101.000	0.3884	0.375
101.500	0.3189	0.3069
102.000	0.2494	0.2393
102.500	0.1799	0.1721

Yield to Maturity for Corporate bond with price = 100.0: 8.000
Yield to Maturity of the Treasury bond with price = 103.795: 7.4876.

The yield spread is simply the difference between the yield to maturity of the riskless bond and the yield to maturity of a corporate bond selling at par. This is calculated as: $8.000 - 7.4876 = 0.5124\%$ and static spread is calculated directly by Equation 2.5.2.

The range of spot rates produces the range of spreads as shown in Table 3.2. The static spreads are those values which need to be added to the spot rates, to correctly price the corporate bond. The yield spreads are the differences in yield to maturity for the corporate bonds and the Treasury bond.

For verification that the static spreads do produce the required cash flows, the value of all cash flows for maturity is calculated using the adjusted spot rates. Table 3.3 shows a sample from the range.

Considering the first two prices: Price 1 shows the addition of 0.912 to the original spot rate is required to make the period rates equal to the adjusted rate, that produces the required cash flows. The table shows that the adjusted rates will produce the required cash flows to equate the original price.

Listing 3.3 shows the application code for this example

```
package FinApps;
import static FinApps.Intr.*;
import static java.lang.Math.*;
import java.util.*;
import static FinApps.PresentValue.*;
import java.text.*;
import java.io.*;
public class Example_3_3 {
   public Example_3_3() {
   }

   public static void main(String[] args) {
```

```
NumberFormat formatter=NumberFormat.getNumberInstance();
formatter.setMaximumFractionDigits(4);
formatter.setMinimumFractionDigits(3);
Spread ss=new Spread();
 double facevalue=100.0;
 double terms=6.0;
 double coupon=8.0;
 double maturity=11.0;
int counter=0;
 double[]testprices={97.5,98.0,98.5,99.0,99.5,100.0,100.5,
                     101.0,101.5,102.0,102.5};
 double[] aa=new double[testprices.length];
 double testspots[]={2.5,2.8,3.0,3.5,3.8,4.0,4.9,5.5,
                    6.5,7.9,8.9};
 double sptest[]=new double[testspots.length];
 double[]newspot=new double[testspots.length];
 double pcstat=ss.spreadsT(testspots,100.0,11,8.0,4.0);
 Tyield ct= new Tyield();
 double yldpc=ct.yieldEstimate(facevalue,terms,coupon,
                              100.0,maturity,0.06);
 double yldtr=ct.yieldEstimate(facevalue,terms,coupon,
                              103.795,maturity,0.06);

 System.out.println("Static spread for 100.0 price ==
                    "+formatter.format(pcstat)+" YTM for the
                    100 cor bond =="+formatter.format(yldpc)+"
                    YTM for the low risk bond ==
                    "+formatter.format(yldtr));
 System.out.println("Static spread for P = 100.0
                    :"+formatter.format(pcstat));
 System.out.println("Yield to Maturity for Corporate Bond with
                    price = 100.0 :"+formatter.format(yldpc));
 System.out.println("Yield to Maturity of the Treasury Bond with
                    price = 103.795 :"+formatter.format(yldtr));
 aa=ss.spreadrateS(testspots,testprices,8.0,11,3.0);
 sptest=ss.spreadrateT(testspots,testprices,8.0,11,2.0);
System.out.println(" Price(%of PAR) Static Spread  Yield Spread");
for(double d:testprices)
{
 Tyield t=new Tyield();
 double yields=t.yieldEstimate(facevalue,terms,coupon,d,
                              maturity,0.02);
 double sprd=aa[counter];
 double stat=sptest[counter];

System.out.println(" "+formatter.format(d)+
                   " "+formatter.format(stat)+
                   " "+formatter.format(sprd));

  counter++;
}
formatter.setMaximumFractionDigits(3);
formatter.setMinimumFractionDigits(3);
int indx=0;
counter=0;
```

```
        System.out.println("Spot : Spread = Adjusted rate
                                Price Original ");
        for(double f:sptest)
        {
        for(double d:testspots)
        {
         System.out.println(" "+formatter.format(d)+
                                 " "+formatter.format(f)+
                                 " "+formatter.format((d+f))+" - "+" - ");
         newspot[counter]=(d+f);
        counter++;
        }
        double newpriced=ss.spotPvannualT(newspot,8.0,2.0);
        System.out.println(" "+formatter.format(newpriced)+
                                 " "+formatter.format(testprices[indx]));
        indx++;
        counter=0;

        }
    }
}
```

LISTING 3.3. Application code for Example 3.3

TABLE 3.3. Verification of spread adjusted rates

Spot:	Spread =	Adjusted	Price	Original	Price 1
2.500	0.912	3.412	–	–	
2.800	0.912	3.712	–	–	
3.000	0.912	3.912	–	–	
3.500	0.912	4.412	–	–	
3.800	0.912	4.712	–	–	
4.000	0.912	4.912	–	–	
4.900	0.912	5.812	–	–	
5.500	0.912	6.412	–	–	
6.500	0.912	7.412	–	–	
7.900	0.912	8.812	–	–	
8.900	0.912	9.812	–	–	
			97.300	97.300	
2.500	0.837	3.337	–	–	Price 2
2.800	0.837	3.637	–	–	
3.000	0.837	3.837	–	–	
3.500	0.837	4.337	–	–	
3.800	0.837	4.637	–	–	
4.000	0.837	4.837	–	–	
4.900	0.837	5.737	–	–	
5.500	0.837	6.337	–	–	
6.500	0.837	7.337	–	–	
7.900	0.837	8.737	–	–	
8.900	0.837	9.737	–	–	
			98.000	98.000	

3.4. Credit Spreads

Whereas static spread reflects the premium that is associated with the general risk amortised over the life of the rate curve, credit spread reflects the further risk associated with maturity length. In calculating credit spread we are concerned with the probability of a bond issuer defaulting within a given period. A simple method is defined which starts with the premise that the price of a corporate bond is equal to an equivalent parameter treasury bond times the probability of the corporate issuer remaining solvent. The probabilities are simply related to the observed prices.

Thus, $P_{corp} = P_{tres} * p_{solv}$ the probability of solvency is $1 - p_{def}$ so,

$$1 - p_{def} = \frac{P_{corp}}{P_{tres}} \tag{3.4.1}$$

This provides us with the single period probability of default. Subsequent or forward probabilities are calculated by similar price ratios and are conditional on the previous period not being in default.

The forward probability of default is:

$$f_{def}^{i} = 1 - \frac{1}{(1 - p_{def}^{i-1})} * \frac{P_{corp}^{i}}{P_{tres}^{i}} \tag{3.4.2}$$

Example 3.4

The following prices are available for five zero coupon bonds:

	1 year	2 year	3 year	4 year	5 year
Treasury	97	95	92	91	89
Corporate	92	89	86	83	79

Compute the probability of default and the forward probabilities of default.
The probability of default is computed by 2.5.3
Probability of default is: 0.0515
The forward probabilities as computed by 2.5.4:

Forward probability of default: period 1: 0.0122
Forward probability of default: period 2: 0.0646
Forward probability of default: period 3: 0.0879
Forward probability of default: period 4: 0.1124

Listing 3.4 shows the application code for this example

```
package FinApps;
import static java.lang.Math.*;
import java.text.*;
import java.io.*;
public class Example_3_4 {
```

```
    public Example_3_4() {
    }
    public static void main(String[] args) {
        NumberFormat formatter=NumberFormat.getNumberInstance();
        formatter.setMaximumFractionDigits(4);
        formatter.setMinimumFractionDigits(3);
        Spread ss=new Spread();
        double[] risky={92,89,86,83,79};
        double[] riskless={97,95,92,91,89};
        double[]risk=new double[risky.length];
        risk=ss.creditS(riskless,risky);
        int i=0;
        System.out.println("Probability of default is:
                                "+formatter.format(risk[i]));
        for(i=1;i<risk.length;i++)
        {
            System.outs.println("Forward probability of default:
period"+i+":"+formatter.format(risk[i]));
        }
    }
}
```

LISTING 3.4. Application code for Example 3.4

The algorithm for computing the various spread measures outlined here are implemented in the class **Spread**. This is shown in Listing 3.5.

Method *spreadsT* take the spot rate data and a single corporate bond price. Using the method *newtraph* from the abstract class *NewtonRaphson* the class computes the static spread of the corporate bond using the implementation of *newtonroot*. This method is overloaded for multiple corporate bonds and will return an array of static spreads for each bond. The method *spreadrateS* takes the spot rates and a corporate bond price and returns the yield spread. The method computes the yield to maturity of the non-treasury bond price and compares this to the YTM for the spot rates. Spot rates are used to derive a riskless price which is the basis of the riskless YTM. Method *spreadrateS* uses a **Tyield** object to compute the yield, this method is overloaded to compute multiple risky (non-treasury) yields.

Spread also contains private methods *spotPvannual* and *spotPvperiod* which calculate annual cash flows or user defined period cash flows. The method *spreadrate* is used to compute the static spread for a nominal term flat spot rate. The method *creditS* computes the probability of default and the forward probability of default (credit spread). Method *creditS* takes the risky and riskless arrays of ordered maturity prices then produces the output array of default probabilities.

```
package FinApps;
import java.util.*;
import static java.lang.Math.*;
import static FinApps.PresentValue.*;
import CoreMath.NewtonRaphson;
import static FinApps.Intr.*;
```

```
import FinApps.Tyield.*;

public class Spread extends NewtonRaphson {
   private double precision=1e-5;
   private int iterations=20;
   public Spread() //default constructor//
   {
      this. terms=2.0;// default twice annual coupon payments
      this.dataperiod=1;//period of spot rate data
      this.facevalue=100.0;//default par
   }
   public Spread(double frequency,int dataterms,double parvalue ) {
      this.terms=frequency;
      this.dataperiod=dataterms;
      this.facevalue=parvalue;
   }
   double terms=0.0;
   int dataperiod=0;
   double facevalue=0.0;
   double periodyield=0.0;
   double nperiods=0.0;
   double periodcoupon=0.0;
   double price=0.0;
   double[]spots;
   double coupon=0.0;
   /** Method computes spread for the annual period rates
   *
   * provides the static spread from the corporate bond price. The amount
      by which each spot needs to be adjusted
   * Assumes annual coupon rate and annual yield estimate
   */
   public double spreadsT(double[]spotrates,double pcorp,
                          double maturity,double couponrate,
                          double estimate) {
      accuracy(precision,iterations);
      spots=spotrates;
      price=pcorp;
      coupon=couponrate;
      return newtraph(estimate);
   }
   /**
   *Calculates the yield spread for YTM of corretly priced risk
      free and arbitrary priced corporatecorporate
   */
   public double spreadrateS(double[]spotrates,double priceval,
                             double couponrate,double maturity,
                             double yieldapprox ) {
      double baseyield=0.0;
      double curveyield=0.0;
      double spotapprox=0.0;
      spots=spotrates;
      coupon=couponrate;
      nperiods=maturity*terms;// number of compounding periods
```

```
    if(((int)maturity*dataperiod)!=spots.length)
    {System.out.println("error: spots data is not ==
                        to the maturity*dataperiods");
    return 0.0;
    }
    periodyield=((yieldapprox/100.0)/terms);
    periodcoupon=(coupon/(terms));
    spotapprox=(((spots[0]+spots[(spots.length-1)])/2.0)/100.0);
                                                     //first guess
    spotapprox=spotapprox/terms;
    price=dataperiod==1?spotPvannual(spots,coupon):
                    spotPvperiod(spots,coupon);
    Tyield c=new Tyield();// create a yield object
    curveyield=c.yieldEstimate(facevalue,6.0,coupon,price,
                            maturity,spotapprox);
    price=priceval;
    Tyield t=new Tyield();
    baseyield=t.yieldEstimate(facevalue,6.0,coupon,price,
                            maturity,periodyield);
    return (abs(baseyield-curveyield));// returns annualised rates
}
/**
*Assumes annual coupon and annual yield with years to maturity
  as input parameters
*assumes coupon and yield is entered as percentage value
*/
public double[] spreadrateS(double[]spotrates,double prices[],
                        double couponrate,double maturity,
                        double yieldapprox ) {
    double curveyield=0.0;
    double spotapprox=0.0;
    spots=spotrates;
    coupon=couponrate;
    double curvest=yieldapprox;
    int index=0;
    double spreads[]=new double[prices.length];
    Tyield t=new Tyield();
    nperiods=maturity*terms;
    periodcoupon=(coupon/(terms));
    price=dataperiod==1?spotPvannual(spots,coupon):
                    spotPvperiod(spots,coupon);
    curveyield=t.yieldEstimate(facevalue,6.0,coupon,price,
                            maturity,(curvest/100.0));
    for(double p:prices) {
        Tyield yld=new Tyield();
        price=p;
        double y=yld.yieldEstimate(facevalue,6.0,coupon,price,
                            maturity,(yieldapprox/100.0));
        spreads[index]=(y-curveyield);
        index++;
    }
    return spreads;
```

```
}
public double[] spreadrateT(double[]spotrates,
                          double prices[],double couponrate,
                          double maturity,double yieldapprox ) {
   spots=spotrates;
   coupon=couponrate;
   int index=0;
   double spreads[]=new double[prices.length];
   accuracy(precision,iterations);
   periodyield=yieldapprox;
   nperiods=maturity*terms;
   periodcoupon=(coupon/(terms));
   for(double p:prices) {
      price=p;
      periodyield=((periodyield/100.0)/terms);
      periodyield=newtraph(periodyield);
      spreads[index]=periodyield;
      index++;
   }
   return spreads;
}
/**
*Method computes the PV for an array of period spots
* and the annual coupon
*periods are user defined
**/
private double spotPvperiod(double[]periodspot,double coupon) {
   double pv=0.0;
   double par=0.0;
   double periodcoupon=0.0;
   double couponadjust=coupon/terms;
   int size=0;
   size=periodspot.length*(int)terms;
   pv= pVonecash(periodspot,couponadjust);
   par=(100.0*exp(-(double)size*log(1.0+
        (periodspot[(periodspot.length-1)]/100.0))));
   return(pv+par);
} /** Method to compute the PV of an array of annual
spots and annual coupon with given annual frequency of compounding
**
**/
private double spotPvannual(double[]periodspot,double coupon ) {
   double pv=0.0;
   double par=0.0;
   if(terms>1.0) {
      int size=0;
      int compfreq=0;
      int index=0;
      compfreq=(int)terms;
      size=periodspot.length*(int)terms;
      double[]periodspotadj=new double[(size)];
      for(double d:periodspot) {
         for (int i=0;i<compfreq;i++) {
            periodspotadj[index]=d/terms;
```

```
            index++;
         }
      }
      double couponadjust=(coupon/terms); //from an annual coupon
                                          //to the period rate
      pv=pVonecash(periodspotadj,couponadjust);
      par=(100.0*exp(-(double)size*log(1.0+
           (periodspotadj[(periodspotadj.length-1)]/100.0))));
      return(pv+par);
   }
   else{
      double couponadjust=(coupon/terms); //from an annual coupon//
                                          //to the period rate
      pv=pVonecash(periodspot,couponadjust);
      par=(100.0*exp(-(double)periodspot.length*log(1.0+
          (periodspot[(periodspot.length-1)]/100.0))))
      return(pv+par);
   }
}

public double spotPvannualT(double[]periodspot,double coupon,
                            double terms ){
   double pv=0.0;
   double par=0.0;
   if(terms>1.0) {
      int size=0;
      int compfreq=0;
      int index=0;
      compfreq=(int)terms;
      size=periodspot.length*(int)terms;
      double[]periodspotadj=new double[(size)];
      for(double d:periodspot) {
         for (int i=0;i<compfreq;i++) {
            periodspotadj[index]=d/terms;
            index++;
         }
      }
      double couponadjust=(coupon/terms); //from an annual coupon
                                          // to the period rate
      pv=pVonecash(periodspotadj,couponadjust);
      par=(100.0*exp(-(double)size*log(1.0+
          (periodspotadj[(periodspotadj.length-1)]/100.0))));
      return(pv+par);
   }
   else {
      double couponadjust=(coupon/terms); //from an annual coupon
                           to the period rate
      pv=pVonecash(periodspot,couponadjust);
      par=(100.0*exp(-(double)periodspot.length*log(1.0+
          (periodspot[(periodspot.length-1)]/100.0))));
   return(pv+par);
   }
}
```

```
    /**Assumes period spot and the yield approximation are period rates
       with period percentages
    *Assumes flat rate spot for entire maturity period
    *assumes the coupon is a coupon percent of par rate.
    */
 public double spreadrate(double periodspot,double priceval,
                          double coupon,double maturity,
                          double yieldapprox ) {
    accuracy(precision,iterations);
    double baseyield=0.0;
    double frequency=0.0;
    price=priceval;
    periods=maturity*terms;
    periodcoupon=coupon;
    periodyield=yieldapprox/100.0;
    periodspot=periodspot/100.0;
    baseyield=newtraph(periodyield);
    return (abs(baseyield-periodspot)*terms*100.0);
                                   //returns annualised spread
 }
 /**
 *credit spread computes probability of default and forward
    prob of default
 *assumes corporate bond zero and treasury zero (riskless)
 */
 public double[] creditS(double[]riskless,double[]risky) {
    int size=riskless.length;
    double[]fdefault=new double[size];
    double[]pdefault=new double[size];
    pdefault[0]=(1.0-(risky[0]/riskless[0]));
    fdefault[0]=pdefault[0];
    for(int i=1;i<size;i++) {
       fdefault[i]=(1.0-(exp(-log(1.0-pdefault[i-1])))*
                  (risky[i]/riskless[i]));
       pdefault[i]=(pdefault[i-1]*fdefault[i]);
    }
    return fdefault;
 }
 public double newtonroot(double spread)
 {int indx=0;
 double[] spotspreads=new double[spots.length];

 for(double d:spots) {
    spotspreads[indx]=(d+spread);
    indx++;
 }
 spread=(spotPvannual(spotspreads,coupon)-price);

 return spread;

 }
}
```

LISTING 3.5. Computation of various spread measures

3.5. Bond Volatility Measures

Bond volatility measures are concerned with the relationship between the price of a bond and the various factors influencing its value. There are many co factors in determining the pricing of a bond, the principal ones being interest rates, coupon terms, maturity and required yield.

A typical bond exhibits a convex characteristic for price with yield. Also for a given bond, all things being equal, the price will tend to par as maturity is reached and for a given yield, the higher the coupon the greater will be the price. Any tool which supports bond market participants therefore needs to deal with coupon rates, maturity lengths and yields.

The class used for basic investigation is **Volatility**. This class contains general accessor methods and methods for calculating the sensitivity of percentage price change to interest rates (*percentVolatilty*) and methods for calculating price value of a basis point (PVBP, also referred to as the dollar value of a basis point-DV01) and yield value of a price change. These are methods; *pVbPoints* and *yieldForPpoint.*

The PVBP measures the change in price for a single percentage point change in the yield (a change of 0.01%). Method *pVbPoints* calculates the value of a basis point by taking the price of the security for its current yield, followed by the price of the security plus or minus the basis point change in current yield. The basic calculations are as in Equation 2.3.3. Method *YieldForPpoint* calculates the yield value for a price change. The method uses public class **Tyield** to calculate the yield values using method *yieldEstimate* .The yield value (YV) of a price change is carried out by calculating the yield at the current price, followed by the yields for plus or minus the point price change. The basic calculation is based on Equation 2.3.4.

Percentage price change measures the sensitivity of % price changes to interest rate changes and is given as:

$$-\frac{\left(\frac{c/2}{\frac{r}{2}}\right)2n-\left(\frac{c/2}{\frac{r}{2}}\right)^{2}\qquad\frac{\left(1+\frac{r}{2}\right)^{2n+1}-\left(1+\frac{r}{2}\right)}{}_{-2nP}}{\left(\frac{c/2}{\frac{r}{2}}\right)\left[\left(1+\frac{r}{2}\right)^{2n+1}-\left(1+\frac{r}{2}\right)\right]+P\left(1+\frac{r}{2}\right)} \tag{3.5.1}$$

Where c/2 is the coupon rate per semi annual period. $\frac{r}{2}$ is the period yield rate. P is the par value and 2n is the number of time periods to maturity, where n is the number of years.

The method *percentVolatilty* implements 3.5.1. This method uses method *pVbPoints* to calculate the PVBP. The current market price is then calculated and used as the divisor to provide the sensitivity measure.

The following sections deal with using class **Volatility**. The class is shown in Listing 3.6.

```
package FinApps;
import static FinApps.Intr.*;
import static java.lang.Math.*;
import java.util.*;
import static FinApps.PresentValue.*;
public class Volatility {
public Volatility() {
  this.facevalue=1000.0;
  this.frequency=2.0;
}
public Volatility(double parvalue,double coupontimes) {
  this.facevalue=parvalue;
  this.frequency=coupontimes;
}
  private double mktprice;
  private double mktpricelow;
  private double mktpricenew;
  private double couponvalue;//couponvalue = par value*annual
                                          //coupon percent/2
  private double facevalue;
  private double frequency;
  private double pv;
  private double par;
  private double relativeprice;
  private double relativepricelow;
  private double upyield;
  private double downyield;
  private double newpriceup;
  private double newpricedown;
  private double currentyield;
  private double currentpvb;
/* Accessor methods */
  private void setInitialYldPp(double yield)
  {
    currentyield=yield;
  }
  public double getInitialPpYld()
  {
    return currentyield;
  }
  private void setPpointpriceup(double price)
  {
    newpriceup=price;
  }
  private void setPpointpricedown(double price)
  {
    newpricedown=price;
  }
  public double getPriceupPp()
  {
```

```
    return newpriceup;
}
public double getPricedownPp()
{
    return newpricedown;
}
private void setdownyieldPp(double yield)
{
    downyield=yield;
}
private void setupyieldPp(double yield)
{
    upyield=yield;
}
public double getUpPp()
{
    return upyield;
}
public double getDownPp()
{
    return downyield;
}
public double getValuePUp()
{
    return (abs(getUpPp()-getInitialPpYld())/100.0);
}
public double getValuePDown()
{
    return (abs(getDownPp()-getInitialPpYld())/100.0);
}
private void setRelativeValue(double price)
{
    relativeprice=price;
}
public double getRelativeValue()
{
    return relativeprice;
}
private void setRelativeValuelow(double price)
{
    relativepricelow=price;
}
public double getRelativeValuelow()
{
    return relativepricelow;
}
private void setCurrentPvb(double price)
{
    currentpvb=price;
}
public double getCurrentPvb()
{
    return currentpvb;
}
```

```
/* Price value of a basis point */
public void pVbPoints(double yield,double yearterm,double coupon,
                     double pointchange)
{
  double yieldval;
  mktprice=Bpricing(yield,yearterm,coupon);
  setCurrentPvb(mktprice);
  yieldval=(yield+(pointchange/100.0));// make basis point
                                        //adjustment higher
  mktpricenew=Bpricing(yieldval,yearterm,coupon);
  setRelativeValue(abs(mktpricenew-mktprice));
  yieldval=(yield-(pointchange/100.0));// make basis point
                                        //adjustment lower
  mktpricelow=Bpricing(yieldval,yearterm,coupon);
  setRelativeValuelow(abs(mktpricelow-mktprice));
}
/* Provides basic bond pricing */
private double Bpricing(double yield,double yearterm,double coupon)
{
  couponvalue=((facevalue*coupon/100)/frequency);
  pv=(couponvalue*pvancert((yield/100.0)/frequency,
      (frequency*yearterm)));
  par=(facevalue*(1.0/pow(1.0+(yield/100.0)/frequency,
      (frequency*yearterm))));
  return(pv+par);
}
public double percentVolatility(double yield,double yearterm,
                               double coupon,double pointchange)
{
  pVbPoints(yield,yearterm,coupon,pointchange);//price value of a
                                               //basis point
  couponvalue=((facevalue*coupon/100)/frequency);
  pv=(couponvalue*pvancert((yield/100.0)/frequency,
      (frequency*yearterm)));
  par=(facevalue*(1.0/pow(1.0+(yield/100.0)/frequency,
      (frequency*yearterm))));
  mktprice=pv+par;
  return((getRelativeValue()/mktprice)*100.0);
}
/**
*Method provides yield values for a percentage point
  change in par value
*Sets via accessor methods the Yield value of a point change,
  the initial yield value prior to applying the point change
*/
public void yieldForPpoint(double couponpercent,double price,
                          double maturity,double estimate,
                          double pointvalue)
{
  double couponterm=12.0/frequency;
  double change=((facevalue/100.0)*pointvalue);
  setPpointpricedown(price-change);
  setPpointpriceup(price+change);
  Tyield CalcBond= new Tyield();
```

```
        setInitialYldPp(CalcBond.yieldEstimate(facevalue,couponterm,
                        couponpercent,price,maturity,estimate));
        setdownyieldPp(abs((CalcBond.yieldEstimate(facevalue,
                       couponterm,couponpercent,getPricedownPp(),
                       maturity,estimate))));
        setupyieldPp(abs((CalcBond.yieldEstimate(facevalue,couponterm,
                     couponpercent,getPriceupPp(),maturity,
                     estimate))));
    }
}
```

LISTING 3.6. Computation of bond volatility measures

3.5.1. *Price Value of a Point*

The PVBP is often used to demonstrate an investor's cash exposure to interest rate movements it also useful in implementing trading strategies. The following exercise demonstrates the application.

Example 3.5

Compute the cash exposure for a range of bonds with coupons of 2,5,8,9 and 10%. Show the cash exposure to the basis point change for yields of 5 and 10% and maturities of 3,5,8,10 and 30 years. Show the difference in $ volatility for a higher yield.

Tables 3.4 and 3.5 show the output data from Example 3.5. The effect of raising or reducing the yield by a single basis point is shown in columns 3 and 4. The cash exposure is per $100 of par.

TABLE 3.4. Price value of a basis point with 5% yield

Coupon	Initial Price	Plus PVBP	Less PVBP	Maturity
2.0	$91.7378	$0.0262	$0.0262	3.0
2.0	$86.8719	$0.0404	$0.0404	5.0
2.0	$80.4175	$0.0576	$0.0577	8.0
2.0	$76.6163	$0.0669	$0.0669	10.0
2.0	$53.6070	$0.1016	$0.1019	30.0
5.0	$100.0000	$0.0275	$0.0275	3.0
5.0	$100.0000	$0.0437	$0.0438	5.0
5.0	$100.0000	$0.0652	$0.0653	8.0
5.0	$100.0000	$0.0779	$0.0780	10.0
5.0	$100.0000	$0.1544	$0.1547	30.0
8.0	$108.2622	$0.0289	$0.0289	3.0
8.0	$113.1281	$0.0471	$0.0472	5.0
8.0	$119.5825	$0.0729	$0.0729	8.0
8.0	$123.3837	$0.0889	$0.0890	10.0
8.0	$146.3630	$0.2071	$0.2076	30.0
9.0	$111.0163	$0.0294	$0.0294	3.0

(Continued)

TABLE 3.4. (*Continued*)

Coupon	Initial Price	Plus PVBP	Less PVBP	Maturity
9.0	$117.5041	$0.0483	$0.0483	5.0
9.0	$126.1100	$0.0754	$0.0755	8.0
9.0	$131.1783	$0.0926	$0.0927	10.0
9.0	$161.8173	$0.2247	$0.2252	30.0
10.0	$113.7703	$0.0298	$0.0298	3.0
10.0	$121.8802	$0.0494	$0.0494	5.0
10.0	$132.6375	$0.0779	$0.0780	8.0
10.0	$138.9729	$0.0963	$0.0964	10.0
10.0	$177.2716	$0.2423	$0.2428	30.0

TABLE 3.5. Price value of a basis point with 10% yield

Coupon	Initial Price	Plus PVBP	Less PVBP	Maturity
2.0	$79.6972	$0.0221	$0.0221	3.0
2.0	$69.1131	$0.0311	$0.0311	5.0
2.0	$56.6489	$0.0387	$0.0388	8.0
2.0	$50.1512	$0.0412	$0.0412	10.0
2.0	$24.2828	$0.0311	$0.0312	30.0
5.0	$87.3108	$0.0233	$0.0234	3.0
5.0	$80.6957	$0.0339	$0.0339	5.0
5.0	$72.9056	$0.0445	$0.0446	8.0
5.0	$68.8445	$0.0491	$0.0491	10.0
5.0	$52.6768	$0.0549	$0.0550	30.0
8.0	$94.9243	$0.0246	$0.0246	3.0
8.0	$92.2783	$0.0367	$0.0367	5.0
8.0	$89.1622	$0.0503	$0.0504	8.0
8.0	$87.5378	$0.0570	$0.0571	10.0
8.0	$81.0707	$0.0787	$0.0788	30.0
9.0	$97.4622	$0.0250	$0.0250	3.0
9.0	$96.1391	$0.0377	$0.0377	5.0
9.0	$94.5811	$0.0522	$0.0523	8.0
9.0	$93.7689	$0.0596	$0.0597	10.0
9.0	$90.5354	$0.0866	$0.0868	30.0
10.0	$100.0000	$0.0254	$0.0254	3.0
10.0	$100.0000	$0.0386	$0.0386	5.0
10.0	$100.0000	$0.0542	$0.0542	8.0
10.0	$100.0000	$0.0623	$0.0623	10.0
10.0	$100.0000	$0.0946	$0.0947	30.0

The graph of Figure 3.1 shows that the 10% yield exhibits less $ volatility and that the 5% yield exhibits greater convexity. The application code for Example 3.5 is shown in Listing 3.7

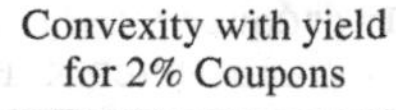

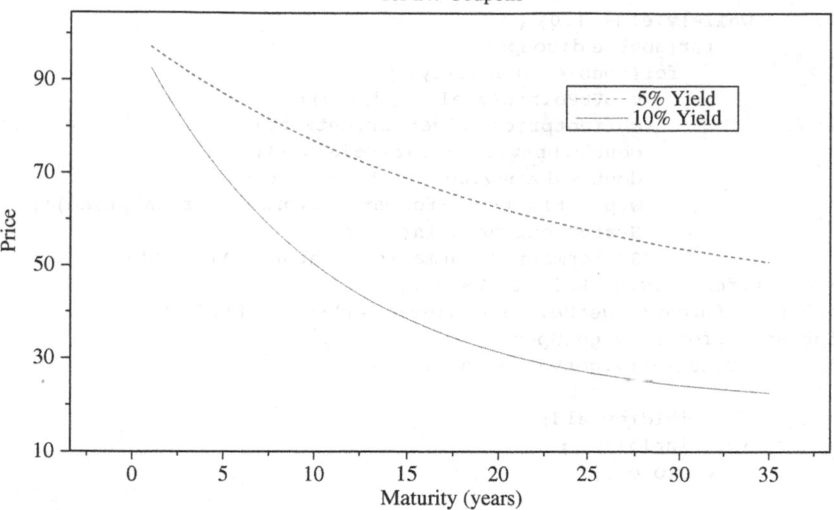

FIGURE 3.1. Convexity v Yield.

```
package FinApps;
import static FinApps.Intr.*;
import static java.lang.Math.*;
import java.util.*;
import static FinApps.PresentValue.*;
import java.text.*;
import java.io.*;
public class Exercise_3_5 {
   /** Creates a new instance of Exercise_3_5 */
   public Exercise_3_5() {
   }
   /**
   * @param args the command line arguments
   */
   public static void main(String[ ] args) {
      NumberFormat formatter=NumberFormat.getNumberInstance();
      formatter.setMaximumFractionDigits(4);
      formatter.setMinimumFractionDigits(4);
      double currentprice;
      double par;
      try{
         PrintWriter w=new PrintWriter(new FileWriter
                                 ("c:\\Ex2_8data.txt"),true);
         double[ ] coups={2.0,5.0,8.0,9.0,10.0};
         double [ ] maturity={3.0,5.0,7.0,15.0,30.0};
         par=100.0;
         double yield=5.0;
         Volatility v=new Volatility(100.0,2.0);
```

```
        System.out.println("Coupon Initial Price Plus PVBP
                              Less PVBP Maturity ");
        while(yield<11.0) {
            for(double d:coups) {
                for(double f:maturity) {
                    v.pVbPoints(yield,f,d,1.0);
                    currentprice =v.getCurrentPvb();
                    double up=v.getRelativeValue();
                    double down=v.getRelativeValuelow();
                    w.println(f+","+formatter.format(currentprice));
                    System.out.println(" "+d+"
                    $"+formatter.format(currentprice)+"    $"+
formatter.format(v.getRelativeValue())+"   $"
+formatter.format(v.getRelativeValuelow())+"    "+f);//+"
"+formatter.format(v.getUpPp())+"
"+formatter.format(v.getValuePUp()));      }
                }
            }yield+=yield;
        }w.println(" ");
            w.close();
        }
        catch(IOException foe) {
            System.out.println(foe);
        }
    }
}
```

LISTING 3.7. Application code for Example 3.5

Referring to Table 3.4, with a 5% yield. The 2% coupon with 3 year maturity has a current price of $91.7378. The PVBP with an increase of a basis point is around 2.6 ¢ per $100 of par, it is the same for a decrease of a basis point. The 2% coupon with 30 year maturity has a PVBP of 10.16 ¢ for an increase in basis point and around 10.19 ¢ for a decrease in a basis point. This reflects greater $ volatility of the longer maturity period. Comparing the first entry in Table 3.5 for a 10% yield, this shows a lower $ volatility, reflecting the higher yield. In general, the data in Table 3.3 will exhibit greater convexity than the data in Table 3.5. This can be seen from the graph of data taken for the 2% coupons from both tables shown in Figure 3.1. The plot shows the difference in convexity with the 2% coupon for 5% and 10% yields.

3.6. Bond Pricing Characteristics

Table 3.6 shows the data for a 20 year option free bond, with a semi annual coupon of 9%. The data consists of prices for a range of required yields. The price column is made up from the discounted coupon cashflows and the discounted face value (par).

TABLE 3.6. Price yield relationship

Yield	PV of coupon	PV of par	Price
1.000	1,627.750	819.139	2,446.889
2.000	1,477.561	671.653	2,149.214
3.000	1,346.213	551.262	1,897.475
4.000	1,230.997	452.890	1,683.887
5.000	1,129.625	372.431	1,502.056
6.000	1,040.165	306.557	1,346.722
7.000	960.978	252.572	1,213.551
8.000	890.675	208.289	1,098.964
9.000	828.071	171.929	1,000.000
10.000	772.159	142.046	914.205
13.000	636.549	80.541	717.089
17.000	509.153	38.266	547.419

Figure 3.2 below shows the price yield characteristic for the data of Table 3.5 .. The graph shows that as required yield increases, bond price decreases. Figure 3.3 shows the graph of present values for the coupon payments (20 years paid semi annually). The 40 payments have a discounted value which also decreases as the required yield increases. The dotted line of Figure 3.3 shows the discounted value of the par value, which also declines as the required yield increases.

The characteristics shown in Figures 3.2 and 3.3 are of use in determining the required yield for an investor in order to achieve a required series of cash flows. The formula of Equation 2.3.3 is used to determine the price for a range of required yields. This is shown in Listing 3.8.

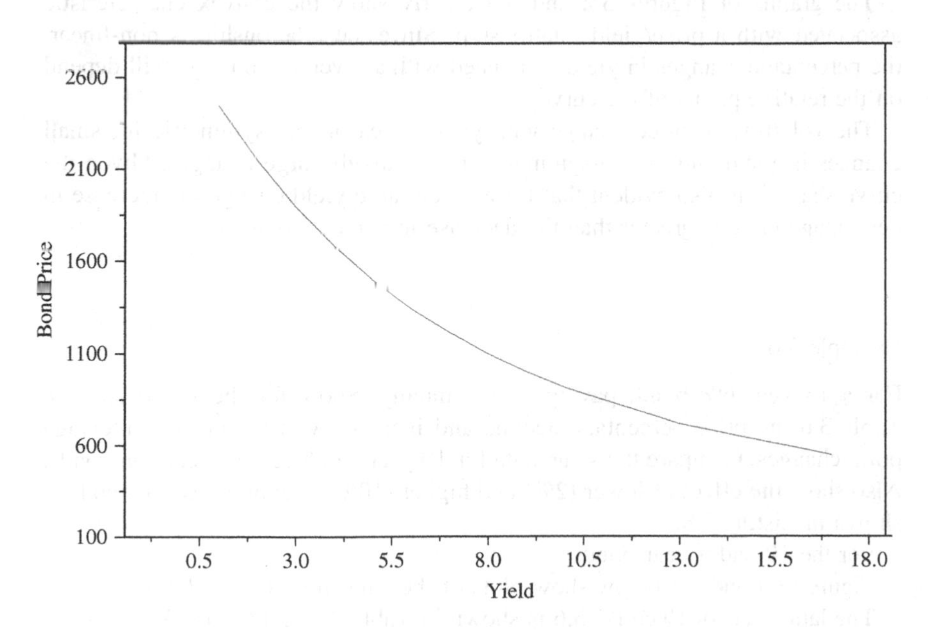

FIGURE 3.2. Data plot for Table 3.5.

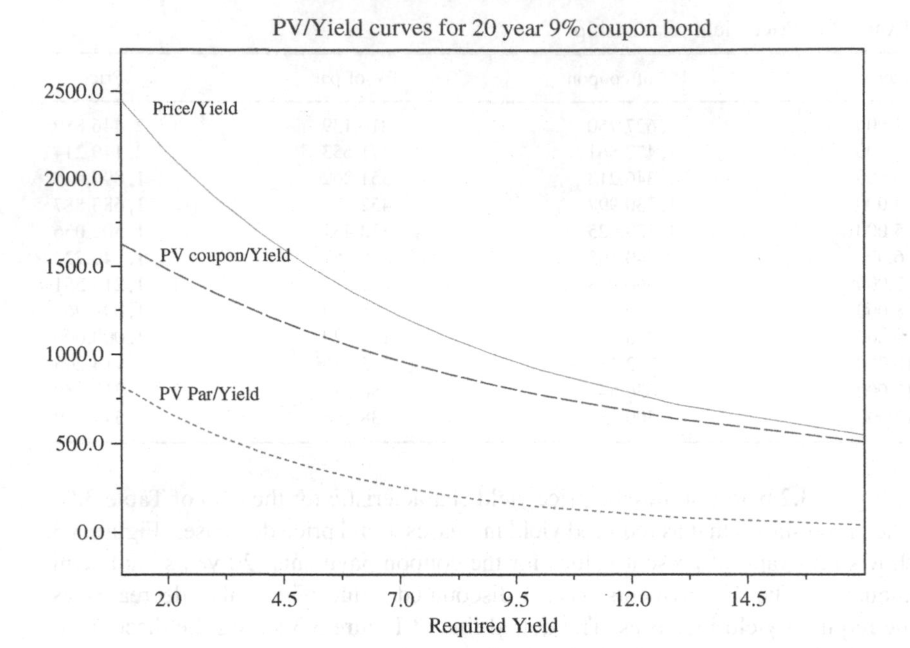

FIGURE 3.3. P.V v Required yield characteristics.

The graphs of Figures 3.2 and 3.3 clearly show the convex characteristic associated with a price/yield relationship. Since the relationship is non-linear, the percentage changes in yield associated with a given price range will depend on the relative points of the curve.

The volatility of price change with yield is reasonably symmetric for small changes in yield and is non-symmetric for relatively large changes. Given the curve shape it is also evident that for a given large yield change the increase in percentage price is greater than the decrease in percentage price.

Example 3.6

For a 15 year 6% bond, paying semi annually. Show for the yield range in Table 3.6 the price percentage decline and increase with the (100) percentage point changes. Compare the same data for 10 year and 5 year 6% coupon bonds. Also show the effect of lower (2%) and higher (10%) coupon rates. The code is shown in Listing 3.8.

For the 10 and 5 year bonds:

Figures 3.4 and 3.5 below show plots of the data from Tables 3.7 & 3.8.

The latter part of Exercise 3.6 is shown in Table 3.9 and Figure 3.6.

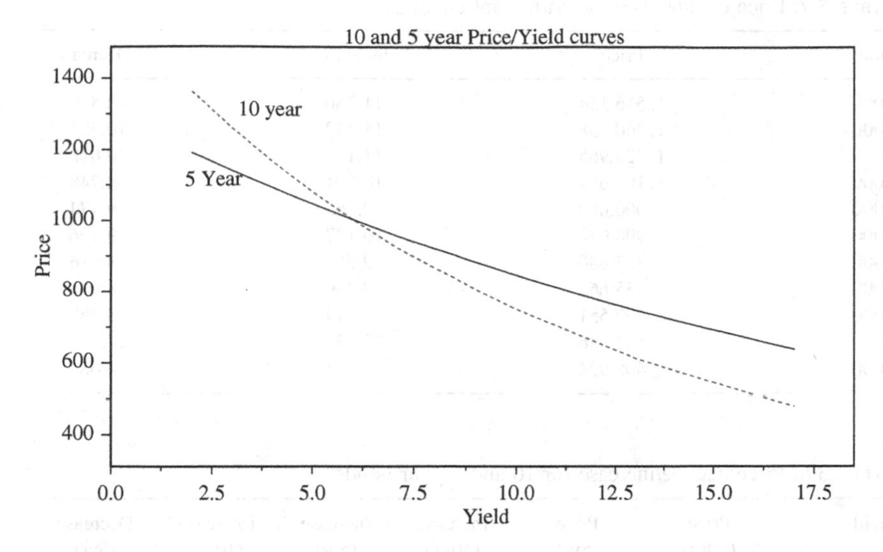

FIGURE 3.4. Price/yield curves for Example 3.6.

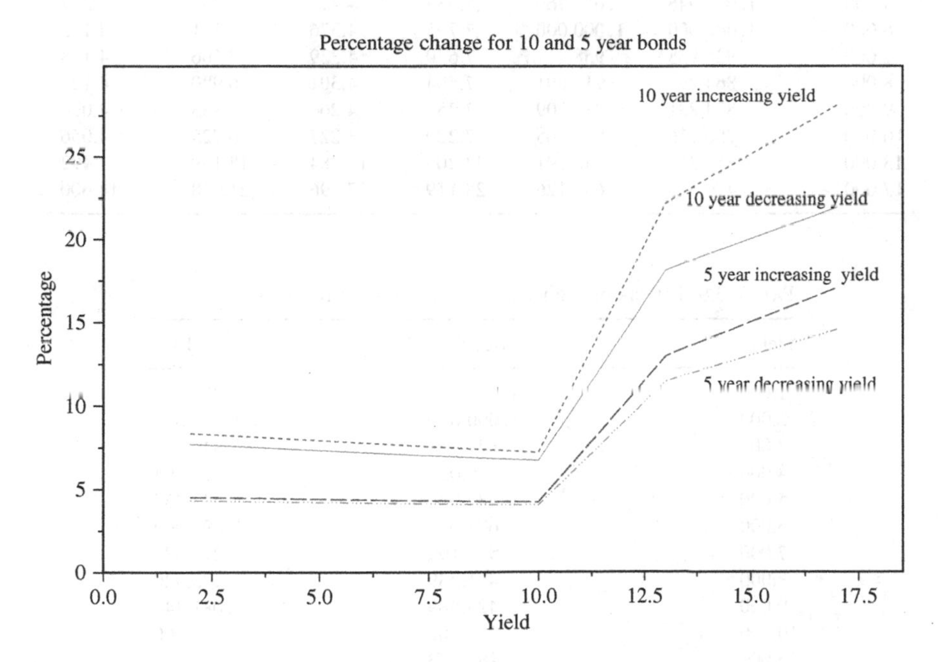

FIGURE 3.5. Percentage changes for Example 3.6.

TABLE 3.7. Price decline/increase with point changes

Yield	Price	Increase	Decrease
2.000	1,516.154	11.786	10.544
3.000	1,360.238	11.462	10.284
4.000	1,223.965	11.134	10.018
5.000	1,104.651	10.801	9.748
6.000	1,000.000	10.465	9.474
7.000	908.040	10.127	9.196
8.000	827.080	9.789	8.916
9.000	755.667	9.450	8.634
10.000	692.551	9.114	8.352
13.000	542.946	27.554	21.602
17.000	408.924	32.775	24.684

TABLE 3.8. Price decline/increase for 10 and 5 year bonds

Yield	Price (10yr)	Price (5yr)	Increase (10yr)	Increase (5yr)	Decrease (10yr)	Decrease (5yr)
2.000	1,360.911	1,189.426	8.360	4.526	7.715	4.330
3.000	1,257.530	1,138.333	8.221	4.488	7.596	4.296
4.000	1,163.514	1,089.826	8.080	4.451	7.476	4.261
5.000	1,077.946	1,043.760	7.938	4.413	7.354	4.227
6.000	1,000.000	1,000.000	7.795	4.376	7.231	4.193
7.000	928.938	958.417	7.650	4.339	7.106	4.158
8.000	864.097	918.891	7.504	4.301	6.980	4.124
9.000	804.881	881.309	7.357	4.264	6.853	4.090
10.000	750.756	845.565	7.209	4.227	6.725	4.056
13.000	614.352	748.391	22.203	12.984	18.169	11.492
17.000	479.516	639.126	28.119	17.096	21.948	14.600

TABLE 3.9. Effects of higher and lower coupon rates

Yield	Price(2%)	Price(10%)
1.000	1,138.970	2,250.732
2.000	1,000.000	2,032.308
3.000	879.921	1,840.554
4.000	776.035	1,671.894
5.000	686.046	1,523.257
6.000	607.991	1,392.009
7.000	540.199	1,275.881
8.000	481.239	1,172.920
9.000	429.889	1,081.444
10.000	385.102	1,000.000
13.000	281.773	804.120
17.000	193.987	623.860

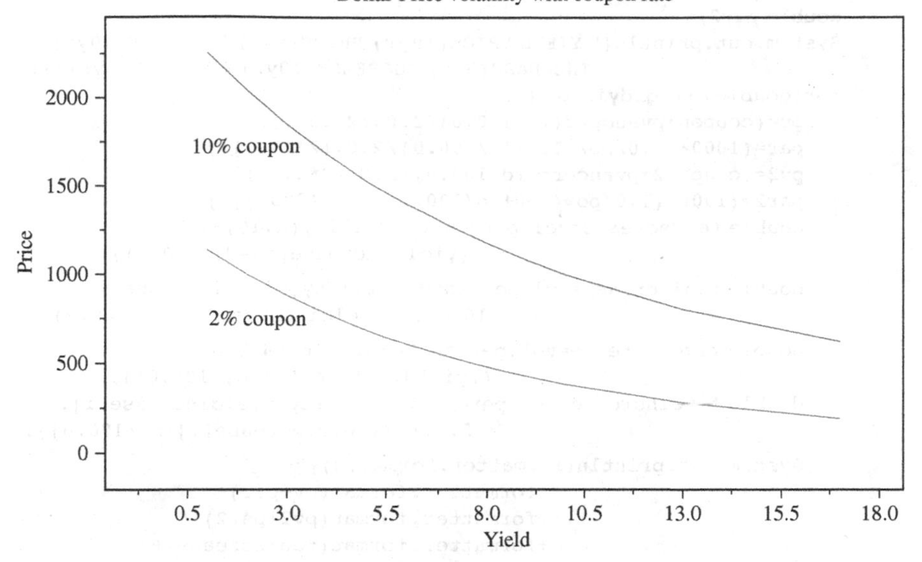

FIGURE 3.6. Volatility for different coupon rates.

```
package FinApps;
import static FinApps.Intr.*;
import static java.lang.Math.*;
import java.util.*;
import static FinApps.PresentValue.*;
import java.text.*;
import java.io.*;
public class Exercise2_9 {
  public Exercise2_9() {
  }
  public static void main(String[] args) {
    //Assumes par value of 1,000
    NumberFormat formatter=NumberFormat.getNumberInstance();
    formatter.setMaximumFractionDigits(3);
    formatter.setMinimumFractionDigits(3);
    Volatility vol=new Volatility();
    double increase;
    double decrease;
    double[] reqrdyield={1,2,3,4,5,6,7,8,9,10,13};
    double[]yieldincrease={2,3,4,5,6,7,8,9,10,13,17};
    double coupon=10;//coupon = par value*annual coupon percent/2..2%
    double coupon2=50;//coupon = par value*annual coupon percent/2..10%
    int i=0;
    double pv;
    double par;
```

```
    double pv2;
    double par2;
    System.out.println(" YIELD PRICE(10yr) PRICE(5yr) INCREASE(10yr)
                        INCREASE(5yr) DECREASE(10yr) DECREASE(5yr)");
    for(double d:reqrdyield) {
       pv=(coupon*pvancert((d/100.0)/2.0,(2*10)));
       par=(1000*(1.0/pow(1.0+(d/100.0)/2.0,(2*10))));
       pv2=(coupon2*pvancert((d/100.0)/2.0,(2*5)));
       par2=(1000*(1.0/pow(1.0+(d/100.0)/2.0,(2*5))));
       double tendecrease=vol.percentVolatility(d,10,6,
                              ((yieldincrease[i]-d)*100.0));
       double tenincrease=vol.percentVolatility(yieldincrease[i],
                              10,6,(-(yieldincrease[i]-d)*100.0));
       double fivedecrease=vol.percentVolatility(d,5,6,
                              ((yieldincrease[i]-d)*100.0));
       double fiveincrease=vol.percentVolatility(yieldincrease[i],
                              5,6,(-(yieldincrease[i]-d)*100.0));
       System.out.println(formatter.format(d)+"
                        "+formatter.format(pv+par)+"
                        "+formatter.format(pv2+par2)+"
                        "+formatter.format(tenincrease)+"
                        "+formatter.format(fiveincrease)+"
                        "+formatter.format(tendecrease)+"
                        "+formatter.format(fivedecrease));
             i++;
    }
    System.out.println(" YIELD PRICE(2%) PRICE(10%)");
          for(double d:reqrdyield) {
       pv=(coupon*pvancert((d/100.0)/2.0,(2*15)));
       par=(1000*(1.0/pow(1.0+(d/100.0)/2.0,(2*15))));
       pv2=(coupon2*pvancert((d/100.0)/2.0,(2*15)));
       par2=(1000*(1.0/pow(1.0+(d/100.0)/2.0,(2*15))));
       double marktprice=(pv+par);
       double marktprice2=(pv2+par2);
       System.out.println(formatter.format(d)+"
                        "+formatter.format(pv+par)+"
                        "+formatter.format(pv2+par2));
    }
  }
}
```

LISTING 3.8. Code implementing Example 3.6

Yield value of a price change is another measure often used to gauge the price volatility of a bond. This measure tracks the change in yield for a given price change. The difference between initial and changed yields per $ change is the ‘yield value for $x’. Since US Treasury bonds and notes are quoted in 32nds, it is common to compute the yield value for a 1/32nd price change.

TABLE 3.10. Yield values for 1/32

Maturity	Initial Price	Yield %	New Price	New Yield %	Value of 1/32	Coupon
1.0	95.3514739	10.0000000	95.3827239	9.9651614	0.0003484	
2.0	91.1351237	10.0000000	91.1663737	9.9812917	0.0001871	
3.0	87.3107698	10.0000000	87.3420198	9.9866193	0.0001338	
4.0	83.8419681	10.0000000	83.8732181	9.9892452	0.0001075	5%
5.0	80.6956627	10.0000000	80.7269127	9.9907896	0.0000921	
10.0	68.8444741	10.0000000	68.8757241	9.9936376	0.0000636	
15.0	61.5688724	10.0000000	61.6001224	9.9943159	0.0000568	
30.0	52.6767762	10.0000000	52.7080262	9.9943182	0.0000568	
1.0	100.0000000	10.0000000	100.0312500	9.9663952	0.0003360	
2.0	100.0000000	10.0000000	100.0312500	9.9823779	0.0001762	
3.0	100.0000000	10.0000000	100.0312500	9.9876888	0.0001231	
4.0	100.0000000	10.0000000	100.0312500	9.9903318	0.0000967	10%
5.0	100.0000000	10.0000000	100.0312500	9.9919076	0.0000809	
10.0	100.0000000	10.0000000	100.0312500	9.9949859	0.0000501	
15.0	100.0000000	9.9999999	100.0312500	9.9959352	0.0000406	
30.0	100.0000000	9.9999999	100.0312500	9.9966991	0.0000330	
1.0	104.6485261	10.0000000	104.6797761	9.9675446	0.0003246	
2.0	108.8648763	10.0000000	108.8961263	9.9833448	0.0001666	
3.0	112.6892302	10.0000000	112.7204802	9.9886000	0.0001140	
4.0	116.1580319	10.0000000	116.1892819	9.9912190	0.0000878	15%
5.0	119.3043373	10.0000000	119.3355873	9.9927835	0.0000722	
10.0	131.1555259	10.0000000	131.1867759	9.9958627	0.0000414	
15.0	138.4311276	10.0000000	138.4623776	9.9968365	0.0000316	
30.0	147.3232238	10.0000000	147.3544738	9.9976739	0.0000233	

Example 3.7

Show the yield value of a 1/32 increase for US Treasury securities with coupon rates of 5%,10% and 15%,priced to yield 10%. Compare the yield value characteristics of these bonds with varying maturities. Assume the par value is $100 and coupons are semi annual.

Table 3.10 shows the data for US bonds with maturities ranging from 1 to 30 years and coupon rates of 5, 10 and 15 percent. The initial price of the bond is shown (priced to yield 10%) and the new price with a 1/32 added. The value of a 1/32 is the difference between yields.

The yield value characteristics are shown in Figure 3.7. From this it can be seen that the yield value of a 1/32 increase, decreases with maturity. This effect is seen for all coupon rates and confirms that the $ volatility is greater for longer maturity lengths.

The code for Example 3.7 is shown in Listing 3.9 below.

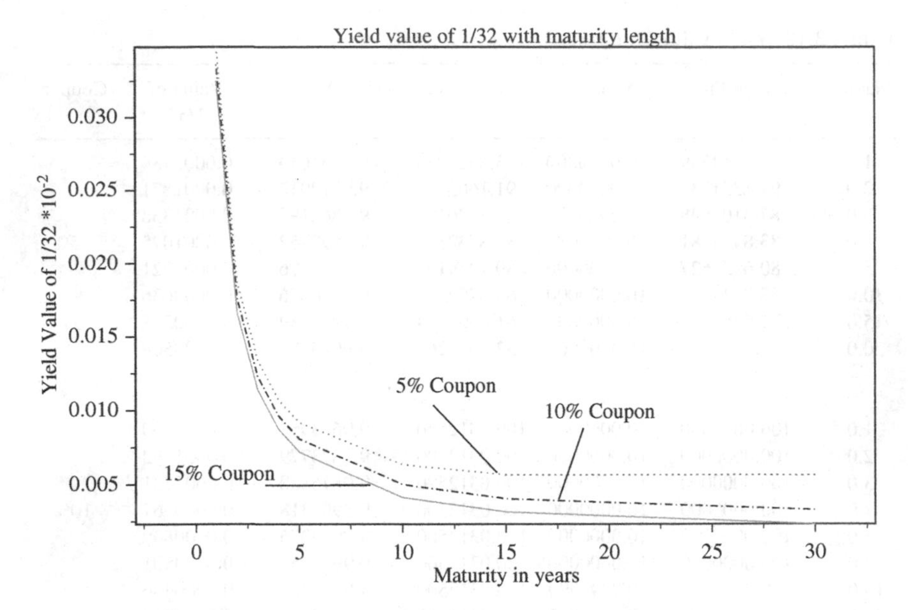

FIGURE 3.7. Yield value of a 1/32.

```
package FinApps;
import static FinApps.Intr.*;
import static java.lang.Math.*;
import java.util.*;
import static FinApps.PresentValue.*;
import java.text.*;
import java.io.*;
public class Exercise2_10 {
   /** Creates a new instance of Exercise2_10 */
   public Exercise2_10() {
   }
   public static void main(String[] args) {
      NumberFormat formatter=NumberFormat.getNumberInstance();
        formatter.setMaximumFractionDigits(7);
        formatter.setMinimumFractionDigits(7);
      Volatility vol=new Volatility();
      double par;
      double currentprice;
      double change;
      double newprice;
      try{
                  PrintWriter w=new PrintWriter(new FileWriter
                     ("c:\\data_for_2_10.txt"),true);
         double[] coups={5.0,10.0,15.0};
         double [] maturity={1.0,2.0,3.0,4.0,5.0,10.0,15.0,30.0};
              par=100.0;
         double yield=0.1;
```

```
        Volatility v=new Volatility(par,2.0);
        System.out.println("Maturity Initial Price Yield %
                              New Price New Yield % Value of 1/32 ");
        for(double d:coups)
        {w.println(d+",");
           for(double f:maturity)
           {
              currentprice=(((pvancert(yield/2.0,f*2))*
                             (d/100.0*par)/2.0)
                             +(par/pow((1+yield/2.0),f*2)));
              v.yieldForPpoint(d,currentprice,f,0.0950,1.0/32.0);
              double up=v.getValuePUp();
              double down=v.getValuePDown();
        System.out.println(" "+f+" "+formatter.format
(currentprice)+" "+ formatter.format(v.getInitialPpYld())+
   " "+formatter.format(v.getPriceupPp())+
   " "+formatter.format(v.getUpPp())+
   " "+formatter.format(v.getValuePUp()));
           w.println(formatter.format(v.getValuePUp()));
           }
        }
   w.println(" ");
              w.close();
   }
   catch(IOException foe)
              {
              System.out.println(foe);
                    }
}
}
```

LISTING 3.9. Application code for Example 3.7

References

1. Fabozi, F. J. *Fixed Income Mathematics Analytical & Statistical Techniques* 3rd edition. Irwin.

4
Duration

Bonds that sell at a premium or par have increased price volatility with longer terms to maturity. The price of a bond is partially based on the reinvestment of coupon receipts at the quoted yield. This holds an inherent risk in not finding a suitable investment vehicle that provides the required yield. As the maturity of the bond increases the reinvestment risk therefore also increases.

With a drop in interest rates the price of a bond will increase so a bondholder will see an immediate appreciation in value but since the reinvestment rate (new yield) is now lower for the reinvestment of coupons the growth in value is slowed. With an increase in interest rates the capital value of a bond will immediately drop but the growth with the higher yield figure will see a more rapid increase in value. All other things being equal there will be some investment duration for the bondholder that equates to indifference for small changes to interest rates. See the book by Sharpe et al for a comprehensive overview.

4.1. Macaulay Duration

The Macaulay duration (proposed by F Macaulay in 1938) is the weighted average of the times to an asset's cash flows. The present value of the cash flows, divided by the asset's price constitutes the weights.

Macaulay duration (MD) for periods is given as:

$$MD \equiv \frac{1}{P}\sum_{t=1}^{n}\frac{tF_t}{(1+y)^t} \tag{4.1.1}$$

Where P is the bond price and F_t is the period cash flow, n the number of periods to maturity, t the time period. For a coupon bond the MD has an addition of the discounted face value so that the MD becomes:

$$MD \equiv \sum\nolimits_{t=1}^{n}\frac{\frac{tF_t}{(1+y)^t}}{P}+\frac{\frac{nV}{(1+y)^n}}{P}. \text{ Where V is the par value.} \tag{4.1.2}$$

MD (in years) $= \frac{MD(periods)}{k}$ where k is the compounding frequency. This converts Equation 4.1.2 to:

$$\frac{\sum_{t=1}^{n} \frac{t}{k} \frac{F}{\left(1+\frac{y}{k}\right)^{t}} + \frac{n}{k} \frac{V}{\left(1+\frac{y}{k}\right)^{n}}}{P} \tag{4.1.3}$$

The MD effectively computes the period cash flow and for each cash flow in the maturity divides by the total discounted cash flow to weight each period as a percentage of the total. The total period cash flow, which is of course, the bond price, also includes accrued interest (the dirty price).

Price volatility is defined as the change in price for a change in yield, this can be represented as:

$$\frac{\frac{\partial Price}{Price}}{\partial Yield} \tag{4.1.4}$$

This relationship is also expressed as the changes in cash flows so $\frac{\frac{\partial P}{P}}{\partial Y} \equiv$ Equation (4.1.5)

From which we can derive:

$$MD = -(1+y)\frac{\frac{\partial P}{P}}{\partial Y} \tag{4.1.6}$$

The equations above only hold if the variables are independent of the yield. Re-arrangement of 4.1.6 gives another ratio, the modified duration. Modified Duration is expressed as:

$$\frac{MD}{(1+y)} = -\frac{\frac{\partial P}{P}}{\partial Y}$$

Thus percentage price change is approximately equal to modified duration times the yield change. The MD is seen as a useful sensitivity measure of price volatility.

The measures derived from Macaulay's original equations are, modified duration, dollar duration and effective duration.

Modified duration is used to approximate the % change in price with a change in the market yield. The dollar duration reflects the $ value for a small change in yield.

The class ***Duration*** contains the method *duration* which takes the yield, years to maturity, par price and coupon rate. The method sets the values for a range of accessor methods; *setMDyears*, *setMDmodyrs*, *setDduration* and *setPerchange*. The accessor methods return the Macaulay duration in years (get*MDyears*), the modified duration in years (get*MDmodyrs*), the dollar duration (*getDolduration*) and percentage change (*getPerchange*). The class is shown in Listing 4.1 below.

```
package FinApps;
import static FinApps.PresentValue.*;
public class Duration {

   /** Creates a new instance of Duration */
   public Duration() {this.frequency=2.0;

   }
   public Duration(double couponfreq) {
        this.frequency=couponfreq;
   }
   private void setMDyears(double mdperiods) {
        this.mdyrs=mdperiods/frequency;
   }
   public double getMDyears() {
        return mdyrs;
   }
   public double getMDmodyrs() {
        return modmdyrs;
   }
   private void setMDmodyrs(double mdyears,double discvalue){
        this.modmdyrs=mdyears/discvalue;
   }
   private void setDduration(double modurationyrs,double price)
   {
        this.dolduration=((modurationyrs*price)/1e4);
   }
   public double getDolduration()
   {
        return dolduration;
   }
   private void setPerchange(double moduration)
   {
        this.percentchange=-moduration;
   }
   public double getPerchange(double basispoints)
   {
        return 100*(percentchange*basispoints);
   }
   private double percentchange;
   private double dolduration;
   private double modmdyrs;
   private double mdyrs;
   private double frequency;

   /** Requires the yield and coupon as a decimal value */
   public double duration(double yield,double period,
                          double parprice,double coupon) {

        double val=0;
        Volatility v=new Volatility(parprice,frequency);
        double bondprice=v.Bpricing((yield*100.0),period,
                                         (coupon*100.0));
        yield=yield/frequency;
        coupon=coupon/frequency;
        int n=(int)(period*frequency);
```

```
        val=(n*(pVs(yield,parprice,n)/bondprice));// face value
                                                 //discounted..
            for(int i=1;i<(n+1);i++) {
          double value=((pVw(yield,(coupon*parprice),
                       i))/bondprice);
          val+=((pVw(yield,(coupon*parprice),i))/bondprice);
              // individual period cash flows
        }
        setMDyears(val);
        setMDmodyrs(getMDyears(),(1+yield));
        setDduration(getMDmodyrs(),bondprice);
        setPerchange(getMDmodyrs());
              return val;
    }
}
```

LISTING 4.1. Computation of duration

Exercise 4.0

Show the percentage price change and PVBP for a security with 6% coupon, 15 years to maturity and priced to yield 8%. Compare the percentage change and PVBP with duration estimates for a range of yield changes from 1 to 500 basis points.

The bond price = 84.6275. Table 4.1 shows the data for calculated and duration estimates. It can be seen that the percentage price changes as measured by volatility calculation agree reasonably with the duration estimates for base point changes up to around 5 basis points. There is a similar agreement with the $ price value of a basis point up to around 10 basis points. After yield increments of around 10 basis points it can be seen that there is an increasing divergence between direct calculation and duration based estimates of price change. The graphs of Figures 4.1 and 4.2 show this divergence rate clearly. The application code is shown in Listing 4.2.

TABLE 4.1. Comparison of calculated percentage change with duration estimates

Basis Points	% Volatility	% Duration Estimate	PVBP	PVBP Estimate
0.000	0.000	−0.000	0.000	0.000
1.000	0.080	−0.080	0.068	0.068
2.000	0.161	−0.161	0.136	0.136
3.000	0.241	−0.241	0.204	0.204
5.000	0.401	−0.402	0.340	0.341
10.000	0.800	−0.805	0.677	0.681
20.000	1.591	−1.609	1.346	1.362
30.000	2.372	−2.414	2.008	2.043
50.000	3.908	−4.024	3.307	3.405
100.000	7.596	−8.047	6.428	6.810
200.000	14.366	−16.094	12.157	13.620
300.000	20.412	−24.141	17.274	20.430
500.000	30.680	−40.235	25.964	34.050

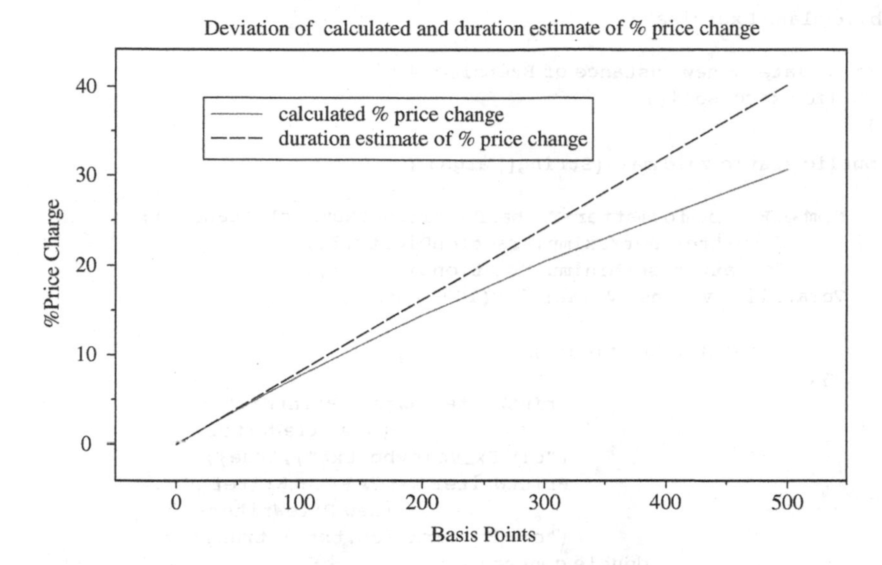

FIGURE 4.1. Deviation characteristics of percentage change for data in Table 4.1. Figure 4.1 shows that as yields get progressively higher the duration estimate tends to continue in a more linear fashion than the directly calculated volatility measure. The difference between both curves show the progressive error of duration estimate of the percentage price change.

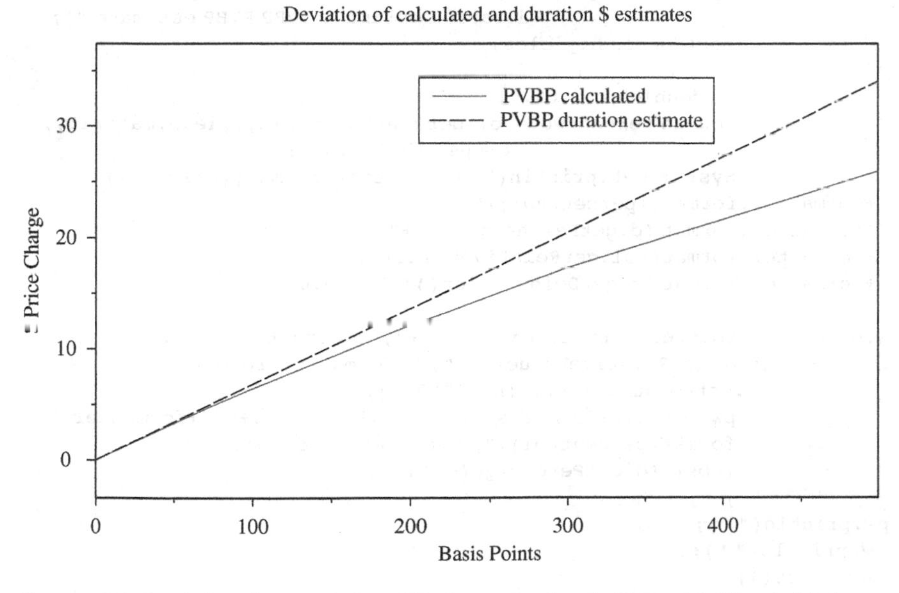

FIGURE 4.2. Deviation characteristics of $ change for data in Table 4.1. Figure 4.2 shows that the $ price value of a basis point change shows an increasing rate of error for duration based estimates. The duration estimate is also linear and does not track the calculated curve of $ price change with basis point increases.

```
public class Exercise_4 {

   /** Creates a new instance of Exercise_4 */
   public Exercise_4() {
   }

   public static void main(String[] args) {

      NumberFormat formatter=NumberFormat.getNumberInstance();
            formatter.setMaximumFractionDigits(3);
            formatter.setMinimumFractionDigits(3);
      Volatility vol=new Volatility(100.0,2.0);

      Duration d=new Duration();
     try{
                                 PrintWriter pw=new PrintWriter
                                                 (new FileWriter
                                 ("c:\\Ex_volpvbp.txt"),true);
                                 PrintWriter w=new PrintWriter
                                                 (new FileWriter
                                 ("c:\\Ex_duration.txt"),true);
                          double coups=8.0;
               double maturity=15;
               double[] yields={0.0,0.01,0.02,0.03,0.05,0.10,0.20,0.30,
                                   0.50,1.0,2.0,3.0,5.0};
                          double par=100.0;
               double yield=10.0;
               d.duration(yield/100.0,15,par,coups/100.0);// set values
                                    //for the base duration calculations
               System.out.println("Basis points %volatility
                                %duration estimate PVBP PVBP estimate ");
               for(double h:yields)
               {
                  double g=h/100.0;
               double percentvol=vol.percentVolatility(yield,maturity,
                                    coups, (h*100.0));
               System.out.println(" "+formatter.format((h*100.0))+ "
"+formatter.format(+percentvol)+"
"+formatter.format (d.getPerchange(g))+"
"+formatter.format(vol.getRelativeValue())+"
"+formatter.format (d.getDolduration()*(h*100.0)));

w.println(formatter.format((h*100.0))+","+formatter.format
          (vol.getRelativeValue())+", "+formatter.format
          (d.getDolduration()*(h*100.0)));
               pw.println(formatter.format((h*100.0))+","+formatter.
               format(percentvol)+", "+formatter.format
               (abs(((d.getPerchange(g))))));
               }
pw.println(" ");
 w.println(" ");
 pw.close();
                 w.close();
 }
 catch(IOException foe)
```

```
            {
            System.out.println(foe);
                    }
    }
}
```

Listing 4.2. Application code for Example 4.0

The duration based estimates for an increase in basis points, at higher yields consistently shows an over estimate in price. As yields get progressively higher the calculated curve deviates from the duration estimates at a greater rate.

4.2. Effective Duration

When a bond cash flow is affected by yield changes the duration measures we have looked at so far are not applicable. The Effective Duration is an approximation method of duration that can be used to measure volatility of yield dependant bonds.

Effective duration is calculated by taking the price of a security with small yield changes up and down from a reference value.

The effective duration is represented as:

$$\frac{P_{-\partial y} - P_{+\partial y}}{P_y(y_+ - y_-)} \tag{4.2.1}$$

Where $P_{-\partial y}$ is the price with yield minus a small amount (y_-) and $P_{+\partial}$ is the price with an equal magnitude increase in yield (y_+). P_y is the price with yield y, the reference value.

If we examine the graphs of percentage and dollar duration in figures 4.1 and 4.2 above, it is clearly seen that the actual prices are non linear whereas the duration estimates are linear. If we take a data set around the base value of 10% and plot $ duration with actual price change, the relationship between duration and actual price change can be seen. The tangent line at the base (10%) point gives the zero change duration. The duration based estimate is linear across the range of yield values whereas the actual price curve is convex. The difference between the tangent line and the actual price curve is the error of estimate, which can be seen to be larger for greater point changes. See figure 4.3.

To improve on our linear estimates we need to account for convexity of the actual price curve. Convexity in periods can be defined as:

$$\frac{1}{P}\frac{\partial^2 P}{\partial y^2} \tag{4.2.2}$$

To represent the price/yield curve the convexity is given as:

$$\frac{1}{P}\left[\sum_{t=1}^{n} t(t+1)\frac{F_t}{(1+y)^{t+2}} + n(n+1)\frac{V}{(1+y)^{n+2}}\right] \tag{4.2.3}$$

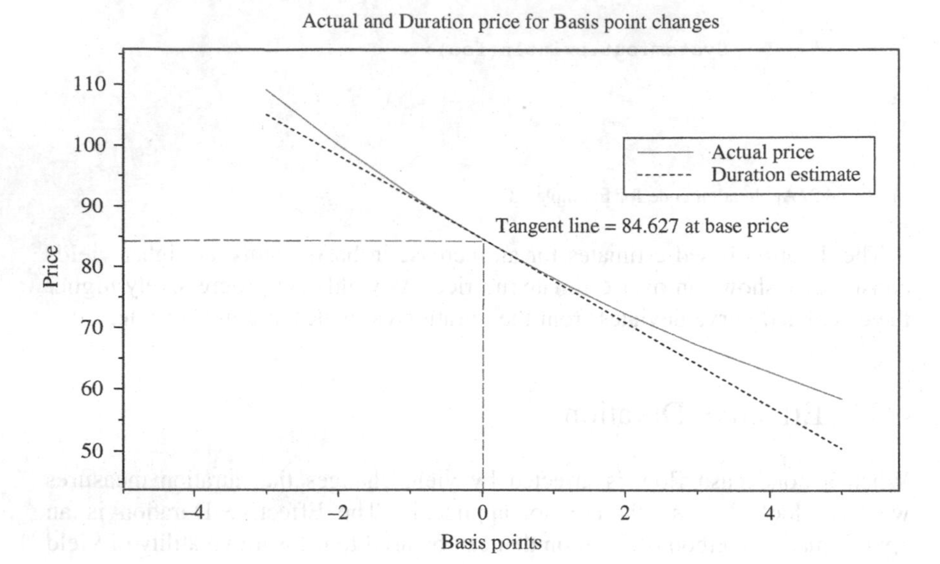

FIGURE 4.3. Actual v Duration price for change in basis points.

To convert the convexity in periods to years the transformation is given as; convexity (periods) = convexity (years)$^*k^2$, where k is the number of periods in a year.

For a small change in yield we have seen that the modified duration is a reasonable first order estimate, so from $\frac{MD}{(1+y)} = -\frac{\partial P}{\frac{P}{\partial Y}}$ we have the approximation $\frac{\Delta P}{P} \approx$ -duration$^*\partial y$. To continue with this approximation we can extend it to a second order Taylor series such that;

$$\frac{\Delta P}{P} \approx \frac{\partial P}{\partial y}\frac{1}{P}\Delta y + \frac{1}{2}\frac{\partial^2 P}{\partial y^2}\frac{1}{P}(\Delta y^2)$$

This is equal to $-duration^*\Delta y + \frac{1}{2}\ {}^*convexity^*(\Delta y^2)$. We will look at convexity and other possible convexity corrections in later chapters where we examine option valuation methods.

References

Sharpe, Alexander, Bailey (1999). *Investments* 6th edition. Prentice Hall.

5 Futures

Futures are agreements between parties to buy or sell assets at a future price. Futures are traded on exchanges such as the Chicago Board of Trade (CBT) or Chicago Mercantile Exchange (CME). Forward contracts are generally traded 'over the counter' by financial institutions or between financial institutions and corporate organizations. There are two positions to a forward contract; the long position agrees to buy at a future date for an agreed price and the short position agrees to sell at the forward date for the delivery price. In a forward contract the price is agreed which makes the forward value of the contract zero.

Although a forward contract has an initial value of zero, the value of a forward contract will change as time progresses to maturity. If the underlying asset decreases in price the short position stands to gain, if the asset price increases the value will change to favour the long position. Thus the delivery price will tend to vary from the forward price. Forward contracts are most useful in hedging strategies.

A futures contract traded at an exchange has a set of processes that control the cash flows of the instrument. The positions are taken through brokers to floor traders, for each client requiring a short position, a corresponding client requiring a long position is identified for the particular asset (say, coffee bean delivery). A participant 'closes out' a position by taking an opposite trade, so that the trade for a short in coffee beans for May delivery is closed out by taking a long position for May delivery. Most of the futures trading is closed out and very rarely is the underlying asset actually taken.

The underlying assets for any futures trading are very wide ranging. Typically trading is done across commodities such as tea, coffee, wheat grain etc and metals such as tin, copper, iron ore and nickel. The assets are well described and well defined in terms of quality and delivery details. The amount and volume of the asset is also related to the asset and its logical possibilities of delivery. The value of wheat contracts for July delivery for instance are specified by the CBT as being No 2 soft red, No 2 hard red winter wheat, No 2 dark Northern wheat or No 1 Northern spring wheat. Quantity is specified as 5,000 bushels.

Pricing for futures contracts involve taking the 'market' price negotiated on the floor of the exchange, when both positions (long and short) have takers. The pricing methods take account of the underlying assets, so oil is priced per barrel, treasury notes and bill futures are quoted in dollars and 32nds(CBT). Price movements are related to the base methods, so oil futures have movements of

$0.01 and bond futures 1/32nd of a dollar. The limits for daily price fluctuations on NYMEX as of early 2006 for light sweet crude oil are:

Trade basis 1,000 US barrels (42,000 gallons) per contract. Limits are, within first two months of contract, $7.50. Thereafter the limit is $7.50. After two months, the limit is $3.0 per barrel, rising to $6.0 per barrel if the limit has been reached in a previous back month. There are also position limits for futures. In the case of NYMEX oil futures the limit is 20,000 net futures but the limit is 1,000 in the last 3 days of trading in the spot month.

The futures exchange manages the contract so that default risk is minimised, this is principally done through 'marking to market' the account on a daily basis. Marking to market adjusts gains and losses of the account based on the end of day settlement prices. The settlement price is the average of the days trading price changes just prior to the ending of daily business.

When an investor buys a futures contract a percentage margin deposit is required, this varies with the asset and the exchange. The contract is started with an initial margin, set by the broker. If by the end of day the price of the futures contract has dropped, the amount of decrease in price is deducted from the margin account and paid to the sellers broker account, the same process applies to price rises, in which case the buyer account is credited with the end of day price rise. When the contract comes to an end the futures price is the price of the last mark to market. Investors are able to withdraw funds from the margin account that is in excess of the original margin. When margins go below a set maintenance level, the investor will be sent a 'margin call' to immediately top up to the initial margin level. If an investor defaults on this margin variation, the position is immediately closed out by the broker.

Short sellers are participants who effectively sell stock which they don't own. The market mechanism involves a broker taking a short order from the investor, the asset has to be borrowed by the broker, from a client who holds the asset, the asset is sold at the market price and the proceeds put into the short sellers account. When the position is closed out, the broker purchases the asset at the market price and replaces the asset to the lending client. The cost of purchase is deducted from the short sellers account. The short seller makes profit if the stock prices declines and a loss if it rises, however there are also considerations of cash flows from the asset, such as dividend payments or coupon receipts. The short seller has to make good interim cash flows that would have been paid from the asset even though the cash flows are not available to the short seller (the asset has been immediately sold). An excellent overview is given in the book by Hull.

5.1. Forward & Futures Pricing

Forward contracts can be broadly classified as being one of three types:

1. The security pays no income, e.g. a zero coupon bond.
2. The asset provides known cash income, e.g. any dividend paying stock.

3. An underlying security which pays a known dividend yield, e.g. a coupon bearing bond.

1. Forward price where the underlying asset pays no income.

For a forward contract with time T to maturity and current time t in years, with $\tau = (T - t)$ being the time remaining in the forward contract and r the risk free rate of interest. The forward price F at time t is given as

$$F = Se^{\tau r} \tag{5.1.1}$$

Where S is the spot price of an asset.

The forward value f of a contract that pays no income is given as:

$$f = S - ke^{-r\tau} \tag{5.1.2}$$

Where K is the delivery price ($K = F$ on contract initiation). The value of f is initially 0. As time passes both the forward value and the forward price will change.

To eliminate arbitrage opportunities, the relationship shown in 5.1.1 must hold. If $F < Se^{r\tau}$ then we can short sell the asset, go long the contract, invest the income from short selling(S) for τ at r rate of interest for a value of $Se^{\tau r}$.At maturity we can pay the long contract delivery price (which is the forward price). The arbitrage profit is therefore $Se^{r\tau} - F$. A similar argument holds for $F > Se^{r\tau}$. In this case there is an arbitrage opportunity for the investor to borrow funds at the risk free rate and buy the asset, go short the contract and at maturity sell the asset for F to close the short position. The asset was purchased for $Se^{r\tau}$, therefore the profit is $F - Se^{r\tau}$.

Example 5.0

A forward contract on a non income stock has a maturity of 5 months. The risk free interest rate is 5% and the spot price is quoted as $900.80. Calculate the forward price for delivery. Show the forward value of going long this contract, month by month to maturity. If immediately on purchase the spot price rises to $910.80. Show the effect this has on the month by month long position to maturity.

Using 5.1.1, the forward price is computed as $919.7628.

Using 5.1.2, the forward values are shown in Figure 5.1 below. The forward value is −18.9628, with zero elapsed time, rising to zero forward value as the months elapse to maturity. Showing that the spot price is equal to the delivery price, discounted continuously for the 5 months period at the 5% interest rate.

The effect of an immediate price rise is shown in Figure 5.2 below. This shows that at around 0.22 years the forward value is zero. This is the point at which the delivery price (919.7628) discounted at 5% is equal to the new price. By closing the contract up to the zero value the opportunity exists to take the forward value as profit

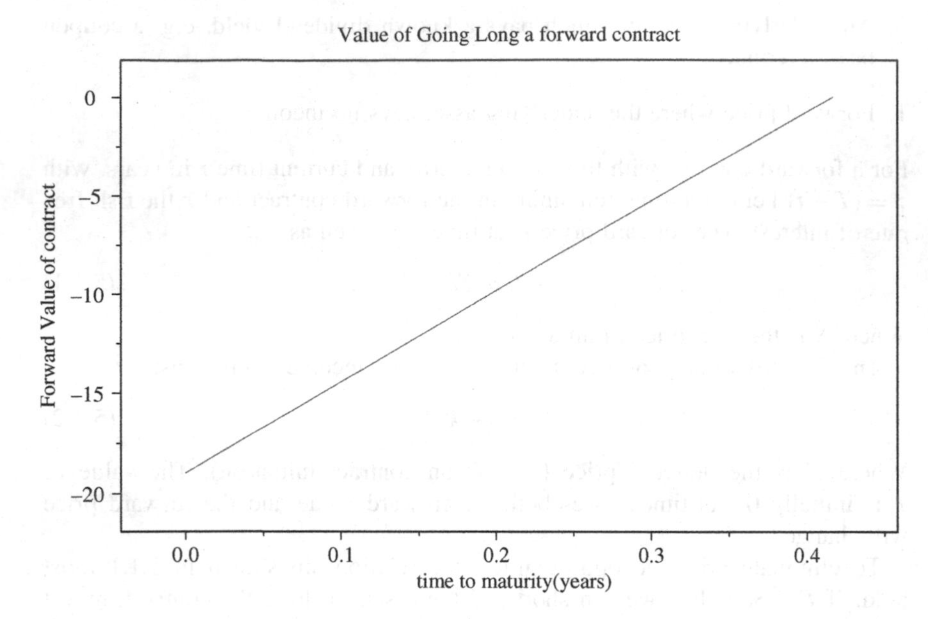

FIGURE 5.1. Long forward contract.

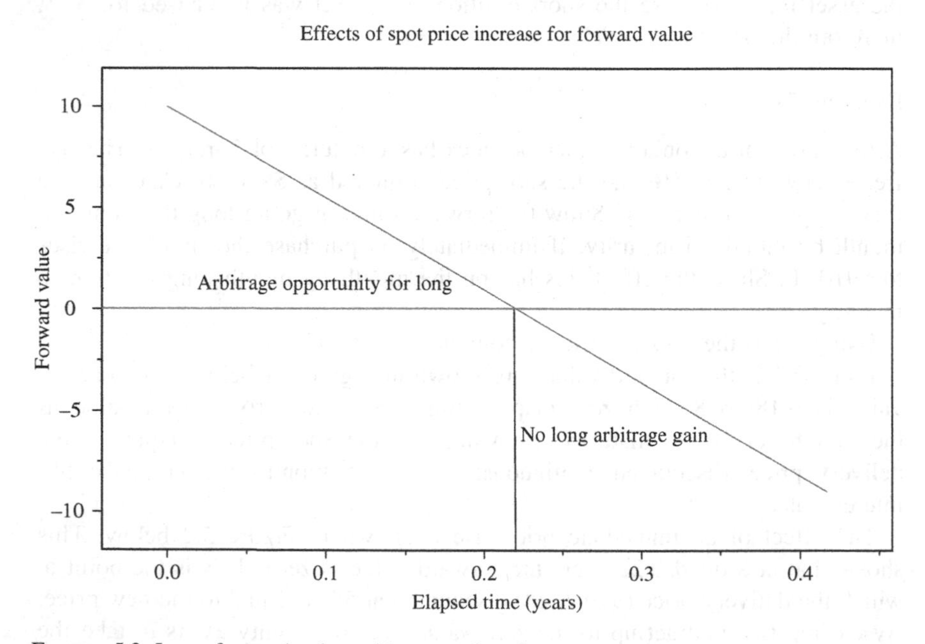

FIGURE 5.2. Long forward value with spot price change.

Listing 5.1 gives the example code

```
package FinApps;
import static FinApps.Intr.*;
import static java.lang.Math.*;
import java.util.*;
import static FinApps.PresentValue.*;
import static FinApps.Forwards.*;
import java.text.*;
import java.io.*;
public class Example_5_0 {

   public Example_5_0() {
   }

   public static void main(String[] args) {
      NumberFormat formatter=NumberFormat.getNumberInstance();
      formatter.setMaximumFractionDigits(4);
      formatter.setMinimumFractionDigits(3);
      double forprice;
      double monthvalue=0.08333;
      double currentime=0.0;
      double delprice=delpriceNoinc(900.80,0.41665,0.05);
      double delpricenew=delpriceNoinc(910.80,0.41665,0.05);
      forprice=fpriceNoinc(900.80,0.41665,0.41665,919.7628,0.05);
      try{
         PrintWriter pw=new PrintWriter(new FileWriter("c:\\
                                          example5_1a.txt"),true);
         PrintWriter w=new PrintWriter(new FileWriter("c:\\
                                         example5_1b.txt"),true);
         System.out.println("Delivery Price == "+formatter.
                          format(delprice));
         System.out.println("Delivery Price NEW == "+formatter.format
                          (delpricenew));
         for(int i=0;i<6;i++) {
            forprice=fpriceNoinc(900.80,0.41665,currentime,
                                delprice,0.05);
            System.out.println("Forward Value == "+formatter.format
                              (forprice));
            pw.println(formatter.format(forprice)+","+formatter.format
                                          (currentime));
            currentime+=monthvalue;
         }
         System.out.println("Delivery Price == "+formatter.format
                          (delprice));
         for(int i=0;i<6;i++) {
            forprice=fpriceNoinc(910.80,0.41665,currentime,
                                delprice,0.05);
            System.out.println("Forward Value == "+formatter.format
                              (forprice));
            w.println(formatter.format(forprice)+","+formatter.format
                                (currentime));
            currentime+=monthvalue;
         }
```

```
        w.println(" ");
        //pw.close();
        w.close();

      } catch(IOException foe) {
        System.out.println(foe);
      }
    }
}
```

LISTING 5.1. Application code for Example 5.0

5.2. Forward Price

Where an asset pays a predictable income (such as coupon bearing bonds) the present value of the known income streams has to be factored in. For the elimination of arbitrage opportunities the relationship between the forward price and spot price has to be:

$$F = (S - I)e^{r\tau} \tag{5.2.1}$$

If $F < (S - I)e^{r\tau}$, the possibility exists for the investor to short the asset and invest the income for τ at the risk free interest rate. By taking out a long contract for F and on maturity buying the asset (at F), the short position is closed for a profit of $(S - I)e^{r\tau} - F$. Similarly if $F > (S - I)e^{r\tau}$, the opportunity exists to borrow S for r interest and τ years, buy the asset and short the forward contract for F. On maturity the short position is closed by selling the asset for an arbitrage profit of $F - (S - I)e^{r\tau}$.

The value of a forward contract paying predictable income is:

$$f = (S - I) - Ke^{-r\tau} \tag{5.2.2}$$

Where I , is the present value of the income stream from the asset and K is the delivery price.

Example 5.1

A security with 10 months to maturity and spot price of € 60.50 has a dividend of € 0.50 payable every three months. Assuming a flat rate of 6%, calculate the forward delivery price. If the security is offered at a delivery price of € 61.0. Would there be an arbitrage opportunity for going short or long the forward? Using 5.2.1, the series of cash flows are:

$$I = 0.50e^{-0.06*3(0.08333)} + 0.50e^{-0.06*6(0.08333)} + 0.50e^{-0.06*9(0.08333)}$$

The income cash flows are: 0.4925, 0.4852, and 0.4779. The sum is 1.455.
The forward price F is $(60.50 - 1.455)e^{0.06*10(0.08333)} = 62.0702$.
The forward value for a delivery price of € 61.0 is given by:

$$f = 60.50 - 1.455 - 61.0e^{-0.06*10(0.08333)}$$

This is 1.0181. The present value of the delivery price is 58.0261. There is therefore an arbitrage opportunity to short the stock and go long the forward contract.

The general formulae of 5.2.1 and 5.2.2 are applicable to non flat interest rates. For example, if we have a 9 months forward contract, for a € 37.50 stock, that pays a dividend of € 2.0 every two months. The forward price is given as:

$$(37.50 - 2.0e^{0.05*2(0.08333)} - 2.0e^{-0.051*4(0.08333)} - 2.0e^{-0.573*6(0.08333)} - 2.0e^{-0.61*8(0.08333)})e^{r_9T}$$

The application code for this example is shown below in Listing 5.2.

```
package FinApps;
import static FinApps.Intr.*;
import static java.lang.Math.*;
import static FinApps.PresentValue.*;
import static FinApps.Forwards.*;
import java.text.*;

public class Example6_2 {
  public Example6_2() {
  }

    public static void main(String[] args) {
    NumberFormat formatter=NumberFormat.getNumberInstance();
    formatter.setMaximumFractionDigits(4);
    formatter.setMinimumFractionDigits(3);
    double forprice=fpriceInc(60.50,0.833,0.0,0.06,3.0,0.50);
    System.out.println("Forward price for income bearing security =
                        "+formatter.
format(forprice));
    double forvalue=fvalueInc(60.50,0.833,0.0,0.06,3.0,0.50,61.0);
    System.out.println("Forward value for income bearing security =
                        "+formatter.format(forvalue));
  }
}
```

LISTING 5.2. Application code for Example 5.1

Example 5.2

A 7 year bond with a price of £800.0 paying a semi annual coupon of 6% has a delivery price of £815.0 with a one year maturity. The continuously compounded period interest rates are 6.5% and 7.0%, respectively. Compute the value of a forward long position.

The period income stream has a value of; $I = 24.0e^{-0.065*0.5} + 24.0e^{-0.065}$

The period cash flows are 23.232 and 22.377 giving 45.609. The forward value is:

$$800.0 - 45.609 - 815.0e^{-0.07} = -5.511.$$

The value of taking a forward long position is −5.511 the value of a short position is +5.511.

Forward price and value, where the underlying asset pays a continuous dividend yield.

A stock paying a continuous dividend is one where the income is a defined percentage of the stock price. The dividend is assumed to be paid continuously throughout the period of the contract. If we assume that the rate q is 8% and the price in period one is 100 units, the dividend is 0.08*100. If the price is adjusted in subsequent periods the dividend is adjusted accordingly. So, if the next period price of the stock is 105 units, the dividend becomes 0.08*104.0. The forward price of a contract with a known dividend yield is:

$$F = Se^{(r-q)\tau} \tag{5.2.3}$$

The forward value of a contract with known dividend yield is given as:

$$f = Se^{-q\tau} - Ke^{-r\tau} \tag{5.2.4}$$

Example 5.3

A 6 months forward contract on an asset paying a continuous dividend yield of 5.0% per annum and a spot price of $29.50 has a delivery price of $31.0. Assuming the risk free interest rate is 10.5%. What are the forward price and the value of a long position on the contract?

Using 5.2.3 the forward price is:

$$F = 29.50e^{(0.105-0.05)0.5} = 30.32250.$$

Using 5.1.6 the forward value is:

$$f = \left(29.50e^{-(0.05*0.5)} - 31.0e^{-(0.105*0.5)}\right) = -0.64284.$$

The value of going long is −$0.64284. The forward price is $30.32250.

The example application code is shown in Listing 5.3.

```
package FinApps;
Import static FinApps.Intr.*;
```

```
import static java.lang.Math.*;
import java.util.*;
import static FinApps.PresentValue.*;
import java.text.*;
import java.io.*;
import static FinApps.Forwards.*;
public class Example5_4 {

  public Example5_4() {
    public static void main(String[] args) {
     NumberFormat formatter=NumberFormat.getNumberInstance();
     formatter.setMaximumFractionDigits(4);
     formatter.setMinimumFractionDigits(3);
      double forprice=fpriceDyld(29.50,0.50,0.0,0.105,0.05);
      double forvalue=fvalueDyld(29.50,0.50,0.0,0.105,0.05,31.0);
      System.out.println(" Long value == "+(formatter.format
                          (forvalue))+"
Forward price : "+(formatter.format(forprice)));
    }
}.
```

LISTING 5.3. Application code for Example 5.3

5.3. Pricing On Different Markets

The arguments posed for prices and values of forward contracts are applicable to futures contracts only when the interest rates are non stochastic. Recall that in futures, marking to market involves the contract being effectively re written at the close of each day's business; this isn't the case for forward contracts. When the interest rates are not constant the differences will matter for long term interest rate sensitive stock. The futures markets are broadly based on stock, stock index, currencies and commodities.

5.3.1. Stock Index

Stock indices track a virtual portfolio of stock. The weighting given to an individual stock is in proportion to the percentage investment in that stock. Stocks can vary over time with the relative percentage of the portfolio. The index movement is normally related to the total volume of the price changes.

Stock index futures were first traded at the Kansas City Board of Trade as the 'value line index futures'. Index futures are traded on a worldwide basis on numerous exchanges. The most prominent are the US markets with the S&P 500, DJIA, NASDAQ, and NYSE. For Europe there is the FTSE 100, CAC 40, DAX etc' and the NIKKEI 225 in Tokyo and Hang Seng in Hong Kong.

The S&P 500 is traded on the CME and is made up from 500 of the prime stocks on the exchange. The fluctuation of the index is measured in ticks, with one tick being equal to 0.05 points. Each tick is traded at $25.0. Each contract for the S&P 500 is traded at a minimum of \$500 times the index. The DJIA is traded on the CBT for \$10.0 times the index, the NIKKEI is traded at \$5.0 times the index on the CME. Note that the NIKKEI is quoted in Yen, not US\$. This means that the equations relating to arbitrage opportunities do not apply, as there are no securities in multiples of \$5 times the index.

An index future can be regarded as dividend paying securities, with the security being the portfolio of stocks making up the index and the dividend being the dividends due to a holder of the stocks. Since the basket of stocks that make up an index are representative of all the major market participants, dividends can be assumed continuous.

By assuming that q is the annual average dividend throughout the contract's life the futures price is:

$$F = Se^{(r-q)\tau} \tag{5.3.1}$$

This is the same as the forward price with a known dividend yield. The value of q is the yield from all of the aggregated stocks. If the yield values are not immediately obvious or there is a degree of variation not acceptable, then the \$ dividend payable during the life of the contract can be used. If the \$ dividend is used then the appropriate futures contract price is:

$$F = (S - I)e^{r\tau} \tag{5.3.2}$$

This is the same as the forward price with known cash income.

Example 5.4

The S&P 500 equity index is at 1103.30. The underlying stocks currently provide a dividend yield of 3.3%. The risk free interest rate is 4.2%. What is the futures price for a 3 months contract?

$$S = 1103.30,\ r = 0.042,\ q = 0.033,\ \tau = 0.25.$$

Using 5.1.7. The futures price is:

$F = 1103.30e^{(0.042-0.033)0.25} = 1105.7852$. The application code is shown in Listing 5.4.

```
package FinApps;
import static FinApps.Intr.*;
import static java.lang.Math.*; import static FinApps.PresentValue.*;
import static FinApps.Forwards.*;
import java.text.*;

public class Example_5_5 {
  public Example_5_5() {
  }
```

```
    public static void main(String[] args) {
     NumberFormat formatter=NumberFormat.getNumberInstance();
     formatter.setMaximumFractionDigits(4);
     formatter.setMinimumFractionDigits(3);

      double futureprice=fpriceDyld(1103.30,0.25,0.0,0.042,0.033);
     System.out.println("Futures price for the index = "+formatter.
                               format(futureprice));
}
    }
```

LISTING 5.4. Application code for Example 5.4

Index Arbitrage
For the case of $F > Se^{(r-q)\tau}$ an investor could buy the stocks which make up the index and short sell futures contracts. For the case of $F < Se^{(r-q)\tau}$ an investor could short the stocks and go long the futures contracts. It is not always necessary to trade the whole of the stock portfolio making up a particular index. An index such as the NYSE composite can be reasonably approximated by taking a representative sub set, since the index is constructed from all of the listed stock.

5.3.2. *Currencies*

If S denotes the foreign exchange rate and K denotes the agreed forward delivery price, with r_f the foreign, risk free rate of interest and r the domestic risk free rate. The forward value of a futures contract can be computed in an analogous way to a futures contract with known dividend yield as:

$$f = Se^{-r_f\tau} - Ke^{-r\tau} \tag{5.3.3}$$

The dividend yield is approximated by r_f since the \$ value is related to the proportion of the underlying foreign currency holding.

The forward price of a futures contract on foreign currency (also known as the equation of interest rate parity) is given as:

$$F = Se^{(r-r_f)\tau} \tag{5.3.4}$$

Currency futures are quoted in the financial press and are generally laid out as in Table 5.1 below, the prices are quoted as \$ value for the equivalent foreign currency. Thus, the rates show that the \$- £ futures price for December is 1.7732 US \$. If we look at the Euro futures prices we can see that the \$-€ futures for March is 0.0002 cents lower than for December, thus suggesting that short term US rates are around 0.08% lower than the Euro risk free rates. It is worth noting that forward rates and spot rates for currencies are normally reported as the foreign currency amounts per \$. So, a forward quote or spot price of \$1.7830 would be a futures quote of 56.08 ¢.

TABLE 5.1. Currency Futures CME

Sep 20 2004.	Open	Sett	Change	High	Low	Est vol.	Open Intr
$-Can $ DEC	0.7692	0.7714	+0.0018	0.7729	0.7666	15,069	77,535
$-Euro € DEC	1.2170	1.2162	−0.0012	1.2173	1.2118	38,729	84,028
$-EURO € MAR	1.2120	1.2160	−0.0012	1.2168	1.2118	4	742
$-STER £ DEC	1.7797	1.7732	−0.0075	1.7810	1.7693	8,783	46,164

Example 5.5

If the current $-£ exchange rate is $1.7830 and UK repo rate is 4.75%, the US rate 1.50%. Compute the 3 months future price.

In this case $S = 1.7830$, $r = 0.015$, $q = 0.0475$. The futures price is therefore 1.7686. The application code is shown in Listing 5.5.

```
package FinApps;
import static FinApps.Forwards.*;
import java.text.*;

public class Example_5_10 {
  public Example_5_10() {
  }
    public static void main(String[] args) {
     NumberFormat formatter=NumberFormat.getNumberInstance();
     formatter.setMaximumFractionDigits(4);
     formatter.setMinimumFractionDigits(3);
      double futureprice=fpriceDyld(1.7830,0.25,0.0,0.015,0.0475);
     System.out.println("Futures price for the currency = "+formatter.
                         format(futureprice));
    }
}
```

LISTING 5.5. Application code for Example 5.5

5.4. Commodity Futures

Commodities come in two varieties, those that are for consumption and those for investment. We can use arbitrage arguments to derive the futures price for investment commodities, but the arbitrage arguments will only provide bounds for deriving the futures price for consumable commodities.

If we ignore the cost of storage, the futures price for an investment commodity is: $F = Se^{r\tau}$. This is the same as the forward price for an asset paying no income. If we do need to consider storage costs, there are two alternative ways of accounting for them. Firstly, we can consider the present value of the storage costs, in which case the futures price can be represented as a forward contract with a predictable income. Secondly, the storage costs can be considered as a proportion of the spot price, in which case the future price can be computed as a forward contract where the asset pays a continuous yield.

For an investment commodity, where U represents the present value of the costs of storage and u represents the cost of storage as a proportion of the spot price. The following formula can be used to compute the futures commodity contracts:

1 For no storage costs

$$F = Se^{r\tau} \tag{5.4.1}$$

2 Where storage costs are reasonably represented by the PV of storage costs throughout the life of the contract

$$F = (S+U)e^{r\tau} \tag{5.4.2}$$

3 If storage costs are determined by a proportion of the commodity price

$$F = Se^{(r+u)\tau} \tag{5.4.3}$$

Precious metals are a good example of a commodity group that is used primarily for investment.

Example 5.6

A 1 year futures contract for platinum has a spot price of $843.0 per oz. the storage cost per year is $6.80 per oz. The risk free rate of interest is 1.5%. What is the futures price for this commodity?

In this case we have a forward price with known income: $S = 843.0,\ r = 0.015,\ \tau = 1.0,\ U = 6.80e^{-0.015*1.0}$. The futures price is $862.540. The example code is shown in Listing 5.6.

```
package FinApps;
import static FinApps.Forwards.*;
import java.text.*;
public class Example_6_11 {

  public Example_6_11() {
  }
  public static void main(String[ ] args) {
     NumberFormat formatter=NumberFormat.getNumberInstance();
     formatter.setMaximumFractionDigits(4);
     formatter.setMinimumFractionDigits(3);
     double futuresprice=fpriceInc(843.0,1.0,0.0,0.015,12.0,-6.80);
     System.out.println("Futures price for Commodity = "+futuresprice);
  }
}
```

LISTING 5.6. Application code for Example 5.6

For a futures commodity which is also consumable (e.g. Oil, cattle, etc') the owners of the asset have an interest in its value for consumption. The usual arbitrage arguments have limited applicability. If the formula of 5.4.2, is examined for an arbitrage opportunity, when the futures price is either greater than or less than the PV of the spot and cost of storage prices. We will see that the equations do not hold. For the case of

$$F > (S+U)e^{r\tau} \tag{5.4.4}$$

The investor can borrow the sum $(S+U)$ at the risk free rate, buy the asset for S and store it for τ at a price U. The investor can then short a futures contract on the asset. The overall effect is that the investor will obtain a profit of $F-(S+U)e^{r\tau}$. If this state of affairs continues for any appreciable time, the spot price will rise in the market and the futures price will equate to the equilibrium state.

In the case of

$$F < (S+U)e^{r\tau} \tag{5.4.5}$$

Some investors who own the asset will find it profitable to buy the futures contract for F, with cash from the sale of the underlying asset that has been invested at the risk free rate. This avoids the cost of storage for τ. This strategy is time limited in that the price will fall on the market and the equation will tend to equilibrium.

The investor in a consumable commodity will not be willing to sell the inventory (it has an intrinsic value in being consumed) so 5.4.5 provides an upper bound on the futures price of a consumable commodity as:

$$F \leq (S+U)e^{r\tau} \tag{5.4.6}$$

If the cost of storage is a proportion of the spot price, the futures price is defined as:

$$F \leq Se^{(r+u)\tau} \tag{5.4.7}$$

If the futures price $F < Se^{(r+u)\tau}$ holds, the implication is that a futures contract does not have a value sufficient to cancel the convenience value of holding the commodity in inventory. If we know the \$ cost of storage (U), the convenience yield y is defined as that yield which satisfies:

$$Fe^{yr} = (S+U)e^{r\tau} \tag{5.4.8}$$

If the cost of storage is a proportion of the spot price (u), then the convenience yield is given as:

$$F = Se^{(r+u-y)\tau} \tag{5.4.9}$$

Since the convenience yield reflects the desirability to have the underlying asset in inventory, it can be viewed as a measure of the market's perception of relative shortage of the commodity in the future. The increasing likelihood of a shortage within the contract duration, the greater will be the value of y.

The relationship between spot and futures prices is largely determined by the cost of carry. The carrying cost is the storage cost plus the interest cost of holding the asset minus the revenue earned from the asset. The carrying charge is given as:

$$C \equiv I + U - D \tag{5.4.10}$$

Where I is the interest cost, D is the cash flows from the asset and U is the storage cost. The convenience yield has a dollar representation as:

$$\$Y = S + C - F \tag{5.4.11}$$

The theoretical basis, defined as the spot price minus the futures price, is therefore $Y - C$. For a stock which pays no dividend there is no storage cost and no cash flows from the stock, therefore the cost of carry is r. For stock that earns an income the cost of carry is $r - q$ since there is a cash flow from the asset. In the case of stock index futures the value of q is determined by the continuous flow of dividend revenue from underlying stock. For currency futures the cost of carry is $r - r_f$, for stock that has associated storage costs the carrying charge is $r + u$.

Listing 5.7 shows the class **Forwards.** This class provides methods to compute the functions described above. Each method has a brief description in the listing.

```
package FinApps;
import static java.lang.Math.*;
import static FinApps.Intr.*;
import static FinApps.PresentValue.*;
public final class Forwards {

  public Forwards() {
  }

  /** method to return the dollar intersest value coefficint
for the term of a repo rate
    *@param term is the term in years (as decimal) greater than 1 day
    *@param reporate the current bank base rate/ federal funds rate
    */
  public static double dollarIntr(double term,double reporate) {
    return reporate*(term/360.0);
  }
  /** method to return the delivery price of a new forward contract
    *@param spotprice is the spot price of the underlying asset
    *@param maturity is the time (in years as a decimal) to maturity
of the contract
    *@param currentime is the start time of the new contract
  */
```

```
  public static double delpriceNoinc(double spotprice,double
maturity, double reporate) {
    return(spotprice*conintr(reporate,maturity));
  }
  public static double fpriceNoinc(double spotprice,double maturity,
double currentime,double deliveryprice,double reporate) {
    return(spotprice-(pVcont(reporate,(maturity-currentime),
deliveryprice)));
  }
  public static double fpriceInc(double spotprice,double maturity,
double currentime,double reporate,double period,double dividend) {
      double income=0.0;
      income= maturity==1.0 ?pVcont(reporate,1.0,dividend):0.0;
//last value
      double limit=0.0;

      limit=(maturity-currentime);//Assumes that later start times
                                 //will floor the pv of dividend payments
      double time =(period/12.0);
      double increment=time;
      while(time<limit) {
        income+=pVcont(reporate,time,dividend);

        time=time+increment;

      }

        return((spotprice-income)*(conintr(reporate,
              (maturity-currentime))));

      }

      public static double fpriceInc(double spotprice,double maturity,
double currentime,
double[] reporate,double period,double dividend) {
        double income=0.0;
        double limit=0.0;
        double forwardprice=0.0;
        limit=(maturity-currentime);//Assumes that later start times
                                    will floor the
pv of dividend payments
        double time =(period/12.0);
        double increment=time;
        for(double r:reporate) {
          income+=pVcont(r,time,dividend);
          time=time+increment;

        }

        return((spotprice-income)*(conintr(reporate[(reporate.
              length-1)], (maturity-currentime))));

      }

      public static double fvalueInc(double spotprice,double
                                      maturity,double
currentime,double reporate,double period,double dividend,double
```

```
deliveryprice) {
          double income=0.0;
          income= maturity==1.0 ?pVcont(reporate,1.0,dividend):
                            0.0;//last value
          double limit=0.0;

          limit=(maturity-currentime);//Assumes that later start
                 times will floor the
pv of dividend payments
          double time =(period/12.0);
          double increment=time;
          while(time<limit) {
            income+=pVcont(reporate,time,dividend);

            time=time+increment;

          }

          return ((spotprice-income)-(deliveryprice*pVcont
                 (reporate, (maturity-currentime))));

        }

        public static double fvalueInc(double spotprice,double
maturity,double currentime,double[ ] reporate,double period,
double dividend,double deliveryprice) {
          double income=0.0;
          double forwardprice=0.0;
          double time =(period/12.0);
          double increment=time;
          for(double r:reporate) {
            income+=pVcont(r,time,dividend);
            time=time+increment;

          }

          return (spotprice-(income+(deliveryprice*pVcont(reporate
                 [(reporate.length-1)], (maturity-currentime)))));
        }
        // Also parity rate calculation

        public static double fvaluegen(double fprice,double
delivprice,double maturity,
double currentime,double reporate)
        {

          return ((fprice-delivprice)*pVcont(reporate,
(maturity-currentime)));
        }
        public static double fpriceDyld(double spotprice,
double maturity,double
currentime,double reporate,double dividendyld) {
          return (spotprice*conintr((reporate-dividendyld),
(maturity-currentime)));
        }
        public static double fvalueDyld(double spotprice,double
maturity,double currentime,double reporate,double dividendyld,
double deliveryprice)
```

```
        {
          return ((fpriceDyld(spotprice,maturity,currentime,
reporate,dividendyld)-(deliveryprice))*pVcont(reporate,
(maturity-currentime)));
        }
    }
```

LISTING 5.7. Computation of futures calculations

References

Hull J. C. (2006). *Options, Futures and Other Derivatives* 6th edition. Prentice Hall.

6 Options

An option grants the holder a right to buy or sell an asset. The value of an option is dependent on the underlying asset value. Because of the indirect derivation of an option's value, options are also referred to as derivatives. A call option is the right to buy an asset for a certain price; a put option is the right to sell an underlying asset at a given price. The price at which a call option is exercised is the strike price. An option gives the holder a choice of exercising the offer, there is no compulsion for the holder to exercise the selling or buying right.

An option can be traded on any asset type. There can be options on a range of almost limitless underlying 'assets'; oil, wheat, fruit juice, interest rates or weather forecasts. Options have a lifetime, for European options the option can only be exercised at the end of its life, for American options the product can be exercised at any time during its life. If an option is embedded it has to be traded with its base asset.

Options are issued by a writer, to own the option a premium is paid to the writer, at exercise an owner (holder) pays the writer a strike price and the option ceases. When a put is exercised the writer pays a strike price to the holder in exchange for the asset and the option ceases. One can take a long or short position in a derivative.

6.1. Option Types

Exchange Traded Options

Options are traded worldwide on international exchanges. Prior to 1973 options were traded as 'over the counter' (OTC). OTC trading is between organisations who trade directly with each other for highly customised products in areas such as currency and interest rates. Exchange based option trading began with the Chicago Board Options Exchange (CBOE), exchanges worldwide now deal with the options market. Broadly the assets being traded for options contracts include stock, index, foreign currencies and futures.

Stock Markets

Stock options are very actively traded at the CBOE the Philadelphia Exchange (PHLX), NYSE and AMEX in the USA. The London International and Financial

Futures Exchange (LIFFE) for the UK. Amongst the most popular traded stocks (for the period 4th April 2005) was American International Group (AIG) at a volume of 22656, Semiconductor Mutual Holdings (SMH) with a volume of 20695 and Siebel Systems with a volume of 16578. A single contract gives the right to buy or sell 100 shares of the stock at the set strike price.

Index Markets

There are a very wide range of index option products available through many exchanges. The two most popular index options in the USA are the S&P500 and S&P100. The S&P500 is a European option, the S&P100 is American. The Philadelphia (PHLX), trading options on the NASDAQ index offer two products: Full size (trading symbol QCX) and mini (trading symbol QCE). The QCX product has an index multiplier of $100.00; the QCE has a multiplier of $10.00. Premium quotes for the QCX are 1 point = $100.00 and has a trading limit of 50,000 contracts on same side with a maximum of 30,000 in near-term month. The QCE has a premium quote of 1 point = $10.00 with a limit of 500,000 contracts and no more than 300,000 in a near-term month. Hedge exceptions are allowed. There is a minimum change in premium allowed for a change under 3 of $5.00 for QCX and $0.50 for QCE. For a change greater than 3. The minimum change is $10.00 for QCX and $1.00 for QCE.

As an example of index settlement, the PHLX QCX product has a strike price of $1900.00. The index at settlement date has a value of 1950. The writer has therefore to pay 1950 − 1900 = 50*$100.00 = $5000. This amount is paid rather than delivering the underlying share index!

Foreign Currency Markets

A major player in the US foreign currency options market is PHLX. The size of a single contract is dependent on the currency being traded. Examples are:

Australian dollar AUS$ $50,000 = 1 contract
UK Sterling £ £31,250 = 1 contract
Euro € €62,500 = 1 contract
Japanese Yen ¥ ¥6.25M = 1 contract.

Futures Option Markets

Futures options are traded extensively, the popular exchanges are LIFFE in the UK and CBOT in the US. Futures options use a futures contract as the underlying asset. The options contract is usually chosen to settle just prior to the options contract maturity date. A put option on a futures contract when exercised entitles the holder to a short position in the contract and an amount equal to the strike price minus the futures price. The holder of a call option is given a long position in the underlying futures contract plus cash equal to the futures price minus the strike price. The futures contracts have zero value (both long and short positions). The futures products encompass the whole range applicable to standard futures trading.

6.2. Option Specifications

Stock Options

A stock option is an American option which conveys the right of the holder (but not the obligation) to buy or sell the underlying stock at a specified price on or prior to a given expiration date. The seller of an option ***is*** obligated to buy or sell the underlying asset to the option buyer at the specified price if required.

Strike Price

The strike price of a stock option is the share price at which the underlying shares will be bought or sold if the option is exercised by the holder against the option writer (seller). Strike prices are listed in increments of $2^{1/2}$, 5 or 10 points, dependent on the underlying market price. Only a few levels around the current market price are traded. A given option is traded with expiration on one of four dates. The closing exchange traded option prices (premiums) are published daily; the option prices for the day are set by active floor trading.

The exercise price (strike price) is the specified share price at which stock can be bought or sold by the option holder. The strike price is initially set at the level near to the current share price. Subsequent strike prices are set at intervals. Where the initial strike price is below $25.00, the additional interval is $2\frac{1}{2}$ points, where the initial price is above $25.00 and up to $200.00, the interval is 5 points. Any initial price over $200.00 attracts a 10 point interval. New strike prices are set when the stock rises to the highest strike price already set or falls to the lowest strike price. As an example XYZ Corporation has a strike price of 13.00, the traded options have strike prices of 11.00, $13^{1/2}$, 16.00, $18^{1/2}$ and 21.00.

The strike price is a fixed option specification unlike the premium which is the daily fluctuating trade price. The option premium price is the price a buyer pays to have the right of exercise. The premium is paid to the option writer. In return for the premium the writer of a call agrees to deliver the underlying asset in return for the strike price. The writer of a put option is obliged to take delivery of the underlying stock at the specified strike price if the right is exercised by the put option holder. Premiums are retained by the writer whether or not the options are exercised. Premiums are quoted per share so a premium of $^7/_8$ equates to a price of $87.50 per option contract (0.875*100 shares).

Expiration Process

Stock options are traded around eight months prior to expiration. A defining feature of the option is the date of expiration. Thus a January call on GM is a call option on General Motors which has an expiry date in January. The final day of trading for an option is the third Friday of the expiration month. Stock options trade on a cyclical basis, trading either on a January, February or March cycle. Each cycle has four expiration dates as shown in Table 6.1 below. For a stock with a January call, this means an option call with a January expiration date. The

TABLE 6.1. Expiration cycles.

January Cycle		February Cycle		March Cycle	
Current Months	Month	Current Months	Month	Current Months	Month
Jan	Jan ***Feb*** Apr Jul	Jan	Jan Feb May ***Aug***	Jan	Jan ***Feb*** Mar Jun
Feb	Feb ***Mar*** Apr Jul	Feb	Feb ***Mar*** May Aug	Feb	Feb Mar Jun ***Sep***
Mar	Mar Apr Jul ***Oct***	Mar	Mar ***Apr*** May Aug	Mar	Mar ***Apr*** Jun Sep
Apr	Apr ***May*** Jul Oct	Apr	Apr May Aug ***Nov***	Apr	Apr ***May*** Jun Sep
May	May ***Jun*** Jul Oct	May	May ***Jun*** Aug Nov	May	May Jun Sep ***Dec***
Jun	Jun Jul Oct ***Jan***	Jun	Jun ***Jul*** Aug Nov	Jun	Jun ***Jul*** Sep Dec
Jul	Jul ***Aug*** Oct Jan	Jul	Jul Aug Nov ***Feb***	Jul	Jul ***Aug*** Sep Dec
Aug	Aug ***Sep*** Oct Jan	Aug	Aug ***Sep*** Nov Feb	Aug	Aug Sep Dec ***Mar***
Sep	Sep Oct Jan ***Apr***	Sep	Sep ***Oct*** Nov Feb	Sep	Sep ***Oct*** Dec Mar
Oct	Oct ***Nov*** Jan Apr	Oct	Oct Nov Feb ***May***	Oct	Oct ***Nov*** Dec Mar
Nov	Nov ***Dec*** Jan Apr	Nov	Nov ***Dec*** Feb May	Nov	Nov Dec Mar ***Jun***
Dec	Dec Jan Apr ***Jul***	Dec	Dec ***Jan*** Feb May	Dec	Dec ***Jan*** Mar Jun

January cycle has January, April, July and October. The February cycle has the months February, May, August and November. March has a sequence of March, June, September and December. While the expiration date for the current month has not been reached, the current month cycle is used. Once the current month expiration date has passed the following options will trade with the next month, next but one month and the next two months in sequence.

6.3. Pricing Specification

Option Contract

The option contract is defined as the style: American or European or Capped, type put or call, number of shares, underlying asset, strike price and expiration. A *class* of option refers to an option of the same type, underlying asset and style. So for instance put options on GM are of the same class and call options on GM are of another class. All options of the same class that have the same number of shares and cover the same underlying asset with strike price and expiration dates are a *series*.

Options are referred to as *in-the-money*, *at-the-money* or *out-of-the-money*. If the strike price of a call option is less than the market price of the underlying security the call is in-the-money, since the holder of the call can buy the stock at a lower than market price. If a put option has a strike price greater than the market price for the underlying asset then the holder is also in-the-money, since the put holder can sell the underlying asset at a price greater than the market price. Where in-the-money options have a positive cash flow to the holder the converse is true for out-of-the money option holders. At-the-money options have a zero cash flow to the holder.

An option is normally exercised when the outcome is of some benefit to the holder. The simple payoff of an option at expiration shows the benefit to the holder.

The payoff of a call at expiration, where the strike price is less than the market price can be represented as:

$C = \max(0, S - X)$ Where C is the call value, X the strike price and S the stock price. The payoff of a put option at expiration can be represented as:

$P = \max(0, X - S)$. It should be noted that the payoff is the simple value of the option at expiration and is not necessarily the profit, which will depend on other factors including premium costs. The relationship between long and short puts and calls is shown in Figures 6.1–6.4 below.

A long call position with a stock price of 80 at expiration and a strike price of 30 has the characteristic shown in Figure 6.1. The payoff will be the maximum of 0 or the difference of the market price minus strike price.

For a short call position on the same stock as in Figure 6.2, the payoff will have a maximum of 0 or minus the difference of the market price minus strike price. This is also: $\min((X - S), 0)$

Going long a put on the same stock as in Figure 6.1 will have a maximum payoff of 0 or the difference of the strike price minus market price.

Going short a put on the same stock as 6.1 will have a payoff of maximum 0 or minus the difference between the strike price, minus market price. This is also: $\min((S - X), 0)$.

The *intrinsic* value of a stock option is the maximum of 0 and the value it would be if exercised immediately. For a put option the intrinsic value is therefore $\max((X\_S), 0)$ and a call option intrinsic value is $\max((S\_X), 0)$. And in-the-money American option is worth at least its intrinsic value (as it can be

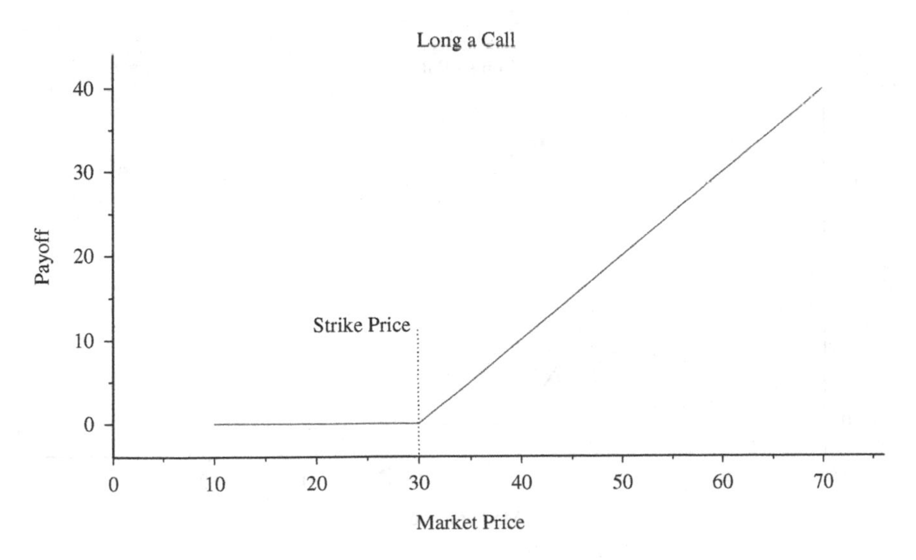

FIGURE 6.1. Long call characteristic.

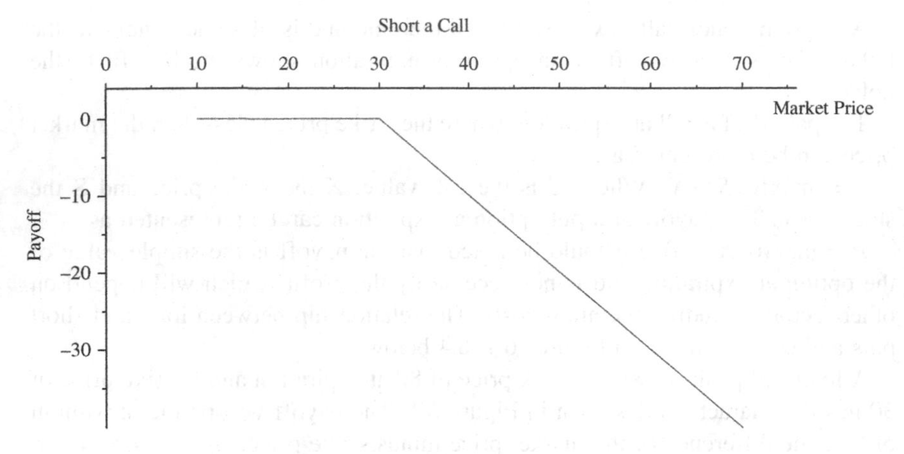

FIGURE 6.2. Short call characteristic.

exercised immediately) also an American option is worth at least as much as an equivalent European option, due to the early exercise feature.

6.3.1. Dividends and Stock Splits

Exchange traded options are not cash dividend protected which means that options are not adjusted for cash dividends. In early OTC trading the amount by which a company declared its cash dividend was deducted on the ex-dividend day from the strike price. The stock price generally falls just after a cash dividend is made, so dividends are detrimental to call options (but conversely enhance put options).

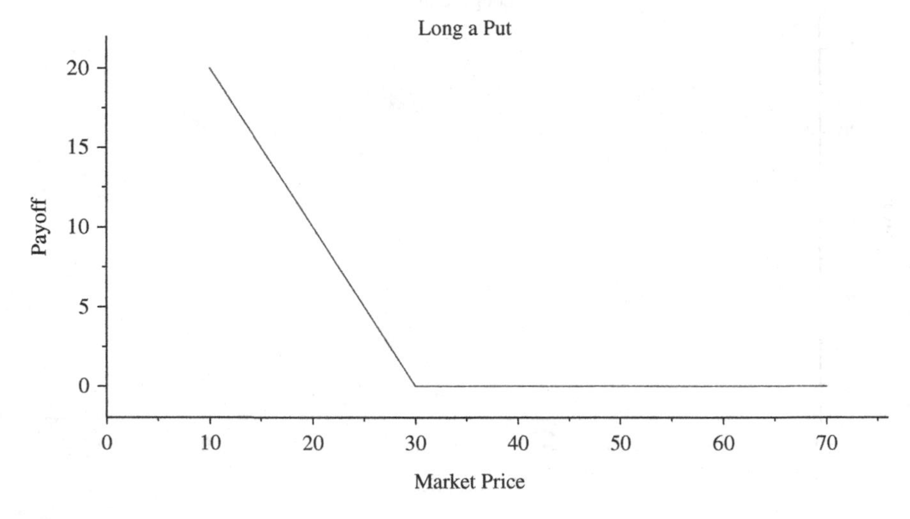

FIGURE 6.3. Long a put characteristic.

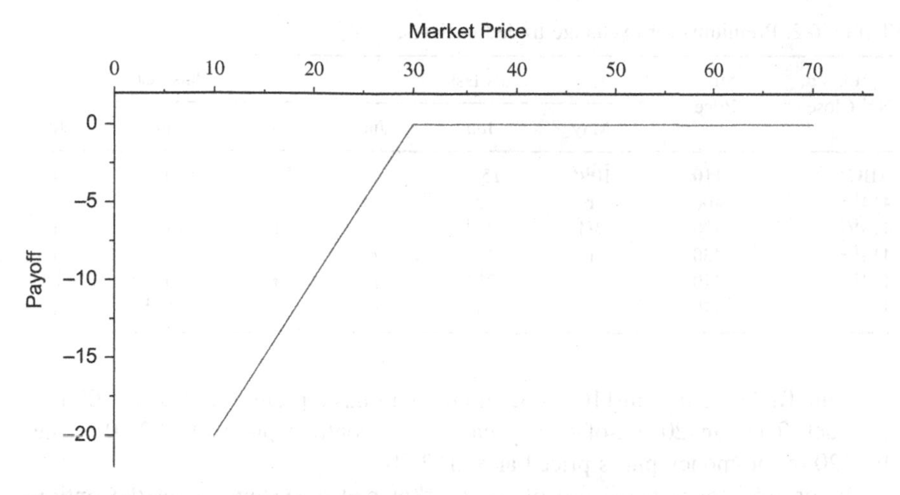

FIGURE 6.4. Short a put characteristic.

Stock splits occur when stock is split into more share issues. In an n-for-m split the strike price is diluted by m/n times the previous value. The number of shares covered by the contract is adjusted by n/m times the previous amount. For example if ABC Inc' has a call option for 100 shares at \$27 per share and then announces a 3 for 1 split. The option would be adjusted to 300 shares at \$9.00 per share.

If there is a stock dividend made then the exchange will adjust the options accordingly. A stock dividend involves the company issuing extra shares to existing shareholders. The stock price can be expected to reduce by an amount proportionate to the percentage stock dividend. The exchange will adjust the option price in much the same way as it does for stock splits. For example ABC Inc' has a put option on 100 shares at \$10.00 per share. The company declares a 10% stock dividend which effectively means issuing 1 share for every 10 held. This is the same as an 11 for 10 stock split so the option would be adjusted to a put for 110 shares at \$9.0909.

6.3.2. *Option Quotes*

Newspapers carry the daily premium prices for exchange traded options. A typical newspaper listing is as outlined in Table 6.2 below.

The first column lists the closing stock price with the second column listing the strike price. Columns three to five list the call premium and columns six to eight list the closing put premium. The premium prices are listed per share and a contract is for 100 shares, so the cost of a single contract is 100*list price. An r indicates that the option is not traded at the exchange and an s indicates that the option is not listed on the exchange.

TABLE 6.2. Premiums for exchange traded options.

Option & NY Close	Strike Price	Calls last			Puts last		
		May	Jun	Jul	May	Jun	Jul
ABC	110	$10^{1/8}$	$15^{1/4}$	r	$1^{1/16}$	r	r
$414^{1/2}$	400	r	r	r	r	4	r
$414^{1/2}$	420	3/4	$8^{1/4}$	r	r	$13^{1/8}$	r
$414^{1/2}$	430	r	$4^{3/4}$	r	r	r	r
$414^{1/2}$	440	r	$2^{3/8}$	r	r	r	$34^{3/8}$
$414^{1/2}$	470	r	s	s	r	$54^{1/4}$	r

From Table 6.2 the Jun110 in-the-money call has a premium of $1525.00 per contract. The Jun420 out-of-money call has a contract price of $825.00. The Jun420 in-the-money put is priced at $1312.50.

Option exchanges make use of the market maker system to set the option prices. Market makers will quote for both the bid and ask price for an option. The difference between ask (market maker's price to sell) and bid (market maker's buying price) is the bid-ask spread and is set at limits determined by the exchange. The limits are: $0.25 for options less than $0.50 and $0.50 for options priced at $0.50 to $10.00, $0.75 for options in the range $10.00 to $20.00 and $1.0 for higher prices. For exchange trading an option holding can be closed out (liquidated) by an offset order to sell the option. The option writer can close out by issuing an offset to buy the option.

6.3.3. Margin Accounts

Option contracts cannot be purchased on a margin account, option writers however operate a margin deposit where the funds are deposited to ensure the writer's ability to pay any liability from an option being exercised. In writing a Covered Call the writer is the owner of the underlying stock used to write the call option. If a covered call is out of the money no margin is needed. If the option is written as a Naked Option there is no offsetting in an underlying stock. If the option is in the money the margin is set at 30% of the stock value plus the in the money amount. If the option is out of the money the margin is 30% of the stock value minus the out of the money value. The following examples show the operation of margin accounts.

Example 6.0 Writing naked options

A participant writes two naked call option contracts. The option price is $8.0, with a strike price of $56.0 and an underlying stock price of $59.0. The margin addition required is therefore:

1. 30% of $59.00 *200 = $3540.0
2. The option is $59.0 – $56.0 = $3.0*200 = $600.0 in the money
3. The price received for the contract is $8.0*200 = $1600.0

$$\$3540.0 + \$600.0 - \$1600.0 = \$2540.0.$$

If the option was a put (out of the money) the margin required would be: $3540.0 – $600.0 – $1600.0 = $1340.0.

Example 6.1 Writing a covered call

An investor wishes to purchase 400 shares of stock on margin and write four call options on the stock. If the market price of the stock is currently $70.00 and the strike price is $66.0, with a premium of $8. How much will the investor need to pay up front?

1. 50% of $70.00*400 = $14000.00.
2. The option is $70.00 – $60.00 = $4.00*400 = $1600.00 in the money.
3. Margin borrowing allowed is $14000.00 – $1600 = $12400.00.
4. Premium received from option is $8*400 = $3200.00.
5. Up front payment is therefore: Cost = $28000.00 – $12400.00 – $3200.00 = $12400.00.

The OCC (Options Clearing Corporation)

OCC is a registered clearing corporation with the Securities and Exchange Commission (SEC) and has a triple A rating with Standard and Poor's Corporation. The OCC guarantees the fulfilment of the writer's liabilities and keeps a record of the long and short positions. All option trades are cleared through an OCC member, the members are collectively responsible for ensuring that sufficient funds are deposited to cover an individual member being in default.

When an option is purchased the buyer must pay the premium by the next business day, this is paid into the OCC account. The writer of an option maintains the margin account with a broker; in turn the broker maintains a margin account with the OCC (or an OCC member if the broker is not a member). The OCC guarantees contract performance of all parties in an exchange traded option. The option holder effectively deals with the 'Corporation' rather than an individual. The OCC is buyer to the seller and seller to the buyer.

6.4. Arbitrage in Option Prices

6.4.1. Main Components of Pricing

There are six main components of stock option pricing:

1. Underlying Stock Price
2. Strike Price
3. Time to expiration

4. Volatility
5. Risk free interest rates
6. Dividends

Underlying Stock Price

An option value is heavily dependant on the underlying market price of the stock. The option is in the money if the price of the stock is in excess of the call strike price. The payoff from a call option is the amount by which the market price is above the strike price. Call options increase in value with the increase in market price of the underlying stock and decrease in value with an increase of strike price. For a put option if the strike price is below the market price the option is in the money. The payoff for a put option is increased as the strike price decreases and the option payoff decreases as the strike price increases. Put options are more valuable as strike price increases and less valuable as market price increases.

The difference between an in the money option's strike price and the market price is the *intrinsic* value of the option. By this definition only in the money options have an intrinsic value.

Time to Expiration

For both put and call options the time to expiration has a direct relationship to value. For any two American options if one considers that the only difference is the expiration time it is intuitively obvious that the option with a longer expiration time has more opportunity to move in the money. As the option with the shorter time approaches expiration the opportunity for moving becomes less. Thus *ceteris parabus* the longer life option is always worth at least as much as the shorter life option. Time value of an option drops very rapidly as it nears expiration.

European options do not necessarily behave in the same way as American options. Since a European option cannot be exercised until expiration, the longer life option has only opportunity to exercise at maturity. If two call options identical in all respects other than time have the underlying stock paying an unexpectedly large dividend shortly after the expiration of the shorter life option and just before the expiration of the longer life option. It could well be the case that the longer life option is disadvantaged by the subsequent lowered stock price, whereas the shorter life option would be more valuable.

Volatility

Volatility refers to the propensity of the underlying stock to move in price. Volatility has a major influence on the premium price, with higher volatility of stock giving higher premiums. The premium price reflects a degree of risk in the option, with higher volatility the risk is perceived as higher for the option to go in- the- money. Since volatility works both ways (changes are both up and down in the stock price) the holder of a put or call option is equally well affected by the volatility.

Interest Rates

The risk free interest rate has a subtle effect on option prices. Generally as interest rates increase the general level of stock prices are expected to increase. However since the risk free rate has increased the discount factor for future cash flows for option holders has increased and the value of future cash flows to the holder is reduced. The combined effect of an increase in stock price and reduced payoff tends to push a put option towards being out-of-the-money. A push option therefore tends to reduce in value as interest rates increase and the push option premium decreases.

For a call option increased risk free interest rates producing higher stock prices makes the option more likely to be in-the-money, the effect of decreased value of future cash flows with a higher discount rate tends not to be as pronounced as the positive push effect of higher stock prices. In general call options increase in value as interest rates increase and the premium of call options increases.

Cash Dividends

Dividends are paid to the stock owner and the amount of dividend paid reduces the immediate value of the stock price in proportion to the dividend magnitude. On the ex-dividend date the stock price is reduced making the likelihood of a call option being out-of-the-money, thus reducing the option premium. The value of put options are pushed in the opposite direction and the tendency is to have a higher put premium. Non cash dividends are usually incorporated through other mechanisms into the option pricing mechanism.

The trade off for various factors affecting the premium rate of options is outlined in Table 6.3.

The trade off effects are either positive on premium rates or negative for puts and calls or in the case of volatility the effect is positive for both. With time to expiration the effect on European option premiums is indeterminate.

Arbitrage Effects in Option Pricing

A riskless arbitrage opportunity exists when, for no initial investment, a nonnegative return is achieved in all circumstances and positive returns under some circumstances. In the theory of efficient markets there should be no riskless arbitrage opportunities. When considering the factors affecting option prices

TABLE 6.3. Trade off for six co factors affecting stock option pricing.

	American Call	European Call	European Put	American Put
Stock Price	pos	pos	neg	neg
Strike Price	neg	neg	pos	pos
Time	pos	variable	variable	pos
Volatility	pos	pos	pos	pos
Interest Rates	pos	pos	neg	neg
Dividends	neg	neg	pos	pos

we will consider the variables in relation to no arbitrage opportunities. As an example an American put cannot take on a price less than its intrinsic value otherwise the option can be purchased and immediately sold taking profit from selling the underlying stock (realising an arbitrage opportunity).

The relationship between option prices will assume a no arbitrage principle and will not depend on other extraneous factors such as volatility or transaction costs such as fees margin deposits etc'. By considering option pricing in this way we can derive some basic principles which will provide a foundation for discovering the trading and pricing behaviours that include considering probabilistic models and other external factors.

6.4.2. *Limits for Pricing*

Relative Option Pricing

There are limits set on the pricing of options if they are to adhere to the no arbitrage principle. Several lemmas have been derived to describe the principles. We will use a standardised notation to explain the option pricing mechanisms.

The notation used is as follows:

S: Stock price
X: The option strike price
τ : Expiration time (T-t)
T: Time of expiration
t: The current time
S_τ: The stock price at time τ
r: Risk free interest rate
P: Put value of an American option
C: Call value of an American option
EP: Put value of a European option
EC: Call value of a European option
σ: Volatility measure of stock prices

Upper Limits

Principle 1.0

An American or European call is never worth more than its underlying stock price:

$$EC \leq S \text{ and } C \leq S.$$

An American put is never worth more than the strike price:

$$P \leq X$$

A European put is never worth less than the present value of the strike price:

$$EP \leq Xe^{-r\tau}$$

This has to hold true otherwise the call value would exceed the stock price and a covered call (see above) would enable an arbitrage profit. If the put value was to be greater than the strike price there would be an opportunity for a riskless profit by writing the option and investing the proceeds until expiration.

Lower limits

The lower limit for a European call option given a non dividend paying stock can be represented by:

$$S - Xe^{-r\tau}$$

The portfolio dominance principle tells us that portfolio A should be more valuable than portfolio B if the payoff from A is at least as good under all circumstances and better under some. This is analogous to the principle of riskless arbitrage opportunities. For the lower limit of a European call:

Principle 2.0

The call option value on a non dividend paying stock is never worth less than its intrinsic value. Consider two portfolios.

Portfolio A: A European call option plus cash equivalent to $Xe^{-r\tau}$
Portfolio B: A single share of the underlying stock.

For portfolio A, assume the cash is invested at the risk free rate, at expiration the value of cash will be X. If the stock price at expiration is greater than the strike price, the option will be exercised and the value of the portfolio will be S_T. If the stock price is less than the strike price the option will expire at no value and the portfolio will be worth X. The portfolio is therefore going to have a value of:

$$\max(S_T, X)$$

Portfolio B is worth the stock price at expiration, S_T. Therefore portfolio A will always be at least as good under all circumstances as portfolio B and better than portfolio B under some circumstances (at expiration). The no arbitrage principle would suggest:

$$EC > S - Xe^{-r\tau}$$

Given that a call option value is always greater than its intrinsic value it follows that in the worst case where the option expires worth nothing, that:

$$EC \geq \max(S - Xe^{-r\tau}, 0) \tag{6.4.1}$$

The lower limits for put options can be derived from the no arbitrage argument in a similar way to the methods applied for call options. Consider the lower limit for a European put on a non dividend paying stock. The limit can be expressed:

$$Xe^{-r\tau} - S$$

Thus, the lower limit is the discounted strike price to expiration minus the stock price. If the value was to be less than this difference an arbitrage opportunity would exist. The arbitrageur could borrow at the risk free rate to buy the option and the stock. At expiration, assuming the stock price is less than the strike price, the stock option is sold, and loan is repaid, leaving a profit of: Strike price – Loan repayment. If the stock price is higher than the strike price at expiration, an arbitrageur will not exercise the option, but sells the stock. The loan can be repaid and a profit of: Stock price – Loan.

If we consider two portfolios one of which contains a European option and a share in the underlying stock and the other cash amount equivalent to the strike price.

Portfolio A: A European put option and a single share of underlying stock.
Portfolio B: A cash amount equivalent to $Xe^{-r\tau}$.

Assume that the stock price at expiration is less than the strike price, $S_{\tau=T} < X$. Portfolio A then becomes worth X. If $S_{\tau=T} > X$, then portfolio A becomes worth the stock price at expiration, $S_{\tau=T}$. This is because the put is worthless at expiration, leaving the single share of stock. Portfolio A is therefore worth:

$$\max(S_{\tau=T}, X)$$

With portfolio B it is assumed that the cash is invested at the risk free rate so that the value of B is worth the strike price at expiration (X). it is therefore true that portfolio A is always worth at least the value of portfolio B and in some circumstances worth more. It therefore holds:

$$EP \geq Xe^{-r\tau} - S$$

The worst case for a put option is to expire worthless therefore the lower limit of the put option value can be represented as:

$$EP \geq \max(Xe^{-r\tau} - S, 0) \tag{6.4.2}$$

Code which implements the basic limit calculations for options is shown in Listing 6.1. below.

```
package FinApps;
import static FinApps.PresentValue.*;
import static java.lang.Math.*;
public class Optionlimits {
```

```
public Optionlimits() {
}
public double lowerlimitCall(double stockprice,double rate,
                              double time,double strikeprice)
{
  return max( (stockprice-pVcont(rate,time,strikeprice)),0);
}
public double[] lowerlimitCall(double[]stockprice,double rate,
                               double time, double[] strikeprice)
{
  int indx=0;
  double[] lowervalues = new double[stockprice.length];
  for(double s:stockprice)
  {
    lowervalues[indx]=max((s-pVcont(rate,time,
    strikeprice[indx])),0);
    indx++;
  }
  return lowervalues;
}
  public double lowerlimitPut(double stockprice,double
                              rate,double time,double strikeprice)
{
  return max( (pVcont(rate,time,strikeprice)-stockprice),0);
}
public double[] lowerlimitPut(double[]stockprice,double rate,double
                              time, double[] strikeprice)
{
  int indx=0;
  double[] lowervalues = new double[stockprice.length];
  for(double s:stockprice)
  {
    lowervalues[indx]=max((pVcont(rate,time,
                         strikeprice[indx])-s),0);
    indx++;
  }
  return lowervalues;
}
}
```

LISTING 6.1. Optionlimits.

The following two examples show the use of **Optionlimits** code to calculate the option limits for European calls and puts.

Example 6.2 Calculation of lower limits of call options

The stock price and theoretical strike price for ABC Inc. has varied over a trading day as follows:

Intra Day Stock 51.0, 50.0, 52.3, 53.6, 51.0.
Proposed Strike 50.0, 49.5, 51.5, 52.5, 50.5.

Assume the risk free interest rate is 13% and the time to expiration is 6 months, with current time = 0. What would the minimum premium per share be if any of the Stock prices had been the closing price for a European call option ?

Using the formula of 6.1. We get:

Call option premium per share == 4.15
Call option premium per share == 3.62
Call option premium per share == 4.04
Call option premium per share == 4.40
Call option premium per share == 3.68

The code in Listing 6.2 is used to get the lower limit for a European call premium calculation.

```
package FinApps;
import java.text.*;
import java.io.*;
import static java.lang.Math.*;

public class Example6_1 {

   public Example6_1() {
   }

   public static void main(String[] args) {
      NumberFormat formatter=NumberFormat.getNumberInstance();
      formatter.setMaximumFractionDigits(2);
      formatter.setMinimumFractionDigits(2);
     Optionlimits ops = new Optionlimits();
     double[] stockprice = {51.0,50.0,52.3,53.6,51.0};
     double[] strikeprice = {50.0,49.5,51.5,52.5,50.5};
     double time = 0.5;
     double interest = 0.13;
     double[] premiums = ops.lowerlimitCall(stockprice,interest,
                                            time,strikeprice);
     for(double pr:premiums)
     {
       System.out.println("Call option premium per share ==
                         "+formatter.format(max(pr,0)));
     }
   }
}
```

LISTING 6.2. Lower premium calculation for a range of stock/strike prices.

Example 6.3 Calculation of lower limits for European put options

The stock price and theoretical strike price for ABC Inc. are as follows:

Inta Day Stock : 38.0,39.0,40.0,45.0,38.0
Theoretical Strike: 40.5, 41.5, 42.5, 47.5, 41.0

The time to expiration is 6 months and the risk free rate of interest is 11%. What would the minimum premium per share be if any of the Stock prices had been the closing price for a European put option?

Using the formula of 6.2 we get:

Put option premium per share == 0.33
Put option premium per share == 0.28
Put option premium per share == 0.23
Put option premium per share == 0.00
Put option premium per share == 0.81

The code in Listing 6.3 is used to get the lower limits for a European put option.

```
package FinApps;
import java.text.*;
import java.io.*;
public class Example6_2 {
  public Example6_2() {
  }
  public static void main(String[] args) {
   NumberFormat formatter=NumberFormat.
                                getNumberInstance();
   formatter.setMaximumFractionDigits(2);
   formatter.setMinimumFractionDigits(2);
   Optionlimits ops = new Optionlimits();
   double[] stockprice = {38.0,39.0,40.0,45.0,38.0};
   double[] strikeprice = {40.5,41.5,42.5,47.5,41.0};
   double time = 0.5;
   double interest = 0.11;
   double[] premiums = ops.lowerlimitPut(stockprice,interest,
                                          time,strikeprice);
   for(double pr:premiums)
   {
    System.out.println("Put option premium per share ==
                         "+formatter.format(pr));
   }  }}
```

Listing 6.3. Application code for Example 6.2

6.5. Early Exercise of American Options

Call Options

We assume that interest rates are always positive and derive the following two principles:

Principle 3.0

An American call option on a non dividend paying stock will not normally be exercised before expiration.

Principle 4.0

An American call option on a dividend paying stock will only be exercised at expiration or just prior to going ex-dividend.

It can be shown that it is non optimal to exercise an American call prior to expiration. If the stock price underlying a call is above the strike price, with some time, say n months prior to expiration. The early exercise of the option would result in the holder paying the strike price, foregoing the interest on the strike price for n months. If the stock was immediately sold the holder would realise the intrinsic value, but a better strategy is to hold the option and short the stock. Further, if the holder of the option simply sold the option rather than exercise it, he or she would realise a profit greater than the intrinsic value. For these reasons it is never optimal to exercise an American option prior to maturity.

Consider two portfolios:

Portfolio A: An American call option plus cash equivalent to the discounted strike price $Xe^{-r\tau}$

Portfolio B: A single share in the underlying stock.

Portfolio A contains cash with a value equal to X at maturity (T). Prior to maturity the value of the cash is $Xe^{-r(T-t)}$ at some time t, with $t < T$, the value of exercising portfolio A is $S - X + Xe^{-r(T-t)}$. If the portfolio remained until exercise at maturity the value would be $S - X + X$ (the value of the option at expiration is max(S,X)). The value of portfolio A is therefore always less than S when t < T, so portfolio A is worth less than portfolio B at early exercise.

If we keep the option to maturity and consider portfolio B, there is a prospect of S < X, but portfolio A at maturity has a value which is max(S,X). Therefore portfolio A is worth at least as much as portfolio B and under some circumstances is worth more than portfolio B. If the option is exercised early then portfolio A will always be less than portfolio B. An American call option on a non dividend paying stock therefore should not be exercised prior to expiration. In this respect An American call option has the same value as a European call option on the same stock.

Put Options

Consider two portfolios:

Portfolio A: An American put option plus a single share in the underlying asset

Portfolio B: Cash equivalent to $Xe^{-r\tau}$.

When the option in portfolio A is exercised at $\tau < T$ the value of portfolio A is worth the strike price, X. This value is greater than the value of portfolio B. At maturity portfolio A is worth $\max(S, X)$. Thus portfolio A is worth at least as much as portfolio B and under some circumstances worth more. The exercise of an American put is more likely as the underlying stock price or stock price volatility decreases and also becomes more attractive as the risk free interest rates increase. The value of an American put increases as shown in Figure 6.5. As interest rates decrease the value increases all other things being equal. Figure 6.5 also shows the price of an American put as X-S.

Figure 6.4 shows the characteristics for a put option with a strike price of \$40.50, with 6 months to expiration. The base interest rate is 10%, with each curve reflecting a decrease of 1% from the base rate. The stock price is linear in the range \$37.50 to \$40.50. The American put is linear and shows the price as (X-S) its intrinsic value.

6.6. Option Convexity

It can be seen from Figure 6.5 that option prices are convex and those on parts of the curve the price rise is higher for a rate decrease than it is lower for an equal rate increase.

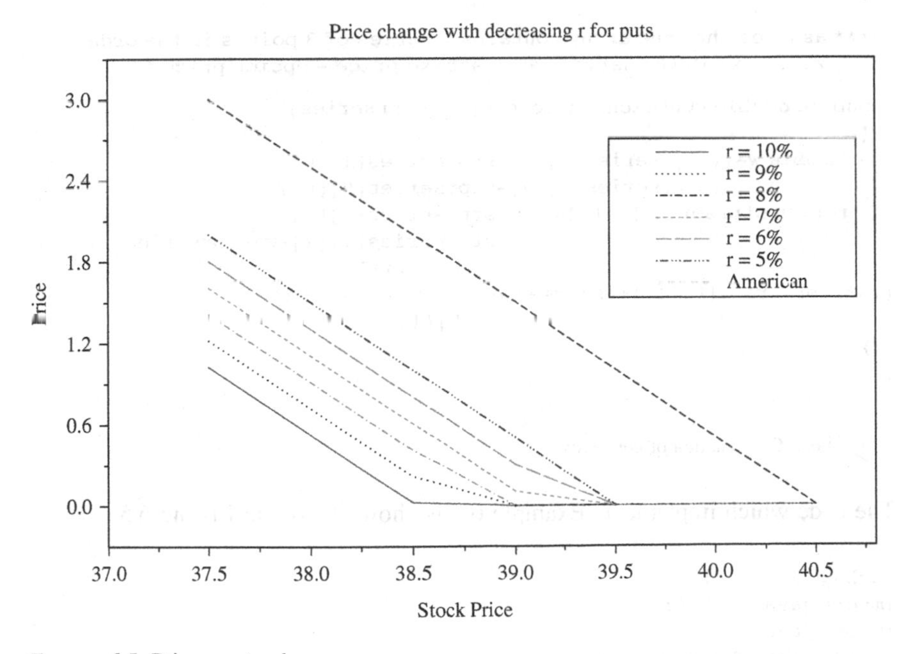

FIGURE 6.5. Price v rate change.

Example 6.4

Consider three strike prices $X_1 < X_2 < X_3$. With weight, $\omega = \frac{X_3 - X_2}{X_3 - X_1}$.

$P_{X_2} \leq \omega P_{X_1} + (1-\omega)P_{X_3}$ and $C_{X_2} \leq \omega C_{X_1} + (1-\omega)C_{X_3}$. As an example consider the ABC Inc June 420,430 and 440 calls from Table 6.1.

$$\omega = \frac{440 - 430}{440 - 420} = 0.5$$

$$C_{X_2}(4.75) \leq 0.5^* C_{X_1}(8.25) + (1.0 - 0.5)C_{X_3}(2.375)$$
$$= 4.75 \leq 5.312$$

Confirming the convexity property.

Listing 6.4 shows a small utility class which will test an option price series for convexity. The class uses a single method *convexcheck* The method uses a two dimensional array to hold the option data series in order. The method first calculates the weighting required and then performs equality checking by implementing the basic equation. The output shows the calculated value and a string representing the equality check.

```
package FinApps;

public class Optconvex {

  public Optconvex() {
  }

  /** assumes that the array contains a series of 3 points in the order
      x1<x2<x3 in the series pairs strike price - option price */

  public double convexcheck(double[][]stprseries)
  {
     double w=((stprseries[2][0]-stprseries[1][0])/
              (stprseries[2][0]-stprseries[0][0]));
    return stprseries[1][1]<= (w*stprseries[0][1]
                              +(stprseries[2][1]-w*stprseries
                                [2][1]))?
((w*stprseries[0][1])+(stprseries[2][1]-w*stprseries
                                [2][1])):0.0;

  }
}
```

Listing 6.4. Computation of convexity

The code which implements Example 6.3 is shown below in Listing 6.5.

```
package FinApps;
import java.text.*;
import java.io.*;
public class Example6_3 {
```

```
public Example6_3() {
}

public static void main(String[] args) {
    NumberFormat formatter=NumberFormat.getNumberInstance();
   formatter.setMaximumFractionDigits(3);
   formatter.setMinimumFractionDigits(3);
   Optconvex conv = new Optconvex();
   double prices[][]={
 {420.0,8.25},
 {430.0,4.75},
 {440.0,2.375}
 };
 double convalue=conv.convexcheck(prices);
 String s=(4.75<=convalue)?"OPTION DATA IS CONVEX : value
          ":"OPTION DATA NOT CONVEX : value ";
 System.out.println(s+convalue);
}
}
```

LISTING 6.5. Application code for Example 6.3

6.7. Put Call Parity

Put-Call parity deals with the relationship between the variables S, X, EC, EP, C, P and r. By analysing the relationships it is possible to value puts in terms of calls and calls (puts) in terms of stock values and strike prices etc.

Consider two portfolios:

Portfolio A: A European call option and a cash amount equivalent to $Xe^{-r\tau}$.
Portfolio B: A single European put plus a single share in the underlying stock.

We have already seen that that at maturity the value of both portfolios is the same; $\max(S, X)$. The value of portfolio A is equivalent to the value of portfolio B so:

$$EC + Xe^{-r\tau} = EP + S \tag{6.7.1}$$

This is the put-call parity relation. This relationship allows us to derive the value of a call with the same exercise price and maturity as a put on the same underlying stock. This relationship is reversible and allows the derivation of a put in terms of call data.

Example 6.5

Given a call price of $3.50 for a European option with the following characteristics:

Stock price of $30.0
Strike price of $28.0
Time to expiration of 6 months
Risk free interest rate of 10%

Calculate the parity put price of the option and show the parity call price given the calculated parity value.

$$3.50 + 28.0e^{-0.1*0.5} - 30.0 = 0.134.$$

The parity call price is therefore:

$$0.134 + 30.0 - 28.0e^{-0.1*0.5} = 3.5.$$

The code for Example 6.4 is shown in Listing 6.6. This makes use of the class **PutCallpar** which is explained in Listing 6.7.

```
package FinApps;
import java.text.*;
import java.io.*;
public class Example6_4 {

  public Example6_4() {
  }

  public static void main(String[] args) {
        NumberFormat formatter=NumberFormat.getNumberInstance();
    formatter.setMaximumFractionDigits(3);
    formatter.setMinimumFractionDigits(3);
     PutCallpar parp= new PutCallpar("put");
        double ansput= parp.europarity(3.5,28.0,30.0,0.1,0.5);
        System.out.println("PUT PRICE IS == "+ansput);
    PutCallpar parc= new PutCallpar("call");
        double anscall= parc.europarity(ansput,28.0,30.0,0.1,0.5);
        System.out.println("CALL PRICE IS == "+anscall);
  }
}
```

LISTING 6.6. European put-call parity.

Although put-call parity is applied to European options there are put call relationships that hold for American options.

Given that $P > EP$, $P > EC + Xe^{-r\tau} - S$ and the observation that a European call has the same value as an American call it follows that:

$$C - P < S - Xe^{-r\tau} \tag{6.7.2}$$

Consider two portfolios:

Portfolio A: EC plus cash equivalent to the strike price X.
Portfolio B: P plus a single share in the underlying stock S.

At maturity portfolio B is worth $\max(S, X)$ and portfolio A is worth: $\max(S, X) + Xe^{-r\tau} - X$.

Portfolio A is worth $Xe^{r\tau} - X$ more than portfolio B. The relationship is therefore:

$EC + X > P + S$, which is also, $C + X > P + S$. By re arrangement we get:

$$(S - X) < (C - P) < (S - Xe^{-r\tau}) \tag{6.7.3}$$

Example 6.6

An American call option with a price of $2.40 has an exercise price of $23.0. The stock price is $21.0, the risk free interest rate is 10% and the time is 3 months. Find the parity put price.

$$S - X = 21.0 - 23.0, S - Xe^{-0.1*0.25} = 21.0 - 22.432 \text{ and } C = 24.0$$

$$\text{So,} \quad 21.0 - 23.0 < 2.40 - P < 22.432 - 23.0$$

Which rearranged gives us the identity:

$$23.0 - 21.0 > P - 24.0 > 22.432 - 21.0$$

$$2.0 > P - 2.40 > 1.432$$

Therefore P is in the range upper limit; $4.40 and lower limit; $3.832.

The code for example 6.5 is shown in Listing 6.7.

```
package FinApps;
import java.text.*;
import java.io.*;
public class Example6_5 {
  public Example6_5() {
  }
  public static void main(String[] args) {
     NumberFormat formatter=NumberFormat.getNumberInstance();
     formatter.setMaximumFractionDigits(3);
     formatter.setMinimumFractionDigits(3);
     PutCallpar par= new PutCallpar("put");
     par.amerparity(2.40,23.0,21.0,0.1,0.25);
     double[]parval=par.getAmerput();
     System.out.println("PUT VALUE UPPER LIMIT == "+parval[0]+"
                          PUT VALUE LOWER LIMIT == "+parval[1]);
  }
}
```

LISTING 6.7. American put call parity code for Example 6.5.

The code in listing 6.8 shows the class **PutCallpar** this class computes the put-call parity for European and American options through the methods *europarity* and *amerparity*. The method *europarity* deals with European call and put options. The class is instantiated through the constructor PutCallpar (). The non default constructor is called with the string "put" or "call" which is passed to the string 'typeoption' . Typeoption is used to define the put or call calculation in the methods. The methods *setAmerput*(double,double) and *getAmerput*() are used by the method *amerparity* to set and give access to the calculated values for American put options. The convenience methods *getAmercall*(double,double) and *getAmercall*() are similarly used to set and give access to American call options.

```
package FinApps;
import static java.lang.Math.*;
import static FinApps.PresentValue.*;

public class PutCallpar {
   public PutCallpar() {
   }
   String typeoption= "call";
   double[] Amerput=new double[2];
   double Amercall;
        public PutCallpar(String type)
   {
      this.typeoption=type;
   }
        public double[] getAmerput()
        {
          return Amerput;
        }
        private void setAmerput(double limitlower,
                                   double limithigher)
        {
          Amerput[0]=limitlower;
          Amerput[1]=limithigher;
        }
        private void setAmercall(double call)
        {
          Amercall=call;
        }
   public double europarity (double optionprice,
                               double strike, double stcprice,
                               double rate,double time)

   {
      double putfrmcall= (optionprice+(pVcont
                          (rate,time,strike))-stcprice);
      double callfrmput=((optionprice+stcprice)-pVcont
                          (rate,time,strike));

      return (typeoption=="put") ? putfrmcall:callfrmput;
}
```

```
public void amerparity (double optionprice,double strike,
                        double stcprice,double rate,double time)
{
  if(typeoption=="put")
  {
  double limit1=abs((stcprice-strike))+optionprice;
  double limit2=abs((stcprice-pVcont(rate,time,
                    strike)))+optionprice;
  setAmerput(limit1,limit2);
  }else
  {
    double callvalue=(optionprice+stcprice)-strike;
    setAmercall(callvalue);
  }
}
}
```

LISTING 6.8. Class **PutCallpar.**

6.8. Strategies

An option strategy involves the taking of a position with one or more options together with the underlying stock and associated borrowing or lending. There are a range of possible option strategies available. The basic strategies are centred on the notion of a portfolio containing an option, stock and some degree of borrowing or lending to finance aspects of the strategy. See Options Clearing Corporation (OCC) guide (2003).

We have already examined the six fundamental options available for a stock namely, long a call, long a put, short a call, short a put, long stock and short stock. Strategies are used to manage the risk reward profile of an investor, some strategies are extremely risky others risk averse or neutral. Where the strategy involves both an option and a security which offers protection to the return, the strategy is a covered option. Covered options are the hedge, the spread and the combination.

Hedge

A hedge combines the option with its underlying stock, so that there is mutual protection against loss. A hedge which combines a long position in the stock with a short position in a call is called a covered call. A protective put is a hedge which combines a long position in the stock with a long put option. A reverse hedge is one that offers the reverse of the covered call or protective put.

By examining the profit v stock price characteristic of an option strategy the behaviour of the payoff can be seen in relation to the underlying stock and option changes with time to expiration. The basic analytical technique is put-call parity. If we consider the basic put-call parity on a dividend paying stock:

$$EP + S = EC + Xe^{-r\tau} + D$$

The value of a long position in a put plus a long position in the underlying stock is the same as a long position in a call plus the discounted cash value of the strike price and an amount equivalent to the discounted dividends. Strategies are characterised as bullish, bearish or neutral in market terms. A bullish view is one where the strategy is geared to a market (stock) rise, whereas the bearish view is based on the market going down. The profit diagram is useful for showing the general characteristic of taking a particular strategy with a range of stock and option prices and aids decision making relative to possible alternatives.

6.8.1. *Hedge with a Protected Put*

Figure 6.6 shows the characteristic of a protected put strategy. The profit diagram shows that a long stock position is negative until the market price is in excess of the price paid. The long put position is positive for values less than the exercise price (minus the option cost). The profit diagram does not take into account the time value of cash initially expended. The investor is therefore bullish concerning the stock

6.8.2. *Reverse Protected Put Hedge*

Figure 6.7 displays the reverse protective put. The characteristic shows opposite behaviour to that shown in Figure 6.5. As the stock price is less than the strike price, a short stock position exhibits greater profit. Towards the strike price, the profit from going short on stock falls and eventually becomes negative. The long call exhibits negative profit to the cost of the option. If the option is exercised greater than the strike price the long call increases in profit. The market outlook is thus bullish for the long call.

6.8.3. *Hedge with a Covered Call*

Figure 6.8 outlines the characteristics of a covered call. A covered call strategy involves writing a call option whilst having bought the underlying stock. If the stock is already owned the strategy is referred to as a 'buy-write'. The position is therefore long in stock and short a call. The strategy is largely neutral or bullish in relation to the market outlook. The strategy offers an opportunity to cover any losses in the underlying stock price, with a decline in stock price. In this case the gain is in any appreciation left in stock value from initial purchase price plus the premium paid for the option. If the price of the stock exceeds the strike price such that the option is exercised then the profit is the premium plus any difference between the original purchase price for stock and the exercise price.

6.8.4. *Reverse Covered Call Hedge*

Figure 6.9 shows the reversed covered call this is short stock and long a call. In the reverse of a covered call, protection is given against a rise in the stock price. The short stock position is most profitable when stock can be sold (short) at a higher price than is currently on the market. As the stock price increases to the

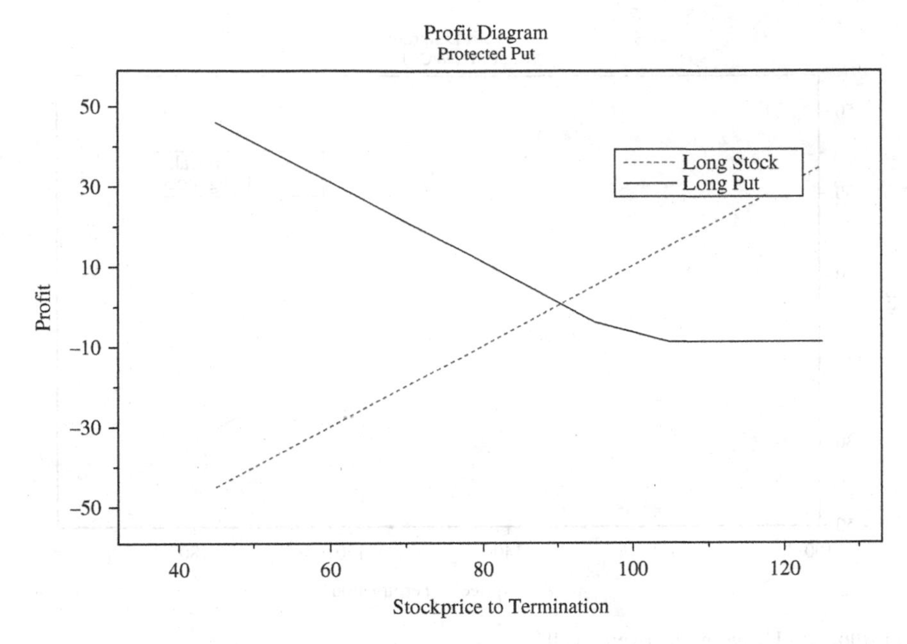

FIGURE 6.6. Hedge with protected put.

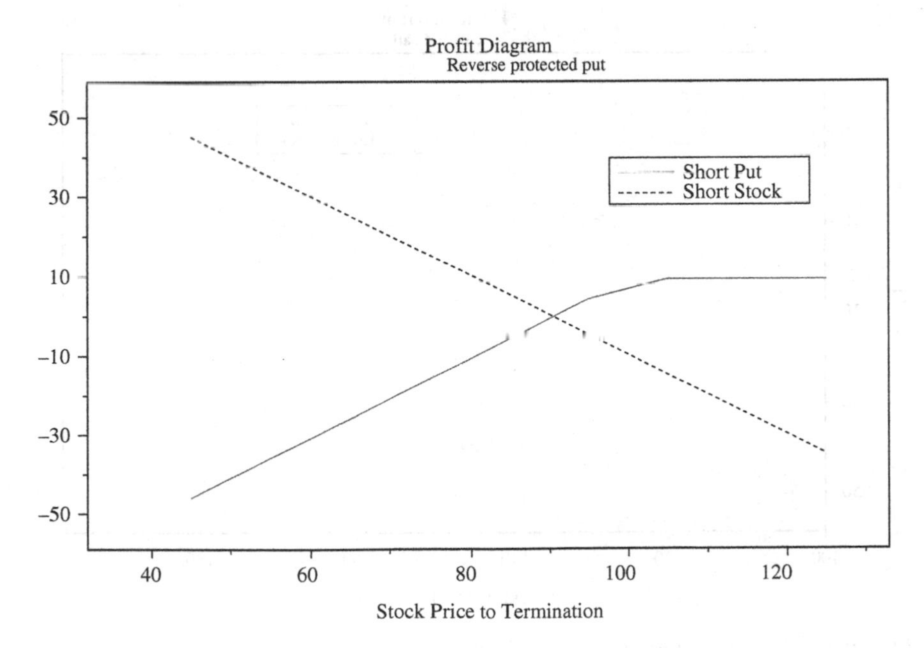

FIGURE 6.7. Reverse protected put hedge.

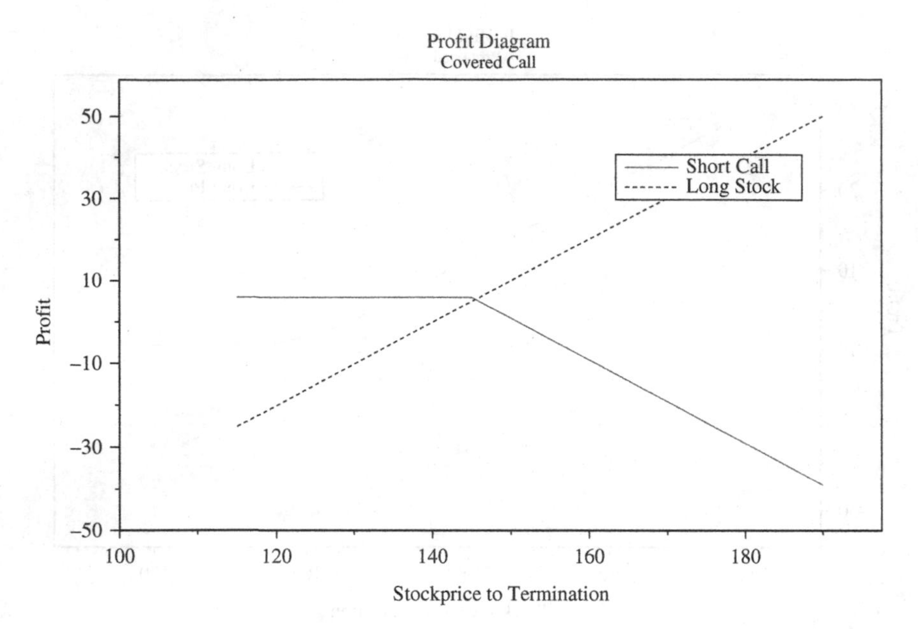

FIGURE 6.8. Hedge with covered call.

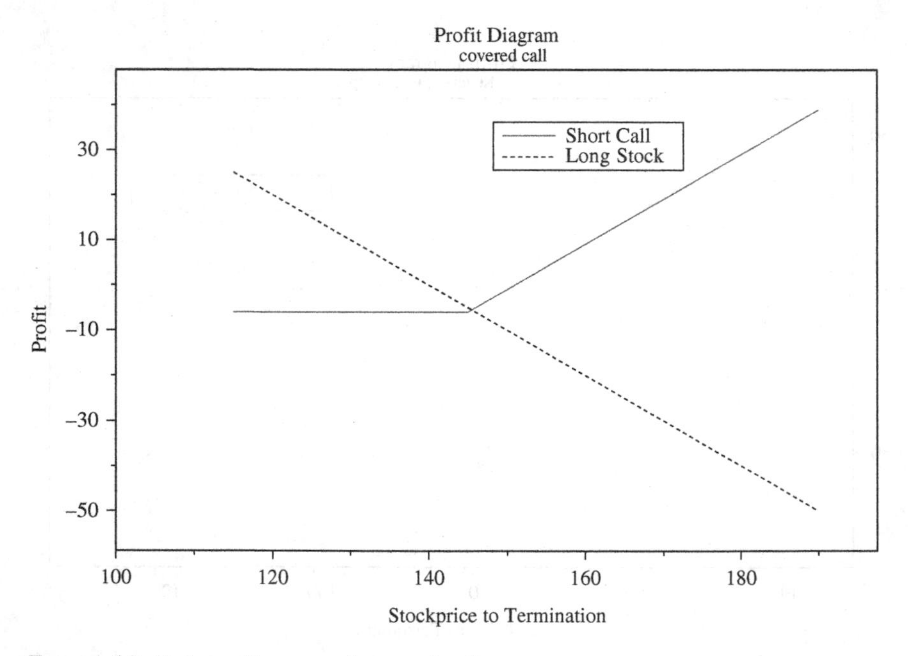

FIGURE 6.9. Hedge with reversed covered call.

short price and beyond the investor can loose substantial amounts of cash. The long call will offer protection against this occurrence.

The protected put defined as: $EC+Xe^{-r\tau}+D$

The reverse protected put defined as: $-EC+Xe^{-r\tau}+D$

Covered call defined as: $Xe^{-r\tau}+D-EP$

Reverse covered call defined as: $EP+Xe^{-r\tau}+D$

Listing 6.9 describes the class **Putcallpos**. This class contains a series of convenience methods to access profit data for the basic hedge strategies. Access methods such as *getShortcallprofits ()* are public methods used by the calling program. The private methods such as *setScallprof ()* are used by the class void methods to store the calculated data. The implementation of the calculations is therefore contained within the class and not available to user view. The calling programs access all data through the get methods.

The void methods *callprof ()* and *putprof ()* perform the basic calculations for put and call option profits. There are two implementations of the methods one for single data points, where a point on the profit diagram is required and multiple data points where a series of profit points are needed for series display.

```
package FinApps;
public class Putcallpos {

   public Putcallpos() {
   }

   double shortcallprof;
   double[]shortcallprofarray;
   double longcallprof;
   double[]longcallprofarray;              // data storage variables //
   double longputprof;
   double[] longputprofarray;
   double shortputprof;
   double[] shortputprofarray;
   private void setScallprof(double scallprofit) {
      this.shortcallprof=scallprofit;
   }

   private void setScallprof(double[]scallprofit) {
      this.shortcallprofarray=scallprofit;
   }
   private void setLcallprof(double lcallprofit) {
      this.longcallprof=lcallprofit;
   }
   private void setLcallprof(double[] lcallprofit) {
                 //Methods that contain local calculation results//
      this.longcallprofarray=lcallprofit;
   }
   private void setLputprof(double lputprofit) {
      this.longputprof=lputprofit;
   }
   private void setLputprof(double[] lputprofit) {
      this.longputprofarray=lputprofit;
   }
   private void setSputprof(double sputprofit) {
```

```
    this.shortputprof=sputprofit;
  }
  private void setSputprof(double[] sputprofit) {
    this.shortputprofarray=sputprofit;
  }
  public double getShortcallprofit() {
    return shortcallprof;
  }
  public double[] getShortcallprofits() {
    return shortcallprofarray;
  }
  public double getLongcallprofit() {
             // Public methods allowing access to the set results//
    return longcallprof;
  }
  public double[] getLongcallprofits() {
    return longcallprofarray;
  }
  public double getLongputprofit() {
    return longputprof;
  }
  public double[] getLongputprofits() {
    return longputprofarray;
  }

  public double getSputprofit() {
    return shortputprof;
  }
  public double[] getSputprofits() {
    return shortputprofarray;
  }
  public void callprof(double callpr,
             double exercisepr,double stockprice) {// single data//

    double shrtcallprofit= stockprice<=exercisepr?callpr:
             (callpr-(stockprice-exercisepr));
    setScallprof(shrtcallprofit);
    double lngcallprofit= stockprice<=exercisepr?-callpr:
             (-callpr+(stockprice-exercisepr));
    setLcallprof(lngcallprofit);
  }
  public void callprof(double callpr,double
             exercisepr,double[] stockprice) {//series data//
    double[]shrtcallprofit=new double[stockprice.length];
    double[]lngcallprofit=new double[stockprice.length];
    int indx=0;
    for(double s:stockprice) {
      shrtcallprofit[indx]= stockprice[indx]<=exercisepr?callpr:
                (callpr-(stockprice[indx]-exercisepr));

      lngcallprofit[indx]= stockprice[indx]<=exercisepr?-callpr:
                (-callpr+(stockprice[indx]-exercisepr));
      indx++;
    }
```

```
        setScallprof(shrtcallprofit);
        setLcallprof(lngcallprofit);

    }

    public void putprof(double putpr, double
                exercisepr, double stockprice) {
        double lngputprofit= stockprice<=exercisepr?
                (-putpr+(exercisepr-stockprice)):-putpr;
        setLputprof(lngputprofit);
        double shrtputprofit=stockprice<=exercisepr?
                (putpr+(exercisepr-stockprice)):putpr;
        setSputprof(shrtputprofit);
    }
    public void putprof(double putpr, double exercisepr,
                        double[] stockprice) {
        double[ ]lngputprofit=new double[stockprice.length];
        double[ ]shrtputprofit=new double[stockprice.length];
        int indx=0;
        for(double s:stockprice) {
            lngputprofit[indx]= stockprice[indx]<=exercisepr?
                    (-putpr+(exercisepr-stockprice[indx])):-putpr;
            shrtputprofit[indx]=stockprice[indx]<=exercisepr?
                    (putpr+(stockprice[indx]-exercisepr)):putpr;
            indx++;
        }
        setLputprof(lngputprofit);

        setSputprof(shrtputprofit);
    }}
```

LISTING 6.9. Computation of basic profit data

6.9. Profit Diagrams

The profit diagram is an investor tool which allows the behaviour of option and stock prices to be assessed when developing a particular strategy. The profit diagram is not inherently analytical; it provides the user with an indication of likely outcomes that might not be intuitively apparent.

We have already seen that using hedge strategies offer protection against undesirable moves in the market. Hedge strategies also offer the possibility of adopting different profit transformations with combinations of stock and option positions. For example the four basic hedge positions shown in Figures 6.1–6.4 actually transform into four basic profit patterns. If we include the cost of stock acquisition (not including any time value) and add this to the basic profit diagrams and sum the characteristics we will get the basic profit diagrams transformed into those shown in Figures 6.10–6.13.

The stock price ranges are:

for call options 115.0,125.0,135.0,145.0,155.0,165.0,175.0,185.0,190.0.

for put options 45.0,55.0,65.0,75.0,85.0,95.0,105.0,115.0,125.0. Assume that the option premium in both cases is 3.50.

The transformations of pattern show that for a covered call the resulting profit behaviour is similar to that of a short European put. For a reverse covered call the pattern is that of a long European put. The pattern transformations with put option profits shows that the protected put is a long European call profile, the reverse protected put has a European short call profile.

The transformations can be seen from the put-call parity relationships namely:

Covered call $(S - EC) = -EP$ (short put) plus cash
Reverse covered call $(EC - S) = EP$ (long put) plus cash
Protected put $(S + EP) = EC$ (long call) plus cash
Reverse protected put $(-(S + EP)) = -EC$ (short call) plus cash.

Listing 6.9 shows the class **Hedgepos.** This class contains the void methods *covercall ()* and *protectedput ()* these two overloaded methods are used to provide the basic profit calculations for the covered call and protected put respectively. Each method instantiates an object from the **Putcallpos** class. The object's from **Putcallpos** are used to access the methods from that class such as *getShortcall-profits()* and *getLongputprofits()*. The *covercall()* and *protectedput()* methods use an iterative loop to perform the profit point calculations and store the results for the array series version. For the single point version only that single point in the profit space is stored.

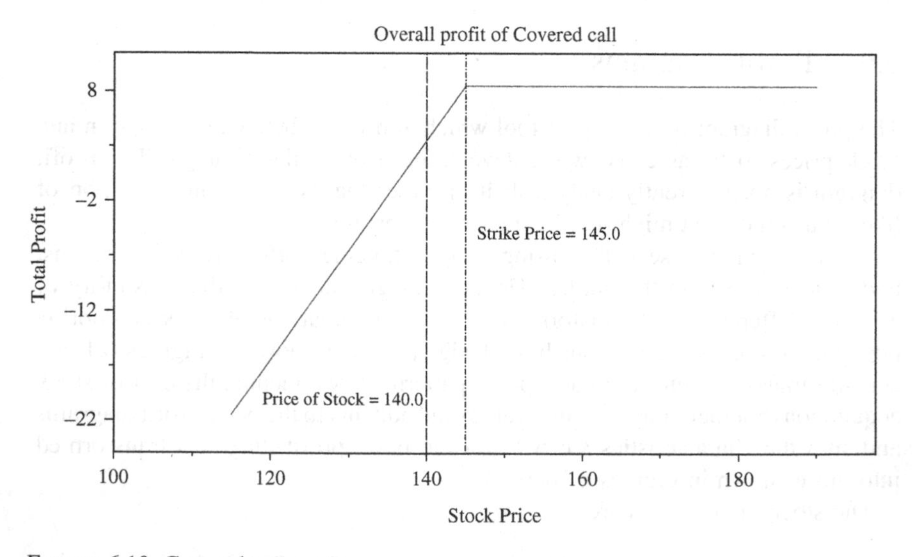

FIGURE 6.10. Covered call profit.

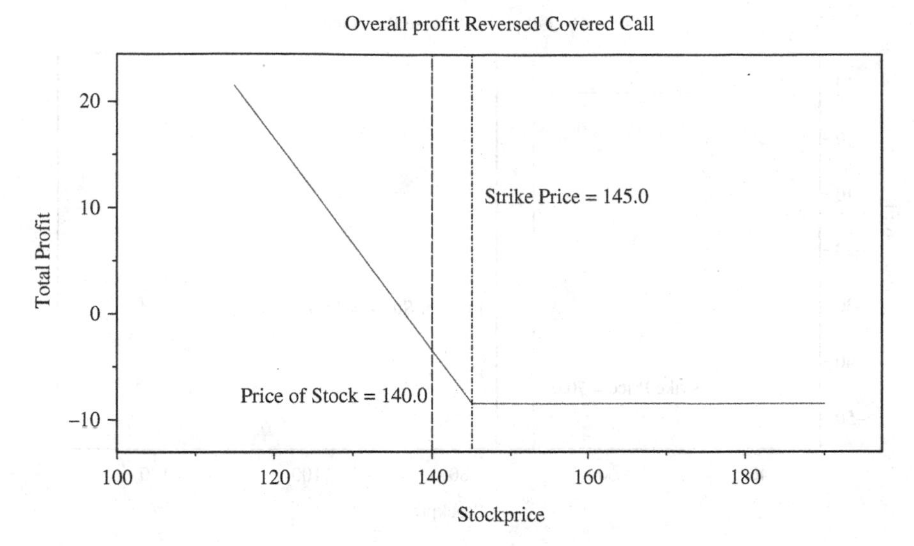

FIGURE 6.11. Reversed covered call profit.

Hedgepos also provides void methods *revcovercall()* and *revprotectedput()* to provide calculations for the reverse hedge. The class has a series of accessor methods for the calling program to retrieve profit data.

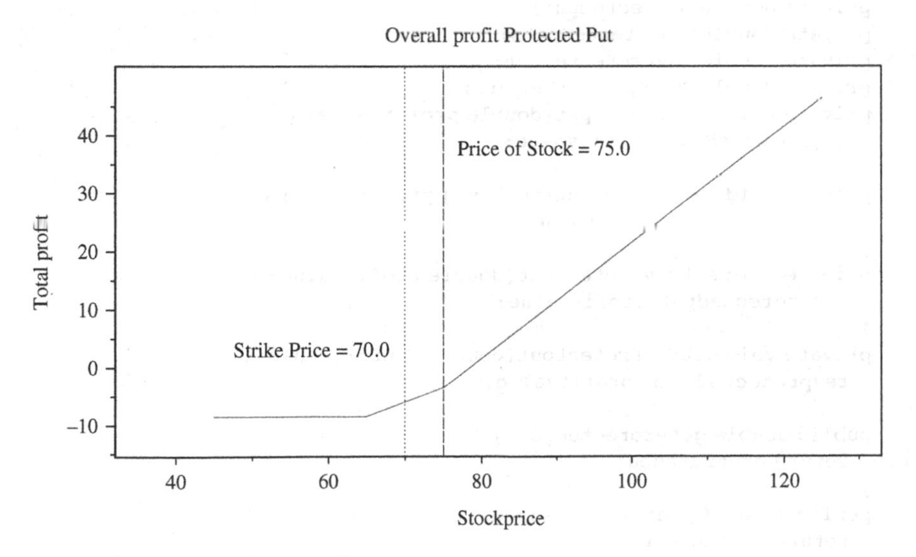

FIGURE 6.12. Protected put profit.

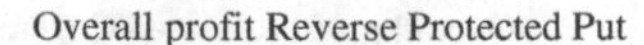

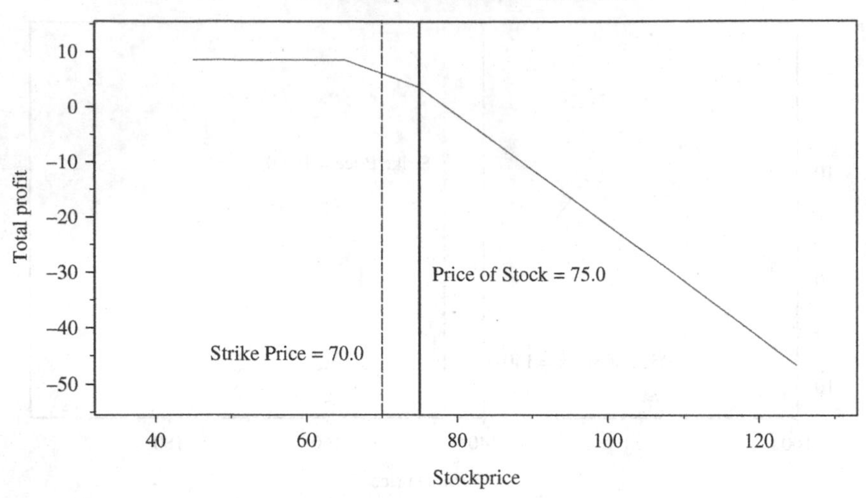

FIGURE 6.13. Reverse protected put profit.

```
package FinApps;

public class Hedgepos {
  public Hedgepos() {
  }
  private double coveredcall;
  private double[]coveredcalls;
  private double revcoveredcall;
  private double[]revcoveredcalls;
  private double protectedput;
  private double[]protectedputs;
  private double revprotectedput;
  private double[]revprotectedputs;
  private void setProtectput(double profitvalue) {
      protectedput=profitvalue;
  }
  private void setProtectput(double[] profitvalue) {
    protectedputs=profitvalue;
  }
  private void setrevProtectput(double profitvalue) {
    revprotectedput=profitvalue;
  }
  private void setrevProtectput(double[] profitvalue) {
    revprotectedputs=profitvalue;
  }
  public double getProtectedput() {
    return protectedput;
  }
  public double[] getProtectedputs() {
    return protectedputs;
  }
```

```
public double getrevProtectedput() {
  return revprotectedput;
}
public double[] getrevProtectedputs() {
  return revprotectedputs;
}
private void setCoveredcall(double profitvalue) {
  coveredcall=profitvalue;
}
private void setrevCoveredcall(double profitvalue) {
  revcoveredcall=profitvalue;
}
public double getCoveredcall() {
  return coveredcall;
}
public double getrevCoveredcall() {
  return revcoveredcall;
}
private void setCoveredcalls(double[]profitvalues) {
  coveredcalls=profitvalues;
}
private void setrevCoveredcalls(double[]profitvalues) {
  revcoveredcalls=profitvalues;
}
public double[] getCoveredcalls() {
  return coveredcalls;
}
public double[] getrevCoveredcalls() {
  return revcoveredcalls;
}

public void covercall(double costofstock,double strike,
double stockprice,double costofoption) {
  Putcallpos p=new Putcallpos();
  p.callprof(costofoption,strike,stockprice);
  double stockprofit=p.getShortcallprofit();
  double profit=costofstock<=stockprice?(-costofstock+stockprice):
            (stockprice-costofstock);
  setCoveredcall(profit+stockprofit);
}
public void covercall(double costofstock,double strike,double[]
          stockprice,double costofoption) {
  Putcallpos p=new Putcallpos();
  int indx=0;
  double[]profits=new double[stockprice.length];
  double[]optionvalues=new double[stockprice.length];
  p.callprof(costofoption,strike,stockprice);
  optionvalues=p.getShortcallprofits();

  for(double s:stockprice) {
    double profit=costofstock<=stockprice[indx]?(-costofstock
              +stockprice[indx]):
(stockprice[indx]-costofstock);
```

```
        profits[indx]=(optionvalues[indx]+profit);
        indx++;
    }
    setCoveredcalls(profits);
}

public void revcovercall(double costofstock,double strike,
                double stockprice,double costofoption) {
    Putcallpos p=new Putcallpos();
    p.callprof(costofoption,strike,stockprice);
    double stockprofit=p.getLongcallprofit();
    double profit=(costofstock-stockprice);
    setrevCoveredcall(profit+stockprofit);
}
public void revcovercall(double costofstock,double strike,double[]
                stockprice,double costofoption) {
    Putcallpos p=new Putcallpos();
    int indx=0;
    double[]profits=new double[stockprice.length];
    double[]optionvalues=new double[stockprice.length];
    p.callprof(costofoption,strike,stockprice);
    optionvalues=p.getLongcallprofits();

    for(double s:stockprice) {
        double profit=(costofstock-stockprice[indx]);

        profits[indx]=(optionvalues[indx]+profit);
        indx++;
    }
    setrevCoveredcalls(profits);
}
public void protectedput(double costofstock,double strike,double
                stockprice,double costofoption) {
    Putcallpos p=new Putcallpos();
    p.callprof(costofoption,strike,stockprice);
    double stockprofit=p.getLongputprofit();
    double profit=costofstock<=stockprice?(-costofstock+stockprice):
                (stockprice-costofstock);
    setProtectput(profit+stockprofit);
}
public void protectedput(double costofstock,double strike,double[]
                        stockprice,double costofoption) {
    Putcallpos p=new Putcallpos();
    int indx=0;
    double[]profits=new double[stockprice.length];
    double[]optionvalues=new double[stockprice.length];
    p.putprof(costofoption,strike,stockprice);
    optionvalues=p.getLongputprofits();

    for(double s:stockprice) {
        double profit=costofstock<=stockprice[indx]?(-costofstock
                +stockprice[indx]):(stockprice[indx]-costofstock);
        profits[indx]=(optionvalues[indx]+profit);
        indx++;
    }
```

```
    setProtectput(profits);
  }

  public void revprotectedput(double costofstock,double strike,double
              stockprice,double costofoption) {
    Putcallpos p=new Putcallpos();
    p.putprof(costofoption,strike,stockprice);
    double stockprofit=p.getSputprofit();
    double profit=(costofstock-stockprice);

    setrevProtectput(profit+stockprofit);
  }
  public void revprotectedput(double costofstock,double strike,
                          double[] stockprice,double costofoption) {
    Putcallpos p=new Putcallpos();
    int indx=0;
    double[]profits=new double[stockprice.length];
    double[]optionvalues=new double[stockprice.length];
    p.putprof(costofoption,strike,stockprice);
    optionvalues=p.getSputprofits();

    for(double s:stockprice) {
      double profit=(costofstock-stockprice[indx]);
      profits[indx]=(optionvalues[indx]+profit);
      indx++;
    }
    setrevProtectput(profits);
  }
}
```

LISTING 6.10. Class Hedgepos provides calculations for hedge profit diagram

References

Options Clearing Corporation, The Equity Options Strategy Guide, April 2003.

7 Modelling Stock Prices

7.1. The Stochastic Process

A stochastic process is one which changes over time in an uncertain way. A discrete stochastic process for a variable is one where the variable can only change at fixed points in time, a continuous time stochastic process is one where the variable is capable of changing at any point in time. A process can have a continuous or discrete variable associated with it. A discrete variable is one that takes on a discrete value within a range, a continuous variable process is one where the variable can take on any value within a range.

The Markov process is a stochastic one where the present value of a variable is the only factor in influencing the prediction of its future value. It is assumed that stock market prices follow a Markov process. Some definitions are useful:

A stochastic process $X = \{X(t)\}$ is a series of random variables over time. $X(t)$ is the state of the process at time t. In a discrete process the state is usually written as a subscript; $X = X_t$. For the continuous process, parentheses are maintained.

The continuous time process has independent increments if for all times, $t_0 < t_1 \ldots < t_n$, the variables $X(t_1) - X(t_0), X(t_2) - X(t_1) \ldots X(t_n) - X(t_{n-1})$ are independent. If the random variables for all times $t_0 < t_1 \ldots < t_n$ have the same distribution for $X(t+s) - X(t)$ the process is stationary and depends only on s. The mean $m_x \equiv E[X(t)]$ is the expected value of the random variable. See the book by Higham for an overview.

7.1.1. Random Walks

A random walk is the behaviour of a variable where the next value is a function of its present value plus a movement with a random probability:

$$X_n = X_{n-1} + \varepsilon_n$$

Figure 7.1 shows the up and down movement of a variable, which can go up with a probability p of 0.5 and down with a probability of $(1 - p) = 0.5$. The up and down movements are cumulative. If the random variable ε is positive the next state is the current state plus an upward movement. If ε is negative the next state is the present state plus a downward movement.

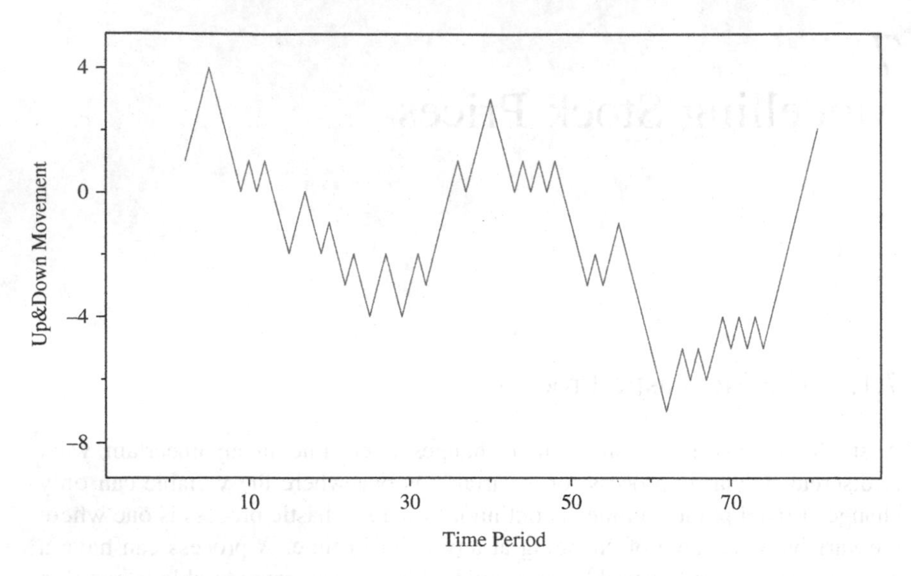

FIGURE 7.1. Random Walk.

A random walk with drift is a random walk with the addition of a term to include the expected value of the change per time period. The random walk with drift is defined as:

$X_n = X_{n-1} + \varepsilon_n + \mu$. Where μ is the expected change per time period.

7.1.2. *Brownian Motion*

Brownian motion is a stochastic process where the random variables are independent of each other. $X(t_n) - X(t_{n-1})$ is normally distributed with a mean $\mu(\Delta x)$ and variance $\sigma^2(\Delta x)$. Brownian motion is described by the mean and variance parameters as a process (μ, σ) with a drift of μ and variance σ^2. Brownian motion with zero drift and unity variance (0, 1) is effectively a symmetric random walk. This is shown in Figure 7.2.

Figure 7.3 shows the value of a stochastic variable following a 20, 10 path of Brownian motion. The central line is the mean value, μ. The line either side of the mean represents a standard deviation of $0.1^* \sqrt{t}$.

7.1.3. *Wiener Process*

Brownian motion with a mean of zero and standard deviation of unity is also known as the Wiener process. Brownian motion expressed in terms of the Wiener process can be represented by:

$$Y(t) = \mu t + \sigma X(t) \tag{7.1.1}$$

Where $X(t)$ is the Wiener process.

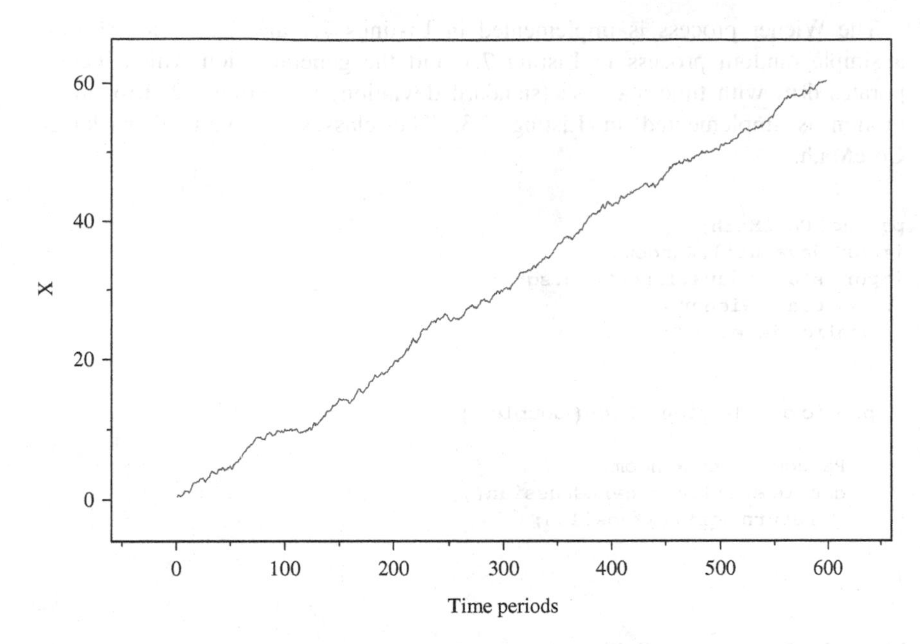

FIGURE 7.2. Drift & Variance for a variable following Brownian Motion.

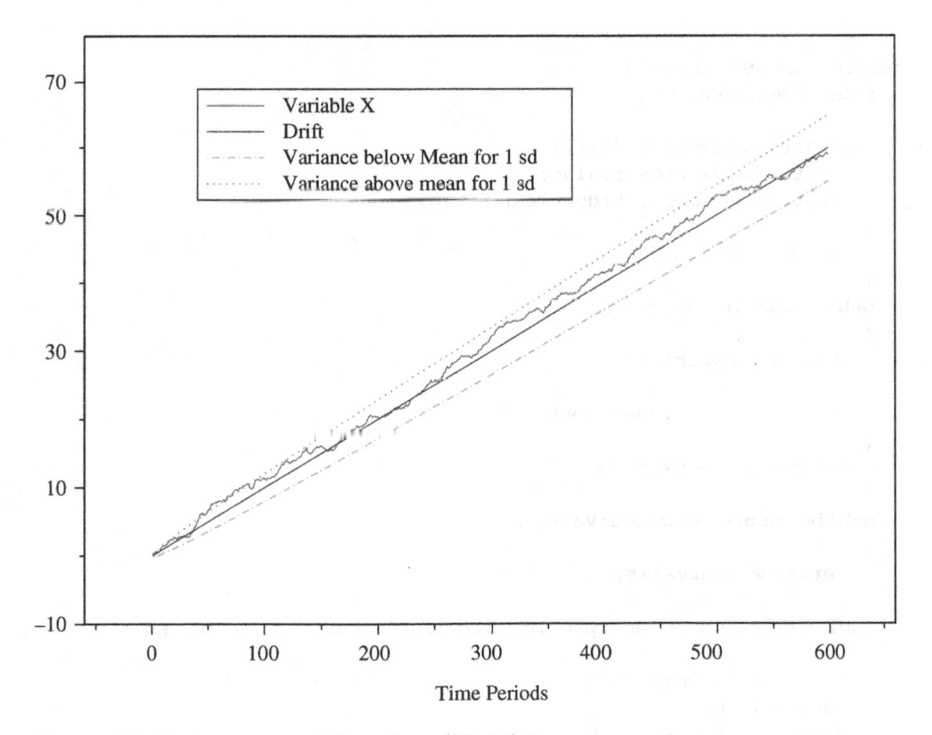

FIGURE 7.3. Variance around Mean for a (20, 10) Brownian motion.

The Wiener process is implemented in Listings 7.1 and 7.2 which shows a simple random process in Listing 7.1 and the generalisation which incorporates drift with time and risk (standard deviation) in Listing 7.2. Brownian motion is implemented in Listing 7.3. The classes are part of package CoreMath.

```
package CoreMath;
import java.util.Random;
import static java.lang.Math.sqrt;
public class Wiener {
   public Wiener() {
   }

   public double wienerProc( double t)
   {
      Random r= new Random();
      double epsilon=r.nextGaussian();
         return sqrt(t)*epsilon;
   }
}
```

LISTING 7.1. Random source for Wiener

```
public class Genwiener {
   public Genwiener() {
   }
    private double constdrift;
    private double wienervalue;
    private void setDrift(double driftval)
   {
      constdrift=driftval;
   }
   public double getDrift()
   {
      return constdrift;
   }
   private void setWiener(double wienval)
   {
      wienervalue=wienval;
   }
   public double getwienerVal()
   {
      return wienervalue;
   }
   public double genWienerproc(double drift, double t, double sd)
   {
      Wiener w=new Wiener();
      double deltaz;
      double driftvalue;
      double deltax;
```

```
        deltaz=w.wienerProc(t);
        setWiener(deltaz);
        driftvalue=drift*t;
        setDrift(driftvalue);

        deltax=(driftvalue+(sd*deltaz));
        return deltax;
    }
}
```

LISTING 7.2. Generalised Wiener process

```
package CoreMath;
import static java.lang.Math.*;

public class Geobrownian {

    public Geobrownian() {
    }
    private double pointdrift;
    private double pointsd;
    private void setDrift(double drift) {
        pointdrift=drift;
    }
    private void setSd(double sd) {
        pointsd=sd;
    }
    public double getpDrift() {
        return pointdrift;
    }
    public double getpSd() {
        return pointsd;
    }
    Genwiener g=new Genwiener();


    public double[][] expBrownian(double mu, double sigma,
                                  double times, int points) {
        double[][] wval=new double[points+1][4];
        wval[0][0]=0.0;
        wval[0][1]=0.0;
        wval[0][2]=(sqrt((exp(0.0)-1)*exp(2*0.0)));
                                    // assumes sd == 1
        double varval;
        double interim=0.0;
        int counter=1;
        double d=points;
        double driftvalues=0.0;
        while(counter<points) {
            varval=(sqrt((exp(counter/d)-1)*exp(2*counter/d)));
            interim=(g.genWienerproc(mu, times,sigma)+interim);
            wval[counter][0]=exp(interim);
            driftvalues=(driftvalues+g.getDrift());
```

```
        wval[counter][1]=exp(driftvalues);
        wval[counter][2]=(wval[counter][1]+varval);
                                                //drift plus variance
        wval[counter][3]=(wval[counter][1]-varval);
                                                //drift minus variance
        counter++;
    }
        return wval;
    }

    public void geoBrownian(double mu,double sigma,double time)
                        //Assumes annual periods/ratios

    {
        Genwiener g=new Genwiener();
        double process=exp(g.genWienerproc(mu,time,sigma));
        setDrift(exp( (g.getDrift())));
        setSd( sqrt( (exp(2.0*mu*(time)
            +pow(sigma, 2.0)*(time))*
            (exp(pow (sigma, 2.0)*(time))-1))));
    }
}
```

LISTING 7.3. Brownian Motion

7.1.4. *Ito Differential*

The Ito differential is a stochastic equation that solves:

$$dX_t = a(X_t, t)dt + b(X_t, t)dz \tag{7.1.2}$$

The equation $dX_t = a_t d_t + b_t dz_t$ is an instance of Brownian motion with an instantaneous drift of a_t and variance b_t^2. The parameters a and b are functions of the variable X, the dz is assumed to be a $\phi(0, dt)$ (normally distributed with mean 0 and variance dt). An equivalent form of 12.1.3 is the Langevin equation:

$$dX_t = a_t dt + b\varepsilon\sqrt{dt} \tag{7.1.3}$$

Where ε is a random drawing from a standardised normal distribution, $\phi(0, 1.0)$. The Ito process is implemented in the class shown in Listing 7.4.

```
package CoreMath;
import static java.lang.Math.*;
import java.util.*;
/**
* Computes the generalised Wiener process where the parameters are
functions of the underlying variable
public class Itoprocess {

   public Itoprocess() {
   }
private double sdchange;
private double meanvalue;
private double changebase;
private void setChange(double changevalue)
{
   changebase=changevalue;
}
public double getBaseval()
{
   return changebase;
}
private void setSd(double sd)
{
   sdchange=sd;
}
public double getSd()
{
   return sdchange;
}
private void setMean(double drift)
{
   meanvalue=drift;
}
public double getMean()
{
   return meanvalue;
}
    /**
    *
    * @param mu mean value
    * @param sigma The variance
    * @param timedelta time periods for each step
    * @param basevalue the starting value
    * @return The change in the base value
    */
   public double itoValue(double mu, double sigma, double
                            timedelta,double basevalue)
   {

       setSd(basevalue*(sigma*sqrt(timedelta)));
       Genwiener g=new Genwiener();
       mu=mu*basevalue;
       sigma=sigma*basevalue;
```

```
        double change=( g.genWienerproc(mu, timedelta, sigma));
        setChange(change);
        setMean(g.getDrift());
        return change;
    }
}
```

LISTING 7.4. Ito Process

Example 7.0

A zero dividend stock has an expected return of 10% per annum with a volatility of 20% per annum. What is the expected return in one week if the current stock price is \$65.0?

Using the Langevin equation for the Ito differential we get:

The change in stock price = −\$1.1649.

This is a random drawing from a normal distribution with mean = \$0.1248, and standard deviation of \$1.8013. The expected return is therefore \$63.835.

The code for this example is shown in Listing 7.5.

```
package FinApps;
import static java.lang.Math.*;
import java.io.*;
import CoreMath.Itoprocess;


public class Example7_0 {


    public Example7_0() {
    }

    public static void main(String[] args) {


        Itoprocess i=new Itoprocess();
        i.itoValue(0.10, 0.20, 0.0192, 65.0);
        System.out.println("NEW Stock price == "+(65.0+i.getBaseval())+
                                "for a change of "+i.getBaseval()+" from a
mean of "+i.getMean()+" and a standard deviation of "+i.getSd());
    }
}
```

LISTING 7.5. Example 7.0

7.2. Lognormal Modelling of Stock Prices

Generally speaking, we are not concerned with the historical behaviour of actual stock price data. This is a natural corrolary of the Markov property which governs stock prices where the past behaviour has no relevance to future behaviour.

The exception to this general rule is in the validation and calibration of stock modelling techniques.

We will examine in the next section, various properties of the lognormal distribution for stock prices. One of the central questions is how realistic are the mathematical models? There are numerous statistical techniques available to the modeller for calibration and verification, one of the simplest is a lognormal based model.

The central purpose behind the use of empirical data is to provide a 'standard' for estimating the input parameters used in price process simulation models. The parameters used in the example above were μ, σ and S_0; the aim of the (Ito) process being to generate a feasible value for the expected return at a future date. We could have used a trivially simple approach which takes the current published price of the stock and taken the last two days and divided by two, to arrive at a mean and variation. This would be highly unlikely to provide us with a 'realistic' appreciation of the particular stock price behaviour. The sample size and distribution for empirical data is an area where there is some debate and we will examine some more technical techniques in later chapters.

A fairly straightforward and reasonably robust method for determining input parameters is the lognormal analysis of empirical price data. The method involves taking a set of historical price data, this can be closing, adjusted closing or some adjusted combination of opening/closing prices and/or ex-dividend adjusted prices. These factors and the period of sample data are largely a matter of experience and judgement as to the appropriateness of the technique for a particular process model (and market).

7.2.1. Handling Empirical Data

The method involves the following steps:

a) Source the price data for a given period, which should be appropriate for the process model
b) Pre-process the data into the appropriate form (in this case lognormal)
c) Statistically analyse the pre-processed data
d) Extract the required statistics (in this case μ and σ)
e) Use the statistics as input parameters to the process model
f) Gauge the simulated output against the empirical data and calibrate the model if needed.

In the straightforward case of deriving input parameters step f is omitted.

We know that prices can be modelled as an instance of Brownian motion with mean μ and variance σ^2, which can be viewed as a variable X following a path with drift μ and a variance influenced by a random drawing from a standardised normal distribution $\phi(0, 1.0)$. If we denote the initial stock price as S_0 at time $t = 0$ and the price S_n at a future time, $t = n$. The t_n can represent any time frame such as trading days, intra day (hourly) or real time, so that $t_{1.5}$ might refer to one and a half trading days. The random movement of a stock price

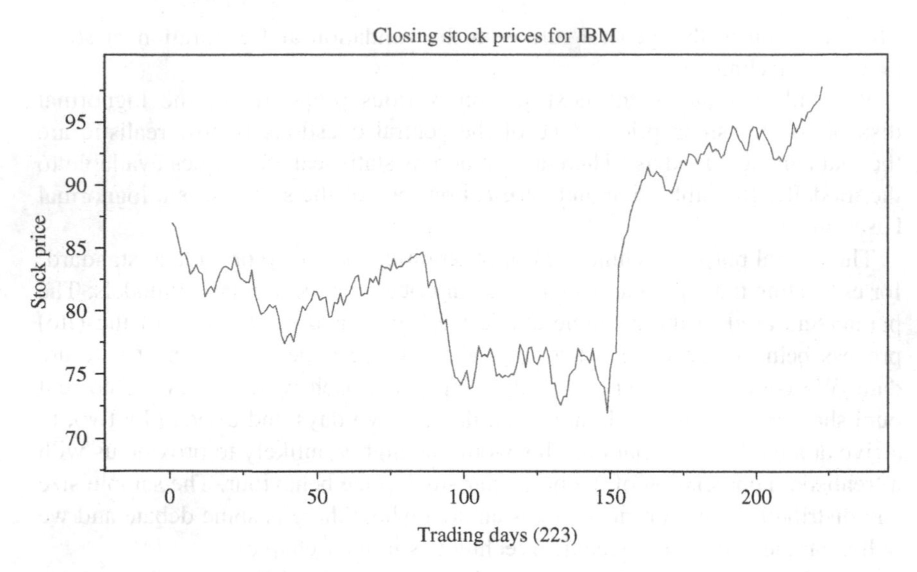

FIGURE 7.4. Step a. Raw data showing closing prices for 223 trading days.

can then be described as $S_0, S_{i=1....,} S_{i=n}$, with $S_i = S_{i-1}X_i$, $1 \leq i \leq n$ where X_i are independently and identically distributed, ie, $X_i : (\mu_i, \sigma_i^2)$. The variable X is said to be lognormal since it is dependent on a random variable ζ such that $\zeta = (\frac{X-\mu}{\sigma})$ and $X = \exp(\mu + \zeta\sigma)$, thus $\log X$ has mean μ and standard deviation σ.

A process simulation that can generate a series of variables $X_{0...}X_n$ which are lognormal can therefore generate a simulation of stock prices with mean μ and standard deviation σ.

As an example, using price data for IBM stock, we will determine the mean and standard deviation of closing price movements. From this we will use a process simulator to generate a series of lognormal price trajectories and compare these to the empirical data.

Step a.

Following the steps outlined above, stock price data from IBM, following closing prices over a period (03.01.2005 to 17.11.2005) of 223 trading days is collected and shown in Figure 7.4.

Following step a, we transform the raw data into a natural log form:

Step b.

From our observation that and $S_i = S_{i-1}{}^*X_i$ and $X_i : \phi(\mu_i, \sigma_i^2)$ from our discussion above that $\log(X)$ is (standardised) normally distributed, then $l_n(\frac{S_i}{S_{i-1}})$, where S_n are raw stock prices, will give us the necessary transformation into a lognormal series of prices.

Step c.

If the series of logarithm transforms are denoted as $Y_1 \ldots Y_n$ IID (independent and identically distributed) normal variates, it is known from the properties of

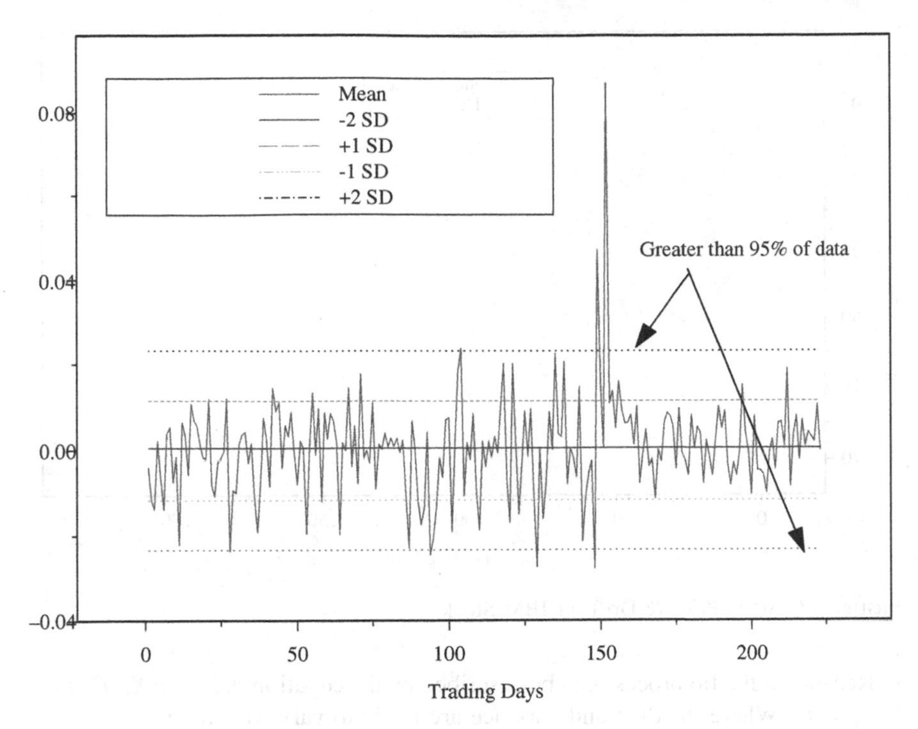

FIGURE 7.5. Log(Prices) mean with +/− 2 SD's.

standard normal statistics that the mean $\hat{Y} = \left(\frac{Y_1 + \ldots + Y_n}{n}\right)$. This gives us an unbiased estimate of μ and $\frac{1}{n-1} * \sum_{i=1}^{n} \left(Y_i - \hat{Y}\right)^2$ gives us an estimate of σ^2.

Figure 7.5 shows the log of stock prices, together with mean and plus/minus two standard deviations for the distribution of log (raw data).

We can see from the graph that more than 95% of data points are within ± 2 standard deviations, which is in reasonable agreement with the theoretical position that around 95% of a normally distributed variable should be within ± 2 Standard Deviations of its mean. If the data points deviated significantly from this observation we would conclude that the data is exhibiting aberrant behaviour and was not suitable for the particular model.

Step d.

We have the drift (μ) and the standard deviation ($\sqrt{\sigma^2}$) of the natural log of closing stock prices.

Figure 7.6 shows the raw stock prices with the calculated drift line.

Step e

The μ and σ which we derived from raw data has been extracted from the logarithmic process, $l_n \frac{S_i}{S_{i-1}}$. The theoretical process followed by the natural log of stock prices can be described by Ito's lemma.

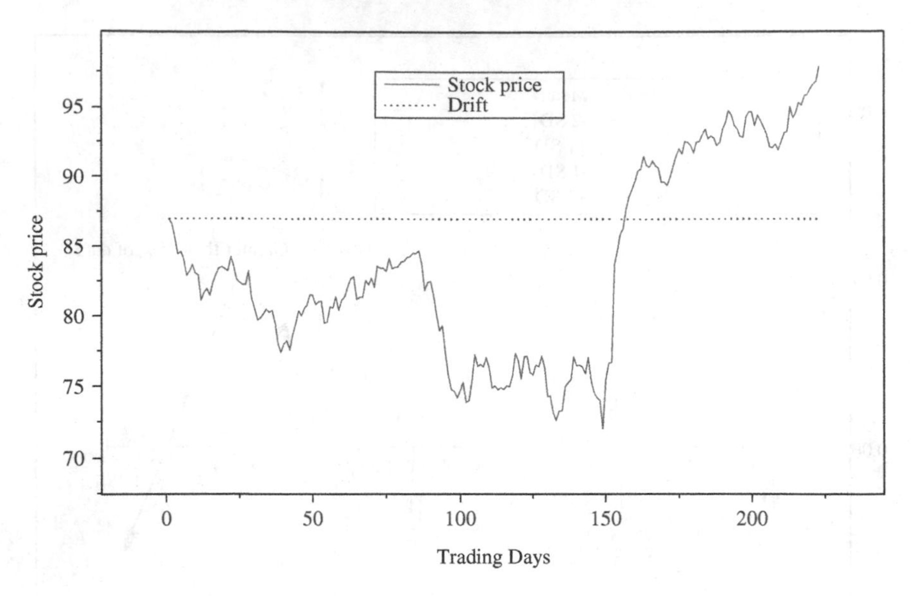

FIGURE 7.6. Stock Price & Drift for IBM Stock.

Recall that the Ito process can be described by the equation $dX_t = a(X_t, t)dt + b(X_t, t)dz$. Where the drift and variance are liable to vary over time.

To model the real situation where a stock price exhibits even variation over the percentage return. We can define σ^2 as the variance of proportional change in stock price and $\sigma^2 \Delta_t$ as the change in time, Δ_t. Therefore $\sigma^2 S^2 \Delta_t$ is the variance of the change in stock price in time Δ_t. This infers that the stock price can be modelled with an Ito process where the instantaneous variance is $\sigma^2 S^2$. The drift rate of actual stock prices is shown to be proportional to the stock price and not constant. For a short time period Δ_t the drift μ is therefore going to be a proportion of the stock price change, so in time Δ_t the increase in stock price will be $S\mu\Delta_t$. Combining the proportional changes in drift and volatility suggests the Ito model which can be expressed as:

$$dS = S\mu dt + S\sigma dz \tag{7.2.1}$$

The proportional change is therefore:

$$\frac{dS}{S} = \mu dt + \sigma dz \tag{7.2.2}$$

If we define Φ as a function of S and t, following 7.1.5,

$$d\Phi = \left(\frac{\partial \Phi}{\partial S} \mu S + \frac{\partial \Phi}{\partial t} + \frac{1}{2} \frac{\partial^2 \Phi}{\partial S^2} \sigma^2 S^2 \right) dt + \frac{\partial \Phi}{\partial S} \sigma S dz \tag{7.2.3}$$

If we define the function as the natural logarithm such that $\Phi = l_n S$. Working through the equation:

$$d\Phi = \left(\frac{1}{S}\mu s + 0 - \frac{1}{2}\frac{1}{S^2}\sigma^2 S^2\right) dt + \frac{1}{S}\sigma S dz.$$

Note:
Since $\Phi = l_n S$, the differentiation of l_n, gives us $\partial\Phi/\partial S = \frac{1}{S}$, $\partial\Phi/\partial t = 0$, $\partial^2\Phi/\partial S^2 = -\frac{1}{S^2}$.

Thus, $d\Phi = \left(\mu - \frac{\sigma^2}{2}\right) dt + \sigma dz$. This indicates that $l_n S$ follows a generalised Wiener process with drift, $(\mu - \sigma^2/2)$ and variance σ^2. The change in $l_n S$ is therefore

$$N : \phi\left((\frac{\mu - \sigma^2}{2})(T - t), \sigma\sqrt{(T - t)} \right) \tag{7.2.4}$$

We have an appropriate stock price model based on 6.1.8 in to which we can insert the empirical parameters, μ and σ. The empirical value for the mean can be used directly in a discrete time model or can be transformed into the continuous time model. Recall that the discrete time model for the process is:

$$\Delta S = \Delta t \mu S + \Delta z \sigma S$$

In this case the μ and σ, which we have taken from the raw data is used. Note that the empirical estimates of σ and μ are based on the daily (not the usual annual) returns, thus $\Delta_t = 1$. If we wish to extrapolate to other time periods, then the daily returns need to be transformed into the appropriate trading period ratios.

To make use of the continuos model we transform the mean to : $\frac{\mu - \sigma^2}{2}$. The choice of model is determined by the particular rate of return we have an interest in.

In the following example we will use a discrete rate of return which will of course have a higher drift rate.

Example 7.1

This example takes the daily stock price returns for 223 trading days. The data chosen are the unadjusted closing prices, which are read into the array data. These data points are transformed into the log ratios using a simple for loop which places the log ratios into the array expon. The array is passed to method *variance()* which is a static method from class **DataDispersion.** The class provides convenience methods *DgetVariance()* and *DgetMean()* which provide the basic statistics from our data points.

The data from IBM, when analysed provide the estimated daily return as 0.000528 and estimated daily return standard deviation of 0.01163.

The class **MonteCprices** provides a simulation process based on the Monte Carlo methods, which are desribed with greater detail in the book by Glasserman. The class **MonteCprices** is shown in Listing 7.6. The class is instantiated as an object 'm' with the constructor passed as 223 (the data set size) the object provides access to the simulation method *simValuep()*. The mean and standard deviation are passed to this method, which returns the simulated data array.

```
/* Calculates the simulated values based on log ratio of raw data
*Produces the daily volatility from raw data anlysis.
*Default time is 1 day
*/
package FinApps;
import static java.lang.Math.*;
import static BaseStats.DataDispersion.*;
import static FinApps.Fileinput.*;
import java.text.*;
import java.io.*;
public class ExampleModel1 {
  public ExampleModel1() {
  }
  public static void main(String[] args) {
    NumberFormat formatter=NumberFormat.getNumberInstance();
    formatter.setMaximumFractionDigits(4);
    formatter.setMinimumFractionDigits(3);

  double[] data=new double[223];
  double[]expon=new double[223];
  double[]monte=new double[223];
  double[]drift=new double[224];
  MonteCprices m= new MonteCprices(223);

  fileinput("c:\\IBM.txt",223);
  data=getFileData();//get raw data
  for(int i=1;i<data.length;i++){
     System.out.println("DATA =="+data[i-1]);
     expon[i-1]=log((data[i]/data[i-1]));
     System.out.println("EXPON data=="+expon[i-1]);
  }
  double s=variance(expon);// variance and mean of the log ratios
  double gsd=sqrt(DgetVariance());
  double gmean=DgetMean();
  double time=1.0;
  double t=(1.0/data.length);
  double So=data[0];
  drift[0]=So;
  int k=1;

   monte=m.simValuep(gmean,gsd,So,time); // Each trading day is 1/223
                                            which is 0.00448

   try{
```

```
    PrintWriter pw=new PrintWriter(new FileWriter("c:\\
                                exampleIBMExSim3.txt"),true);
    PrintWriter w=new PrintWriter(new FileWriter("c:\\
                                exampleIBMExModel1RawA.txt"),true);
    PrintWriter dw=new PrintWriter(new FileWriter("c:\\
                                exampleIBMExModeldriftA.txt"),true);
    for(double d2:monte )
  {


      drift[k]=(So*(exp(k*t*gmean)));
      System.out.println(" Simulated data "+formatter.format(exp(d2))
                             +"DRIFT POINT MOVEMENT =
"+formatter.format(drift[k-1])+"RAW DATA FROM SOURCE = "+data[k-1]);
      pw.println((exp(d2))+",");
      w.println(data[k-1]+",");
      dw.println(drift[k-1]+",");
  k++;
}
  pw.println(" ");
  w.println(" ");
  dw.println(" ");
    w.close();
    pw.close();
    dw.close();
    } catch(IOException foe) {
    System.out.println(foe);
  }
    }
}
```

LISTING 7.6. Computation of Monte Carlo simulations

7.2.2. Simulation with Monte Carlo

Listing 7.7 shows the class **MonteCprices**. This class provides methods to generate simulated stock price movements based on the input parameters, mean, standard deviation, time and initial stock price.

The class uses two imported classes from package **CoreMath**; **Itoprocess** and **Genwiener**. These two classes provide the fundamental calculations required to generate the data based on our random process models.

This class is instantiated with the constructor **MonteCprices(int n)** This constructor sets the value for creation of the return array size. The three methods *simValue(), simValuep() and simValueln()* provide the basic process handling functionality. Both *simValue()* and *simValuep()* provide a process based on the discrete model and *simValueln()* provides a continuos time model. The convenience methods *getValuesim(), getValuesimln()* and *getValuesimp()* give access to the final value of the simulated run.

```
package FinApps;
import java.util.*;
```

```
import static java.lang.Math.exp;
import static java.lang.Math.log;
import static java.lang.Math.sqrt;
import CoreMath.Itoprocess;
import CoreMath.Genwiener;
public class MonteCprices {

  public MonteCprices() {
  }
  public MonteCprices(int n)
  {
    iterations=n;
  }
  public double getValuesim()
  {
    return finalvalue;
  }
  public double getValuesimln()
  {
    return finalvalueln;
  }
  public double getValuesimp()
  {
    return finalvaluep;
  }


  private int iterations;
  private double finalvalue;
  private double finalvalueln;
  private double finalvaluep;
  public double[] simValue(double mean,double sd,double
                              initialvalue,double time)
  {


    double[] simvalues=new double[iterations];
    Itoprocess ito=new Itoprocess();
     for(int i=0;i<iterations;i++)
     {
  simvalues[i]=ito.itoValue(mean, sd, time, initialvalue);
  initialvalue=initialvalue+simvalues[i];
     }
     finalvalue=initialvalue;
     return simvalues;//returns the changes from period to period
  }
  public double[] simValueP(double mean,double sd,double
                              initialvalue,double time)
  {
    double[] simvalues=new double[iterations];
    Itoprocess ito=new Itoprocess();
     double change;
     for(int i=0;i<iterations;i++)
     {simvalues[i]=initialvalue;
    change=ito.itoValue(mean, sd, time, initialvalue);
```

```
    initialvalue=initialvalue+change;
      }
     finalvalue=initialvalue;
     return simvalues;//returns the new price from period to period
   }
   public double[] simValueln(double mean,double sd,double
                                    initialvalue,double time)//continuos time
   {
     double[] simvalues=new double[iterations];
     double so=initialvalue;
     initialvalue=log(initialvalue);
      Genwiener g=new Genwiener();
     mean=((mean-(sd*sd))/2.0);
     sd=(sqrt(time)*sd);


      for(int i=0;i<iterations;i++)
      {simvalues[i]=initialvalue;
       double change=g.genWienerproc(mean, time, sd);// period to
                                                     //period change
      initialvalue=(change+initialvalue);
      }
      finalvalueln=exp(initialvalue);
      return simvalues;//returns the new prices from period to period
   }
}
```

LISTING 7.7. Class **MonteCprices**

Step f
Figure 7.7 shows a simulated run for IBM stock prices. The graph shows that the simulated price (in this particular run) trajectory quickly runs from the drift line.

Figure 7.8 Shows that each simulated run differs according to the random nature of each trajectory.

Figure 7.9 depicts the variation of empirical price data for IBM stock around the mean (drift). Once again the simulated run quickly deviates from the mean line in a random fashion.

Figure 7.10 displays the result of two simulations shown with the empirical data. In this case the convergence between simulated data and empirical data is quite striking, with both runs being closely matched to the drift. Although this graph displays a seductive similarity between the simulated and empirical data, it is not, in any real sense a 'better' series of simulations. The statistical properties of the stochastic process necessarily involve a random range (within the normal distribution) of variation around the drift.

Figure 7.11 exhibits the convergent nature of using multiple runs to obtain a reasonable approximation for the lognormal process. What needs to be kept in mind is that any one of the trajectories IS a reasonable path for a stock price that has a random variation around a trend and is therefore a reasonable model for determining a simulated value.

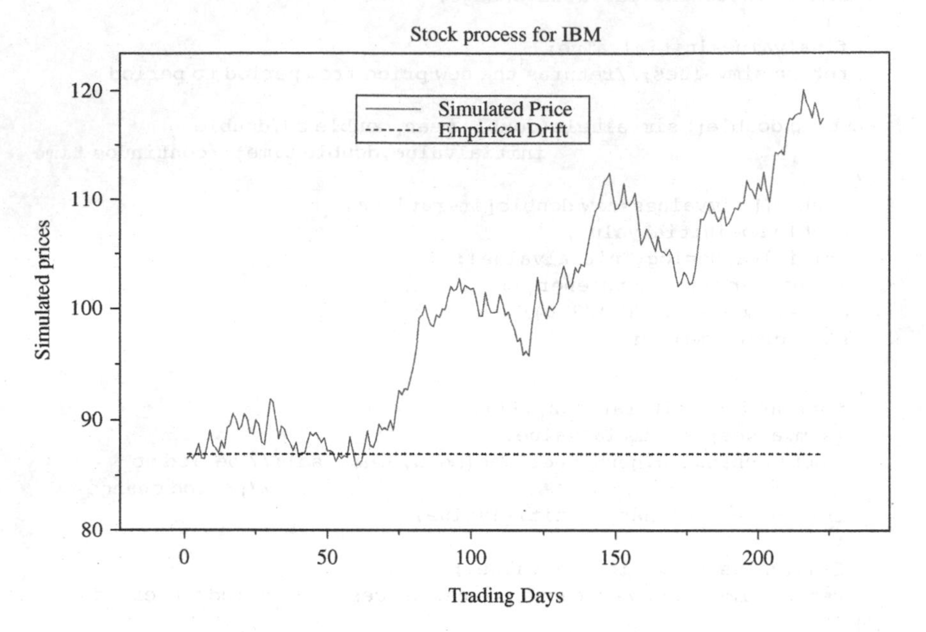

FIGURE 7.7. Price simulation.

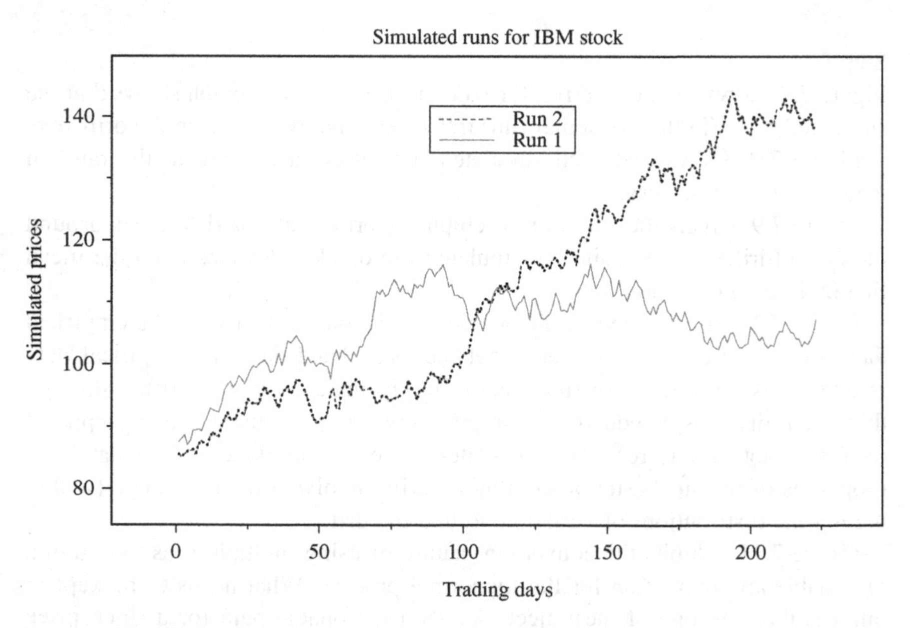

FIGURE 7.8. Simulated series.

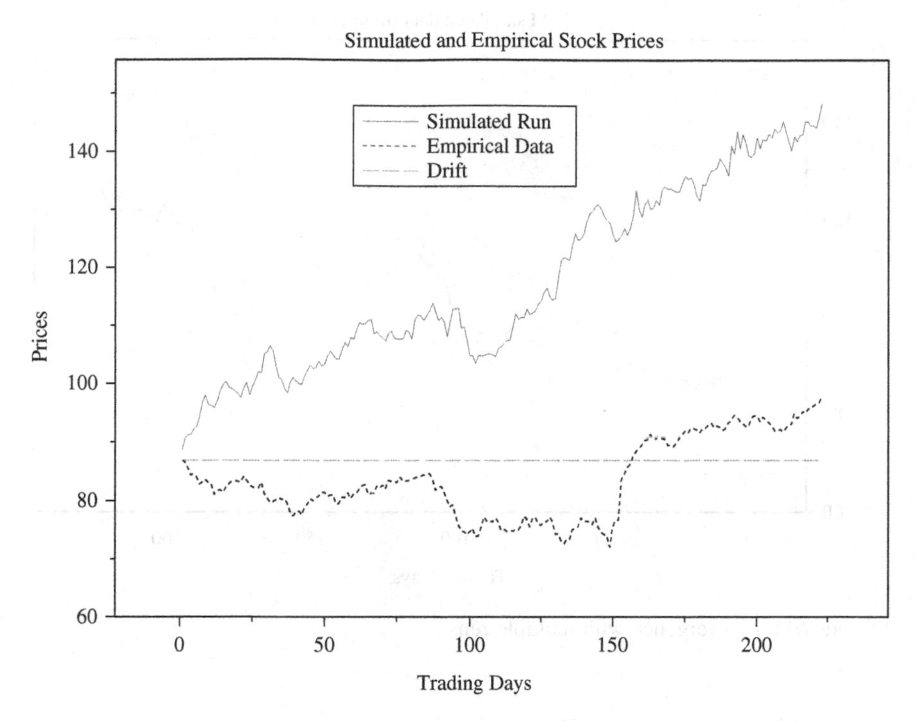

FIGURE 7.9. Comparison of empirical and simulated runs.

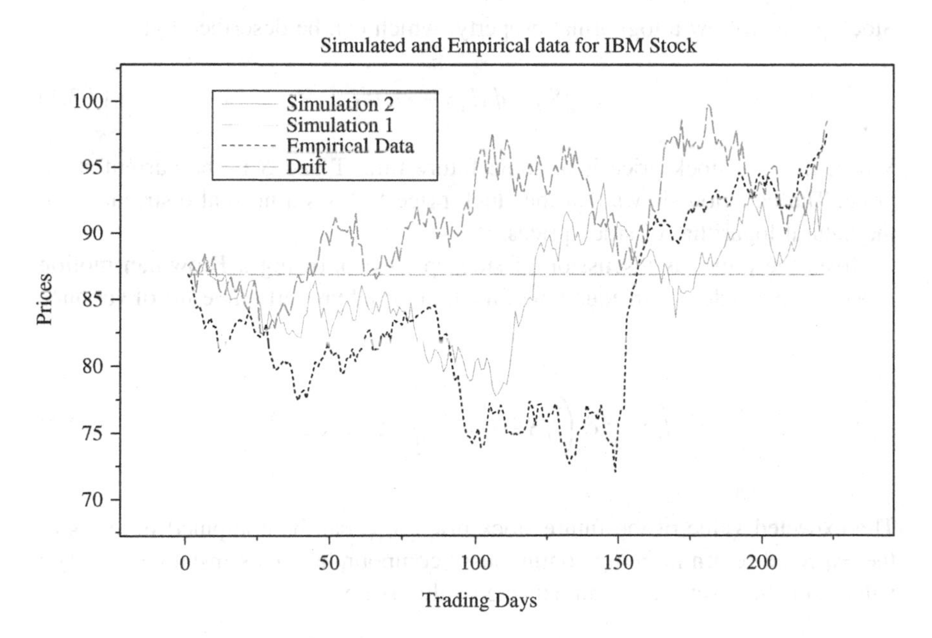

FIGURE 7.10. Comparison of some simulated runs with empirical data and drift line.

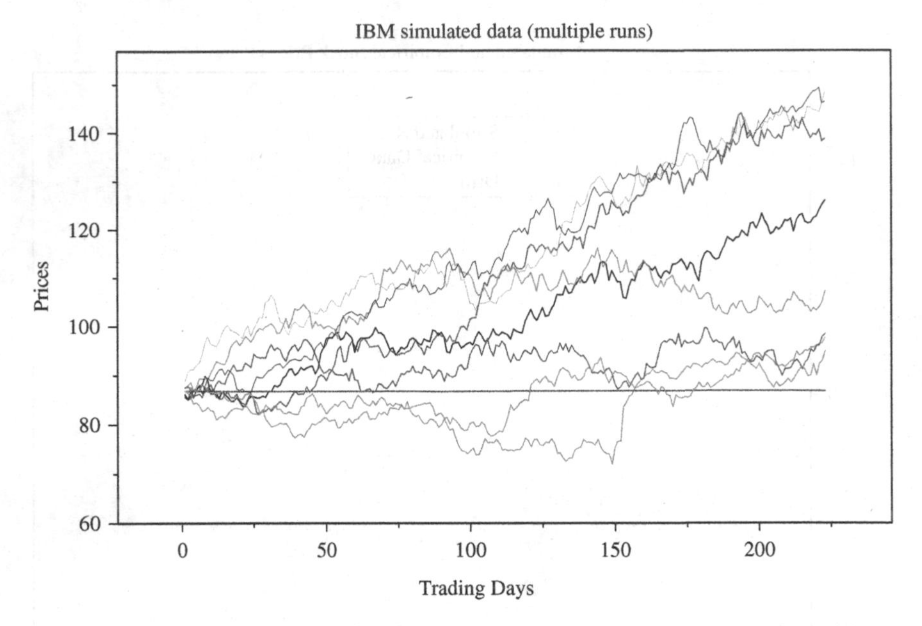

FIGURE 7.11. Convergence with multiple runs.

7.3. The Lognormal Property

Stock prices follow a lognormal property, which can be described by:

$$l_n S_{(T)} : \phi(l_n S + m, \sigma) \tag{7.3.1}$$

Where $S_{(T)}$ the stock price is at some future time T and S is the current stock price. The equation shows that the stock price follows a normal distribution of the natural logarithm of stock prices.

Given the previous discussion of stock prices following a Brownian motion process, we can derive an equation of the lognormal property in terms of variance and time as:

$$l_n S_{\Delta t} : \phi\left(l_n S + (m - \frac{\sigma^2}{2})\Delta t, \sigma\sqrt{\Delta t}\right) \tag{7.3.2}$$

The expected value of the future stock price $S_{(T)}$, can be computed in terms of the expected return m being continuously compounded or its instantaneous (μ) value. For the former case, an expected value is given by:

$$E(S_T) = Se^{mT+\sigma^2/2} \tag{7.3.3}$$

For the latter case the expected return is given by:

$$E(S_{(T)}) = Se^{\mu t} \quad (7.3.4)$$

In either case the variance is given by:

$$Var(S_{(T)}) = e^{2mt+\sigma^2 t}(e^{\sigma^2 t} - 1) \quad (7.3.5)$$

If we assume that the changes in stock price over a short time frame exhibit properties of a normal distribution and define the mean, $\mu\Delta t$ as the percentage change in stock price in a time Δt, with standard deviation $\sigma\sqrt{\Delta t}$. We can derive the change in stock price as:

$$\frac{\Delta S}{S} : \phi(\mu\Delta t, \sigma\sqrt{\Delta t}) \quad (7.3.6)$$

$\frac{\Delta S}{S}$ is the proportional change in stock price for a short time Δt with $\mu\Delta t$ the expected value and variance $\sigma^2\Delta t \equiv \sigma\sqrt{\Delta t}$. The expression $\phi(\mu\Delta t, \sigma\sqrt{\Delta t})$ denotes the expected value and variance following a normal distribution. Following from this we can show that

$$l_n S_{\Delta t} - l_n S : \phi\left((\mu - \frac{\sigma^2}{2})\Delta t, \sigma\sqrt{\Delta t}\right)$$

This provides the rearrangement:

$$l_n \frac{S_{\Delta t}}{S} : \phi\left((\mu - \frac{\sigma^2}{2})\Delta t, \sigma\sqrt{\Delta t}\right) \quad (7.3.7)$$

and

$$l_n S_{\Delta t} : \phi\left(l_n S + (\mu - \frac{\sigma^2}{2})\Delta t, \sigma\sqrt{\Delta t}\right) \quad (7.3.8)$$

The distinction between continuous and instantaneous time for the mean is important. The end result will differ. For some situations the results will differ very significantly and can have an impact on the statutory requirements for financial reporting, which in some countries is stringent (e.g. Sarbanes Oxley).

The stochastic model for stock prices, $\delta S/S = \mu dt + \sigma dz$, where $dz = \xi\sqrt{\Delta_t}$. Recall that ξ is drawn from a standardised normal distribution, $\xi : \phi(0, 1)$. The discrete term μdt is the expected value of the change in stock price $\Delta S/S$ which is a proportional return. Thus:

$\Delta S/S : \phi(\mu\Delta_t, \sigma\sqrt{\Delta_t})$ Shows that the percentage returns per Δ_t is μ_t and μ is the instantaneous expected rate of return with variance σ^2.

For the continuosly compounded rate of return, we can define the annualised compound rate as θ, so that $S_{\Delta t} = Se^{\theta\Delta t}$. From the property of natural logarithms:

$S_{\Delta t} - S = e^{\theta \Delta t}$, using the natural log, $\log S_{\Delta t} - \log S = \log(e^{\theta \Delta t})$ this gives $\log(\frac{S_{\Delta t}}{S}) = \theta \Delta t$, which is $\theta = \frac{1}{\Delta t} \log(\frac{S_{\Delta t}}{S})$. We know from 7.3.4 this implies:

$$\theta : \phi \left(\mu - \frac{\sigma^2}{2}, \frac{\sigma}{\sqrt{\Delta t}} \right) \tag{7.3.9}$$

So, the continuosly compounded rate of return is normally distributed with mean $\frac{\mu - \sigma^2}{2}$ and standard deviation of $\frac{\sigma}{\sqrt{\Delta t}}$. The expected rate of return is therefore $\frac{\mu - \sigma^2}{2}$.

Example 7.2 Probability distribution of stock prices

This example shows the distribution of expected returns based on the lognormal property of stock prices.
Part A.
Assume that a stock has an initial price of £51.0 and an expected return of 11% per annum. The stock exhibits volatility of 19% per annum. We wish to know the distribution of this stock in the next quartrer (3 months). For a 90% confidence level what is the distribution range of stock prices? What is the probability of the stock price being at least £53.50?
Part B.
Finally, if the expected rate of return is continuously compounded, what is the probability of the return over the next year?.

Using 7.3.9 we have:

$\ln S_{0.25} : \phi \left(\ln 51.0 + (\frac{0.11 - 0.19^2}{2})*0.25, 0.19\sqrt{0.25} \right)$, Which gives us,

$\ln S_{0.25} : \phi(3.9318 + (0.11 - 0.01805)*0.25, 0.19*0.5)$. Thus the stock price is distributed normally; $l_n S_{0.25} : (3.9548, 0.095)$.

The 90% probability is that the stock is within plus/minus 1.28 standard deviations of its mean value. Therefore,

$$3.8332 < l_n S_{0.25} < 4.0764.$$

This can be written,

$$e^{3.8332} < S_{0.25} < e^{4.0764} = £46.210 < S_{0.25} < 58.93$$

In three months there is a 90% probability of the stock price being between £46.21 and £58.93.

The probability of the stock price being at least £53.50 is calculated by using the relationship as shown above from 7.3.9. The cumulative distribution function (CDF) is then used to calculate the probability of this value. The CDF is used to provide probabilities of a variable occurring when distributed as a normal variate, based on the mean and standard deviation of the sample. Given our example and using 7.3.9, the parameters are:

$$m = l_n S_0 + (\mu - \frac{\sigma^2}{2})t \tag{7.3.10}$$

$$sd = \sigma\sqrt{t} \tag{7.3.11}$$

The CDF of, x, $\varphi(x)$ can be computed using the error function as:

$$\varphi(x) = \frac{1}{2}\left[1 + errf\left(\frac{x}{\sqrt{2}}\right)\right] \tag{7.3.12}$$

Where x is the normalised variable. Recall that a variable ψ is normalised by transforming as;

$$\psi = (x - \mu/\sigma)$$

Thus our variable x is transformed by

$$x = (l_n(x) - m/sd)$$

Using our actual data

$$\varphi(x) = \frac{1}{2}\left[1 + errf\left(\frac{\log(53.50) - \frac{3.9548}{0.095}}{\sqrt{2}}\right)\right]$$

$$\varphi(x) = 0.603$$

The probability of achieving £53.50 is therefore 0.603. This is a 0.26 standard deviation of the mean (allowing for rounding error) giving an increase of around £1.31.

The expected rate of return (from 7.3.10) over one year is (assuming a 90% confidence level) normally distributed with mean $(\mu - \frac{\sigma^2}{2})$ and standard deviation $\sigma/\sqrt{\delta t}$. There is therefore a 90% chance that the return will be between −15.15% and 33.54%.

Listing 7.8 shows the code for Example 7.2

```
package FinApps;
import CoreMath.Inversenorm;

public class Examplelognorm_1 {

    public Examplelognorm_1() {
  }
  public static void main(String[] args) {
      double[] rangevalues=new double[2];
      double[]rets=new double[2];
      Inversenorm inv=new Inversenorm();
```

```
    double conflevel=inv.InverseNormal(0.90);// get the x factor
                                             //for the chosen
                                             //confidence level

    Lnormprice l=new Lnormprice(conflevel);
   l.logprice(51.0,53.50,0.11,0.19,0.25);
   l.returnrate(0.11,0.19,1.0);
   rangevalues=l.getRange();
   rets=l.getRetrange();
    System.out.println("lnormices"+l.getAverage()+"sd=="
   +l.getSd()+"pDF=="+l.getPdf()+"cdf=="+l.getCdf());
    System.out.println(" RANGE low ="+rangevalues[0]
   +" HIGHER=="+rangevalues[1]);
  System.out.println(" RET RANGE low ="+rets[0]
   +" RET HIGHER=="+rets[1]);

  }
}
```

LISTING 7.8. Application code for Example 7.2

The code in Listing 7.2 makes use of class **Lnormprice.** This class implements the lognormal and rate of return calculations. The method *logprice ()* is passed the values from Example 7.2 part A. The method *returnrate()* is passed the values from part B. The object ‘l’ is instantiated by the constructor for **Lnormprice** being passed a value conflevel, this being the chosen confidence level.

The numeric value of the variable conflevel is returnd from the object ‘inv’, this is instantiated from the class **Inversenorm** which calculates the inverse CDF (i.e. the value of x which gives probability $\varphi(x)$), $\varphi(x)^{-1}$. The method *InverseNormal()* provides the inverse calculation. Class **Inversenorm** is listed at the end of this chapter as Listing A1.

Listing 7.9 shows the class **Lnormprice**

```
package FinApps;
import static java.lang.Math.*;
import CoreMath.Csmallnumber;
import BaseStats.Probnorm;
public class Lnormprice {
   public Lnormprice() {conflevel=1.0;
   }
   public Lnormprice(double confidence) {conflevel=confidence;
   }
   private double conflevel;
   private double pdf;
   private double cdf;
   private double vaverage;
   private double vsd;
   private double[] range= new double[2];
   private double[] retrange=new double[2];
   private void setPdf(double pdfvalue) {
      pdf=pdfvalue;
   }
```

```
private void setCdf(double cdfvalue)//P(X>x)
{
   cdf=cdfvalue;
}
public double getPdf() {
   return pdf;
}
public double getCdf() {
   return cdf;
}
private void setVaverage(double average) {
   vaverage=average;
}
private void setSd(double sd) {
   vsd=sd;
}
public double getAverage() {
   return vaverage;
}
public double getSd() {
   return vsd;
}
public double[] getRange() {
   return range;
}
public double[]getRetrange() {
   return retrange;
}
public void logprice(double So,double St, double mulog, double
                        sdlog, double t) {
   Probnorm p=new Probnorm();
   double meanval=(log(So)+((mulog-(pow(sdlog,2.0)*0.5))*t));
   setSd((sdlog*sqrt(t)));
                       //sets a variance value
   setVaverage(meanval);
   double sdlevel=(getSd()*conflevel);
   range[0]=exp((getAverage()-sdlevel));
   range[1]=exp((getAverage()+sdlevel));
   setCdf(p.ncDisfnc((log(St)-getAverage())/getSd()));
   double divisor=0.0;
   divisor=(sqrt(2*PI));
   divisor=(1.0/(divisor*getSd()*St));

   Double testval=new Double(divisor);
   divisor= testval.isInfinite()?Csmallnumber.
                           getSmallnumber():divisor;

   setPdf( floorvalue( (exp(-0.5*pow(((log(St)-getAverage())/
        getSd()),2)))*divisor));

}
public void returnrate(double exreturn,double volatility,
                           double time) {
   double mean=(exreturn-(pow(volatility,2.0)*0.5));
   double sd=(volatility/sqrt(time));
```

```
        retrange[0]=((mean-(conflevel*sd))*100.0);
        retrange[1]=((mean+(conflevel*sd))*100.0);
    }

    public double floorvalue(double x) {
        return abs(x)<Csmallnumber.getSmallnumber()?Csmallnumber.
        getSmallnumber():x;
    }
}
```

Listing 7.9. Computation of lognormal and return rates

Lnormprice consists of two process methods; *logprice()* and *returnrate()*. The first method implements a lognormal distribution calculation following 7.3.9. The mean is available from the convenience method *getAverage()*, and standard deviation from the method *getSd()*. The mean is constructed using the variable *So*, which is the initial stock price. The variable *St* is the target stock price which is evaluated in relation to the sample mean.

Method *returnrate()* implements 7.3.10. and provides the solution as a range within the chosen confidence level, which is available as the convenience method *getRetrange()*.

The CDF is implemented in the argument to method *setCdf()*, which is passed a return value from the object 'p'. Object 'p' is instantiated from the class **Probnorm**. This class uses method *ncDisfnc()* to calculate the normal CDF.

```
import package CoreMath;
BaseStats.Probnorm;
import java.util.*;
import static java.lang.Math.*;
public class Inversenorm extends NewtonRaphson{
    public Inversenorm() {accuracy(1e-9,20);//optimum values for
                                              xup to 5
    }
    Probnorm p=new Probnorm();
    private double target=0.0;
    public double InverseNormal( double uvalue)//Probability
                                                //between 0 and 1.0
    {
        double xval=0.0;
        target=uvalue;
        if(target==0.5)
            return 0.0;
        if(uvalue<0.5) {
            uvalue=(1.0-uvalue);
            xval=-sqrt(abs(-1.6*log(1.0004-pow((1.0-2.0*uvalue),2))));
            return newtraph(xval);
        }else{
            xval=sqrt(abs(-1.6*log(1.0004-pow((1.0-2.0*uvalue),2))));}
        return newtraph(xval);
```

```
    }
    public double newtonroot(double rootinput) {
        return (target-p.ncDisfnc(rootinput));
    }
}
```

LISTING 7.10. Class Inversnorm in Package CoreMath

References

Higham, D. J. "An Introduction to Financial Option Valuation Mathematics," *Stochastics and Computation*. Cambridge.

Glasserman, P. (2004). *Monte Carlo Methods in Financial Engineering*. Springer.

8
The Binomial Model

Stock price variations are binomial in a short time period. From the lognormal property of stock prices it can be seen that the expected value of a stock price with instantaneous value is as shown in 7.3.1. The equation can be modified to accommodate a risk neutral environment, where the expected return μ is replaced by the risk free interest rate r. From the observation that the change in stock price has a variance of $S^2\sigma^2\Delta_t$ we can construct a binomial model of stock price changes over short time intervals with the following properties.

8.1. Stock Price

A stock is considered over a time frame N. The time frame is split into a series of very short periods Δ_t. At the end of each time period the stock can rise or fall in price. This is shown in the diagram below:

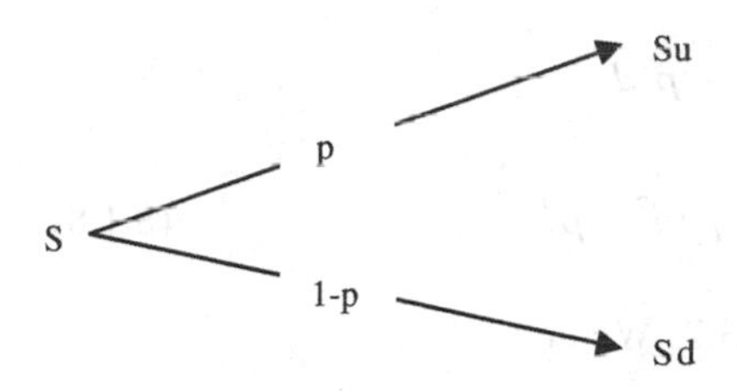

The stock price S can rise to Su with probability p and fall down to Sd with probability (1-p).

The variables we have to deal with are probability p and the amounts u and d. These parameters are constrained to operate within the lognormal properties of stock prices. From 7.3.1 we see the stock value can be given as:

$$Se^{r\Delta_t} = pSu + (1-p)Sd \tag{8.1.1}$$

From the definition of variance as

$$S^2\sigma^2\Delta_t = E(pS^2u^2 + (1-p)S^2d^2) - [E(S^2(pu + (1-p)d)^2)]$$

This reduces to:

$$\sigma^2 \Delta_t = pu^2 + (1-p)d^2 - [pu + (1-p)d]^2 \tag{8.1.2}$$

Both 7.3.3 and 7.3.4 and the identity that $u^*d = 1$ provide the constraints for calculating p, u and d.

8.1.1. Cox Ross Rubinstein (CRR) Model

The conditions provide for parameters with the following values:

$$p = \frac{e^{r\Delta_t} - d}{u - d} \tag{8.1.3}$$

$$u = e^{-\sigma\sqrt{\Delta_t}} \tag{8.1.4}$$

$$d = e^{\sigma\sqrt{\Delta_t}} \tag{8.1.5}$$

$$a = e^{(r-y)\Delta_t} \tag{8.1.6}$$

In the Cox, Ross and Rubinstein (CRR) model it is also a condition to include the constraint of $u = \frac{1}{e^{-\sigma\sqrt{\Delta_t}}} = \frac{1}{d}$ in the calculation of each node in the binomial tree. The equations are accurate for very small values of Δ_t. Because p is a value between 0 and 1, it can be seen as a probability measure, the value y is an expected yield term for the asset. In a risk free environment the return from stock is the risk free rate r, therefore y is considered to be zero. The variable a is termed the growth factor.

For an arbitrary value of δ_t parameters are more accurately described in terms of the lognormal property for the mean and variance of stock prices. In this case the appropriate calculations are:

$$\mu = pu + (1-p)d \tag{8.1.7}$$

$$\sigma^2 = pu^2 + (1-p)d^2 - \mu^2 \tag{8.1.8}$$

From the lognormal property and 7.3.1, 7.3.2. We get:

$$\mu = e^{r\delta_t} \tag{8.1.9}$$

$$\sigma^2 = \mu^2(e^{\sigma^2\delta_t} - 1) \tag{8.1.10}$$

$$u = \frac{(\mu^2 + \sigma^2 + 1)\sqrt{(\mu^2 + \sigma^2 + 1)^2 - 4\mu^2}}{2\mu} \tag{8.1.11}$$

$$p = \frac{\mu - d}{u - d} \tag{8.1.12}$$

8.1.2. Binomial Tree

The binomial model involves the construction of a tree with each node representing a movement in stock price. Figure 8.1 shows a binomial tree for four time periods. There are five possible stock prices at the end of four time periods. At 1 Δ_t we have two possibilities, Su, Sd. At $2\Delta_t$ we have three possible prices (from four movements), Su^2, Sd^2, S. S represents a convergence of Sud and Sdu, this reduces to S. From the identity $u = \frac{1}{d}$, each node is evaluated, so that the movement from $Sd^2u = Sd * \frac{1}{u} * u = Sd$. In general for any time $i\Delta_t$ there are $i+1$ possible prices.

If we denote the time to maturity for a stock as N this is divided into Δ_t discrete intervals the stock price takes on a range of values given as:

$$Su^j d^{i-j} \tag{8.1.13}$$

From the binomial coefficient $\binom{n}{j}$ we can find the number of possible ways to achieve j successes from n trials where $j = 0 \ldots$n. Thus for $n = 4, j = 1$ there are 4 possibilities. $u_1d_1d_2d_3 \equiv d_1d_2d_3u_1 \equiv d_1u_1d_2d_3 \equiv d_1d_2u_1d_3$. For $n = 4, j = 2$ there are 6. In general, the binomial cumulative distribution defines the probability of achieving a target stock price. In the case of determining the cumulative probability of achieving a value at least equal to a particular node,

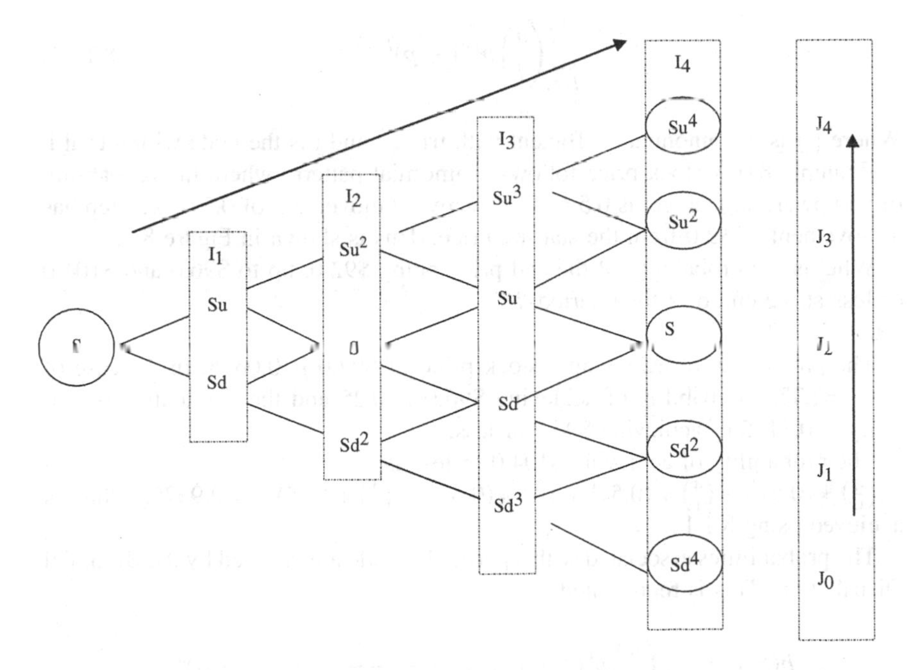

FIGURE 8.1. Binomial tree for five time periods.

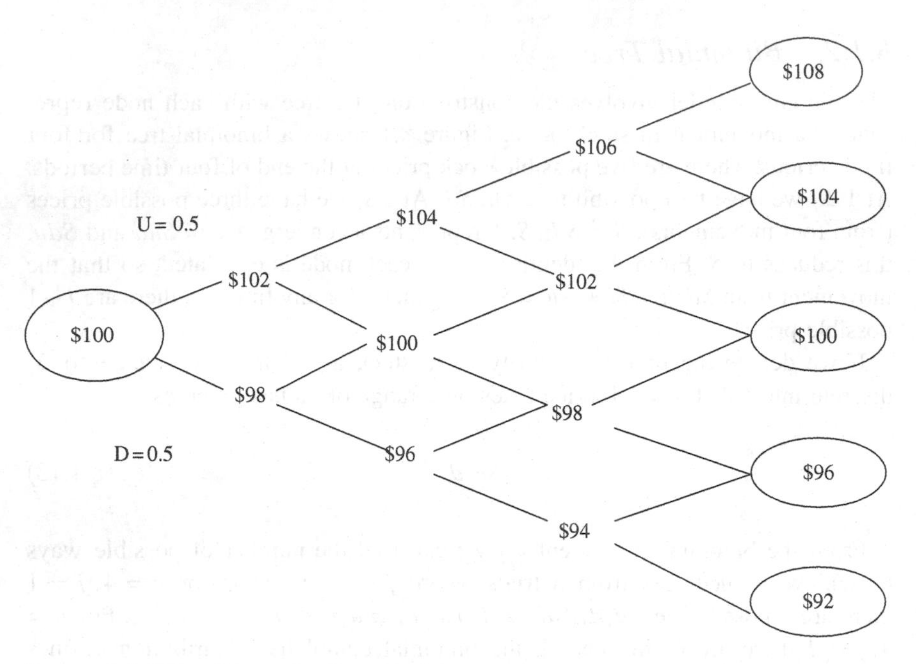

FIGURE 8.2. Binomial tree for Example 8.0.

the following formula is used:

$$\sum_{j=n}^{i} \binom{n}{j} p^j (1-p)^{i-j} \tag{8.1.14}$$

Where $\binom{i}{j}$ is the binomial coefficient with trials i and j is the node within trial i.

Example 8.0 A stock price follows a binomial pattern, where the probability of an upward movement is 0.5 and a downward movement of 0.5. Each step has a movement of $2.0 from the starting price. This is shown in Figure 8.2.

What is the probability of the end price being $92.0, up to $96.0 and $104.0 or less at the end of 4 time periods?

Example 8.0:

The probability of achieving a stock price of $92.0 is 0.0625. In the case of n=4, j=1.The probability of achieving $96.0 is 0.25 and the cumulative probability is 0.3125 of achieving $96.0 or less.

The probability of achieving $104.0 or less is
$\binom{4}{0} * (0.5)^4 + \binom{4}{1} * (0.5)^4 + \binom{4}{2} * (0.5)^4 + \binom{4}{1} * (0.5)^4 = 0.9375$. This is achieved using 8.1.14.

The probabilities associated with a particular node are denoted by the Binomial Distribution. This is represented as:

$$b(k; n, p) \equiv \binom{n}{k} p^k (1-p)^{n-k} = \left(\frac{n!}{k!(n-k)!}\right) p^k (1-p)^{n-k}$$

Where $b(k; n, p)$ is the probability of achieving k successes in n trials (another way of saying k tails in the tossing of a fair coin n times) for the probability p of achieving k.

Figure 8.3 shows the path of probabilities on the binomial tree. To reach node H we need two shortest paths, which give \$100-\$102-\$100-\$98 and \$100-\$98-\$96-\$98. Both routes are given by $(0.5)^4$. This sums to 0.25. Similarly for node G the two shortest paths give an associated probability $(0.5)^3$ this sums to 0.25. If we sum all of the node probabilities for a given column (e.g. \$106+\$102+\$98+\$94) it will cumulate to 1.0.

Example 8.1

A stock has an initial price of £50.0. The risk-free rate is 6% per annum and the stock has an annual volatility of 20%. Show the price distribution for the last month as a binomial tree. Assuming growth is over a complete year.

This example requires the generation of binomial parameters to use in calculating the values of stock prices at each node in the tree.

$$u = e^{0.20\sqrt{0.0833}} = 1.059$$

$$d = e^{-0.20\sqrt{0.0833}} = 0.994$$

$$a = e^{(0.06-0.0)0.0833} = 1.005$$

The variable p is therefore $p = \frac{a-d}{u-d} = \frac{1.005-0.994}{1.059-0.994} = 0.529$

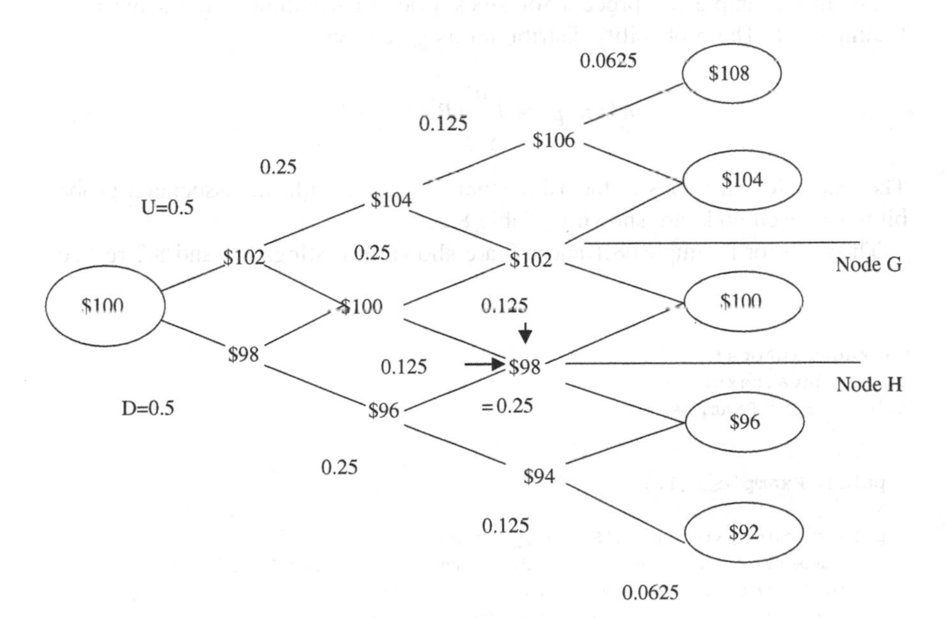

FIGURE 8.3. Binomial with associated probabilities.

TABLE 8.1. Prices for each node (0 to 12) for the last month in the 12 month sequence

PRICE AT EACH NODE:
25.012
28.072
31.508
35.364
39.691
44.548
50.00
56.119
62.986
70.694
79.345
89.055
99.953

From 8.1.14 the stock price will take a value: $S*1.059^{i}*0.994^{i-j}$. Where i is the step and j is the node within each step of the tree. Table 8.1 below shows the prices for each node at the year end (12th step)
Example 8.2
A stock has an initial price of $32.50. If the risk free rate is 10% and the stock has a volatility of 30%. Show the price and probability distribution for the end of month 10.

For this example the process for stock price distribution is the same as in Example 8.1. The probability distribution is given by:

$$b(k; n, p) \equiv \binom{n}{k} P^{k}(1-p)^{n-k}$$

The prices for all nodes at the 10th month together with the associated probabilities at each node are shown in Table 8.2.

The code for Examples 8.1 and 8.2 are shown in Listings 8.1 and 8.2 respectively.

```
package FinApps;
import java.text.*;
public class Example_8_1 {

  public Example_8_1() {
  }
  public static void main(String[] args) {
     NumberFormat formatter=NumberFormat.getNumberInstance();
     formatter.setMaximumFractionDigits(3);
     formatter.setMinimumFractionDigits(2);
     double[] pricetree=new double[13];
```

TABLE 8.2. Price and probability values for Binomial tree on month 10

Probability	Price	Sum of P
0.001	13.673	0.001
0.006	16.258	0.007
0.031	19.331	0.038
0.093	22.986	0.132
0.182	27.332	0.313
0.243	32.50	0.556
0.225	38.645	0.781
0.143	45.951	0.924
0.06	54.639	0.984
0.015	64.97	0.998
0.002	77.254	1.00

```
      Binomparams bp=new Binomparams();
      bp.binomodel(0.0833,0.06,0.20,0.0);
      Binomprice b=new Binomprice(bp.getU(),bp.getD(),bp.getP());
      System.out.printIn("Parameter Values are:"
                        +"Up :"+formatter.format(bp.getU())
                        +" Down :"+formatter.format (bp.getD())
                        +" P :"+formatter.format(bp.getP())
                        + "Growth :"+formatter.format (bp.getG()));
      pricetree=b.binomTprice(12,12,50.0);
      System.out.printIn(" PRICE AT EACH NODE :");
      for(double prices:pricetree) {

           System.out.printIn(formatter.format(prices));
      }

  }
}
```

LISTING 8.1. Application code for Example 8.1

```
package FinApps;
import static java.lang.Math.*;
import java.text.*;
public class Example {

  public Example() {
  }
  public static void main(String[] args) {
     double[][]probprices=new double[11][2];
     NumberFormat formatter=NumberFormat.getNumberInstance();
     formatter.setMaximumFractionDigits(3);
     formatter.setMinimumFractionDigits(2);
     Binomparams bp=new Binomparams();
     bp.binomodel(0.0833,0.10,0.30,0.0);
```

```
    Binomprice b=new Binomprice(bp.getU(),bp.getD(),bp.getP());
    //Binomprice b=new Binomprice(1.5,0.5,0.70);
    //probprices=b.binodeVals(5,4,50.0);
    probprices=b.binodeVals(10,11,32.50);
    double summer=0.0;
    System.out.printIn("  PROBABILITY   PRICE   SUM of P");
    for(int i=0;i<probprices.length;i++) {
        summer+=(exp(probprices[i][0]));
        System.out.printIn("    "+formatter.format
                            (exp(probprices[i][0]))+"
"+formatter.format(exp(probprices[i][1]))+"    "+formatter.
                    format(+summer));

    }
  }
}
```

Listing 8.2. Application code for Example 8.2

The class **Binomparams** is shown in Listing 8.3. This class computes the basic parameters from equations

$$p = \frac{e^{r\Delta_t} - d}{u - d} \quad , u = e^{\sigma\sqrt{\Delta_t}} \quad , d = e^{-\sigma\sqrt{\Delta_t}} \quad , a = e^{(r-y)\Delta_t}$$

```
package FinApps;
import static java.lang.Math.*;
import static CoreMath.Csmallnumber.*;
public class Binomparams {

  public Binomparams() {
  }
  private double p;
  private double u;
  private double d;
  private double g;
  private void setU(double uval) {
     u=uval;
  }
  private void setD(double dval) {
     d=dval;
  }
  private void setP(double pval) {
     p=pval;
  }
  private void setG(double gval) {
     g=gval;
  }
  public double getU() {
     return u;
  }
```

```
      public double getD() {
         return d;
      }
      public double getP() {
         return p;
      }
      public double getG() {
         return g;
      }
      public void binomodel(double time,double rate, double sigma,
                    double yield)//time is as fraction of the rate period
      {
         setG(exp((rate-yield)*time));
         setU(exp(sigma*sqrt(time)));
         setD(floorvalue(exp(-sigma*sqrt(time))));
         setP(floorvalue((getG()-getD())/(getU()-getD())));

      }
      private double floorvalue(double x) {
         return abs(x)<getSmallnumber()?getSmallnumber():x;
      }
}
```

LISTING 8.3. Computation of Cox, Ross and Rubinstein parameters

Class **Binomprice** is responsible for the computation of the binomial tree with associated probabilities. The imported CoreMath classes are listed in the Appendix .

```
package FinApps;
import static CoreMath.Function.*;
import CoreMath.Ibeta.*;
import static java.lang.Math.*;
public class Binomprice {

   public Binomprice() {
   }

   public Binomprice(double u, double d, double p) {
      this.upvalue=u;
      this.downvalue=d;
      this.prob=p;
   }
   private double upvalue=0.0;
   private double downvalue=0.0;
   private double prob=0.0;

   public double[] binomprob(int n,int nodes,int k,double probability)
                           {//altering nodes from n goes from 0 to node,
                           k is starting point from 0

      double[]nodeprobs=new double[nodes-(k-1)];
      nodeprobs[0]=(log(((binom(n,k)*pow(probability, k))
                 *pow((1.0-probability),(n-k)))));
```

```
    int h=1;

    for (int j=k+1;j<(nodes+1);j++)// from 1 to k inclusive....
                                    //Does the strip and each node in it
    {
      nodeprobs[h]=(nodeprobs[h-1]+(log(probability*(n-j+1))
                                    -log((1-probability)*j)));

      h++;
        }
    return nodeprobs;
  }
  public double[] binomTprice(int n,int nodes,int h,double price) {
    double stripsum=0.0;
    double[]nodeprices=new double[nodes-(h-1)];
    int k=0;

    for (int j=h;j<(nodes+1);j++)// from h to nodes inclusive....
                                 //Does the strip n and each node in it
    {
      nodeprices[k]=(price*pow(upvalue,j)*pow(downvalue,(n-j)));

      k++;

    }
    return nodeprices;

  }
  public double[][] binodeVals(int n,int nodes,int h,double price) {

    int k=0;
    double[][] nodeval=new double[nodes-(h-1)][2];

    nodeval[k][0]=log(((binom(n,h)*pow(prob,h))
                                    *pow((1.0-prob), (n-h))));
    nodeval[k][1]=(log(price*pow(upvalue,h))+log(pow
                    (downvalue,(n-h))));
    k=1;
    for (int j=h+1;j<(nodes+1);j++)// from 1 to k inclusive....
                                    //Does the strip and each node in it
    {
      nodeval[k][0]=(nodeval[k-1][0]+(log(prob*(n-j+1))
                                    -log((1-prob)*j)));
      nodeval[k][1]=(log(price*pow(upvalue,j))
                    +log(pow(downvalue,(n-j))));
      k++;

    }
    return nodeval;
  }

  public double cumbinomDistL(double k, double n, double x)
                              {//Does the basic probability for that node
    CoreMath.Ibeta m= new CoreMath.Ibeta();
    double inverseprob;
    inverseprob=m.betai(n-k,k+1.0, 1.0-x);
```

```
      // For direct computation
      return inverseprob;

   }

   public double cumbinomDist( double k, double n, double x) {
      CoreMath.Ibeta j= new CoreMath.Ibeta();
      double cumprob;
      cumprob=j.betai(k, (n-k+1.0), x);

      return cumprob;
   }
}
```

LISTING 8.4. Computation of the Binomial Tree

8.2. Trees for American & European Pricing

The binomial tree can be used to price European and some American options. The price of an American option is complicated by the ability to exercise at any time prior to expiration. Recall from our discussion of options that the payoff from an American option is $\max((x-s),0)$ at any time during the life of the option. If we model the option life as a binomial tree, with each node being evaluated as the maximum payoff, the model will show optimum exercise points.

Using the same terminology as the examples for stock price evolution. The payoff for an option can be represented as:

$$\max(0, (X - S_0 u^j d^{i-j})) \tag{8.2.1}$$

where X is the option strike price.

Since we can now represent each node of the tree as the payoff from that node. The tree effectively represents the lifetime value of each time step. The basic CRI tree method for constructing the values is used to store the end node values, the standard backward induction method for constructing prices towards the staring value are then used to construct a full tree of option prices.

Using 8.2.1 to construct the node prices and the associated probability of reaching that node as the up and down probabilities of the nodes ahead.

The value of the end nodes is first constructed by 8.2.1, the preceding nodes are then constructed by:

$$p_{i,j} = \max(\max(X - Su^i d^{j-i}, 0), [p^* P_{j+1,i+1} + (1-p)^* P_{j+1,i}] \tag{8.2.2}$$

By constructing the binomial tree with these values, each node accounts for all of the possible prior exercise points. Listing 8.5 gives the implementation of the American tree class.

```
package FinApps;

public class Amertreeop {

   public Amertreeop(double time,double rate,double yield,
                     double volatility) {
     t=time;
     r=rate;
     q=yield;
     sigma=volatility;
  }
  private double t;
  private double r;
  private double q;
  private double sigma;

  public double amerCall(int n, int nodes, int h, double price,
                          double strike)
  {
     double[] pricetree=new double[n+1];
     Binomparams bp=new Binomparams();
     t=(t/n);
     bp.binomodel(t,r,sigma,q);
     Binomprice b=new Binomprice(bp.getU(),bp.getD(),bp.getP(),
       t,r,1);// 1 is American
     pricetree=b.binomTprice(n,nodes,h,price,strike,1);// non zero
       //is a cal
      double amps =pricetree[0];
     return amps;
  }
  public double amerPut(int n, int nodes, int h, double price,
                                         double strike)

  {
     double[] pricetree=new double[n+1];
     Binomparams bp=new Binomparams();
     t=(t/n);
     bp.binomodel(t,r,sigma,q);
     Binomprice b=new Binomprice(bp.getU(),bp.getD(),bp.getP(),
       t,r,1); // 1 is American
     pricetree=b.binomTprice(n,nodes,h,price,strike,0);
       // zero is a put
     return pricetree[0];
  }
}
```

LISTING 8.5. Computation of an American Tree

Example 8.4 Valuing an American option

Consider an American opton on a stock with six months to expiry. The stock price is \$110.0, the strike price is \$100.0. The risk free rate is 7% and the option volatility is 27%. Evaluate the put option price with five time steps over the option lifetime.

The input parameters are: S=110, X=100, T=0.5, r=0.07, y=0.0, volatility = 0.27. Time steps (n) = 5. The values for nodes= n, and we wish to evaluate the whole tree so start from h=0.

The put value = $3.054. Listing 8.6 shows the code for implementing Example 8.4.

```
package FinApps;
import java.text.*;
public class Example_8_4 {

   public Example_8_4() {
   }
    public static void main(String[] args) {
     NumberFormat formatter=NumberFormat.getNumberInstance();
     formatter.setMaximumFractionDigits(3);
     formatter.setMinimumFractionDigits(2);
     Amertreeop e=new Amertreeop(0.5,0.07,0.00,0.28);
     double price=e.amerPut(5,5,0,110.0,100.0);
     System.out.printIn(" Put price=="+price);
   }

}
```

LISTING 8.6. Application code for Example 8.4

Reference

Lyuu Yuh-Dah (2002). *Financial Engineering & Computation Principle, Mathematics, Algorithms*. Cambridge University Press.

9 Analytical Option Pricing Methods

9.1. Black-Scholes-Merton

The Black-Scholes and Merton (1973) (more usually refered to as the shortened 'Black Scholes') model of option pricing is based on the risk free hedge of a portfolio consisting of the underlying asset and cash.

The model uses the lognormal property of stock prices and assumes a Geometric Brownian Motion process.

The Black Scholes equation is of the form:

$$dS = S\ \mu dt + S\ \sigma\ dz \tag{9.1.1}$$

Recall this is following an Ito process.

If we define Φ as the price of a derivative which is contingent on the variable S, the price will be some function of S and t.

$$d\Phi = \left(\frac{\partial \Phi}{\partial S}\mu S + \frac{\partial \Phi}{\partial t} + \frac{1}{2}\frac{\partial^2 \Phi}{\partial S^2}\sigma^2 S^2\right) dt + \frac{\partial \Phi}{\partial S}\sigma S dz \tag{9.1.2}$$

Also recall that the discrete versions are given as:

$$\Delta S = \Delta t\ \mu S + \Delta z \sigma\ S \tag{9.1.3}$$

$$\Delta \Phi = \left(\frac{\partial \Phi}{\partial S}\mu S + \frac{\partial \Phi}{\partial t} + \frac{1}{2}\frac{\partial^2 \Phi}{\partial S^2}\sigma^2 S^2\right) \Delta t + \frac{\partial \Phi}{\partial S}\sigma S\ \Delta z \tag{9.1.4}$$

Where the Δ S and Φ are changes in the stock price within a short period of time Δt.

Following from Ito's lemma the underlying process for S and Φ are the same, namely the uncertainty associated with the Wiener process ($\Delta\Phi$, ΔS). The uncertainty affects both S and Φ as the stock price movement is subject to a random fluctuation ($\Delta z = \varepsilon\sqrt{\Delta t}$).

The Black Scholes equation is satisfied by the price of a derivative which is dependent on the underlying non dividend paying stock. Given that in a very short (instantaneous) time the stock price has perfect correlation with the derivative price, positions in the stock and derivative can be established so that the return is the riskless rate. Riskless because the source of uncertainty (risk) is the same.

By constructing a portfolio which holds appropriate positions in a derivative and its underlying stock, which eliminates the Winer process we can build the riskless portfolio. By going short the derivative and long a stock amount $\frac{\partial \Phi}{\partial S} S$. The portfolio value is:

$$\prod = -\Phi + \frac{\partial \Phi}{\partial S} S \tag{9.1.5}$$

The value of this portfolio is only true for small time changes. The change in value for change in time is given by:

$$\Delta \prod = -\Delta \Phi + \frac{\partial \Phi}{\partial S} \Delta S \tag{9.1.6}$$

Thus $\Delta \prod = \left(-\left(\left(\frac{\partial \Phi}{\partial S} \mu S + \frac{\partial \Phi}{\partial t} + \frac{1}{2} \frac{\partial^2 \Phi}{\partial S^2} \sigma^2 S^2 \right) \Delta t + \frac{\partial \Phi}{\partial S} \sigma S \Delta z \right) + \frac{\partial \Phi}{\partial S} (\Delta t \ \mu S + \Delta z \sigma S) \right)$

This gives us.

$$\Delta \prod = \left(-\frac{\partial \Phi}{\partial t} - \frac{1}{2} \frac{\partial^2 \Phi}{\partial S^2} \sigma^2 S^2 \right) \Delta t \tag{9.1.7}$$

The rate of return for this portfolio (assuming an arbitrage free environment) is the short term riskless rate. Therefore in a short time:

$$\Delta \prod = r \prod \Delta t \tag{9.1.8}$$

By substitution from 9.1.4 , 9.1.6 and 9.1.8.

$$\left(-\frac{\partial \Phi}{\partial t} - \frac{1}{2} \frac{\partial^2 \Phi}{\partial S^2} \sigma^2 S^2 \right) \Delta t = r \left(\Phi - \frac{\partial \Phi}{\partial S} S \right) \Delta t$$

Giving:

$$r\Phi = \frac{\partial \Phi}{\partial t} + rS \frac{\partial \Phi}{\partial S} + \frac{1}{2} \sigma^2 S^2 \frac{\partial^2 \Phi}{\partial S^2} \tag{9.1.9}$$

This is the Black Scholes differential equation. There are a range of solutions to the equation which depend on the particular derivative using the underlying stock variable. If we have a tradeable derivative then it will be possible to solve it using the Black Scholes equation. If we cannot solve the equation using Black scholes then there will be arbitrage opportunities from the derivative.

If we consider an example from forward contracts. The value of a forward contract is given as:

$$f = S - ke^{-r(T-t)}$$

Where S is the stock price , k is the delivery price and r the risk free rate. If we consider parameters: $S = \$905.80$, $k = \$919.7328$, $r = 5\%$ and time is 5 months to maturity ($t = 0.41665$). The value of the contract is \$5.00.

Substituting the values of the derivative into the Black Scholes equation we get:

$$rf = \frac{\partial f}{\partial t}(S,t) + rS\frac{\partial f}{\partial S} + \frac{1}{2}\sigma^2 S^2 \frac{\partial^2 f}{\partial S^2}(S,t)$$

This is the partial differential equation that the price of a forward contract satisfies.

The first term becomes $-rke^{-r(T-t)} = -0.05*919.7628*e^{-0.05*0.41665}$. The second term becomes 1.0 and the third term 0.0. This evaluates to: – 45.04 + 45.29 = 0.25. Looking at $rf = 0.05*5.0 = 0.25$. This shows that the equation is satisfied.

The only variables considered in the Black Scholes equation are stock price, volatility, time, forward price and risk free rate. There are no risk preferences such as expected return or probability of return. The Black Scholes model depends on the investor having no risk preferences. If it is assumed that investors are risk neutral, then we can assume that the expected return is the risk free rate.

9.2. Pricing with Black-Scholes

The basic stock option model allows pricing of an option where the underlying asset pays no dividend.

The prices of a basic European call and put option which will satisfy the Black-Scholes equation are given by:

$$c = SN(d_1) - Xe^{-rT}N(d_2) \tag{9.2.1}$$

$$p = Xe^{-rT}N(-d_2) - SN(-d_1) \tag{9.2.2}$$

Where:

$$d_1 = \frac{\ln(S/X) + (r + \sigma^2/2)T}{\sigma\sqrt{T}} \tag{9.2.3}$$

$$d_2 = \frac{\ln(S/X) + (r - \sigma^2/2)T}{\sigma\sqrt{T}} = d_1 - \sigma\sqrt{T} \tag{9.2.4}$$

S = Stock Price
X = Strike Price of the option
R = Risk free rate
T = Time to expiration (annual)
σ = Underling asset volatility (from relative price change)

The function $N(x)$ is the cumulative normal distribution function.

The formulae for call and put options can be deduced intuitively from the basic process of a European call. Recall that the expected value is

$$\hat{E}[\max(S - X, 0)]$$

A European call option at expiration will have its expected value discounted at the risk free rate so,

$$c = e^{-rT}\hat{E}.$$

The expected return can be interpreted in terms of the probability of the stock price being greater than the strike price at maturity, or zero. The second term is the probability of the strike price being paid. This can be written:

$$c = e^{-rT}[SN(d_1)e^{rT} - XN(d_2)] \tag{9.2.5}$$

9.2.1. *Pricing without Dividends*

Example 9.0 EUROPEAN OPTION WITHOUT DIVIDENDS

A European call option has an underlying stock with a current price of \$ 55.0. The strike price is \$60.0. There is three months to expiry. The risk free interest rate is 9% and the volatility is 26%.

$$d_1 = \frac{\ln(S/X) + (r + \sigma^2/2)T}{\sigma\sqrt{T}} = \frac{\ln(55/60) + (0.09 + 0.26^2/2)0.25}{0.26\sqrt{0.25}}$$

$$d_2 = \frac{\ln(S/X) + (r - \sigma^2/2)T}{\sigma\sqrt{T}} = d_1 - \sigma\sqrt{T} = -0.412 - 0.26\sqrt{0.25}$$

$d_1 = -0.412$. $d_2 = -0.542$.

$N\ (d_1) = 0.340$. $N(d_2) = 0.293$.

$c = SN\ (d_1) - Xe^{-rT}N\ (d_2) = 55N(d_1) - 60e^{-0.09*0.25}N\ (d_2) = 1.65$

$p = Xe^{-rT}N\ (-d_2) - SN\ (-d_1) = 60e^{-0.09*0.25}N\ (-d_2) - 55N\ (-d_1) = 4.767$

Thus the price of a European call is \$1.65. The stock price would need to increase by \$6.65 to break even (ignoring time value of money). The price of the European put is \$4.767; this requires no change in the stock price to be slightly in the money.

9.2.2. *Effects of Dividends*

Dividends payable on the underlying stock of an option can be split into short term and long term. The short term dividends are usually regarded as cash dividends where the lump sum payments are regular and known. For long term options the dividend is expressed as a dividend yield. The dividend on a stock is generally assumed to be paid on the ex-dividend date. The amount by which the stock price reduces is not always the same as the dividend cash payment; it is sometimes a percentage range of the dividend to discount from the stock value.

A European option can be priced based on the two components of the stock price when dividends are expected. The first component is the riskless aspect of known dividend payments over time. This component is discounted at the riskless rate over the time periods for which the dividends are paid. The risky component is the normal risk associated with the stock price, in the Black Scholes model with the associated volatility.

Example 9.1 EUROPEAN OPTION WITH DIVIDENDS
Consider a European call option with payments due in n periods. The stock price will take into account the discounted cash flows for each period:

$$S_0 = S - P_1 e^{-rt_1} - P_2 e^{-rt_2} \ldots - P_n e^{-rt_n}.$$

Where P is the dividend payment for periods 1 ... n and t is the time period as a fraction of the annual time. The interest rate is the annual risk free value.

Take the dividend payments to be two five and eight months. The current stock price is £70, the strike price is £66. The time to expiration is 10 months, with the risk free rate being 7% and volatility of 13%. A dividend of £2.0 is paid on each due date.

The stock price, discounted will be :

$$70.0 - (2.0e^{-0.1666*0.07} + 2.0e^{-0.1466*0.07} + 2.0e^{-0.6666*0.07})$$

$$70 - (1.976 + 1.979 + 1.908) = 64.137.$$

Using 9.2.3 and 9.2.4

$$d_1 = \frac{\ln(64.137/66.0) + (0.07 + 0.13^2/2)0.833}{0.13\sqrt{0.833}} = 0.309$$

$$d_2 = d_1 - \sigma\sqrt{T} = 0.190$$

$$N(d_1) = 0.621$$

$$N(d_2) = 0.575$$

$$c = 64.137{}^*N\ (d_1) - 66{}^*e^{-0.07*0.833}N\ (d_2) = 4.020.$$

$$p = 66{}^*e^{-0.07*0.833}N\ (-d_2) - 64.137N\ (-d_1) = 3.519$$

The call price for this option is £4.020 and the put is £3.519. Once the dividend payments have been discounted from stock prices, the basic Black Scholes formulae apply.

9.2.3. *Options Paying a Yield*

There are two choices available in pricing an option providing a known dividend yield. Since paying a dividend causes the stock price to reduce by an amount equivalent to the yield rate.

If there is a dividend yield q, the stock price grows from $S_{t=0}$ to $S_{t=T}$ at time T. This value incorporates the discount of the stock price at the yield rate. If the stock pays no dividend the stock would grow at a greater rate $S_{t=T}e^{qT}$. Therefore we can view the stock price growth from either starting at S_0, including the growth rate q, or starting at S_0e^{-qT}, paying no dividend. The Black-Scholes-Merton model incorporates the yield adjustment to the basic Black Scholes model by taking the stock price as one providing no dividend. The current stock price is then adjusted from S_0 to S_0e^{-qT}.

By adopting a convention of reducing the stock price to accommodate known yield, we have an alteration to the bounds for put call parity. The lower bounds on a put are now $p = (Xe^{-rT} - S_0e^{-qT})$ and 9.2.2 becomes

$$p \geq \max(Xe^{-rt} - S_0e^{-qT}, 0) \tag{9.2.6}$$

Similarly the lower bounds for a call become $c = (S_0e^{-qT} - Xe^{-rT})$ and 9.2.1 becomes

$$c \geq \max(S_0e^{-qT} - Xe^{-rT}, 0) \tag{9.2.7}$$

Put call parity is then represented as:

$$c + Xe^{-rT} = p + S_0e^{-qT} \tag{9.2.8}$$

The formulas for call and put are then:

$$c = S_0e^{-qT}N(d_1) - Xe^{-rT}N(d_2) \tag{9.2.9}$$

$$p = Xe^{-rT}N(-d_2) - S_0e^{-qT}N(-d_1) \tag{9.2.10}$$

$$d_1 = \frac{\ln(S_0e^{-qT}/X) + (r + \sigma^2/2)T}{\sigma\sqrt{T}} = \frac{\ln(S/X) + (r - q + \sigma^2/2)T}{\sigma\sqrt{T}}. \tag{9.2.11}$$

(Recall that $\ln(\frac{S_0e^{-qT}}{X}) = \ln(\frac{S_0}{X}) - qT$).

$$d_2 = \frac{\ln(S_0/X) + (r - q - \sigma^2/2)T}{\sigma\sqrt{T}} = d_1 - \sigma\sqrt{T} \tag{9.2.12}$$

Equations 9.26 and 9.27 fit into the Black Scholes differential equation as:

$$rf = \frac{\partial f}{\partial t}(r-q)S\frac{\partial f}{\partial S} + \frac{1}{2}\sigma^2 S^2 \frac{\partial^2 f}{\partial S^2} \tag{9.2.13}$$

Example 9.2 EUROPEAN OPTION WITH YIELD

A European put option has 6 months to maturity. The stock price is \$102.0 with a strike price of \$96.0,volatility of 18%, yield of 4%. The risk free rate is 10%.
Using the adjusted stock values:

$$d_1 = \frac{\ln(S_0/X) + (r - q + \sigma^2/2)T}{\sigma\sqrt{T}}$$

$$d_1 = \frac{\ln(102.0/96.0) + (0.10 - 0.04 + 0.18^2/2)0.5}{0.18\sqrt{0.5}} = 0.6826$$

$$d_2 = \frac{ln(S_0/X) + (r - q - \sigma^2/2)T}{\sigma\sqrt{T}}$$

$$d_2 = \frac{ln(102.0/96.0) + (0.10 - 0.04 - 0.18^2/2)0.5}{0.18\sqrt{0.5}}$$

$$= d_1 - \sigma\sqrt{T} = 0.6826 - 0.18\sqrt{0.5} = 0.555$$

The put value is therefore $p = Xe^{-rT}N(-d_2) - S_0e^{-qT}N(-d_1)$

$$p = 96.0e^{-0.10*0.5}N(-0.555) - 102.0e^{-0.04*0.5}N(-0.6826) = 0.8227$$

9.2.4. Stock Index Options

We have already looked at index trading through the use of forward contracts. Stock indices are also traded as options on the major markets. The CBOE trades options on the Dow Jones (DJX), which is an index on the industrial average. It also trades the S&P 500 (SPX), Nasdaq 100 (NDX) amongst others.Typically an index contract is 100 times the quoted closing index. The options have cash settlement. The CBOE trade LEAPS which are 'long- term equity anticipation securities', these can have maturities for up to three years.

The stock which underlies the particular index will pay dividends on the ex-dividened date. The various international markets have different convention dates so calculations have to be adjusted to the domain. In the USA ex-dividened dates 'typically' occur in the first of the month for February, May, August and November. The yield is averaged for the underlying stock and is used to construct the q value.

The lower bounds for index options are the same as those for trading known dividend yields, so 9.2.7 and 9.2.8 apply. Equations 9.2.9 and 9.2.10 are used to value index options.

Example 9.3 VALUATION OF STOCK INDICES

A European call on the S&P 500 is 3 months from maturity. The index is \$975. The risk free rate is 9% and the index volatility is 22%. Dividend yields of 0.2%, 0.2% and 0.25% are expected over the next 3 months. The exercise price is \$940. The total yield estimate over the option life is 0.65%; this is 2.6% per annum.

With input parameters:

$$S_0 = 975, X = 940, \sigma = 0.22, T = 0.25, r = 0.09, q = 0.026$$

Gives:

$$d_1 = 0.5327$$
$$d_2 = 0.4227$$
$$N(d_1) = 0.702$$
$$N(d_2) = 0.663$$
$$c = 70.83.$$

The contract cost is \$7,830.0.

9.2.5. *Options on Futures*

The Black (1976) formula is an adjustment to the basic Black Scholes formula. It is used to price a European futures option based on the contract price F. The Black-76 model can also used to price bond options in the interest rate market. The basic assumption is that the futures price follows a lognormal process. This is the same underlying assumption used in pricing stock options. This allows us to use the same formula with the spot price being replaced by the forward price.

The call option on an underlying future can be described in terms of a stock providing a continuous dividend yield (9.2.9 and 9.2.10), with the dividend yield q being replaced by r and S by F. This is given as:

$$c = Fe^{-rt}N(x) - Xe^{-rt}N(x - \sigma\sqrt{t}) \tag{9.2.14}$$

$$p = Xe^{-rt}N(-x + \sigma\sqrt{t}) - Fe^{-rt}N(-x) \tag{9.2.15}$$

With

$$x \equiv \left(\frac{\ln(F/X) + (\sigma^2/2)t}{\sigma\sqrt{t}}\right)$$

In terms of our previous calculation of d_1 and d_2:

$$c = e^{-rt}(FN(d_1) - XN(d_2)) \tag{9.2.16}$$

The put option is:

$$p = e^{-rt}(XN(-d_2) - FN(-d_1)) \quad (9.2.17)$$

Where

$$d_1 = \frac{\ln(F_0/X) + (\sigma^2/2)T}{\sigma\sqrt{T}}$$

$$d_2 = \frac{In(F_0/X) - (\sigma^2/2)T}{\sigma\sqrt{T}} = d_1 - \sigma\sqrt{T}$$

Example 9.4 VALUATION OF FUTURES OPTION

A European option on Brent blend has 6 months to expiry. The futures price is \$21.0. The exercise price is \$21.0. The risk free rate is 10%, the futures volatility is 27%.

$$d_1 = \frac{\ln(21.0/21.0) + 0.10^2 80.50/2}{0.10\sqrt{0.50}} = 0.0954$$

$$d_2 = d_1 - \sigma\sqrt{T} = 0.0954 - 0.27\sqrt{0.50} = -0.0954$$

$$N(d_1) = 0.5380,\ N(d_2) = 0.4619$$

$$p = e^{-0.10^{*}0.50}(21.0^{*}N(-d_2) - 21.0N(-d_1)) = \$1.519$$

$$c = e^{-0.10^{*}0.50}(21.0^{*}N(d_1) - 21.0^{*}N(d_2)) = \$1.519$$

9.2.6. Currency Options

The Merton model described earlier can be used to value currency options. The modified model is based on the changes attributed to Garman & Kohlhagen (1983).

The Merton model used a Black Scholes base model with modifications to include a known dividend yield. Garman and Kohlhagen (1983) further developed this model to value currency options, where the yield variable is replaced by the risk free rate of the foreign currency.

Recall that a foreign currency can be regarded as a stock which pays a yield equivalent to the risk free rate of the country of the currency (r_f). The strike price of the option is paid a yield at the national, risk free rate of the trade. The bounds for a put and call option are therefore the same as in 9.2.7 and 9.2.8, respectively. The stock value, is in this instance the spot exchange rate. The bounds are:

$p \geq \max(Xe^{-rT} - S_0e^{-r_fT}, 0)$ for the put option and $c = (S_0e^{-r_fT} - Xe^{-rT})$ for the call option.

The pricing formulas are:

$$c = Se^{-r_f T} N(d_1) - Xe^{-rT} N(d_2) \tag{9.2.18}$$

$$p = Xe^{-rT} N(-d_2) - Se^{-r_f T} N(-d_1) \tag{9.2.19}$$

From the observation that the forward rate for a currency with domestic rate r and foreign spot rate r_f for a maturity T is given by

$$F_0 = S_0 e^{(r-r_f)T}$$

The put and call can be represented as:

$$p = e^{-rT}(XN(-d_2) - F_0 N(-d_1)) \tag{9.2.20}$$

$$c = e^{-rT}(F_0 N(d_1) - XN(d_2)) \tag{9.2.21}$$

Where

$$d_1 = \frac{\ln(F_0/X) + (\sigma^2/2)T}{\sigma\sqrt{T}}$$
$$d_2 = \frac{ln(F_0/X) - (\sigma^2/2)T}{\sigma\sqrt{T}} = d_1 - \sigma\sqrt{T}$$

The alternative representation using the direct computation in terms of $S_0 e^{(r-r_f)T}$ is:

$$d_1 = \frac{\ln(S_0/X) + (r - r_f + \sigma^2/2)T}{\sigma\sqrt{T}}$$

$$d_2 = \frac{ln(S_0/X) + (r - r_f - \sigma^2/2)T}{\sigma\sqrt{T}} = d_1 - \sigma\sqrt{T}$$

Example 9.5 VALUATION OF CURRENCY OPTION

A European call on the € has six months to expiry. The \$US to € exchange rate is 1.206. The strike is 1.240. The US risk free rate is 4.28%, the euro rate is 3%. The volatility is 13% per annum.

$$d_1 = \frac{\ln(1.206/1.240) + (0.0428 - 0.03 + 0.13^2/2)0.5}{0.13\sqrt{0.5}} = 0.2896$$

$$d_2 = \frac{In(1.206/1.240) + (0.0428 - 0.03 - 0.13^2/2)0.5}{0.13\sqrt{0.5}} = d_1 - \sigma\sqrt{T} = 0.1977$$

$$N(d_1) = 0.6139$$
$$N(d_2) = 0.5783$$

$$\text{Using, } c = Se^{-r_f T} N(d_1) - Xe^{-rT} N(d_2)$$

$$c = 1.206e^{-0.03*0.5} N(d_1) - 1.240e^{-0.0428*0.5} N(d_2) = 0.0600$$

The code implementing methods to deal with the calculation of Black Scholes models for European options is in class **Blackscholecp**. This class is listed in full at the end of Chapter 10.

Listing 9.1. Shows the code to run each of the Examples 9.0–9.5

```
package FinApps;
public class Example_9Bschole {
  /** Creates a new instance of Example_14Bschole */
  public Example_9Bschole() {
  }
  public static void main(String[] args) {
     Blackscholecp b=new Blackscholecp(0.09);
     b.bscholEprice(55.0,60.0,0.26,0.25,0.09);
  */
     /* Example 9.1*/
     /* Dividend = 2/12+5/12+8/12
      *Stock price=70-(1.976+1.979+1.908)=64.137
     Blackscholecp b=new Blackscholecp(0.07);
     b.bscholEprice(64.137,66.0,0.13,0.833,0.07);
      */
     /* Example 9.2
     Blackscholecp b=new Blackscholecp(0.04);
     b.bscholEprice(102.0,96.0,0.18,0.50,0.10);
      */
     /*Example 9.3
      *q=0.2+0.2+0.25=0.65 *4=annual= 2.6%==0.026
     Blackscholecp b=new Blackscholecp(0.026);
     b.bscholEprice(375.0,340.0,0.22,0.25,0.03);
      */
     /*Example 9.4
      *
      * Blackscholecp b=new Blackscholecp(-0.1);
      * b.bscholEprice(21.0,21.0,0.27,0.6,0.10);
      *
      */
     /*Example 9.5
      */
     Blackscholecp b=new Blackscholecp(0.03);
     b.bscholEprice(1.26,1.24,0.13,0.50,0.0428);
     double c=b.getCalle();
```

```
        double put=b.getPute();
        System.out.println("CALL == "+c+" PUT == "+put);

    }
    }
```

LISTING 9.1. Application code for Examples 9.0 to 9.5

9.3. Analytical Approximations for American Options

An American call option should never be exercised early if there are no dividend payments.It has been shown that immediately prioir to the final ex-dividend date is the most likely time for early exercise.

For an American put option it is often optimal to exercise prior to expiration. Any time that a put option is deep in the money, the option should be exercised. Because an American put can be optimally exercised early it always has a greater value than the corresponding European option.

American options can be priced using numerical procedures such as Binomial trees etc, the analytical models of the previous chapter are in general not applicable. However analytical approximations have been developed which give reasonably satisfactory results. These analytical approximations are based on quadrature methods or generalized curve fitting techiniques. The methods therefore tend to involve a series of fixed co-factors in forming equations.

9.3.1. Roll Geske Whaley (RGW) Approximation

The Roll Geske & Whaley (1981) approximation is based on a Black Schole's model where the stock price is modified to take account of known dividends. This approach is reasonable, in the short term where the amount of a dividend can be known with a degree of certainty . For longer terms to dividend payments, models based on percentage dividends offer better results.

When considering an option price the dividends can be viewed as the 'riskless' component, combined with the 'risky' component of stock price and volatility. A basic Black model for estimating an American call is to price the option once, using the discounted stock price; $S = S_0 - De^{-rt_n}$, where the dividend is discounted to the ex-dividend date. The option is then priced as a normal European. The highest of the option prices gives the American approximation. The RGW model refines this approach to take account of a single dividend payment and uses the approximated price which corresponds to the risky period $(T - t)$ of the option life. An iterative procedure is needed to obtain this critical stock price.

If the dividend is less than or equal to the discounted strike for the risky period then exercise is not optimal. So, if;

$$D \leq X(1 - e^{-r(T-t)})$$

Then the option is priced at the European price with the stock price being adjusted to:

$$S = S_0 - De^{-rt_n}.$$

The RGW formula has the following form: (9.3.1)

$$c = (S - De^{-rt})N(b_1) + (S - De^{-rt})M\left(a_1, -b_1; \sqrt{\frac{t}{T}}\right)$$

$$-Xe^{-rt}M\left(a_2, -b_2; -\sqrt{\frac{t}{T}}\right) - (X - D)e^{-rt}N(b_2)$$

Where the variables are:

$$a_1 = \frac{\ln((S - De^{-rt})/X) + (r + \sigma^2/2)T}{\sigma\sqrt{T}}, \; a_2 = a_1 - \sigma\sqrt{T} \quad (9.3.2)$$

$$b_1 = \frac{\ln((S - De^{-rt})/ib) + (r + \sigma^2/2)t}{\sigma\sqrt{t}}, \; b_2 = b_1 - \sigma\sqrt{t} \quad (9.3.3)$$

$N(d)$ is the cumulative normal distribution and $M(a, b; \rho)$ is the cumulative bivariate normal distribution with limits a, b and correlation ρ. The term ib is the critical ex-dividend stock price which iteratively solves:

$$c(ib, X, T - t) = ib + D - X = 91.245 \quad (9.3.4)$$

Example 9.6

Consider an American call option on a stock which is due to pay a dividend of \$4.50 in three months. The stock price is \$90.0, the risk free rate is 5% and the option strike price is \$93.0. Time to expiration is four months and the volatility is 32%.

Using: $a_1 = \dfrac{\ln((S - De^{-rt})/X) + (r + \sigma^2/2)T}{\sigma\sqrt{T}}$

$$a_1 = \frac{\ln((90.0 - 4.50e^{-0.05*0.25})/93.0) + (0.05 + 0.32^2/2)0.333}{0.32\sqrt{0.333}} = -0.269$$

$$a_2 = a_1 - \sigma\sqrt{T} = (-0.269) - 0.32^*\sqrt{0.333} = -0.453$$

The stock price that is solved iteratively:

$$c(ib, X, T-t) = ib + D - X = 91.245$$

Variables b_1, b_2:

$$b_1 = \frac{\ln((S - De^{-rt})/ib) + (r + \sigma^2/2)t}{\sigma\sqrt{t}} = \frac{\ln((90.0 - 4.50^* e^{-0.05^*0.25})/91.245}{0.32^*\sqrt{0.25}}$$

$$= -0.244$$

$$b_2 = b_1 - \sigma\sqrt{t} = (-0.244) - 0.32^*\sqrt{0.25} = -0.404$$

$$M\left(a_1, -b_1; -\sqrt{\frac{t}{\sigma}}\right) = 0.0758$$

$$M\left(a_2, -b_2; -\sqrt{\frac{t}{\sigma}}\right) = 0.0671$$

$$N(b_1) = N(-0.244) = 0.403, \quad N(b_2) = N(-0.404) = 0.343$$

The call value is therefore:

$$c = (S - De^{-rt})N(b_1) + (S - De^{-rt})M\left(a_1, -b_1; \sqrt{\frac{t}{T}}\right)$$

$$-Xe^{-rT}M\left(a_2, -b_2; -\sqrt{\frac{t}{T}}\right) - (X - D)e^{-rt}N(b_2)$$

$$c = (90.0 - 4.50e^{-0.05^*0.25}) * 0.403 + (90.0 - 4.50^* e^{-0.05^*0.25})^* 0.0758$$

$$-93.0^* e^{-0.05^*0.333*}0.0671 - (93.0 - 4.50)e^{-0.05^*0.25*}0.343 = 4.892$$

The American call is priced at $4.892.

Using the adjusted stock price as; $S_0 = S - De^{-rt}$. The equivalent European price for the call option is $3.976 .

The code for RGW is shown in Listing 9.2.

```
package FinApps;
import CoreMath.IntervalBisection;
import static java.lang.Math.*;
import BaseStats.Bivnorm;
import BaseStats.Probnorm;
public class Americrgw extends IntervalBisection{
   public Americrgw() {
   }
```

```
public Americrgw(double dividendval, double divitime) {
   dividend=dividendval;
   divtime=divitime;
   super.setiterations(13);
   super.setprecisionvalue(1e-6);
}
double dividend=1.0;
double divtime=1.0;
double stockprice;
double strikeprice;
double rate;
double time;
double volatility;

public double computeFunction(double rootinput) {
   double solution=0.0;

   double c=americanCall(rootinput,strikeprice,volatility,
                        (time-divtime),rate);
   solution=((c-rootinput)+(strikeprice-dividend));
      return solution;
}
private double americanCall(double s, double x, double sigma,
                        double t,double r) {
   Blackscholecp bp =new Blackscholecp(r);
   bp.bscholEprice(s,x,sigma,t,r);
   return bp.getCalle();
}
public double amCall(double s, double x, double sigma,double t,
                  double r, double low, double high) {
   stockprice=s;
   strikeprice=x;
   volatility=sigma;
   time=t;
   rate=r;
   double callvalue;
   if(dividend<=(x*(1.0-exp(-r*(t-divtime))))) {
      s=(s-(dividend*exp(-r*t)));
      Blackscholecp bp =new Blackscholecp(r);
      bp.bscholEprice(s,x,sigma,t,r);
      return bp.getCalle();
   }
   Probnorm pn=new Probnorm();
   double si=evaluateRoot(low,high);
   double a1=(((log((s-dividend*exp(-rate*divtime))/x)+((rate+
   (sigma*sigma)/2)*t))/(sigma*sqrt(t))));
   double a2=(a1-(sigma*sqrt(t)));
   double b1= (((log((s-dividend*exp(-r*divtime))/si)+((rate+
   (sigma*sigma)/2)*divtime))/(sigma*sqrt(divtime))));
   double b2=(b1-(sigma*sqrt(divtime)));
   double norm1=pn.ncDisfnc(b1);
   double norm2=pn.ncDisfnc(b2);
   double m1=Bivnorm.bivar_params.evalArgs(a1,-(b1),
         -(sqrt(divtime/t)));
```

```
        double m2=Bivnorm.bivar_params.evalArgs(a2,-(b2),
                -(sqrt(divtime/t)));
        callvalue=((s-dividend*exp(-r*divtime))*norm1+
                (s-dividend*exp(-r*divtime))
        *m1-x*exp(-r*t)*m2-(x-dividend)*exp(-r*divtime)*norm2);
        return si;
    }
    public static void main(String[] args) {
        // Requires a reasonable idea of the buest guess for the
        // stockprice with that dividend
        Americrgw am = new Americrgw(4.0,0.25);
        // higher dividend = lower the low
        value, lower the divi, higher the upper guess
        double ret= am.amCall(80.0,82.0,0.30,0.3333,0.06,79.0,89.0);
        System.out.println("MIDVALUE IS =="+ret);

    }
}
```

LISTING 9.2. Option Pricing with RGW

9.3.2. Bjerksund and Stensland (B&S) Approximation

The Bjerksund and Stensland (B&S) (1993) method is a useful approximation for pricing a range of options, including futures and currencies.

The solution for pricing a call is given by:

$$\begin{aligned} C = \alpha S^{\beta} - \alpha\phi(S, T, \beta, ib, ib) + \phi(S, T, 1, ib, ib) - \phi(S, T, 1, X, ib) \\ - x\phi(S, T, 0, ib, ib) + X\phi(S, T, 0, X, ib) \end{aligned} \tag{9.3.5}$$

The parameters are given as:

$$\alpha = (ib - X)ib^{-\beta}$$

$$\beta = \left(0.5 - \frac{b}{\sigma^2}\right) + \sqrt{\left(\frac{b}{\sigma^2} - 0.5\right)^2 + 2\frac{r}{\sigma^2}}$$

The function $\phi(S, T, \gamma, \eta, ib)$ has a variable list given by:

$$\phi(S, T, \gamma, \eta, ib) = e^{\lambda} S^{\gamma} \left[N(\varepsilon) - \left(\frac{ib}{S}\right)^{k} N\left(\varepsilon - \frac{2\ln(ib/S)}{\sigma\sqrt{T}}\right)\right]$$

$$\lambda = [-r + \gamma b + 0.5\gamma(\gamma - 1)\sigma^2]T$$

$$\varepsilon = -\frac{\ln(S/\eta) + [b + (\gamma - 0.5)\sigma^2]T}{\sigma\sqrt{T}}$$

$$k = \frac{2b}{\sigma^2} + (2\gamma - 1)$$

The price *ib* at which the option is optimal for exercise is:

$ib = B_0 + (\beta_\infty - B_0)(1 - e^{h(T)})$, where $h(T) = -(bT + 2\sigma\sqrt{T})\left(\frac{B_0}{\beta_\infty - B_0}\right)$ and

$$B_\infty = \frac{\beta}{\beta - 1}X$$

$$B_0 = \max\left[X, \left(\frac{r}{r-b}\right)X\right]$$

For pricing a put, the put call transformation suggested is:

$$p(S, X, T, r, b, \sigma) = c(S, X, r-b, -b, \sigma)$$

Example 9.7

An American call option has nine months to expiry. The underlying stock price is £82.0, with a volatility of 32%, the option strike price is £78.0. The risk free rate is 5% and the annual yield from the stock is 9%. Calculate the price of an American call.

$$\beta = \left(0.5 - \frac{b}{\sigma^2}\right) + \sqrt{\left(\frac{b}{\sigma^2} - 0.5\right)^2 + 2\frac{r}{\sigma^2}} =$$

$$\beta = \left(0.5 - \frac{-0.04}{0.32^2}\right) + \sqrt{\left(\frac{-0.04}{0.32^2} - 0.5\right)^2 + 2\frac{0.05}{0.32^2}} = 2.20$$

The critical price *ib* is calculated as:

$$B_\infty = \frac{\beta}{\beta - 1}X = \frac{2.20}{2.20 - 1} * 78.0 = 141.884$$

$$B_0 = \max\left[X, \left(\frac{r}{r-b}\right)X\right] = B_0 = \max\left[78.0, \left(\frac{0.05}{0.05 - (-0.04)}\right) * 78.0\right]$$

$$= 78.0$$

$$h(T) = -(bT + 2\sigma\sqrt{T})\left(\frac{B_0}{\beta_\infty - B_0}\right)$$

$$= -(0.04{*}0.75 + 2{*}0.32\sqrt{0.75})\left(\frac{78.0}{141.884 - 78.0}\right) = -0.640$$

$$ib = B_0 + (\beta_\infty - B_0)(1 - e^{h(T)}) = 78.0 + (141.884 - 78.0)(1 - e^{-0.640*0.75})$$

$$= 108.201$$

$$\alpha = (ib - X)ib^{-\beta} = (108.201 - 78.0)108.201^{-2.20}$$

$$= 9.164E - 4$$

The call value is given by the polynomial:

$$C = \alpha S^{\beta} - \alpha\phi(S, T, \beta, ib, ib) + \phi(S, T, 1, ib, ib) - \phi(S, T, 1, X, ib) \\ - X\phi(S, T, 0, ib, ib) + X\phi(S, T, 0, X, ib)$$

which resolves to:

$$\begin{aligned} C &= 9.164E - 4{*}82.0^{2.20} - 9.164E - 4\phi(82.0, 0.75, 2.20, 108.201, 108.201) \\ &\quad + \phi(82.0, 0.75, 1, 108.201, 108.201) \\ &\quad - 78.0\phi(82.0, 0.75, 0, 108.201, 108.201) \\ &\quad + 78.0\phi(82.0, 0.75, 0, 78.0, 108.201) \\ &= 9.515 \end{aligned}$$

The price of an American call option is 9.515.

The code for computing the B&S approximation is shown in Listing 9.3.

```
/*An approximatiom from Bjerksund & Stensland (1993)
* Americbs.java
*/

package FinApps;
import static java.lang.Math.*;
import BaseStats.Probnorm;
public class Americbs {

     public Americbs() {
    }
    public Americbs(double carryrate) {
       crate=carryrate;

    }
    double brate;
    double crate=0.0;
    double beta=0.0;
    double time;
    double rate;
    double volatility;
    double ib;
    public double amerBs(double s, double x, double sigma,
                           double t, double r, int pc) {
       brate=crate<0.0?0.0:(brate=crate!=r?(r-crate):r);
       return pc<0?(amBs(x,s,sigma,t,(r-brate))):
                              amBs(s,x,sigma,t,r);
    }
    private double amBs(double s, double x, double sigma,
                         double t, double r) {

       if(brate>=r) {
          Blackscholecp b=new Blackscholecp();
          b.bscholEprice(s,x,sigma,t,r);
          return b.getCalle();
```

```
    }
    time=t;
    rate=r;
    volatility=sigma;

    double alpha=params(s,x,sigma,t,r);// gets a
    return s>=ib?alpha: ((alpha*(pow(s,beta)))-
    (alpha*phi(s,t,beta,ib,ib))+ (phi(s,t,1,ib,ib))-
    (phi(s,t,1,x,ib))-(x*phi(s,t,0,ib,ib))+
    (x*phi(s,t,0,x,ib)));

}
private double params(double s,double x,double sigma,double t,
                     double r) {
    beta=((0.5-(brate/(sigma*sigma)))
    +sqrt((pow(((brate/(sigma*sigma))- 0.5),2)
    +(2.0*(r/(sigma*sigma))))));
    double betinf=((beta/(beta-1.0))*x);
    double bo=(max(x,(r/(r-brate))*x));
    double h=(-(brate*t+(2.0*sigma*sqrt(t)))*(bo/(betinf-bo)));
    ib=(bo+(betinf-bo)*(1.0-exp(h)));
    double a=((ib-x)*pow(ib,-beta));
    return s>=ib?(s-x):a;
}
private double phi(double s,double t,double gamma,double eta,
                   double iota) {
    double sigma=volatility;
    double r=rate;
    Probnorm p=new Probnorm();
    double lambda=(-r+gamma*brate+0.5*gamma*(gamma-1.0)
                   *(sigma*sigma))*t;
    double epsilon=-(log(s/eta)+(brate+(gamma-0.5)
                   *(sigma*sigma))*t)/(sigma*sqrt(t));
    double kappa=2.0*brate/(sigma*sigma)+(2.0*gamma-1.0);
    double retval=exp(lambda)*pow(s,gamma)*(p.ncDisfnc(epsilon)
    -pow((ib/s),kappa)*p.ncDisfnc(epsilon- 2*log(ib/s)
    /(sigma*sqrt(t))));
    return retval;

}

public static void main(String[] args) {
    Americbs a=new Americbs(0.09);
    double retc=a.amerBS(82.0,78.0,0.32,0.75,0.05,1);
        System.out.println("Value ="+retc);
}}
```

LISTING 9.3. B&S approximation

9.3.3. *Quadratic Approximation (Barone-Adesi Whaley Derivation)*

Quadratic approximation can be used to value stock indices, currencies, futures and stock carrying a constant dividend yield. The BAW (1987) model is a quadratic approximation which incorporates a cost of carry term. The basic analytical model is based on the Black Scholes differential equation. If the difference between a European and American option price is ω the standard equation can be shown as:

$$\frac{\partial\omega}{\partial t}+(r-q)S\frac{\partial\omega}{\partial S}+0.5\sigma^2S^2\frac{\partial^2\overline{\omega}}{\partial S^2}=r\omega$$

If we define a few parameters:

$$h(T)=1-e^{-rT}$$

$$\alpha=2r/\sigma^2$$

$$\beta=\frac{2(r-q)}{\sigma^2}$$

Also, if we have a function g of the variables S, h
$\omega=h(T)g(S,h)$ and the general formula can be written:

$$S^2\frac{\partial^2 g}{\partial S^2}+\beta S\frac{\partial g}{\partial S}-\frac{\alpha}{h}g-(1-h)\alpha\frac{\partial g}{\partial h}=0$$

Ignoring the final term which is small when T is large ($(1-h)$ becomes small) and when t is small $\frac{\partial g}{\partial h}$ is also small. We get: $S^2\frac{\partial^2 g}{\partial S^2}+\beta S\frac{\partial g}{\partial s}-\frac{\alpha}{h}g=0$.

If we denote an American call as $C(S,T)$ and a European call as $c(S,T)$ it can be shown that:

$$C(S,T)=c(S,T)+A_2\left(\frac{S}{S^*}\right)^{\gamma 2}\quad \text{if } S<S^*$$

$$C(S,T)=S-X \text{ if } S\geq S^* \tag{9.3.6}$$

S^* is the critical price at which the stock option should be exercised, it is the price which satisfies the following equation:

$$S^*-X=c(S^*,T)+\left\{1-e^{(b-r)T}N\left[d_1(S^*)\right]\right\}S^*/\gamma_2 \tag{9.3.7}$$

where $b=(r-q)$. This equation is solved for the stock price estimate by using an iterative method (Newton Raphson or Interval Bisection).

The variables used in 9.3.6 are derived from:

$$\gamma_2=\left[-(\beta-1)+\sqrt{(\beta-1)^2+\frac{4\alpha}{h}}\right]0.5$$

$$A_2=-\left(\frac{S^*}{\gamma_2}\right)\{1-e^{(b-r)T}N[d_1(S^*)]\}$$

For a put option the critical price is solved iteratively for the price below which the option should be exercised. The valuation for a put is :

$$P(S,T)=p(S,T)+A_1\left(\frac{S}{S^{**}}\right)^{\gamma 1}\quad if,\ S<S^{**}$$

$$P(S,T)=(X-S)\ if,\ S\leq S^{**}$$

The critical price is given by:

$$X - S^{**} = p(S^{**}, T) - \{1 - e^{(b-r)T} N[-d_1(S^{**})]\} S^{**}/\gamma_1$$

$$\gamma_1 = \left[-(\beta - 1) - \sqrt{(\beta - 1)^2 + \frac{4\alpha}{h}} \right] 0.5$$

$$A_1 = -\left(\frac{S^{**}}{\gamma_1}\right) \{1 - e^{(b-r)T} N[-d_1(S^{**})]\}$$

In both cases the general form for d_1 is:

$$d_1(S) = \frac{\ln(S/X) + (b + \sigma^2/2)T}{\sigma\sqrt{T}}$$

The technique used to derive a critical price can be any of the iterative root finding methods. Generally it is more efficient to use the Newton Raphson method, however in circumstances where the solution equation is difficult to present in a differentiable form, interval bisection or secant methods can be useful.

The Newton Raphson method requires a seed value to place the target value within reasonable range of the equation zero. The same approach can be used to provide base values for low and high parameters used in interval bisection procedures.

A seed value suggested by Barone-Adesi Whaley is:

$$S_1^* = X + [S^*(\infty) - X][1 - e^{h_2}]$$

$$h_2 = -\left(bT + 2\sigma\sqrt{T}\right)\left[\frac{X}{S^*(\infty) - X}\right]$$

where

$$S^*(\infty) = \frac{X}{1 - 2\left[-(\beta - 1) + \sqrt{(\beta - 1)^2 + 4\alpha}\right]^{-1}}$$

$$S_1^{**} = S^{**}(\infty) + [X - S^{**}(\infty)]e^{h_1}$$

$$h_1 = (bT - 2\sigma\sqrt{T})\left[\frac{X}{X - S^{**}(\infty)}\right]$$

$$S^{**}(\infty) = \frac{X}{1 - 2\left[-(\beta - 1) - \sqrt{(-1)^2\beta + 4\alpha}\right]^{-1}}$$

$S(\infty)$ is the critical price for time to expiration being infinite.

Example 9.8

A commodity option has a strike price of \$100.0. The underlying stock price is \$90.0 with a volatility of 15%. The option has 1.2 months to expiry (0.1 years) and the risk free annual rate is 10%. The cost of carry is zero.

The variables are:

$$\beta = \frac{2(r-q)}{\sigma^2} = \frac{2(0.0)}{0.15^2} = 0.0, \ \alpha = 2r/\sigma^2 = 2{*}0.10/0.15^2 = 8.888,$$
$$h(T) = 1 - e^{-rT} = 1 - e^{-0.10*0.1} = 0.00995.$$

The valuation formulas are:

$$\gamma_2 = \left[-(\beta - 1) + \sqrt{(\beta-1)^2 + \frac{4\alpha}{h}} \right] 0.5$$
$$= \left[-(0-1) + \sqrt{(0-1)^2 + \frac{48.888}{0.00995}} \right] 0.5 = 30.384$$
$$S^* - X = c(S^*, T) + \left\{1 - e^{(b-r)T} N\left[d_1(S^*)\right]\right\} S^*/\gamma_2$$
$$\equiv (108.436 - 100) - 6.7137 + \{1 - e^{(-0.10)*0.1} N\left[d_1(105.51)\right]\}$$

Thus the critical price is 105.517.

$$A_2 = -\left(\frac{S^*}{\gamma_2}\right)\{1 - e^{(b-r)T} N\left[d_1(S^*)\right]\} =$$
$$-\left(\frac{105.517}{30.384}\right)\{1 - e^{(-0.10)*0.1} N\left[d_1(105.517)\right]\} = 0.0183$$

The price of an American call option is therefore:

$$C(S,T) = c(S,T) + A_2 \left(\frac{S}{S^*}\right)^{\gamma_2} = 0.2049 + 0.0183^* \left(\frac{90}{105.517}\right)^{30.384}$$
$$= 0.0206$$

The class **Americbaw** which implements the Barone-Adesi Whaley formula is shown in Listing 9.4. Appended below is the Example 9.8 code.

```
package FinApps;
import static java.lang.Math.*;
import BaseStats.Probnorm;
import CoreMath.IntervalBisection;
public class Americbaw extends IntervalBisection {

    public Americbaw() {
    }

    public Americbaw(double carryrate) {
```

```
        crate=carryrate;
        super.setiterations(30);
        super.setprecisionvalue(1e-6);
    }
    double crate=0.0;
    double brate=0.0;
    double d1;
    double q2;
    double su;
    double strike;
    double rate;
    double volatility;
    double time;
    double stockprice;
    double european;
    double n;
    double m;
    double amcall;
    double K;
    double ss;
    int v;
    Blackscholecp bp =new Blackscholecp(crate);
    Probnorm p=new Probnorm();
    public double computeFunction(double rootvalue) {
        double stockvalue=0.0;
        bp.bscholEprice(rootvalue,strike,volatility,time,rate);
        double ds=(log(rootvalue/strike)+(brate+
        (((volatility*volatility)*0.5)
        *time))/(volatility*sqrt(time)));
        double c=bp.getCalle();
        stockvalue=((ss-strike)-(c+((1.0-exp(brate-rate)*time)
        *p.ncDisfnc(ds))*(rootvalue/q2)));
        return stockvalue;
    }
double Si(double s,double x, double sigma, double t, double r) {
        brate=crate==0.0?0.0:(brate=crate!=r?(r-crate):r);
        stockprice=s;
        strike=x;
        volatility=sigma;
        time=t;
        rate=r;
        Blackscholecp bps =new Blackscholecp(crate);
        bps.bscholEprice(s,x,sigma,t,r);// Black '76 model
        double european=bps.getCalle();
        if(brate>=r) {

            return bps.getCalle();
        }
        Probnorm p=new Probnorm();
        ss=startvalue(s,x,sigma,t,r);
                                        //get an initial point for range
        K=(1.0-exp(-r*t));
        double k=2.0*r/((sigma*sigma)*(1.0-exp(-r*t)));
        q2=((-(n-1)+sqrt((n-1*n-1)+4.0*k))*0.5);
```

```
        double sev=evaluateRoot((ss-15.0),(ss+15.0));
        double endval=((log(sev/strike)+(brate+
                       (((volatility*volatility)*0.5)
                       *time))/(volatility*sqrt(time))));
        double a2=((sev/q2)*(1.0-exp((brate-rate)*time))
                   *p.ncDisfnc(endval));
        amcall=s<=sev?(european+(a2*pow((s/sev),q2))):(s-x);
        return amcall;
    }
    private double startvalue(double s, double x, double sigma,
                              double t, double r) {
        m=(2*r/(sigma*sigma));
        n=(2*brate/(sigma*sigma));
        double q2u=(-(n-1.0)+sqrt((n-1.0)*(n-1.0)+4.0*m))*0.5;
        double su=x/(1.0-1.0/q2u);
        double h2=-(brate*t+2.0*sigma*sqrt(t))*x/(su-x);
        double si=x+(su-x)*(1.0-exp(h2));
        return si;
    }
    public double amCall(double s, double x, double sigma,
                         double t, double r) {
        return Si(s,x,sigma,t,r);
    }
    public double amPut(double s, double x, double sigma,
                        double t, double r) {
        return Si(x,s,sigma,t,r-brate);
    }

    public static void main(String[] args) {
        Americbaw a=new Americbaw(0.0);
        double retc=a.amCall(90.0,100.0,0.15,0.10,0.10);
        System.out.println(retc);
    }
    }
```

LISTING 9.4. Barone-Adesi Whaley quadratic approximation

References

Barone-Adesi, G. and R. E. Whaley (1987). "Efficient Analytic Approximation of American Options Values," *Journal of Finance*, 42(2), 301–320.

Bjerksund, P. and G. Stensland (1993). "Closed-Form Approximation of American Options," *Scandinavian Journal of Management*, 9, 87–99.

Black, F. (1976). "The Pricing of Commodity Contracts," *Journal of Financial Economics*, 3, 167–179.

Black, F. and M. Scholes (1973). "The Pricing of Options and Corporate Liabilities," *Journal of Political Economy*, 81, 637–654.

Whaley, R. E. (1981). "On the Valuation of American Call Options on Stocks with Known Dividends," *Journal of Financial Economics*, 9, 207–211.

Vasicek, O. (1977). "An Equilibrium Characterization of the Term Structure," *Journal of Financial Economics*, 5, 177–88.

Garman, M. B. and S. W. Kohlhagen (1983): "Foreign Currency Option Values," *Journal of International Money and Finance*, 2, 231–237.

10 Sensitivity Measures (The 'Greeks')

10.1. The Black-Scholes Pde

The B-S Partial differential equation; $r\Phi = \frac{\partial \Phi}{\partial t} + rS\frac{\partial \Phi}{\partial S} + \frac{1}{2}\sigma^2 S^2 \frac{\partial^2 \Phi}{\partial S^2}$, has a number of solutions that depend on the particular derivative that is defined by S as the underlying variable. The partial elements of the equation provide the derivative sensitivity measures. Recall that the Black Scholes formula, which can be written:

$$c = SN(x) - Xe^{-rt}N(x - \sigma\sqrt{t}) \tag{10.1.1}$$

$$p = Xe^{-rt}N(-x + \sigma\sqrt{t}) - SN(-x) \tag{10.1.2}$$

Where

$$x = \frac{\ln\left(\frac{S}{X}\right) + \left(r + \frac{\sigma^2}{2}\right)t}{\sigma\sqrt{t}} \tag{10.1.3}$$

For convenience we have written the generalised formulae with the equivalent as in 9.1.10 to 9.2.4. A succinct coverage is given in Haug (1998).

10.2. Delta Sensitivity

Delta is the option price sensitivity to small changes in the underlying asset:

$$\Delta_{call} = \partial c / \partial S \tag{10.2.1}$$

$$\Delta_{call} = \frac{\partial c}{\partial S} = N(x) > 0 \tag{10.2.2}$$

The delta of a European put is therefore:

$$\Delta_{put} = \frac{\partial p}{\partial S} = (N(x) - 1) < 0 \tag{10.2.3}$$

Which can be written:

$$\Delta_{call} = \partial c/\partial S = e^{-rt}N(d_1) > 0 \qquad (10.2.4)$$

$$\Delta_{put} = \partial p/\partial S = -e^{-rt}N(-d_1) < 0 \qquad (10.2.5)$$

This can be represented in the general form:

$$\Delta_{call} = \partial c/\partial S = e^{(b-r)T}N(d_1) > 0 \qquad (10.2.6)$$

$$\Delta_{put} = \partial p/\partial S = e^{(b-r)T}(N(d_1) - 1) < 0 \qquad (10.2.7)$$

A portfolio which has a total delta of zero is said to be delta-neutral. A delta-neutral position is one which is largely unaffected by small changes to the underlying asset. 10.2.1 shows the delta is defined as a rate of change for option price with respect to an underlying asset.

The curve is plotted with parameters:

$$S_0 = 8\ldots 90, X = 55, \sigma = 0.25, T = 0.876, r = 0.08$$

Figure 10.1 shows the characteristic curve for the delta as a function of stock price. Using Black Scholes the riskless portfolio would consist of a hedge containing:

−1 Option
+Δ Shares of the underlying stock.

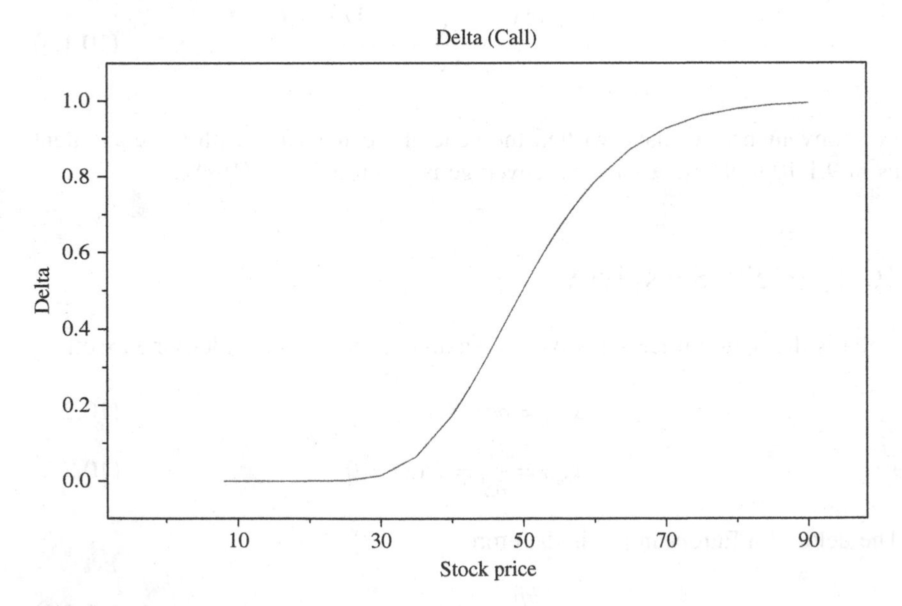

FIGURE 10.1. Delta call for a range of stock prices.

This is the Delta neutral position. From 10.2.6, if the stock is non dividend paying, the delta reduces to: $\Delta = N(d_1)$. For a short position in a call, the delta neutrality requires a long position in the underlying stock, which is an amount $N(d_1)$.

For a Delta put option 10.2.7 shows that for a non dividend paying stock the Delta is negative; $N(d_1) - 1$. Thus a short position in the option requires a short position in the underlying stock and a long option position requires a long stock position.The characteristic is plotted from the following parameters:

$$S_0 = 8 \ldots 90, X = 55, \sigma = 0.25, T = 0.876, r = 0.08$$

This is shown in Figure 10.2.

Example 10.0

Consider a futures option with six months to expiry. The futures price is £110.0, the strike price £106.0 and the risk free rate is 10%. The volatility is 32%. How sensitive are the call and put options to the futures price change?

Using :

$$\Delta_{call} = \partial c / \partial S = e^{(b-r)T} N(d_1)$$

$$d_1 = \frac{\ln(110/106) + (0 + 0.32^2/2)0.5}{0.32\sqrt{0.5}} = 0.2768$$

$$N(d_1) = 0.609$$

$$\Delta_c = e^{(0.0-0.10)0.5} N(d_1) = 0.5793$$

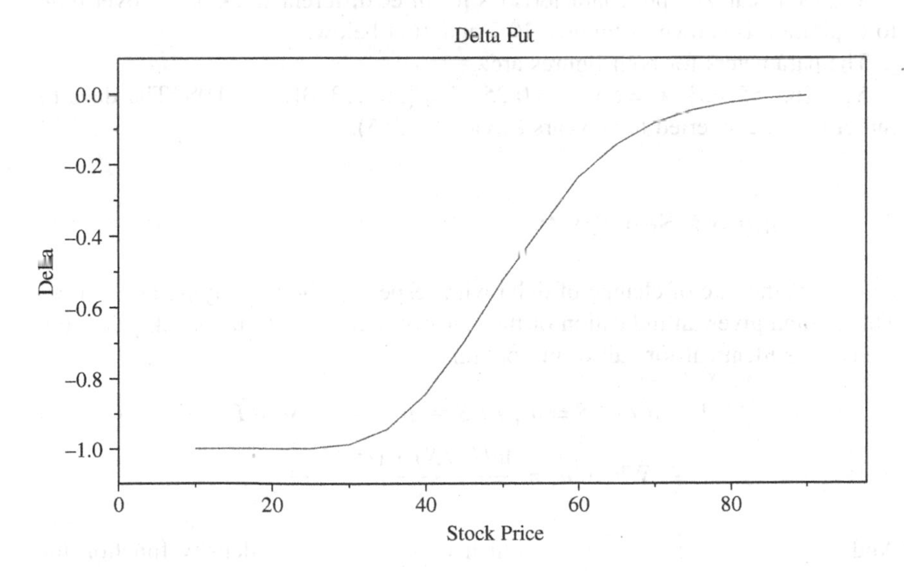

FIGURE 10.2. Delta put for a range of stock prices.

For the put, using:

$$\Delta_{put} = \partial p/\partial S = e^{(b-r)T}(N(d_1) - 1)$$

$$\Delta_p = e^{(0.0-0.10)^{0.5}}(N(d_1) - 1) = -0.3718$$

Example 10.1

Consider the Delta on a currency option.

A US financial institution sells an eight month put option for £2,000,000. The exchange rate is 1.7100 and the strike price is 1.7000. The US risk free rate is 4% and the UK rate is 6%, with a volatility of 9% for Sterling. What is required to make the portfolio Delta neutral?

Using:

$$\Delta_{put} = \partial p/\partial S = e^{-r_f T}(N(d_1) - 1)$$

$$d_1 = \frac{\ln(1.7100/1.7000) + (0.04 - 0.06 + 0.09^2/2)0.666}{0.09\sqrt{0.666}} = -0.0647$$

$$N(d_1) = 0.4741$$

$$\Delta_p = e^{-0.06*0.666}(N(d_1) - 1) = -0.5052$$

The Delta shows that when the exchange rate alters by ΔS, the put option price changes by 50.52%. The Delta of the total option is £1010400 (50.52 % of £2,000,000). The bank will therefore have to short this amount of Sterling to remain Delta neutral.

The Delta call and put characteristics for three different stock prices over time to expiration is shown in Figures 10.3 and 10.4 below.

The parameters for both figures are:

$S_0 = \{65, 55, 45\}$, $X = 55$, $\sigma = 0.25$, $T = \{20 \ldots 320\}$, $r = 0.08$. The days to maturity are converted to % years (divide by 365).

10.3. Gamma Sensitivity

Gamma is the rate of change of delta with respect to the underlying asset price. The gamma gives an indication of the sensitivity of delta to the stock price; the measure is identical for call or put options.

$$\Gamma = \partial^2 c/\partial^2 S = \partial^2 p/\partial^2 S \equiv n(d_1)e^{-rT}/S\sigma\sqrt{T} \tag{10.3.1}$$

$$\text{Where } d_1 = \frac{\ln(F_0/X) + (\sigma^2/2)T}{\sigma\sqrt{T}}$$

And $n(d_1) = (1/\sqrt{2\pi})e^{-d_1^2/2} > 0$ which is the probability density function for the standard normal distribution.

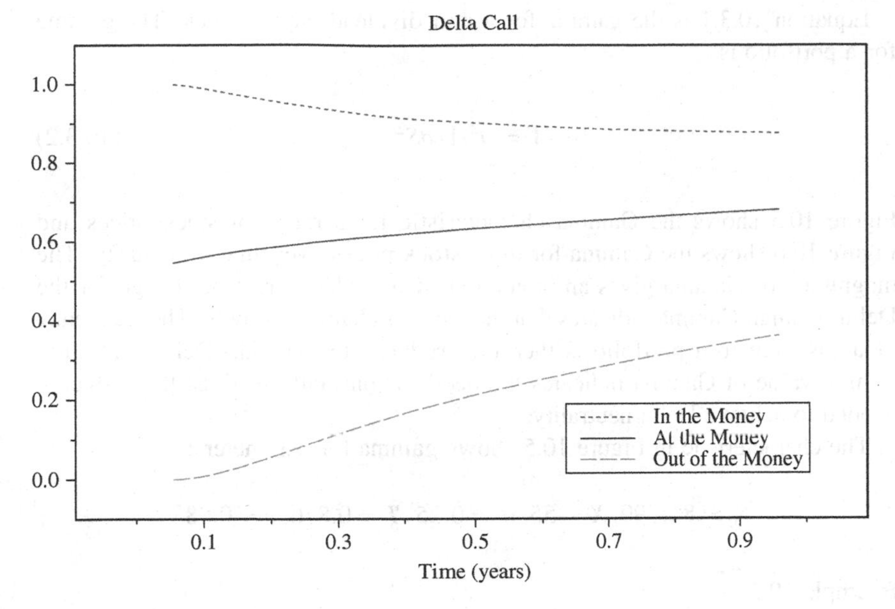

FIGURE 10.3. Delta call for time to expiration with three stock prices.

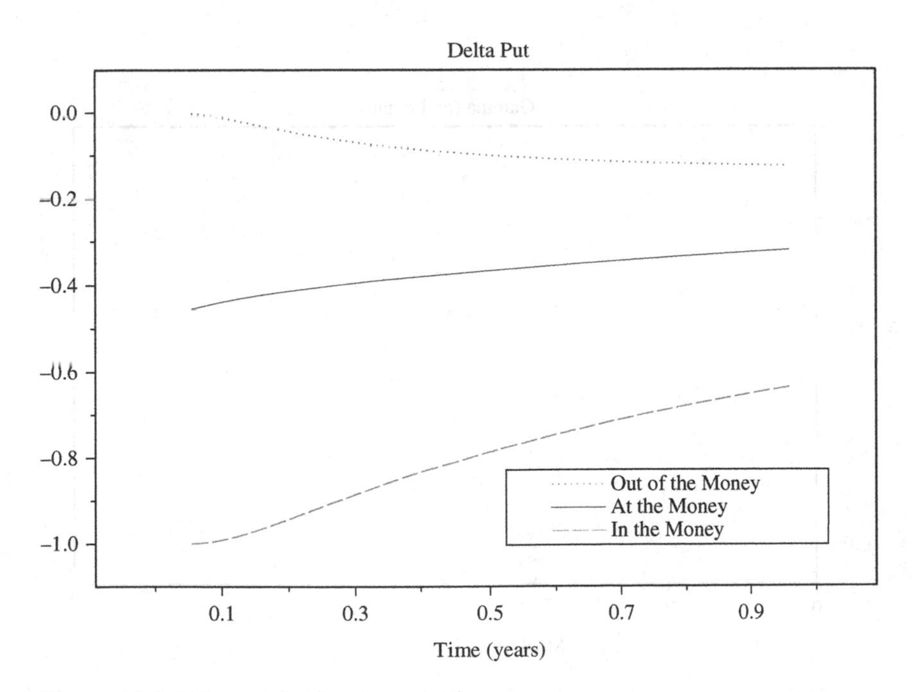

FIGURE 10.4. Delta put for time to expiration.

Equation 10.3.1 is the gamma for a non dividend paying stock. The gamma for a portfolio is:

$$\Gamma = \partial^2\Pi/\partial S^2 \tag{10.3.2}$$

Figure 10.5 shows the Gamma characteristic for a range of stock prices and Figure 10.6 shows the Gamma for three stock prices over time to maturity. The magnitude of Gamma gives an indication of the relative rate of change for the Delta. A small Gamma indicates that the Delta is changing slowly. The frequency of adjustment to a portfolio is therefore reduced to maintain Delta neutrality. A high value of Gamma indicates the need to constantly readjust the hedgeing needed to achieve Delta neutrality.

The characteristic in Figure 10.5 shows gamma for parameters:

$$S_0 = 8\ldots 90, X = 55, \sigma = 0.25, T = 0.876, r = 0.08$$

Example 10.2

A stock option with 9 months to expiry has a strike price of \$50.0. The stock price is \$41.0 with a volatility of 32%. The risk free interest rate is 10%. What is the option Gamma?

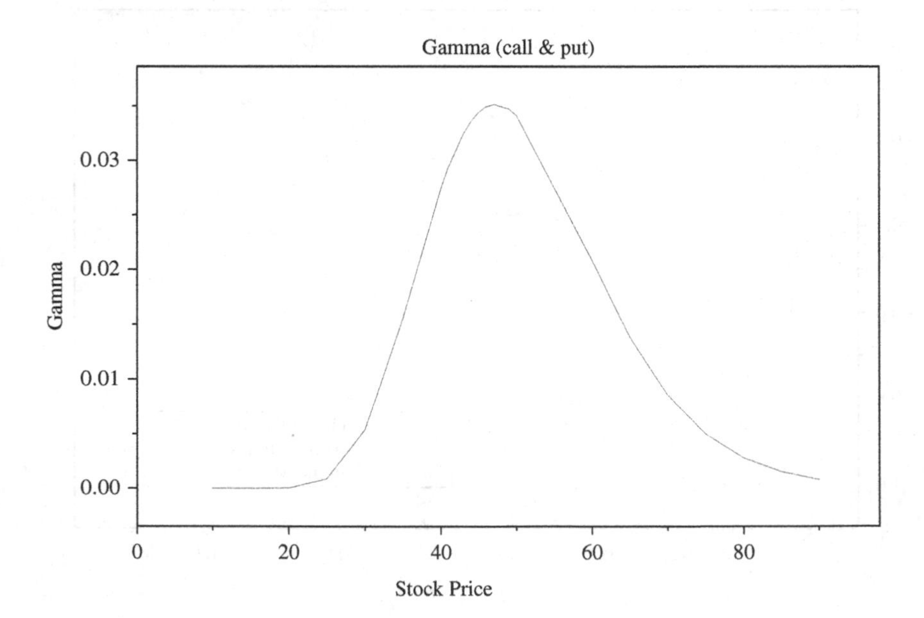

FIGURE 10.5. Gamma for a range of stock prices.

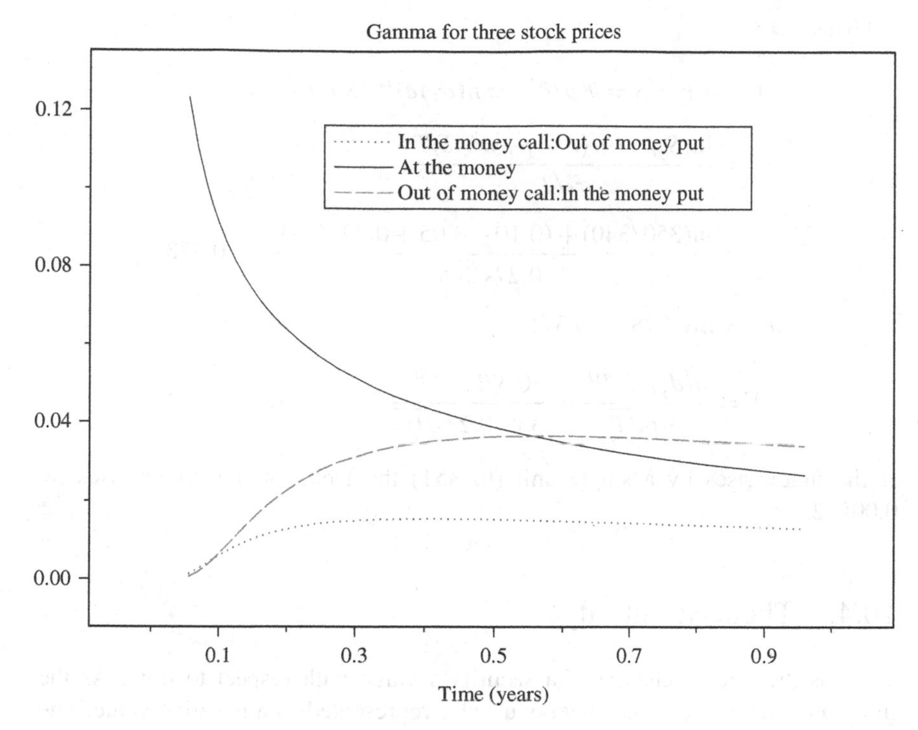

FIGURE 10.6. Gamma for three stock prices.

Using: $\Gamma = \partial^2 c/\partial^2 S = \partial^2 p/\partial^2 S \equiv n(d_1)e^{-rT}/S\sigma\sqrt{T}$

$$d_1 = \frac{\ln(S_0/X) + (r + \sigma^2/2)T}{\sigma\sqrt{T}} = \frac{\ln(41/50) + (0.10 + 0.32^2/2)0.75}{0.32\sqrt{0.75}} = -0.9647$$

$$n(d_1) = n(-0.9647) = 0.250$$

$$\Gamma = \frac{n(d_1)e^{(b-r)T}}{S\sigma\sqrt{T}} = \frac{0.250e^{0^*T}}{41^*0.32\sqrt{0.75}} = 0.022.$$

The Γ is 0.022. Thus for an increase in stock price to \$50.0, the Delta of the option increases by 0.198.

Figure 10.6 shows Gamma for three stock prices with time to maturity. The parameters are:

$S_0 = \{65, 55, 45\}$, $X = 55$, $\sigma = 0.25$, $T = \{20 \ldots 320\}$, $r = 0.08$

Example 10.3

A put option on a stock index has 6 months to expiry. The index is currently at 350, the strike price is 340. The risk free rate is 10% per annum, the stock yield is 5% and the index volatility is 27%. Calculate the Γ of the index option.

Using:

$$\Gamma = \partial^2 c/\partial^2 S = \partial^2 p/\partial^2 S \equiv n(d_1)e^{-rT}/S\sigma\sqrt{T}$$

$$d_1 = \frac{\ln(S_0/X) + (r - q + \sigma^2/2)T}{\sigma\sqrt{T}}$$

$$= \frac{\ln(350/340) + (0.10 - 0.05 + 0.27^2/2)0.5}{0.27\sqrt{0.5}} = 0.378$$

$$n(d_1) = n(0.378) = 0.371$$

$$\Gamma = \frac{n(d_1)e^{(b-r)T}}{S\sigma\sqrt{T}} = \frac{0.371\mathrm{e}^{0.05*0.5}}{350*0.27\sqrt{0.5}} = 0.00542$$

If the index rises by a single unit (to 351) the Delta of the option rises by 0.00542.

10.4. Theta Sensitivity

Theta is the rate of change of a security's value with respect to time. As the time to maturity decreases, theta is usually represented as a negative value. The theta of a portfolio is given as:

$$\Theta \equiv -\partial\Pi/\partial r \tag{10.4.1}$$

The theta for a call on a non dividend paying stock can be represented as:

$$\Theta = -\frac{Sn(d_1)\sigma}{2\sqrt{t}} - rXe^{-rt}N(d_1 - \sigma\sqrt{t}) > 0 \tag{10.4.2}$$

This is also represented in our general formula as:

$$\Theta_c = -\partial c/\partial T = -\frac{Se^{(b-r)T}n(d_1)\sigma}{2\sqrt{T}} - (b-r)Se^{(b-r)T}N(d_1) - rXe^{-rT}N(d_2) < 0 \tag{10.4.3}$$

The theta for a put:

$$\Theta = -\frac{Sn(d_1)\sigma}{2\sqrt{t}} + rXe^{-rt}N(-d_1 + \sigma\sqrt{t}) \tag{10.4.4}$$

This in our general formula is:

$$\Theta_p = -\partial p/\partial T = -\frac{Se^{(b-r)T}n(d_1)\sigma}{2\sqrt{T}} + (b-r)Se^{(b-r)T}N(-d_1) + Xe^{-rT}N(-d_2) \tag{10.4.5}$$

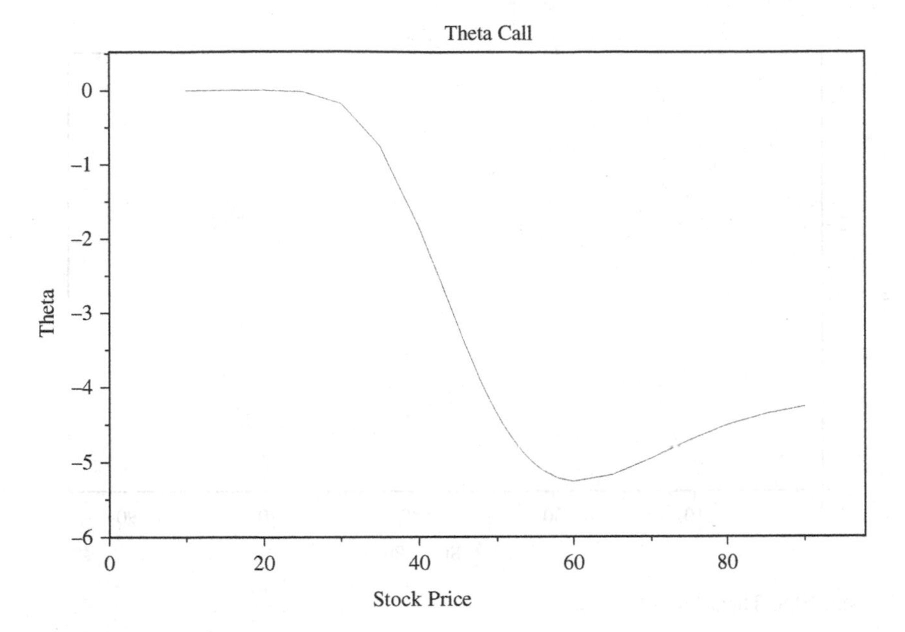

FIGURE 10.7. Theta for a call option.

Theta is usually quoted for days, so the translation into fractions of the year, take into account the calendar year or trading year (recall the trading year consists of 252 days).

Theta is usually a negative value, as depicted in Figure 10.7. The value of an option tends to reduce with a reduction in time to maturity. When the stock price is 8 (from Figure 10.7) the Theta is nearly zero, as the stock price approaches at-the-money (between 50 and 60 in Figure 10.7), the Theta reaches a large negative value. After passing the at-the-money point Theta rises generally following the increasing stock price tending to $-rXe^{-rT}$.

with parameters: $S_0 = 8 \ldots 90, X = 55, \sigma = 0.25, T = 0.876, r = 0.08$

Figure 10.7 shows the characteristic for Theta a put option. Here the early stock prices are associated with a positive value of Theta which gradually decaysto a negative value. Figures 10.8–10.9 show the characteristics for a call and put option with respect to time to maturity (calendar days).

Figure 10.8 Theta put with parameters: $S_0 = 8 \ldots 90, X = 55, \sigma = 0.25,$ $T = 0.876, r = 0.08$

Example 10.4

Consider a European put on a stock index. The current price is €450 with a strike price of €430. The time to expiration is 1 month with the risk free rate being 8% . The yield is 6% with the index volatility at 22%. Compute the trading day Theta.

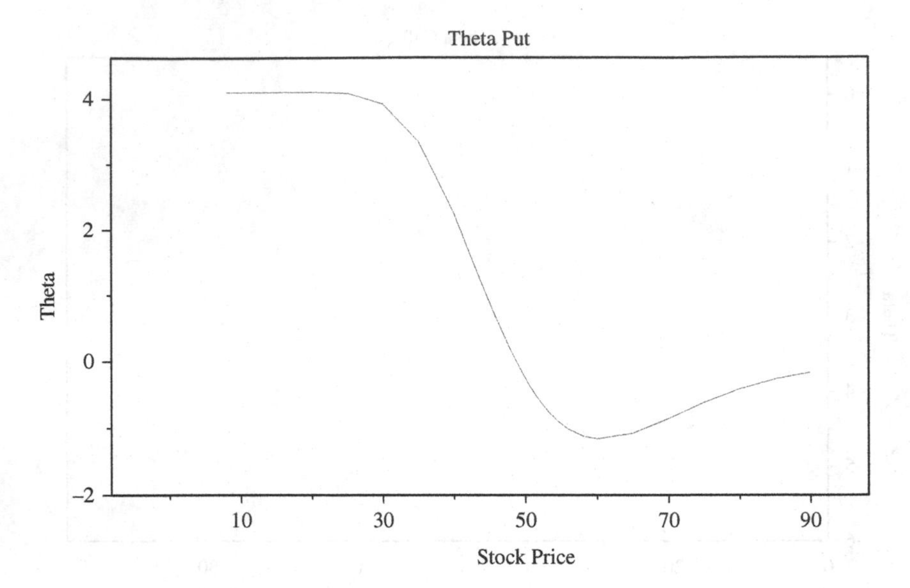

FIGURE 10.8. Theta Put.

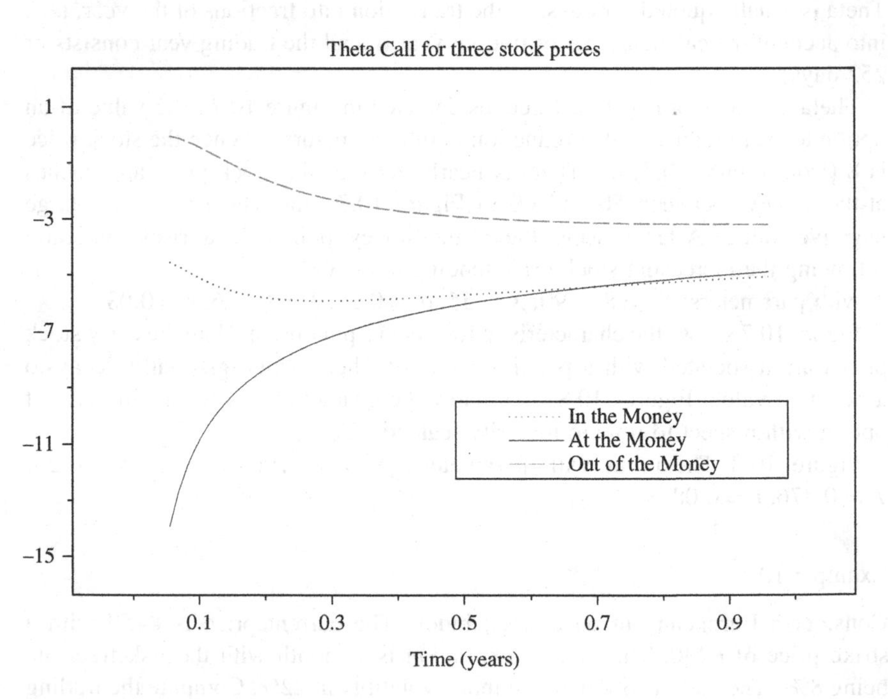

FIGURE 10.9. Theta (call option) for three stock prices with time to maturity.

$$\text{Using}: \Theta_p = -\partial p/\partial T = -\frac{Se^{(b-r)T}n(d_1)\sigma}{2\sqrt{T}} + (b-r)Se^{(b-r)T}N(-d_1)$$
$$- Xe^{-rT}N(-d_2)$$

$$d_1 = \frac{\ln(S_0/X) + (r - q + \sigma^2/2)T}{\sigma\sqrt{T}}$$
$$= \frac{\ln(450/430) + (0.08 - 0.06 + 0.22^2/2)0.08333}{0.22\sqrt{0.08333}} = 0.773$$
$$d_2 = d_1 - \sigma\sqrt{T} = 0.773 - 0.22^*\sqrt{0.08333} = 0.7103$$

$$n(d_1) = n(0.773) = 0.295$$
$$N(-d_1) = N(-0.773) = 0.2195$$
$$N(-d_2) = N(-0.7103) = 0.2387$$

$$\Theta_p = -\frac{450e^{(0.08-0.06)^*0.08333} {}^*0.295^*0.22}{2^*\sqrt{0.0833}} + -0.06^*450e^{-0.06^*0.08333} {}^*0.2195$$
$$+ 0.08^*430e^{-0.08^*0.08333} {}^*0.2387 = -48.19.$$

The Theta for a one trading day time delay is −48.19/252 = −0.191. For a calendar day the Theta would be −0.132.

Figure 10.9 shows the Theta with a range of stock prices where the input parameters are:

$$S_0 = \{65, 55, 45\}, X = 55, \sigma = 0.25, T = \{20 \ldots 320\}, r = 0.08$$

Figure 10.10 Theta put with: $S_0 = \{65, 55, 45\}, X = 55, \sigma = 0.25, T = \{20 \ldots 320\}, r = 0.08$

10.5. Vega Sensitivity

Vega sensitivity measures the rate of change of a derivative's value with respect to the volatility of the underlying asset. The Vega is equal for put and call.

The Vega for a portfolio is:

$$\Lambda \equiv \partial\Pi/\partial\sigma \tag{10.5.1}$$

The Vega for a European call or put is :

$$S\sqrt{t}n(d_1) > 0 \tag{10.5.2}$$

In our general formula.

$$\Lambda = \partial c/\partial\sigma = \partial p/\partial\sigma = Se^{(b-r)T}n(d_1)\sqrt{T} > 0 \tag{10.5.3}$$

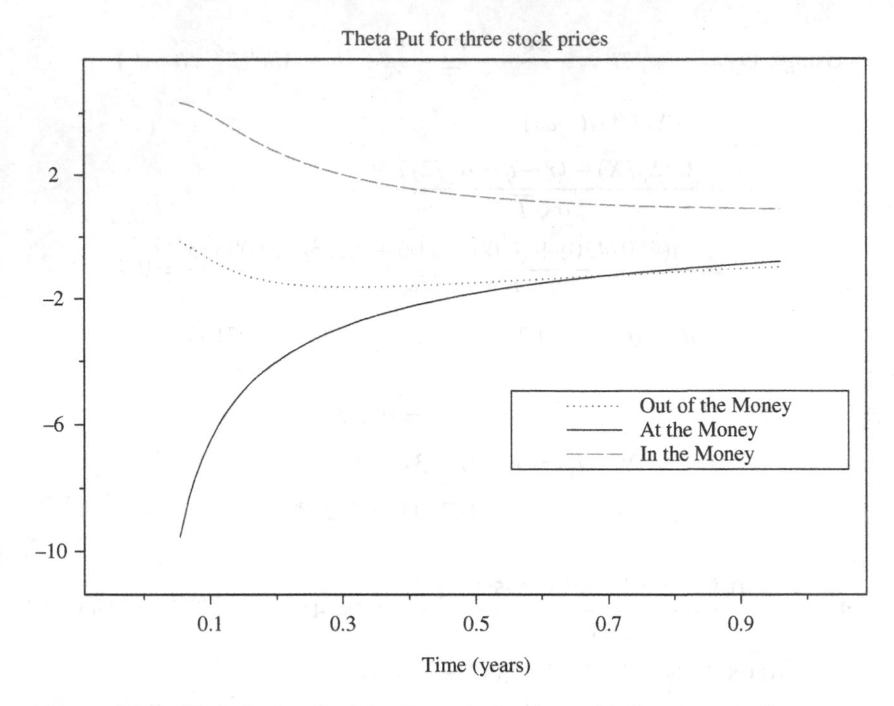

FIGURE 10.10. Theta (put option) for three stock prices with time to maturity.

Vega measures the sensitivity of an option to changes in volatility. So far we have considered that volatility is constant in using other sensitivity measures, in reality the volatility tends to change constantly. Thus the value of an option will change with alterations to stock price, time and volatility. The Black Scholes model requires a constant value for volatility and it would be more technically correct to use a stochastic model for calculating a measure based on changing volatility. The empirical evidence however supports the validity of using a Black Scholes based model such as the Vega.

When considering a portfolio of derivatives, the Vega will increase to a high value when the portfolio is sensitive to volatility changes. If the portfolio is relatively unaffected by volatility changes, the Vega will be low.

Figures 10.11 and 10.12 show that the Vega achieves a high value coincidental with higher option prices.

Figure 10.11 shows Vega for a range of stock prices with parameters:

$$S_0 = 8 \dots 90, X = 55, \sigma = 0.25, T = 0.876, r = 0.08$$

Example 10.5

A stock option has 9 months to expiration. The risk free rate is 9%, the volatility is 27% and the current stock price is £50. If the option strike price is £56. What would be the effect of a 5% rise in volatility?

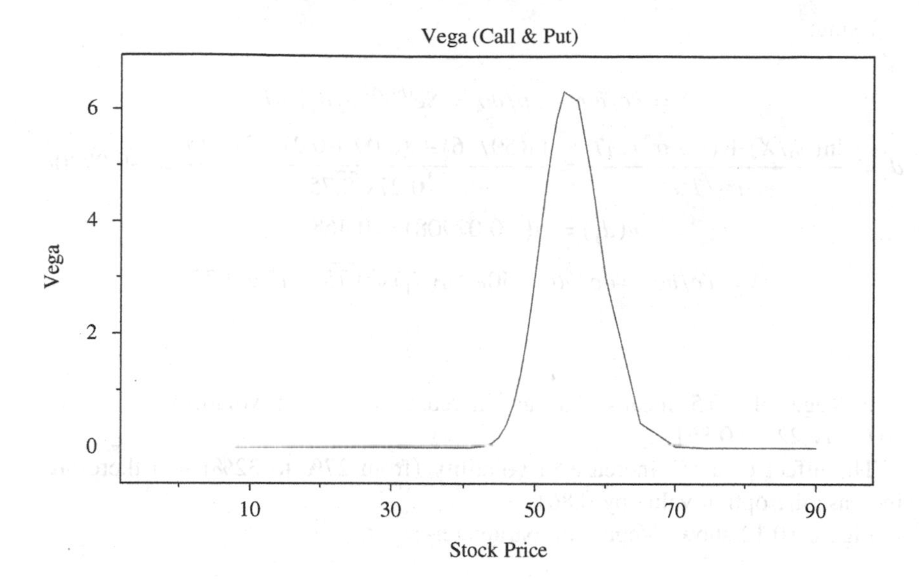

FIGURE 10.11. Vega.

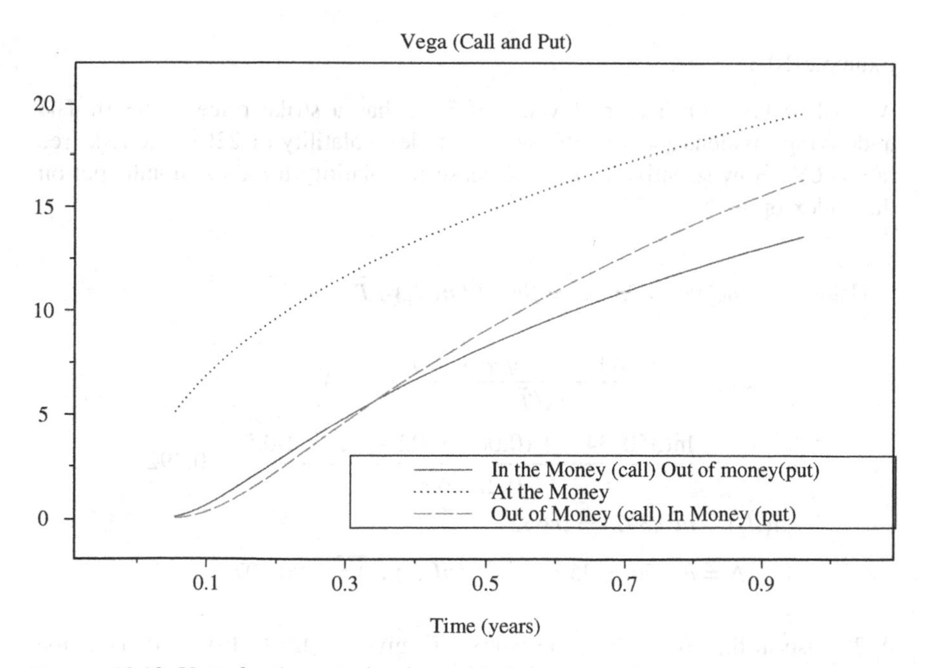

FIGURE 10.12. Vega for three stock prices with time to maturity.

Using:

$$\Lambda = \partial c/\partial\sigma = \partial p/\partial\sigma = Se^{(b-r)T} n(d_1)\sqrt{T}$$

$$d_1 = \frac{\ln(S_0/X) + (r + \sigma^2/2)T}{\sigma\sqrt{T}} = \frac{\ln(50/56) + (0.09 + 0.27^2/2)0.75}{0.27\sqrt{0.75}} = -0.07908$$

$$n(d_1) = n(-0.07908) = 0.468$$

$$\Lambda = \partial c/\partial\sigma = \partial p/\partial\sigma = 50e^{0*} n(d_1)\sqrt{0.75} = 17.22077.$$

The Vega of 10.5 means that an increase of 5% in volatility will give $(0.05*17.22) = 0.861$

The effect of a 5% increase in volatility (from 27% to 32%) will therefore increase the option value by 0.861.

Figure 10.12 shows Vega with parameters:

$$S_0 = \{65, 55, 45\}, X = 55, \sigma = 0.25, T = \{20 \ldots 320\}, r = 0.08$$

Figure 10.12 agrees with the intuitive observation that a higher Vega is equivalent to a higher option price and a higher volatility is associated with longer times to maturity.

Example 10.6

A stock index with a current value of \$350 has a strike price of \$340. The underlying dividend yield is 2%, with an index volatility of 22%. The risk free rate is 6%. How sensitive is a 2% increase in volatility for a six months put on this index option?

Using: $\Lambda = \partial c/\partial\sigma = \partial p/\partial\sigma = Se^{(b-r)T} n(d_1)\sqrt{T}$

$$d_1 = \frac{\ln(S_0/X) + (r - q + \sigma^2/2)T}{\sigma\sqrt{T}}$$

$$= \frac{\ln(350/340) + (0.06 - 0.02 + 0.22^2/2)0.5}{0.22\sqrt{0.5}} = 0.392$$

$$n(d_1) = n(0.392) = 0.652$$

$$\Lambda = \partial p/\partial\sigma = 350e^{-0.02*0.5*} n(d_1)\sqrt{0.5} = 90.497$$

A 2% rise in the volatility of the index will give: (0.02*90.49)=1.80. Thus the put option value will rise by 1.80.

10.6. Rho Sensitivity

The Rho of a derivative is the sensitivity to a small change in the risk free interest rate.

For a portfolio:

$$\rho \equiv \partial\pi/\partial r \tag{10.6.1}$$

The Rho for call and put:

$$\text{Call (non dividend paying)} \quad \rho = Xte^{-rt}N(d_1 - \sigma\sqrt{t}) > 0 \tag{10.6.2}$$

$$\text{Put (non dividend paying)} \quad \rho = -Xte^{-rt}N(-d_1 + \sigma\sqrt{t}) < 0 \tag{10.6.3}$$

In the general form; calls

$$\rho = \partial c/\partial r = TXe^{(b-r)T}N(d_2) > 0, b \neq 0 \tag{10.6.4}$$

$$\rho = \partial c/\partial r = -Tc < 0, b = 0 \tag{10.6.5}$$

In the general form; puts

$$\rho = \partial p/\partial r = -TXe^{(b-r)T}N(-d_2) < 0, b \neq 0 \tag{10.6.6}$$

$$\rho = \partial p/\partial r = -Tp < 0, b = 0 \tag{10.6.7}$$

The Rho for a range of stock prices is shown in Figures 10.13 and 10.14. As the option tends to at the money, Vega rises. At the beginning prices Vega shows that the option is either deep in or deep out of the money.

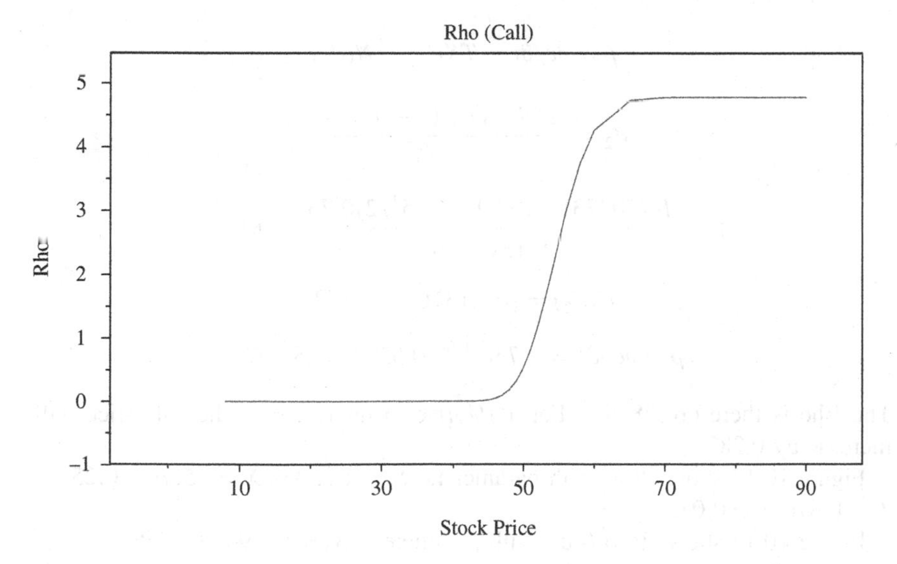

FIGURE 10.13. Rho.

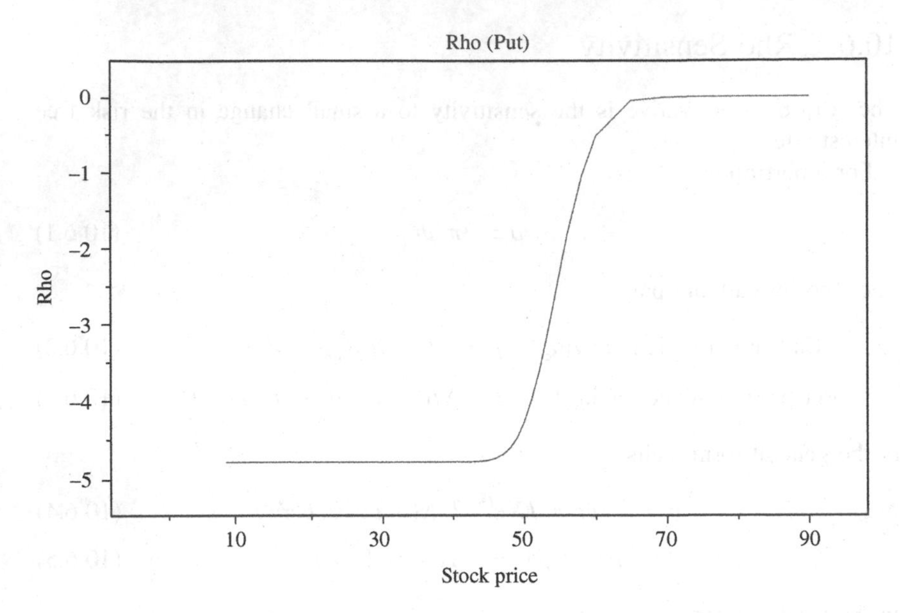

FIGURE 10.14. Rho (Put).

Example 10.7

A European call on a stock option with a price of \$70.0 has a strike price of \$73.0. The risk free rate is 9% and the volatility of the stock is 15%. Calculate the Vega for a 9 months option.

Using:

$$\rho = \partial c/\partial r = TXe^{(b-r)T}N(d_2)$$

$$d_2 = \frac{ln(S_0/X) + (r - \sigma^2/2)T}{\sigma\sqrt{T}}$$

$$d_2 = \frac{ln(70/73) + (0.09 - 0.15^2/2)0.75}{0.15\sqrt{0.75}} = 0.1316$$

$$N(d_2) = N(0.1316) = 0.552$$

$$\rho = \partial c/\partial r = 1^{*}73e^{-0.09*}0.5523 = 28.267$$

The Rho is therefore 28.267. For a 1% rise in interest rate, the call price will increase by 0.282.

Figure 10.13 shows Rho with parameters: $S_0 = 8\ldots90$, $X = 55$, $\sigma = 0.25$, $T = 0.876$, $r = 0.08$.

Figure 10.14 shows Rho (put) with parameters: $S_0 = 8\ldots90$, $X = 55$, $\sigma = 0.25$, $T = 0.876$, $r = 0.08$.

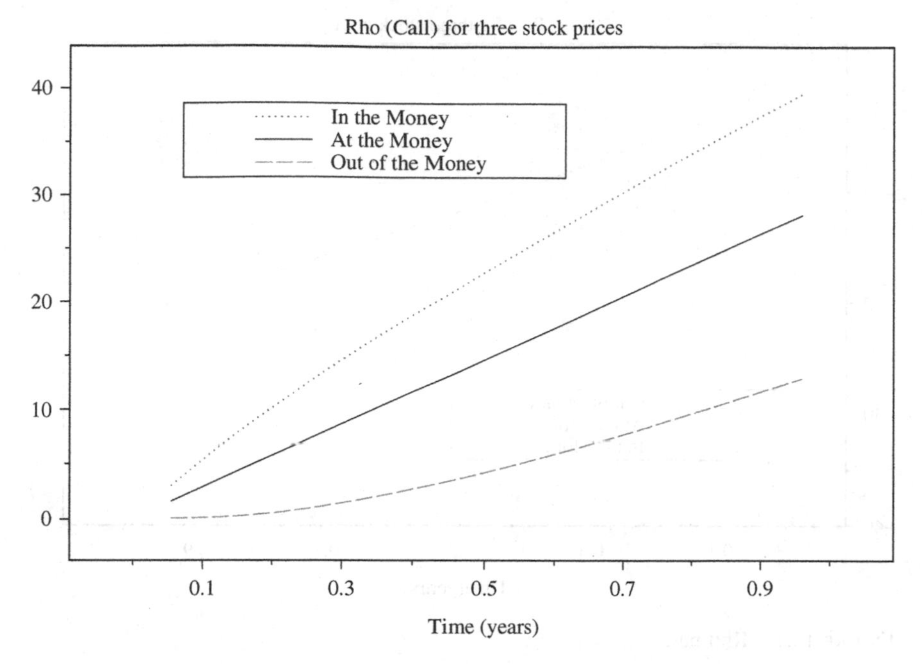

FIGURE 10.15. Rho (call) for three stock prices with increasing time to maturity.

Figure 10.15 shows the Rho for a range of stock prices with parameters:

$$S_0 = \{65, 55, 45\}, X = 55, \sigma = 0.25, T = \{20 \ldots 320\}, r = 0.08$$

Figure 10.16 Rho (Put) for three stock prices with increasing time to maturity, with parameters:

$$S_0 = \{65, 55, 45\}, X = 55, \sigma = 0.25, T = \{20 \ldots 320\}, r = 0.08$$

10.7. Option Extensions

Useful extensions to gauge sensitivity are the cost of carry and elasticity measures.

10.7.1. Elasticity

The elasticity measure is an extension of Delta in that it gives an indication of the sensitivity in percentage terms to a percentage change in the underlying asset price.

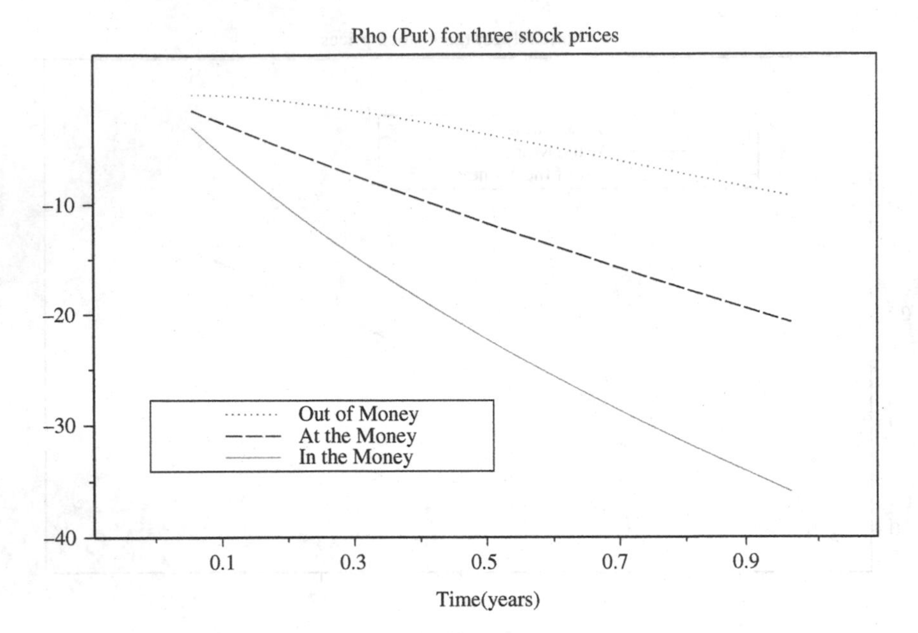

FIGURE 10.16. Rho put.

$$\varepsilon_{call} = \Delta_c \frac{S}{c} = e^{(b-r)T} N(d_1) \frac{S}{c} > 1 \quad (10.7.1)$$

$$\varepsilon_{put} = \Delta_p \frac{S}{p} = e^{(b-r)T} [N(d_1) - 1] \frac{S}{P} < 0 \quad (10.7.2)$$

Example 10.8

Using the data from Example 15.0. calculate the elasticity of a call and put option.

For the call:

$$\varepsilon_{call} = \Delta_c \frac{S}{c} = e^{(b-r)T} N(d_1) \frac{S}{c}$$

$$d_1 = \frac{\ln(S_0/X) + (r + \sigma^2/2)T}{\sigma\sqrt{T}}$$

$$d_1 = \frac{\ln(110/106) + (0.10 + 0.32^2/2)0.5}{0.32\sqrt{0.5}} = 0.497$$

$$\varepsilon_{call} = \Delta_c \frac{S}{c} = e^{(b-r)T} N(d_1) \frac{S}{c} = e^{-0.10*0.5*} N(d_1) \frac{110}{14.77} = 5.14$$

For the put:

$$\varepsilon_{put} = \Delta_p \frac{S}{p} = e^{(b-r)T} [N(d_1) - 1] \frac{S}{p}$$

$$d_1 = \frac{\ln(S_0/X) + (r + \sigma^2/2)T}{\sigma\sqrt{T}}$$

$$d_1 = \frac{\ln(110/106) + (0.10 + 0.32^2/2)0.5}{0.32\sqrt{0.5}} = 0.497$$

$$\varepsilon_{put} = \Delta_p \frac{S}{p} = e^{(b-r)T}[N(d_1) - 1]\frac{S}{p} = e^{-0.10*0.5*}(0.497 - 1)^* \frac{110}{5.601} = -6.07$$

10.7.2. Cost of Carry

The cost of carry refers to the storage costs and interest paid (if any) from the finance of an option's underlying asset. A non dividend paying option has no cost therefore the cost of carry is simply r. For a stock index the cost of carry will be the cost (r) minus the yield from the underlying asset (q) The cost of carry b is therefore, $b = r - q$. For a currency the cost of carry is, $b = r - r_f$. , for a commodity with a utility cost of u the cost of carry is, $b = r - q + u$.

The cost of carry is computed using:

$$\partial c/\partial b = TSe^{(b-r)T}N(d_1) > 0 \quad (10.7.3)$$

$$\partial c/\partial b = TSe^{(b-r)T}N(-d_1) > 0 \quad (10.7.4)$$

Where

$$d_1 = \frac{\ln(S_0/X) + (b + \sigma^2/2)T}{\sigma\sqrt{T}}$$

Example 10.9

A stock index option with three months to expiry has a strike price of £290. The index price is £300 with a volatility of 11% and returns a yield of 4%. The risk free rate is 8%. What is the sensitivity to cost of carry for a call and put option?

For a call:

$$\partial c/\partial b = Tse^{(b-r)T}N(d_1)$$

$$d_1 = \frac{\ln(300/290) + (0.04 + 0.11^2/2)0.25}{0.11\sqrt{0.25}} = 0.8257$$

$$N(d_1) = N(0.8257) = 0.795$$

$$\partial c/\partial b = TSe^{(b-r)T}N(d_1) = 0.25^*300e^{(0.04-0.08)^*0.25*}0.795 = 59.07$$

For a put:

$$\partial p/\partial b = -TSe^{(b-r)T}N(-d_1)$$

$$d_1 = \frac{\ln(300/290) + (0.04 + 0.11^2/2)0.25}{0.11\sqrt{0.25}} = 0.8257$$

$$N(-d_1) = N(-0.8257) = 0.2044$$

$$\partial p/\partial b = -TSe^{(b-r)T}N(-d_1) = -0.25{*}300e^{(0.04-0.08)*0.25}{*}0.2044 = -15.183$$

The 'Greeks' are implemented in the class **Blackscholecp.** The implementation of sensitivity measures in this class is a fairly arbitrary decision. The choice of having this class containing methods to compute both Black-Scholes and the sensitivity measures is biased towards the fact that the measures are based on Black-Scholes basic options. The sensitivity measures could well be implemented as a separate class. The 'Greeks' are shown in Listing 10.1.

```
package FinApps;
import BaseStats.Probnorm;
import static java.lang.Math.*;
public final class Blackscholecp {
   public Blackscholecp() {
}
   *For carryrate=rate. The black Scholes basic model.
         For carryrate a zero value. Black 1976 futures
   * For carryrate != rate Gives cont yield model
   */
   public Blackscholecp(double carryrate) {
      this.crate=carryrate;
   }
   private double crate=0.0;
   private double brate=0.0;
   private double d1=0.0;
   private double d2=0.0;
   private double callprice=0.0;
   private double putprice=0.0;
   private double deltac=0.0;
   private double deltap=0.0;
   private double gamma=0.0;
   private double vega=0.0;
   private double thetac=0.0;
   private double thetap=0.0;
   private double rhoc=0.0;
   private double rhop=0.0;
   private double elasticityc=0.0;
   private double elasticityp=0.0;
   private double carryc=0.0;
   private double carryp=0.0;
   public double getCalle() {
      return callprice;
   }
```

```
public double getPute() {
  return putprice;
}
private void setcalle(double call) {
  callprice=call;
}
private void setpute(double put) {
  putprice=put;
}
public void bscholEprice(double sprice, double strike,
    double volatility,double time, double rate) {

  Probnorm p=new Probnorm();
  dvalues(sprice,strike,volatility,time,rate);
  double probd1=0.0;
  double probd2=0.0;
  probd1=p.ncDisfnc(d1);
  probd2=p.ncDisfnc(d2);
  double densityfunc=p.npdfDisfnc(d1);
  double densityfunc2=p.npdfDisfnc(d2);
  setcalle(((sprice*exp((brate-rate)*time))
                    *probd1)-((strike*exp(-rate
                    **time))*probd2));
  setpute(((strike*exp(-rate*time))
        *p.ncDisfnc(-d2))-(sprice*exp((brate-rate)
        *time))*p.ncDisfnc(-d1));
  deltac=(exp((brate-rate)*time)*probd1);
  deltap=(exp((brate-rate)*time)*(probd1-1));
  gamma=((densityfunc*(exp((brate-rate)*time)))
         /(sprice*volatility*sqrt(time)));
  vega=((sprice*exp((brate-rate)*time))
         *densityfunc*sqrt(time));
  double thetaterm1=((brate-rate)*
        (sprice*exp((brate-rate)*time)*probd1));
  double thetaterm2=(rate*(strike*exp(-rate*time))*densityfunc2);
  thetac=(((-(sprice*exp((brate-rate)*time))
             *densityfunc*volatility)/(2*sqrt(time)))
             -thetaterm1-thetaterm2);

}
private void dvalues(double sprice,double strike,
    double volatility,double time, double rate) {
  brate=crate==0.0?0.0:(brate=crate!=rate?
                   (rate-crate):rate);
  d1=((log(sprice/strike)+(brate+(volatility*volatility)
                    *0.5)*time)/(volatility*sqrt(time)));
  d2=(d1-(volatility*sqrt(time)));

}
public void setDelta(double sprice,double strike,
    double volatility,double time, double rate)
{
  Probnorm p=new Probnorm();
  dvalues(sprice,strike,volatility,time,rate);
  double probd1=0.0;
```

```
      double probd2=0.0;
      probd1=p.ncDisfnc(d1);
      probd2=p.ncDisfnc(d2);
      deltac=(exp((brate-rate)*time)*probd1);
      deltap=(exp((brate-rate)*time)*(probd1-1));
        }
   public void setGamma(double sprice,double strike,
        double volatility,double time, double rate)
   {
      Probnorm p=new Probnorm();
      dvalues(sprice,strike,volatility,time,rate);
      double probd1=0.0;
      probd1=p.ncDisfnc(d1);
      double densityfunc=p.npdfDisfnc(d1);
      gamma=((densityfunc*(exp((brate-rate)*time)))
             /(sprice*volatility*sqrt(time)));
        }
   public void setVega(double sprice,double strike,
        double volatility,double time, double rate)
   {
      Probnorm p=new Probnorm();
      dvalues(sprice,strike,volatility,time,rate);
      double probd1=0.0;
      probd1=p.ncDisfnc(d1);
      double densityfunc=p.npdfDisfnc(d1);
      vega=((sprice*exp((brate-rate)*time))
             *densityfunc*sqrt(time));

      }
   public void setTheta(double sprice,double strike,
        double volatility,double time, double rate)
{
      Probnorm p=new Probnorm();
      dvalues(sprice,strike,volatility,time,rate);
      double probd1=p.ncDisfnc(d1);
      double probd2=p.ncDisfnc(d2);
      double probd3=p.ncDisfnc(-d1);
      double probd4=p.ncDisfnc(-d2);
      double densityfunc=p.npdfDisfnc(d1);
      double densityfunc2=p.npdfDisfnc(d2);
      double thetaterm1=((brate-rate)*(sprice*exp
                        ((brate-rate)*time)*probd1));
      double thetaterm2=(rate*(strike*exp
                        (-rate*time))*probd2);
      double thetaterm3=((-(sprice*exp((brate-rate)*time))
                       *densityfunc*volatility)/(2*sqrt(time)));
      thetac=(thetaterm3-(thetaterm1)-(thetaterm2));
      double thetaterma=((brate-rate)*(sprice*exp
                        ((brate-rate)*time)*probd3));
      double thetatermb=(rate*(strike*exp
                        (-rate*time))*probd4);
      double thetatermc=((-(sprice*exp((brate-rate)*time))
             *densityfunc*volatility)/(2*sqrt(time)));
      thetap=(thetatermc+(thetaterma)+(thetatermb));
```

```
}
public void setRho(double sprice,double
        strike,double volatility,double time, double rate)
{

      Probnorm p=new Probnorm();
   dvalues(sprice,strike,volatility,time,rate);
   if(brate!=0.0)
   {
      rhoc=(time*strike*exp(-rate*time)*p.ncDisfnc(d2));
      rhop=(-time*strike*exp(-rate*time)*p.ncDisfnc(-d2));
   }
   else
   {
      bscholEprice(sprice,strike,volatility,time,rate);
      rhoc=(-time*getCalle());
      rhop=(-time*getPute());
   }

}
public void setElstic(double sprice,double strike,
       double volatility,double time, double rate)
{
   bscholEprice(sprice,strike,volatility,time,rate);
   setDelta(sprice,strike,volatility,time,rate);
   elasticityc=(getDeltac()*(sprice/getCalle()));
   elasticityp=(getDeltap()*(sprice/getPute()));

}
public void setCarry(double sprice,double
        strike,double volatility,double time, double rate)
{
      Probnorm p=new Probnorm();
   dvalues(sprice,strike,volatility,time,rate);
   carryc=(time*(sprice*exp((brate-rate)
                *time))*p.ncDisfnc(d1));
   carryp=(-time*(sprice*exp((brate-rate)*time))
                              *p.ncDisfnc(-d1));

   }
public double getDeltac() {
   if(deltac>0.0)
   return deltac;
   else
      throw new RuntimeException("INCORRECT DELTA
                                    PARAMETRS"+deltac);
}
public double getDeltap() {

   return deltap;
}
public double getGamma() {
   if(gamma>0.0)
   return gamma;
   else
```

```
            throw new RuntimeException("INCORRECT GAMMA
                                       PARAMETRS"+gamma);
    }
    public double getThetac() {
        return thetac;
    }
    public double getThetap() {
        return thetap;
    }
    public double getVega() {
        return vega;
    }
    public double getRhoc()
    {
        return rhoc;
    }
    public double getRhop()
    {
        return rhop;
    }
    public double getElasticc()
    {
        return elasticityc;
    }
    public double getElasticp()
    {
        return elasticityp;
    }
    public double getCarryc()
    {
        return carryc;
    }
    public double getCarryp()
    {
        return carryp;
    }
}
```

LISTING 10.1. Implementation of class to compute Black scholes valuation and option sensitivities

References

Haug, E. G. (1998). *The Complete Guide to Option Pricing Formulas*. McGraw Hill.

11
Interest Rate Derivatives

The valuation models we have looked at so far have made the assumption that interest rates are constant in the lifetime of the option. The expected payoff from an asset is at the risk free rate and the risk free rate is used to discount the future payoff values. The principle of risk neutrality is an underlying assumption of the basic Black Scholes model. An issue arises when the interest rates during an asset's life, follow a stochastic process. We will see later that the Black Scholes model is still consistent, even in a stochastic interest rate environment, when due account is made of the risk effect from the stochastic rates.

11.1. Market Price of Risk

The market price of risk defines a value above the risk free return for an asset. If we assume that all assets exist in a risk environment we can account for risk free returns as being an environment in which the market price of risk is equal to zero.

If we have an option whose price is dependent on a single underlying variable S and assume that it follows the Ito process (see 13.1.2):

$$dS/S = mdt + sdz$$

The variables m, Z are the growth and volatility respectively of the variable S, which depend on S and t only. The variable S does not need to be the price of an asset and does not necessarily have to be traded (e.g. temperatures or interest rates). Given two derivatives $f_1(S, t), f_2(S, t)$ with the condition that they are dependent only on S and t and:

$$\frac{df_i}{f_i} = \mu_i dt + \sigma_1 dz$$

Then,

$$\frac{df_1}{f_1} = \mu_1 dt + \sigma_1 dz \text{ and } \frac{df_2}{f_2} = \mu_2 dt + \sigma_2 dz.$$

Where

$\mu_1, \mu_2, \sigma_1, \sigma_2$ are functions of S and t the term dz represents uncertainty in the process and is the same uncertainty for both derivatives. Both prices share the same Wiener process (dz).

If we construct a portfolio of $\sigma_2 f_2$ and $-\sigma_1 f_1$ this portfolio becomes instantaneously riskless by:

$$\sigma_2 f_2 df_1 - \sigma_1 f_1 df_2 = \sigma_2 f_2 f_1(\mu_1 dt + \sigma_1 dz) - \sigma_1 f_1 f_2(\mu_2 dt + \sigma_2 dz)$$
$$= (\sigma_2 f_2 f_1 \mu_1 - \sigma_1 f_1 f_2 \mu_2)dt$$

Thus,

$$\Delta \prod = (\sigma_2 f_2 f_1 \mu_1 - \sigma f_1 f_2 \mu_2)\Delta t$$

Given that the portfolio has now eliminated risk, it now earns the risk free rate; $\Delta \prod = r \prod \Delta t$.

Combining this with the previous equations gives:

$$\mu_1 \sigma_2 - \mu_2 \sigma_1 = r\sigma_2 - r\sigma_1$$
$$\frac{\mu_1 - r}{\sigma_1} = \frac{\mu_2 - r}{\sigma_2} \tag{11.1.1}$$

The left hand side and right hand side of 15.1.32 show that they depend on the parameters of f_i and therefore S and t and not on the derivative f. If we define $\frac{\mu_1 - r}{\sigma_1} = \frac{\mu_2 - r}{\sigma_2} = \lambda$, then λ is the market price of risk of S. The market price of risk is independent of the derivative.

For any derivative which is dependent on an underlying variable :

$$\frac{df}{f} = \mu dt + \sigma dz \tag{11.1.2}$$

and

$$\mu - r/\sigma = \lambda \tag{11.1.3}$$

Equation 11.1.3 applies to *any* derivative which depends on the same variables S and t. By rearranging 11.1.3 $\lambda\sigma = \mu - r$, we can view the left hand side as the proportion of risk that S contributes multiplied by the price of that risk. The right hand side refers to the amount of excess return required to offset the risk.

If we consider the values of μ and σ in terms of Ito's lemma, the partial differential equation becomes:

$$\mu = \frac{1}{f}\left(\frac{\partial f}{\partial t} + \mu S \frac{\partial f}{\partial S} + \frac{1}{2}\sigma^2 S^2 \frac{\partial^2 f}{\partial S^2}\right), \sigma = \frac{\sigma S}{f}\frac{\partial f}{\partial S}$$
$$\frac{\partial f}{\partial t} + (\mu - \lambda\sigma)S\frac{\partial f}{\partial S} + \frac{1}{2}\sigma^2 S^2 \frac{\partial^2 f}{\partial S^2} = rf \tag{11.1.4}$$

The appearance of μ infers that risk is part of the investors strategy and the derivative valuation needs to take account of the market price of risk. If the asset is a traded security the assumption is that the market price of risk is 0 and $\mu = r$. The valuation of a derivative can be computed using the discounted value of the derivative at the riskless rate if the process is now:

$$\frac{dS}{S} = (\mu - \lambda\sigma)dt + \sigma dz \tag{11.1.5}$$

dz is now the Wiener process of the risk neutral measure.

If there are several variables which can affect the market price of risk for a particular derivative, then each of the individual variables have to be incorporated into the single risk measure.

If we have n variables

$$S_0, S_{1,\dots} S_n$$

which follow the stochastic process

$$dS_i/S_i = m_i dt + s_i dz_i$$

for i=0..n. The m_i, S_i terms are expected growth and volatility rates which may be functions of the underlying asset for any security therefore which is dependent on the variables:

$$\frac{df}{f} = \mu dt + \sum_{i=0}^{n} \sigma_i dz_i \tag{11.1.6}$$

μ is the expected return from the security (whereas m is the expected growth of the variable). The σ_i, dz_i terms are the risk contributed by the ith variable to this return. It can further be shown that:

$$\mu - r = \sum_{i=0}^{n} \lambda_i \sigma_i \tag{11.1.7}$$

Where λ_i is the market price of risk for the S_i

11.2. Martingales

A Martingale is a zero drift stochastic process with the following properties:

Given a variable X it is a Martingale if the stochastic process $\{X(t), t \geq 0\}$ holds that,

$$E[|S(t)|], < \infty, t \geq 0 \text{ and}$$
$$E[S(t)|S(u), 0 \leq u \leq x] = S(x)$$

For a discrete time

$$E\left[S_{n+1}|S_1, S_2, \ldots S_n\right] = S_n$$

Thus the expected value of a Martingale at any future time is its expected value today. A variable follows a Martingale process if $dS = \sigma dz$, the variable σ can also be stochastic with dependencies on S.

The Martingale process follows $E(S_T) = S_0$

The Equivalent Martingale measure

If we have two stocks with prices A and B respectively, being dependent on a single underlying risk.

Let

$$\theta = \frac{A}{B}$$

where θ is the relative price measure which measures A in terms of units of B, with B termed the numeraire. An equivalent Martingale (θ) is a measure that defines the market price of risk in terms of the numeraire, the security B can be thought of as a proxy measure. The market price of risk is taken as the volatility of the numeraire, thereafter for any prices that A might take on, θ is Martingale for the ratio A/B and all securities A.

If the market price of risk is σ_B it can be shown that

$$d\left(\frac{A}{B}\right)(\sigma_A - \sigma_B)\frac{A}{B}dz \tag{11.2.1}$$

If the market price of risk is defined with respect to B then the market is said to be forward risk neutral with respect to B. Thus,

$$A_0 = B_0 E_B(A_T/B_T) \tag{11.2.2}$$

where the expected value is forward risk neutral with respect to B.

The equivalent Martingale measure can be based on whatever is appropriate for the domain, such as the money market account, zero coupon bond prices etc. In terms of interest rates the numeraire is often taken as the price of a bond.

Consider the Black Scholes model which we have used to date in a risk neutral environment. The model can be adjusted to incorporate the effects of stochastic interest rates by using the forward risk neutral process. The effect of this is to adjust our basic Black Scholes model to incorporate the zero coupon rate in place of the risk free rate, the general argument is as follows:

If we take the market price (M_p) of a zero coupon bond as the numeraire such that $M_p(t, T)$ is the price at time t of a bond that pays 1 unit at time T. If we are forward risk neutral with respect to M_p then

$$A_0 = M_p(0, T)E_T(A_T).$$

Given a variable, x (any variable which is not an interest rate) that has a forward contract on it with time to maturity T. The price of this forward contract is:

$$A_0 = M_p(t, T)(E_T(x_t) - X).$$

The forward price F is the strike price at which $A_0 = 0$. Therefore, $F = E_T(X_T)$. This infers that the forward price of a variable is the expected future spot price where the variable is risk neutral with respect to $M_p(t, T)$.

If we consider the situation where the variable is an interest rate. The interest rate when a zero coupon bond is used as the numeraire is represented as:

$$R(t, \tau) \tag{11.2.3}$$

where $\tau = T_2 - T_1$ and the annualised time period is $((T_2 - T_1)$ eg, for $\tau = 0.25$, the compounding period is quarterly. The forward price is given as:

$$Mp(t, T_2)/Mp(t, T_1)$$

Given that the forward interest rate is the implied rate of the forward zero coupon bond rate it can be shown that:

$$R(0, T_1, T_2) = E_{T_2}(R(T_1, T_1, T_2)) \tag{11.2.4}$$

where E_{T_2} is forward risk neutral with respect to $M_p(t, T_2)$. The forward interest rate (15.1.42) is therefore the future interest rate in a risk neutral environment with respect to a zero coupon bond maturing at T_2.

Now consider a European call option on a non dividend paying stock which expires at time T. The option price is

$$c = M_p(0, T)E_T[\max(S_T - X), 0]$$

Assuming that the stock price is distributed lognormally and that the standard deviation is given as ς it can be shown that

$$E_T[\max(S_T - X), 0] = E_T(S_T)N(d_1) - XN(d_2)$$

For,

$$d_1 = \frac{\ln[E_T(S_T)/X] + \varsigma^2/2}{\varsigma}$$

$$d_2 = \frac{\ln[E_T(S_T)/X] - \varsigma^2/2}{\varsigma}$$

From the observation that $E_T(S_T)$ is the forward stock price for maturity at time T. Thus we have

$$E_T(S_T) = S_0 e^{RT} \tag{11.2.5}$$

We now have

$$c = S_0 N(d_1) - Xe^{-RT} N(d_2)$$

for

$$d_1 = \frac{\ln(S_0/X + RT + \varsigma^2/2)}{\varsigma}$$
$$d_2 = \frac{\ln(S_0/X + RT - \varsigma^2/2)}{\varsigma}$$

If stock price volatility σ is defined as, $\sigma\sqrt{T} = \varsigma$, then d_1 and d_2 can be written in the well known form:

$$d_1 = \frac{\ln(S_0/X) + (R + \sigma^2/2)T}{\sigma\sqrt{T}}$$
$$d_2 = \frac{\ln(S_0/X) + (R - \sigma^2/2)T}{\sigma\sqrt{T}}$$

which is the standard Black Scholes equation with the continuously compounded risk free rate being replaced by the forward interest rate risk neutral with respect to a zero coupon bond. Therefore we can use the standard Black Scholes method when using stochastic interest rates.

11.3. Interest Rate Caps & Floors

A rate cap is an option to offer protection against rises in the prevailing interest rate. A common cap is to protect against rises in the LIBOR rate. The floating note rate is set at the current LIBOR rate, at periodic times the LIBOR rate is reset and the interest rate on the note reflects this. A cap option would offer insurance against the rise by having a strike rate which is the value of the capped interest rate.

Suppose we have a cap of 5% and a notional value of £1,000,000. If the tenor (times between reset periods) is 0.25 years and the cap life is 2 years, the cap offers protection against rises above 5%, as follows:

LIBOR at 5%; 0.25*0.05*1, 000, 000. = 12, 500 interest payable. LIBOR rises by 1% to 6%;= 0.25*0.06*1, 000, 000. = 15000. The cap would offer 2,500 worth of protection.

An interest rate cap is made up of a number of caplets, which are individual rate options. The cap is therefore a portfolio of caplets, which have the following properties:

For a cap with life (time to expiry) T made from tenors $t_1, t_2, \ldots t_n$, $(t_{n+1} = T)$. The cap rate is X_r and r is the interest rate between t_m and $t_m + 1$ seen from time t_m. The notional sum (note value) is V. The payoff is

$$V\ \delta_m \max(r - X_r, 0) \tag{11.3.1}$$

where, $\delta_m = (t_{m+1} - t_m)$.

The cap is therefore a portfolio of call options on the LIBOR rate at time t_m with payoff at time t_{m+1}. Equation 11.3.1 describes such a a call option or caplet.

A floor is the converse of the cap. The floor offers protection against the floating rate note falling below the strike value. A floor is defined as a portfolio of floorlets. The individual floorlet options are defined as:

$$V\delta_m \max(X_r - r, 0) \tag{11.3.2}$$

Thus a floor is a portfolio of European style put options on interest rates.

An interest rate cap can also be seen as a portfolio of put options on zero coupon bonds. The payoff is defined as:

$$\max\left[V - \frac{V(1 + X_r\delta_m)}{1 + x_r\delta_m}, 0\right] \tag{11.3.3}$$

where, $\frac{V(1+X_r\delta_m)}{1+x_r\delta_m}$ is the value of a zero coupon bond at t_m that pays at time $t_m + 1$.

A cap is the portfolio of caplets which is defined as:

$$Cap = \sum_{i=1}^{n} caplet_i$$

The caplet value is given as:

$$V\ \delta_m(e^{-rt})[F_m N(d_1) - X_n], \text{ where } T = t_m + 1 \tag{11.3.4}$$

where

$$d_1 = \frac{\ln(F_m/X_r) + (\delta_m^2/2)t_m}{\delta_m\sqrt{t_m}}$$

$$d_2 = \frac{\ln(F_m/X_r) - (\delta_m^2/2)t_m}{\delta_m\sqrt{t_m}}$$

The notional and forward prices should be adjusted to take account of the period basis ie, 360 or 365 days. So that:

$$V = (d/Basis) \text{ and}$$
$$F = (d/Basis)$$

where d is the number of days in the forward rate period.
The floor value is given by:

$$Floor = \sum_{i=1}^{n} floorlet_i$$

The floorlet is given as:

$$V\ \delta_m(e^{-rT})[X_r N\ (-d_2) - F_m N\ (-d_1)], \text{ where } T = t_{m+1}$$

Each caplet in the portfolio is valued separately using 11.3.4. Often spot volatilities are used for each caplet period, in some circumstances flat volatilities are used by brokers. The flat volatilities are adjusted for the life of the caplet.

Example 11.0

Consider a contract to cap LIBOR on a notional £50,000 at 7.6% per annum, for 120 days beginning in one year. The LIBOR curve is flat at 6.9% with a forward rate volatility of 23%. The zero rate is 6.45%.
The caplet price is:

$$d_1 = \frac{\ln(F_m/X_r) + (\delta_m^2/2)t_m}{\delta_m\sqrt{t_m}} = \frac{\ln(0.069/0.076) + (0.23^2/2)1.0}{0.23\sqrt{1.0}} = -0.3051$$

$$d_2 = \frac{\ln(F_m/X_r) - (\delta_m^2/2)t_m}{\delta_m\sqrt{t_m}} = \frac{\ln(0.069/0.076) - (0.23^2/2)1.0}{0.23\sqrt{1.0}} = -0.5351$$

(Alternatively, $d_2 - d_1 - \sigma_m\sqrt{t_m} = -0.5351$).

$$V\ \delta_m(e^{-rt})[F_m N(d_1) - X_n] =$$

$$50,000{*}0.23(e^{-0.0645{*}1.3333})[0.069{*}N(-0.3051) - 0.076N(-0.5351)] = 56.697$$

The caplet price is £56.70.

11.4. Swap Options

Swap options (swaptions) are options offered on interest rate swaps. Swaptions are either payer swaptions, where the holder has the right (not the obligation) to pay at a fixed rate and receive at the floating rate.Or, receiver swaptions, where the holder has the right (not the obligation) to receive the fixed rate and pay the floating interest rate.

Swaptions have many uses as financial instruments, a typical use is to offer a guarantee to an organisation that the level of interest payments against loan capital at a future date will not exceed a fixed level. To achieve this the mechanism has to offer protection against uncertainty (risk) in the floating rate.

As an example suppose a company is planning to take out a 10 year floating rate loan, in three months time. The company perceives a comparative advantage in exchanging the floating rate for a 10 year fixed rate, that should not execeed some level say, 7%. A swaption giving the right to a 3 month LIBOR at a fixed rate of 7% for the loan term would be an attractive option. If the fixed to float exchange rate in three months time exceeds 7%, the option would be exercise, if the exchange rate is less, the option would expire and the company would enter a regular swap at the more advantageous rate.The advantage of a swaption over other instruments for forward rate guarantees is that the holder has no obligation to exercise.

European swap options can be valued using the Black (1976) model with adjustments. Referring to our analysis of the equivalent Martingale measures.If we take an annuity as the numeraire, the swap at time T with payments at times $T_1, T_2, \ldots T_n$, (giving an annuity $A(t) = \sum_{m=0}^{N-1} (T_{m+1} - T_m)^* M_p(t, T))$ is the forward swap rate $(r_s)^* A(t)$. We are therefore in an environment which is forward risk neutral with respect to the zero coupon bond price. The forward interest rate is equal to the expected forward rate. The value is therefore the present value of the annuity times the ratio of swap price to the annuity value.

Thus the swaption value is:

$$VA\left[E_A(r_T)N(d_1) - r_x N(d_2)\right]$$

If V is the notional value and m is the number of payments per annum and n the number of years in the option life. The swaption value will be:

$$\sum_{i=1}^{m^* n} V/m^* M_p(0, T_i)\left[r_{s0} N(d_1) - r_x N(d_2)\right] \qquad (11.4.1)$$

Given that an annuity pays $1/m^* \sum_{i=1}^{mn} M_p(0, T_i)$ and $M_p(0, T_i) = (e^{-rT_i})$ The swaption values are:

$$VA\left[r_f\ N(d_1) - r_x\ N(d_2)\right] \qquad (11.4.2)$$

for a payer swaption.

$$VA\left[r_X\ N(-d_2) - r_f\ N(-d_1)\right] \qquad (11.4.3)$$

for receiver swaption. Where:

$$d_1 = \frac{\ln(r_f/r_X) + (\sigma^2/2)T}{\sigma\sqrt{T}}$$

$$d_2 = \frac{\ln(r_f/r_X) + (\sigma^2/2)T}{\sigma\sqrt{T}} = d_1 - \sigma\sqrt{T}$$

For computational efficiency the leading value $1/m^* \sum_{i=1}^{mn} M_p(0, T_i)$ can be rewritten as:

$$\left[\frac{1 - \frac{1}{(1+\frac{r_f}{m})^{mn}}}{r_f}\right] e^{-rT} \tag{11.4.4}$$

Equation 11.4.2 can now be rewritten for a payer swaption as:

$$\left[\frac{1 - \frac{1}{(1+\frac{r_f}{m})^{mn}}}{r_f}\right] e^{-rT}\left[r_f\ N(d_1) - r_X\ N(d_2)\right] \tag{11.4.5}$$

For a receiver swaption the value is given by:

$$\left[\frac{1 - \frac{1}{(1+\frac{r_f}{m})^{mn}}}{r_f}\right] e^{-rT}\left[r_x N(-d_2) - r_f N(-d_1)\right] \tag{11.4.6}$$

Equation 11.4.5 provides the benchmark model for swaptions.

Example 11.1

Consider a five year swap, which starts in three years time and ends in eight years (5 year swaption) that gives the holder a right to pay 7%. The forward swap volatility is 21%, the risk free rate is 6.2% and the forward swap rate is 6.5%. The principal sum is $100.0. The value of this payer swaption is:

$$d_1 = \frac{\ln(r_f/r_x) + (\sigma^2/2)T}{\sigma\sqrt{T}} = d_1 = \frac{1n(0.065/0.07) + (0.21^2/2)^*3.0}{0.21\sqrt{3.0}} = -0.218$$

$$d_2 = \frac{\ln(r_f/r_x) - (\sigma^2/2)T}{\sigma\sqrt{T}} = d_1 - \sigma\sqrt{T} = -0.0218 - 0.21\sqrt{3.0} = -0.3856$$

$$N(d_1) = 0.491$$

$$N(d_2) = 0.349$$

$$\left[\frac{1-\frac{1}{(1+\frac{r_f}{m})^{mn}}}{r_f}\right] e^{-rT}\left[r_f N(d_1) - r_x N(d_2)\right]$$

$$= 4.2111^* e^{-0.062^*3.0*}\left[0.065^*0.491 - 0.70^*0.349\right] = 0.02601$$

The up front value of the swaption is therefore 2.6% of the notional value, which gives $2.60.

11.4.1. Adjusting Rates for Convexity

The relationship between bond prices and yield have been examined in an earlier chapter where we looked at duration. The Macaulay duration was seen as a reasonable method to account for the non-linear characteristic of the price/yield relationship. A more rigorous approach is now examined, where the assumption that the expected value of a rate is assumed to be exactly its forward rate is challenged.

If we consider a derivative that has its future payoff determined by the zero coupon bond yield at a future time. We have assumed that the future payoff is given by $S_t - X_t$, where X gives us a final value of 0. In looking at interest rate products, we have used the assumption that the forward interest rate is the forward rate determined by the zero bond rate. The bond yield at some future time is dependent on the futures bond price. We know that the bond price/yield characteristic is non linear, so we should accommodate the non-linearity by adjusting the forward interest rates in a non linear way.

When using the zero coupon bond as the numeraire a method which more correctly relates the expected bond yield when the forward bond price is equal to the expected bond price. It can be shown that:

$$E_t(y_t) = y_0 - \frac{1}{2} y_0^2 \sigma_y^2 T \frac{G''(y_0)}{G'(y_0)} \tag{11.4.7}$$

Where y_0 is the forward bond yield and the expected yield is $y_0 - \frac{1}{2} y_0^2 \sigma_y^2 T \frac{G''(y_0)}{G'(y_0)}$, thus the convexity adjustment is

$$-\frac{1}{2} y_0^2 \sigma_y^2 T \frac{G''(y_0)}{G'(y_0)} \tag{11.4.8}$$

Where G' and G'' are the first and second partial derivatives with respect to G, which is the function relating bond value to forward yield.

If p is the fixed swap value (or bond value) and y the forward yield:

$$-\frac{1}{2}\frac{\frac{\partial^2 p}{\partial y^2}}{\frac{\partial p}{\partial y}}y^2(e^{\sigma^2 T}-1) \tag{11.4.9}$$

Example 11.2

A derivative has a single payment in four years time. The payoff is based on the yield of a three year swap with forward yield of 6.57% and volatility of 20%.What is the adjusted rate?

The fixed side value is given as:

$$p=\frac{r_f}{(1+r_f)}+\frac{r_f}{(1+r_f)}+\frac{1+r_f}{(1+r_f)}$$

The first partial derivative:

$$\partial p/\partial r_f=-\frac{r_f}{(1+r_f)^2}-\frac{2r_f}{(1+r_f)^3}-\frac{3r_f}{(1+r_f)^4}=-\frac{0.0657}{(1+0.0657)^2}-\frac{2*0.0657}{(1+0.0657)^3}-\frac{3*0.0657}{(1+0.0657)^4}=-2.645$$

The second partial derivative:

$$\partial^2 p/\partial r^2{}_f=\frac{2r_f}{(1+r_f)^3}+\frac{6r_f}{(1+r_f)^4}+\frac{12r_f}{(1+r_f)^5}=\frac{2*0.0657}{(1+0.0657)^3}+\frac{6*0.0657}{(1+0.0657)^4}+\frac{12*0.0657}{(1+0.0657)^5}=9.717$$

The convexity adjustment is given by:

$$-\frac{1}{2}\frac{9.717}{-2.645}*0.0657^2(e^{0.20^2*4}-1)=0.001375$$

For an adjustment of 0.001375 added to the initial forward yield of 0.0657 gives an adjusted yield of 0.0670. The initial forward yield of 6.57% is now 6.7%.

11.4.2. Zero Coupon Bond as the Asset

Convexity is also a factor in interest rate derivatives that use the forward interest rate of the zero coupon bond. Consider an instrument which pays cash flows at time T_0 (for time τ) where this is between times T_1 and T_0. Effectively paying in advance. The cash flow will be $V^* r_T{}^* \tau$, where $\tau = T_1 - T_0$. The yield is r_T for the compounding period τ.

Example 11.3

An instrument has a payoff in four years time based on the one year zero coupon bond with a face value of \$500.0. The flat yield is 10% with a volatility of 21.5%

$$p = \frac{1}{(1+.10)^4}$$

$$\partial p/\partial r_f = -0.909$$

$$\partial^2 p/\partial {r_f}^2 = 1.652$$

The convexity adjustment is:

$$r_f + \left[-\frac{1}{2} \frac{\frac{\partial^2 p}{\partial y^2}}{\frac{\partial p}{\partial y}} y^2 (e^{\sigma^2 T} - 1) \right] = 0.10 + 0.001846 = 0.1018$$

The bond price unadjusted value is:

$$p = \frac{1}{(1+.10)^4} = 0.683 * 500.0 * 1 = \$34.1506$$

The adjusted price is now:

$$0.683 * 500.0 * 1 * \left[r_f + -\frac{1}{2} \frac{\frac{\partial^2 p}{\partial y^2}}{\frac{\partial p}{\partial y}} y^2 (e^{\sigma^2 T} - 1) \right] = \$34.78$$

There is a difference of around \$0.63 between the unadjusted and convexity adjusted price.

11.4.3. Valuation of Bond Options

A typical bond option is the option to buy or sell a bond at a given date for a particular price. If we can assume the bond price follows a lognormal characteristic then the formulae for Black's 76 model apply.

The call price is:

$$c = M_p(0, T)\,[FN(d_1) - XN(d_2)] \quad (11.4.10)$$

The put price is:

$$p = M_p(0, T)\,[XN(-d_2) - FN(-d_1)] \quad (11.4.11)$$

Where F is the forward price of the bond at the option expiration and variables d_1, d_2 are given by:

$$d_1 = \frac{\ln(F/X) + (\sigma^2/2)T}{\sigma\sqrt{T}},$$

$$d_2 = d_1 - \sigma\sqrt{T}.$$

This basic model is only suitable for short term bond pricing as the volatility (uncertainty) of a bond at time zero is zero and then increases as time to maturity increases. At maturity volatility is once again zero. The concave nature of volatility is however not taken into account with this model. The Black model assumes a linear rise in volatility with maturity, so pricing with this basic model should remain under the quasi-linear rising portion of the concave characteristic.

For a coupon bond the price can be approximated by :

$$F = \frac{B_0 - I}{M_p(0, T)} \tag{11.4.12}$$

Where B_0 is the bond price at time zero and I is the present value of all coupon payments during the option life.

When pricing the bond option choosing the base price as either dirty or clean should be consistent for the strike price. The prices used in 11.3.14 and 11.3.15 assume cash prices rather than quoted prices. If we use quoted prices the formulae should be adjusted accordingly and the strike price have accrued interest added (see Chapter 2 and accrual conventions).

Example 11.4

A European put option has 6 months to expiry with a strike price of £187.0 on a bond with a forward price of £188.0 at expiration. The forward volatility is 5%, the risk free rate is 5.65%. Give the option value.

$$d_1 = \frac{\log(188.0/187.0) + (0.05^2/2)*0.5}{0.05\sqrt{0.5}} = 0.1685$$

$$d_2 = 0.1685 - \sqrt{0.5} = 0.1331$$

$$N(-d_1) = N(-0.1685) = 0.483$$

$$N(-d_2) = N(-0.1331) = 0.447$$

$put = e^{-0.05*0.5} * [188.0*0.447 - 187.0*0.433] = 2.11$

The put value of the option is £2.11.

Example 11.5

A European 9 months call option on a 10 year bond with a face value of $1,000.0 has a strike price of $1,000.0. The current cash price for the bond is $985.0. The

9 month risk free rate is 10% and the forward volatility is 8.5%. The bond pays a semi annual rate of 10%. Coupon's are payable in 2 and 8 months time. The 2 month risk free rate is 9.2% and the 8 month rate is 9.55. Give the option value. Also derive the put option price.

The discounted coupon payments are:

$$I = 50^* e^{-0.1666^*0.092} + 50^* e^{-0.75^*0.95} = 96.17$$

The forward bond price is given by 11.3.16

$$F = \frac{B_0 - I}{M_p(0, T)} = \frac{985.0 - 96.17}{e^{-0.10^*0.75}} = 958.0$$

$$d_1 = \frac{\log(958.0/1000.0) + (0.085^2/2)^*0.75}{0.085\sqrt{0.75}} = -0.545$$

$$d_2 = d_1 - \sigma\sqrt{T} = -0.545 - 0.085\sqrt{0.75} = -0.6189$$

The call value is given by:

$$c = M_p(0, T)\,[FN(d_1) - XN(d_2)] = e^{-0.10^*0.75}\,[958.0^*0.292 - 1000.0^*0.267]$$
$$= 11.602$$

Thus, the price of this call option is \$11.60. The cost of a put option would be:

$$p = M_p(0, T)\,[XN(-d_2) - FN(-d_1)] = e^{-0.10^*0.75}\,[1000.0^*0.732 - 958.0^*0.707]$$
$$= 50.51$$

Therefore a put option will cost \$50.51.

11.5. Short Rate Modelling

Short rate modelling is based on the notion of an interest rate (r) taken over an infinitely short time span. So called equilibrium models are built around a process model of the short rate. The process model is examined in relation to products such as bonds and options that are dependent on a risk neutral environment. There are a multitude of factors which can effect the eventual trajectory of interest rates in the real world. For a risk neutral environment however we ignore extraneous factors and view the process in terms of the constraints offered by risk neutrality.

The simplest process approach is to examine equilibrium in terms of a single factor such as the uncertainty that we can measure in a process for r. A single factor equilibrium model therefore looks at a single reference such as Brownian motion to explain the rate process.

The usual model for a risk neutral process that describes the short rate behaviour is the Ito process:

$$dr = m(r)dt + \sigma(r)dz$$

There are three familiar models based on the Ito process, which offer a single factor term structure model, these are reviewed below:

11.5.1. Rendleman and Bartter

The basic terms in this model are;

$$m(r) = \mu r, \sigma(r) = \sigma r$$

The model assumes that the short rate is lognormal and follows a geometric Brownian motion described by

$$dr = \mu r dt + \sigma r dz \tag{11.5.1}$$

where μ is the constant drift of the instantaneous change in rate and σ is the instantaneous variance of the change in rate. The process is, in this view, the same as that for a change in stock price. The interest rate does not in practice tend to follow the same observed pattern as a stock price. Interest rates exhibit 'mean reversion'. When interest rates rise the tendency is for the short rate to pull the rate towards a more negative value and towards a reversion (overall mean level). When rates fall the opposite reversion tends the short rate towards a more positive level. Models which cannot account for mean reversion are therefore considered less robust than those which do account for the phenomenon.

The Rendleman and Bartter model, is usually implemented as a binomial tree. The parameters used are:

$u = e^{\sigma\sqrt{\Delta t}}$, for the up movement and
$d = e^{-\sigma\sqrt{\Delta t}}$ for the down movement
And probability of an up movement $p = \frac{e^{\mu\Delta t} - d}{u - d}$
To implement this model see Chapter 8.

11.5.2. The Vasicek Model

The **?** model is one which addresses the issue of mean reversion. The model is a yield based one which assumes that interest rates are normal. The fundamental process is:

$$m(r) = a(\theta - r), \sigma(r) = \sigma$$

The Vasicek, risk neutral process is described by:

$$dr = a(\theta - r)dt + \sigma dz \tag{11.5.2}$$

where a, θ, σ are all constants. The short rate is pulled to a mean reversion level θ at a rate a.

If we consider the money market account as numeraire, the expected value of a derivative, paying f_T at time T is given by:

$$E\left[e^{-r_a(T-t)} f_T\right]$$

where r_a is the average interest rate over the period $T-t$. If we consider the expectation in terms of the price of a zero coupon bond over the same time periods;

$$M_p(t,T) = E\left[e^{-r_a(T-t)} f_T\right] \quad (11.5.3)$$

The Vasicek model provides an equivalent as:

$$M_p(t,T) = A(t,T)e^{-B(t,T)r(t)} \quad (11.5.4)$$

where $r(t)$ is the value of r at time t. Also,

$$B(t,T) = \frac{1-e^{-a(T-t)}}{a} \quad (11.5.5)$$

$$A(t,T) = e\left[\frac{(B(t,T)-T+t)(a^2\theta-\sigma^2/2)}{a^2} - \frac{\sigma^2 B(t,T)^2}{4a}\right] \quad (11.5.6)$$

Considering again the price of a zero coupon bond being represented by the expectation of the derivative (11.4.3).

Take $R(t,T)$ as the continuously compounded rate over the same time, then for the same time periods, $M_p(t,T) = e^{-R(t,T)(T-t)}$. Which can be rewritten: $R(t,T) = \frac{1}{T-t}\ln M_p(t,T)$. Thus the term structure for interest rates can be derived from r and its risk free process which can be described by:

$$R(t,T) = -\frac{1}{T-t}\ln E\left[e^{-r_a(T-t)}\right] \quad (11.5.7)$$

In Vasicek's model if we put

$$a = 0,\ B(t,T) = T-t,\ A(t,T) = e(\sigma^2(T-t)^3/6$$

we have:

$$R(t,T) = \frac{1}{T-t}\ln A(t,T) + \frac{1}{T-t}B(t,T)r(t) \quad (11.5.8)$$

The entire term structure can therefore be fully described once suitable values for a, θ, σ and r are chosen.

For pricing zero coupon European bonds the Vasicek model can be used with the following equations:

$$c = P(t, \tau)N(h) - XP(t, T)N(h - \sigma p) \tag{11.5.9}$$

$$p = XP(t, T)N(-h + \sigma p) - P(t, \tau)N(-h) \tag{11.5.10}$$

$$h = \frac{1}{\sigma p} \ln\left[\frac{P(t, \tau)}{P(t, T)X}\right] + \frac{\sigma p}{2} \tag{11.5.11}$$

$$\sigma p = B(T, \tau)\sqrt{\frac{\sigma^2(1 - e^{-2a(T-t)})}{2a}} \tag{11.5.12}$$

11.5.3. Cox Ingersoll Ross (C.I.R) Model

The fundamental process is:

$$m(r) = a(\theta - r), \sigma(r) = \sigma\sqrt{r}$$

The risk neutral process from their model is:

$$dr = a(\theta - r)dt + \sigma\sqrt{dz} \tag{11.5.13}$$

The behaviour of the CIR one factor model is very similar to the Vasicek model and the forming equations are similar.

$$M_p(t, T) = A(t, T)e^{-B(t,T)\gamma}$$

However both $B(t, T)$ and $A(t, T)$ are derived with different characteristics. The variable γ is also not a direct dependence on $r(t)$.

The values are derived as:

$$B(t, T) = \frac{2(e^{\gamma(T-t)} - 1)}{(\gamma + a)(e^{\gamma(T-t)} - 1) + 2\gamma} \tag{11.5.14}$$

$$A(t, T) = \left[\frac{2\gamma e^{(a+\gamma)(T-t)/2}}{(\gamma + a)(e^{\gamma(T-t)} - 1) + 2\gamma}\right]^{2a\theta/\sigma^2} \tag{11.5.15}$$

$$\gamma = \sqrt{a^2 + 2\sigma^2} \tag{11.5.16}$$

See Cox, Ingersoll and Ross (1985).

11.6. Arbitrage Free Models

The equilibrium models do not fit current term structures for interest rates. The fit from these models can be made reasonably accurate with the correct fitting of the characteristic from choosing appropriate forming parameters. The drift in an equilibrium model is not usually dependent on time (although see the CIR model). A no arbitrage model is designed to be consistent with the current rate term structure; the input to the model is the current term structure, whereas the output from an equilibrium model is the estimated current term structure. Equilibrium models can have a time dependency added to convert them to no-arbitrage.

11.6.1. The Ho and Lee Model

The Ho and Lee (1986) model is a yield based no arbitrage one. The model assumes a normally distributed short term rate. The short rate drift is time dependent and is therefore arbitrage free with respect to its input (prices). The original Ho and Lee model was based on a binomial tree of bond prices. The two input parameters were short rate standard deviation and the market price of risk for the short rate. The basic model is not adjusted for mean reversion.

The forming equation is:

$$dr = \theta(t)dt + \sigma dz \tag{11.6.1}$$

where $\theta(t)$ is time dependent drift and σ (standard deviation of the short rate) is a constant. The instantaneous short rate standard deviation refelects the choice of θ, which is selected on the basis of it being a fit of the initial structure. θ can be computed by:

$$\theta(t) = F_t(0, t) + \sigma^2 t \tag{11.6.2}$$

where $F(0, t)$ is the instantaneous forward rate at maturity t. Zero coupon bonds and European options can be computed analytically using the Ho and Lee model. The price at time t of a discount bond maturing at time T is given as:

$$P(t, T) = A(t, T)e^{-r(t)(T-t)}$$

and

$$\ln A(t, T) = \ln\left(\frac{P(0, T)}{P(0, t)}\right) + (T-t)\frac{\partial \ln P(0, t)}{\partial t} - \frac{1}{2}\sigma^2 t(T-t)^2$$

The term, $\frac{\partial \ln P(0,t)}{\partial t}$ can be replaced with a term for the instantaneous forward rate with maturity t viewed at point 0. From 15.1.77 this is, $F(0, t)$. The bond price is then expressed by:

$$\ln A(t, T) = \ln\left(\frac{P(0, T)}{P(0, t)}\right) + (T-t)F(0, t) - \frac{1}{2}\sigma^2 t(T-t)^2 \tag{11.6.3}$$

The Ho-Lee formulae for European zero coupon option prices are:

$$c = P(t, \tau)N(h) - X(t, T)N(h - \sigma p) \tag{11.6.4}$$

$$p = XP(t, T)N(h - \sigma p) - P(t, \tau)N(h) \tag{11.6.5}$$

where

$$\sigma p = \sigma(\tau - T)\sqrt{T - t}.$$

$$h = \frac{1}{\sigma p}\ln\left[\frac{P(t, \tau)}{P(t, T)X}\right] + \frac{\sigma p}{2}$$

11.6.2. Hull and White Model

The Hull and White (1990) model is similar to the Ho and Lee model with the addition of mean reversion. The model extends the Vasicek model by:

Extending $dr = a(\theta - r)dt + \sigma dz$, to; $dr = a\left[\frac{\theta(t)}{a} - r\right]dt + \sigma dz$.

The basic term $\theta(t)$ can be computed from:

$$\theta(t) = F_t(0, t) + aF(0, t) + \frac{\sigma^2}{2a}(1 - e^{-2at})$$

The bond price at time T is given by:

$$P(t, T) = A(t, T)e^{-B(t,T)r(t)} \tag{11.6.6}$$

where,

$$B(t, T) = \frac{1 - e^{-a(T-t)}}{a} \tag{11.6.7}$$

$$\ln A(t, T) = \ln\left[\frac{P(0, T)}{P(0, t)}\right] - B(t, T)\frac{\partial p(0, t)}{\partial t} - \frac{v(t, T)^2}{2} \tag{11.6.8}$$

and

$$v(t, T)^2 = \frac{1}{2a^3}\sigma^2(e^{-aT} - e^{-at})^2(e^{2at} - 1)$$

The partial derivative can be replaced with a term for the instantaneous forward rate:

$$\ln A(t, T) = \ln\left[\frac{P(0, T)}{P(0, t)}\right] + B(t, T)F(0, t) - \frac{1}{4a^3}\sigma^2(e^{-aT} - e^{-at})(e^{2at} - 1) \tag{11.6.9}$$

A European option on a zero coupon bond maturing at time T is given by:

$$c = P(0, \tau)N(h) - XP(0, T)N((h - v(T, \tau)) \tag{11.6.10}$$

$$p = XP(0, T)N(-h + v(T, \tau)) - P(0, \tau)N(-h) \tag{11.6.11}$$

where

$$h = \frac{1}{v(T, \tau)}\ln\left[\frac{P(0, \tau)}{P(0, T)X}\right] + \frac{v(T, \tau)}{2}$$

The Vasicek, Ho-Lee and Hull-White models can be computed by a general purpose set of methods that require minimal modifications to implement an individual model.

If we take an initial time as 0, then the following formulae will provide the basis of a general purpose set of algorithms.

For a call option at time 0 that matures in time T:

$$c = SP(0, T)N(h) - XP(0, T)N(h - \sigma p) \tag{11.6.12}$$

where S is the bond face value and:

$$h = \frac{1}{\sigma p} \ln \frac{SP(0, T_m}{XP(0, T)} + \frac{\sigma p}{2} \tag{11.6.13}$$

The put price is given as:

$$XP(0, T)N(-h + \sigma p) - SP(0, T_m)N(-h) \tag{11.6.14}$$

For the Ho and Lee model:

$$\sigma p = \sigma(T_m - T)\sqrt{T}$$

and for the Vasicek/Hull-White models:

$$\sigma p = \frac{\sigma}{a}\left[1 - e^{-a(T_m - T)}\right]\sqrt{\frac{1 - e^{-2aT}}{2a}}$$

Example 11.6

A European call option with three years to expiry has strike price of $100.0 and volatility of 4%. The mean reverting level is 9.5% and the mean reverting rate is 5.5%. The bond has a face value of $110.0 and a four year maturity. The risk free rate is 8.5%.

$$B(t, T) \equiv B(0, 2) = \frac{1 - e^{-0.055(3-0)}}{0.055} = 2.7655$$

$$B(T, \tau) \equiv B(3, 4) = \frac{1 - e^{-0.055(4-3)}}{0.055} = 0.9729$$

$$B(t, \tau) \equiv B(0, 4) = \frac{1 - e^{-0.055(4-0)}}{0.055} = 3.9505$$

$$A(t, T) \equiv A(0, 3) = \exp\left[\frac{(B(0, 3) - 3 + 0)(0.055^2 * 0.095 - 0.035^2/2)}{0.055^2} - \frac{0.035^2 B(0, 3)^2}{4 * 0.055}\right] = 0.9842$$

$$A(t, \tau) \equiv A(0, 4) = \exp\left[\frac{(B(0, 4) - 3 + 0)(0.055^2 * 0.095 - 0.035^2/2)}{0.055^2} - \frac{0.035^2 B(0, 4)^2}{4 * 0.055}\right] = 0.9759$$

$$P(t,T) \equiv P(0,3) = A(0,3)e^{-B(0,3)^*0.085} = 0.7780$$

$$P(t,\tau) \equiv P(0,4) = A(0,4)e^{-B(0,4)^*0.085} = 0.7192$$

$$\sigma p = B(t,\tau)\frac{\sigma}{a}\left[1 - e^{-a(T_m - T)}\right]\sqrt{\frac{1-e^{-2aT}}{2a}} \equiv$$

$$B(3,4)\sqrt{\frac{\sigma^2(1-e^{-2a(T-t)})}{2a}} = 0.9729^*\sqrt{\frac{0.035^2(1-e^{-2^*0.055(3-0)})}{2^*0.055}} = 0.06221$$

$$h = \frac{1}{\sigma p}\ln\frac{sP(0,T_m)}{XP(0,T)} + \frac{\sigma p}{2} \equiv \frac{1}{0.06221}\ln\frac{110^*P(0.4)}{100^*P(0,3)} + \frac{0.06221}{2} = 0.2996$$

$$c = (F^*P(0,4)N(h) - X^*P(0,3)N(h-\sigma p)) = 2.672.$$

The call option price is $2.67.

Listing 11.1 gives the class **Vasiceckop** which implements the Vasiceck algorithm. The code which runs Example 11.10 is appended.

```
package FinApps;
import static java.lang.Math.*;
import BaseStats.Probnorm;
public class Vasicekop {

  public Vasicekop(double meanrev,double revlevel,
                double volatility, double starttime) {
    a=meanrev;
    theta=revlevel;
    sigma=volatility;
    start=starttime;
  }
  private double a;
  private double theta;
  private double sigma;
  private double start;
  private double pstart;
  private double pmat;
  private double h;
  private double hw;
  private double bondvol;
  private double bondvolw;
  private double btstart;
  private double bmaturity;
  double bexpiry;
```

```
    private void vasiParams(double f, double x, double rate,
                            double time,double tmaturity ) {

      btstart=((1.0-exp(-a*(time-start)))/a);
      bexpiry=((1.0-exp(-a*(tmaturity-time)))/a);
      bmaturity=((1.0-exp(-a*(tmaturity-start)))/a);
      double startat1=((btstart-time+start)*(((a*a)*theta)
                      -((sigma*sigma)*0.5))/(a*a));
      double startat2=(((sigma*sigma)*(btstart*btstart))/(4*a));
      double starta=exp(startat1-startat2);
      double matat1=((bmaturity-tmaturity+start)*(((a*a)*theta)
                      -((sigma*sigma)*0.5))/(a*a));
      double matat2=(((sigma*sigma)*(bmaturity*bmaturity))/(4*a));
      double mata=exp(matat1-matat2);
      pstart=(starta*exp(-btstart*rate));
      pmat=(mata*exp(-bmaturity*rate));
      bondvol=(bexpiry*(sqrt((sigma*sigma)
                      *(1.0-exp(-2*a*(time-start)))/(2*a))));
           h=((1.0/bondvol)*log((pmat*f)/(pstart*x))+(bondvol*0.5));
      }
    public double vasiCall(double f, double x, double rate,
                           double time,double tmaturity ) {
      Probnorm p=new Probnorm();
      vasiParams(f,x,rate,time,tmaturity);
           return ((f*pmat*p.ncDisfnc(h))
                  -(x*pstart*p.ncDisfnc(h-bondvol)));
    }
    public double vasiPut(double f, double x, double rate,
                          double time,double tmaturity ) {
      Probnorm p=new Probnorm();
      vasiParams(f,x,rate,time,tmaturity);
           return ((x*pstart*p.ncDisfnc(-h+bondvol))
                  -(f*pmat*p.ncDisfnc(-h)));
    }
    public static void main(String[] args) {

      Vasicekop v= new Vasicekop(0.055,0.095,0.04,0.0);
      double returnvalue=v.vasiCall(110.0,100.0,0.085,3.0,4.0);
      System.out.println(" CALL =="+returnvalue);
    }
}
```

LISTING 11.1. Implementation of class to compute Vasiceck algorithm

References

Black, F. (1976). "The Pricing of Commodity Contracts". *Journal of Financial Economics*, 3, 167–179.

Vasicek, O. (1977). "An Equilibrium Characterization of the Term Structure". *Journal of Financial Economics*, 5, 177–188.

Cox, J. C., J. E. Ingersoll, and S. A. Ross (1985). "A Theory of the Term Structure of Interest Rates". *Econometrica*, 53(2), 385–407.

Hull, J. and A. White (1990). "Pricing Interest Rate Derivative Securities". *Review of Financial Studies*, 3, 573–592.

12
Conditional Options

Conditional options are variations on standard ('plain vanilla') options, where the ultimate payoff is derived from a model condition imposed on the underlying or the value of another asset class. Conditional options can be considered as the basic subset of exotic options. Exotic options are a class of derivative that have been developed to meet the needs of particular trading environments. Standard put and call options (and some exotic types) are traded on major exchanges with prices and volatilities being quoted, to reflect demand and risk. In the more sophisticated OTC market many derivatives are specifically designed to address a very particular set of circumstances, these exotic products range from being reasonably simple options based on the well understood Black-Scholes process to complex options, which assume a different economy to the Black-Scholes one. In this chapter and those which follow we will only be considering the case of products that have assets which follow Geometric Brownian motion and subsequently the option generally follows a Black-Scholes type process.

This and subsequent chapter's largely follows the sequence of option taxonomy as originally described by Rubinstein & Reiner in their series of papers from 1991 to 1992. The papers are referenced where appropriate. This taxonomy and further extensions provided by the excellent guide from Haug (1998) largely influence the sequence of discussion and presentation.

12.1. Executive Stock Options

Executive stock options are a vehicle used to attract and retain key individuals in an organisation. Typically stock options are offered to an employee, where the options can be exercised only at some future date. The rationale is to tie in the employee with a call option to gain in some future time. If the employee leaves within a specified timeframe (the so-called vesting period), the option is cancelled. After the vesting period the option can be exercised at any time within its available maturity time. If the employee leaves the organisation after the vesting period, the option is immediately exercised, if it is in-the-money. Options that are out-of-the-money are forfeit. Executive options are also widely used to 'incentivise' manager's to achieve a high shareholder worth to an organisation. Options can be tied to the stock price performance and paid when a given performance barrier is exceeded. Options used in this way are part of compensation packages and have to be valued according to regulatory guidelines.

An executive stock option cannot be resold, but needs to be exercised so that the executive is required to sell the underlying company stock. The valuation of an executive stock option can therefore be based on a model of the behaviour for an early exercise decision.

The executive option suggested by Jennergren and Naslund (1993) takes into account the behaviour of the employee in leaving within the vesting period as a probability measure (the annual'jump rate'). The jump rate probability measure is used to modify a standard Black-Scholes formula for valuing this executive option.

The basic formula used is:

$$c = e^{-\lambda T}\left(Se^{(b-r)T}N(d_1) - Xe^{-rT}N(d_2)\right) \qquad (12.1.1)$$

$$P = e^{-\lambda T}\left(Xe^{-rT}N(-d_2) - Se^{(b-r)T}N(-d_1)\right) \qquad (12.1.2)$$

Where λ is the annual jump rate and ;

$$d_1 = \frac{\ln(S/X) + (b = \sigma^2/2)T}{\sigma\sqrt{T}}$$

$$d_2 = d_1 - \sigma\sqrt{T}$$

Example 12.0

A three year to maturity executive stock option is to be issued when the stock price is £50.0, the strike price is £60.0 and the underlying asset volatility is 30%.The stock pays an annual dividend of 3% and the risk-free rate is 5%. If the annual jump rate is 12% what is the value?

$$S = 50.0, X = 60.0, T = 3.0, r = 0.05, q = 0.03, \sigma = 0.30, \lambda = 0.12.$$

Using:

$$d_1 = \frac{\log(S/X) + (b + \sigma^2/2)T}{\sigma\sqrt{T}}$$

$$= \frac{\log(50.0/60.0) + (0.02 + 0.30^2/2)*3.0}{0.30*\sqrt{3.0}}$$

$$= 0.0243$$

$$d_2 = d_1 - \sigma\sqrt{T} = 0.0243 - 0.30*\sqrt{3.0} = -0.495$$

$$N(d_1) = N(0.02430) = 0.509, N(d_2) = N(-0.4950) = 0.310$$

$$e^{-\lambda T} = e^{-0.12*3.0} = 0.6976$$

$$c = e^{-\lambda T}\left(Se^{(b-r)T}N(d_1) - Xe^{-rT}N(d_2)\right)$$

$$C = 0.6976^*\left[50.0^* e^{(0.02-0.05)*3.0^*}0.509 - 60.0^* e^{-0.05*3.0^*}0.310\right] = 5.0737$$

The value is £5.07.

The value given above is the same as

$$e^{\lambda T^*}C_{BS} = 0.6976^*7.272 = 5.0737.$$

The term in brackets of 12.1.1, is the standard Black-Scholes call formula. The code for implementing executive stock options is shown in Listing 12.1

```
package FinApps;
import static java.lang.Math.*;
import BaseStats.Probnorm;
import java.text.*;
public class Execoption {
   public Execoption(double jrate) {
      jump=jrate;
   }
   private double jump;
   private double callprice;
    private double callpricefm;
   private double putprice;
   public double getExcall() {
      return callprice;
   }
   public double getExcallfm() {
      return callpricefm;
   }
   public double getExput() {
      return putprice;
   }
   private void setCall(double call) {
      callprice=call;
   }
   private void setCallfm(double call) {
      callpricefm=call;
   }
   private void setPut(double put) {
      putprice=put;
   }

   public void execOpt(double s, double x, double volatility,
                          double time, double rate, double yield) {
                                //Jennergren & Naslund (1993)
      Blackscholecp b=new Blackscholecp(yield);
      b.bscholEprice(s,x,volatility,time,rate);
      setCall((exp(-jump*time)*(b.getCalle())));
      setPut((exp(-jump*time)*(b.getPute())));
   }
```

```
public void execOptfm(double s1, double s2, double r,
                      double sig1,double sig2,double time) {
                      // after Fischer- Margrabe (1978) for index
                      //-linked compensation
    Probnorm p = new Probnorm();
    double sigs=((sig1*sig1)-2.0*jump*sig1*sig2+(sig2*sig2));
    double sigma=sqrt(sigs);
    double d1=((log(s1/s2)+(sigs*time))/(sigma*sqrt(time)));
    double d2=(d1-sigma*sqrt(time));
    double n=p.ncDisfnc(d1);
    double n2=p.ncDisfnc(d2);
    double c=(s1*n-s2*n2);
    setCallfm(c);
}

public static void main(String[] args) {
    Execoption e=new Execoption(0.12);
    e.execOpt(50.0,60.0,0.30,3.0,0.05,0.03);
     System.out.println("ANS=="+e.getExcall());
}
}
```

LISTING 12.1. Executive Stock Option

12.1.1. Forward Start Option

An executive stock option can also be represented as a forward start option since there is commitment to granting an at-the-money option at some future time.

A forward start option will be at-the-money or out/in-the-money at a given time in the future. If we consider a forward start European call option, which is at-the-money and has maturity T and starts at t. If the asset price at t is denoted as S_1 and at time zero is S_0 the value of the option will be proportional to the asset price S_1/S_0, times the call price at time zero (for an option with maturity $(T-t)$).

Rubinstein (1990) has developed a formula based on the above, where the strike price is represented by a constant (α) times the asset price ratio. If the option is at -the-money (as for an executive option) the constant is set to unity, if the option is in-the-money, the constant is set less than unity and for out-of-the-money, it is set at greater than unity.

The forward start option formulae are:

$$c = Se^{(b-r)t}\left(e^{(b-r)(T-t)}N(d_1) - \alpha e^{-r(T-t)}N(d_2)\right) \tag{12.1.3}$$

$$P = Se^{(b-r)t}\left(\alpha e^{-r(T-t)}N(-d_2) - e^{(b-r)(T-t)}N(-d_1)\right) \tag{12.1.4}$$

where

$$d_1 = \frac{\ln(1/\alpha) + (b+\sigma^2/2)(T-t)}{\sigma\sqrt{T-t}}$$

and

$$d_2 = d_1 - \sigma\sqrt{T-t}$$

Example 12.1

What is the value of a call option with a forward start in four months time assuming the option starts at-the-money and has a maturity of one year, the stock price is \$50.0, the risk free rate is 6% and the continuous dividend yield is 3%, with an expected volatility of 25% ?

$$S = 50, T = 1.0, r = 0.06, b = 0.06 - 0.03 = 0.03, \sigma = 0.25, \alpha = 1.0$$

$$d_1 = \frac{\ln(1/\alpha) + (b + \sigma^2/2)(T-t)}{\sigma\sqrt{T-t}}$$

$$= \frac{\ln(1/1) + (0.03 + 0.25^2/2)(1 - 0.333)}{0.25\sqrt{1-0.333}}$$

$$= 0.20$$

$$d_2 = d_1 - \sigma\sqrt{T-t} = 0.5792 - 0.25^*\sqrt{1-0.333} = -0.004$$

$$N(d_1) = N(0.20) = 0.4207$$

$$N(d_2) = N(-0.004) = 0.5016$$

$$c = Se^{(b-r)t}\left(e^{(b-r)(T-t)}N(d_1) - \alpha e^{-r(T-t)}N(d_2)\right) =$$

$$c = 50e^{(0.03-0.06)0.333}\left(e^{(0.03-0.06)(1.0-0.333)}N(d_1) - 1.0e^{-0.06(1.0-0.333)}N(d_2)\right)$$

$$= 4.4057$$

The call option is worth \$4.40.

```
package FinApps;
import BaseStats.Probnorm;
import static java.lang.Math.*;
public class Forstartop {

   public Forstartop(double carryrate) {
      crate=carryrate;
   }
    public double getCalle() {
      return callprice;
   }
   public double getPute() {
      return putprice;
   }
   private void setcalle(double call) {
      callprice=call;
   }
   private void setpute(double put) {
```

```
        putprice=put;
    }
    private double crate=0.0;
    private double brate=0.0;
    private double d1=0.0;
    private double d2=0.0;
    private double callprice=0.0;
    private double putprice=0.0;
    private void dvalues(double sprice,double alpha,double volatility,
                        double time, double tmaturity,double rate) {
        brate=crate<0.0?0.0:(brate=crate!=rate?(rate-crate):rate);
        d1=((log(1.0/alpha)+(brate+(volatility*volatility)*0.5)*
             (tmaturity-time))
             /(volatility*sqrt(tmaturity-time)));
        d2=(d1-(volatility*sqrt(tmaturity-time)));
    }
    public void fstartOp(double sprice,double alpha,double volatility,
                        double time,double tmaturity, double rate) {

        Probnorm p=new Probnorm();
        dvalues(sprice,alpha,volatility,time,tmaturity,rate);
        double probd1=0.0;
        double probd2=0.0;
        double probdn1=0.0;
        double probdn2=0.0;
        probd1=p.ncDisfnc(d1);
        probd2=p.ncDisfnc(d2);
        probdn1=p.ncDisfnc(-d1);
        probdn2=p.ncDisfnc(-d2);
        setcalle(sprice*exp((brate-rate)*time)*
                ((exp((brate-rate)*(tmaturity-time))*probd1)
                -(alpha*exp(-rate*(tmaturity-time))*probd2)));
        setpute(sprice*exp((brate-rate)*time)*
                ((alpha*exp((-rate)*(tmaturity-time))*probdn2)
                -(exp((brate-rate)*(tmaturity-time))*probdn1)));

}
```

LISTING 12.2. Forward start option

12.1.2. Indexed Stock Options

An indexed stock option is an executive 'compensation' device, where the package is related not to the nominal fixed value strike price of the underlying company stock; rather it is tied to the performance of the organisation's share value in relation to the exchange index. In this way the company's performance is measured against the index as a whole. Schnusenberg & McDaniel (2000) suggest a valuation methodology based on the Fischer-Margrabe formulae for pricing calls on index bonds (Fischer 1978) and the pricing of an option to exchange one option for another (Margrabe 1978). (The latter option is covered in a following section.) The model does not take account of stock paying any dividends.

The indexed stock option model has applicability to both executive compensation schemes and for a more rigorous compliance with standards accounting practices for company reporting. Using a standard Black-Scholes pricing method, executive indexed options are shown to be consistently overvalued, resulting in accounting anomalies with financial reporting when the options are exercised. The overvaluation of indexed options when using a standard Black-Scholes method is very susceptible to an accelerating rate of increase with time to maturity.

From a shareholder's perspective the advantage of tying executive compensation to an index is that the option only pays when the company's stock outperforms the rest of the index. From the executive perspective, a payout is still possible if a bear market, reduces the index value, but the company performance is still ahead of a falling market.

The valuation formula is given by a call:

$$C(S,X,T) = SN(d_1) - XN(d_2) \tag{12.1.5}$$

where:

$$d_1 = \frac{\log(S/X) + \sigma^2 T}{\sigma\sqrt{T}} \tag{12.1.6}$$

$$d_2 = d_1 - \sigma\sqrt{T} \tag{12.1.7}$$

$$\sigma^2 = \sigma_S^2 - 2\rho_{SX}\sigma_X\sigma_S + \sigma_X^2 \tag{12.1.8}$$

$N(.)$ is the standard normal density function.

S The current market price of company stock

X The current value of the indexed-exercise price: given by the 'fair' determination of exercise price = a % of the index value.

T The number of years to maturity of the stock option.

σ_X The instantaneous standard deviation of the underlying index; solved from an iterative procedure on the Black-Scholes call option valuation of the index.

σ_S The instantaneous standard deviation of the company common stock; solved from an iterative procedure on the Black-Scholes call option valuation of the stock.

σ The instantaneous proportional standard deviation of the change in stock to exercise price.

ρ Is the instantaneous correlation coefficient between stock and exercise price.

Listing 12.3 shows class ***Execoption***, which provides two methods to compute valuations.

Exercise 12.2

An executive stock option is to be issued, based on the current stock price of ABC company as a percentage of the XYZ 500 index. The current price of ABC stock is $164.0 which is around 8% of the XYZ index. The XYZ is currently at 2,000.0. The strike is therefore 2,000/8 = $160.0. The risk-free rate (taken as the current one year T-Bill yield) is 5%. The value on the day, of a Black-Scholes standard call on the option stock implies an instantaneous volatility for the stock to be 0.45. Similarly a call option on the XYZ implies a volatility of 0.20. Based on historical daily price comparisons for the previous year, the correlation between ABC and XYZ stock is found to be 0.90.

1. What is the standard Black Scholes value of this stock option?
2. What is the correct valuation?

The standard Black-Scholes valuation with:

$$S = 164.0, X = 160.0, r = 0.05, \sigma_S = 0.45, T = 1.0$$

is, $34.426.

The index-strike adjusted valuation is given as:

$$\sigma^2 = \sigma_S^2 - 2\rho_{SX}\sigma_X\sigma_S + \sigma_X^2 = 0.45_S^2 - 2{*}0.90_{SX}{*}0.20_x{*}0.45_S + 0.20_X^2 = 0.0805$$

$$d_1 = \frac{\log(S/X) + \sigma^2 T}{\sigma\sqrt{T}} = \frac{\log(164.0/160.0) + 0.0805{*}1.0}{0.2837{*}\sqrt{1.0}} = 0.3707$$

$$d_2 = d_1 - \sigma\sqrt{T} = 0.3707 - 0.2837{*}\sqrt{1.0} = 0.0870$$

$$C(S, X, T) = SN(d_1) - XN(d_2) = 164.0{*}0.644 - 160.0{*}0.534 = 20.164$$

The ‘correct’ price is $20.164 showing that the Black-Scholes standard valuation for an executive stock option is overvalued by around 70%.

The effect of time to maturity of the option is shown in Figure 12.1.

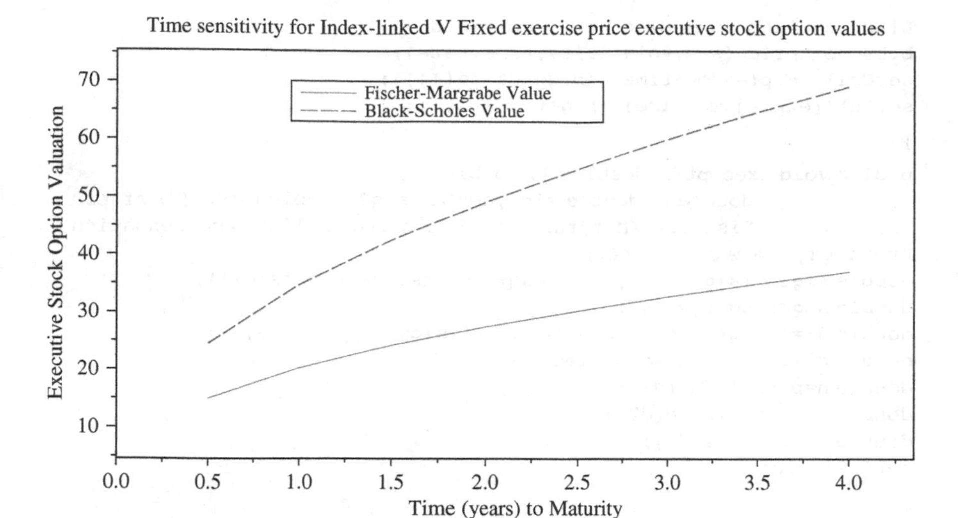

FIGURE 12.1. Simple chooser option value.

```
import static package FinApps;
java.lang.Math.*;
import BaseStats.Probnorm;
public class Execoption {
public Execoption(double jrate) {
jump=jrate;
}
private double jump;
private double callprice;
private double callpricefm;
private double putprice;
public double getExcall() {
return callprice;
}
public double getExcallfm() {
return callpricefm;
}
public double getExput() {
return putprice;
}
private void setCall(double call) {
callprice=call;
}
private void setCallfm(double call) {
callpricefm=call;
}
private void setPut(double put) {
putprice=put;
}

public void execOpt(double s, double x, double volatility, double time,
            double rate, double yield) {//Jennergren & Naslund (1993)
```

```
Blackscholecp b=new Blackscholecp(yield);
b.bscholEprice(s,x,volatility,time,rate);
setCall((exp(-jump*time)*(b.getCalle())));
setPut((exp(-jump*time)*(b.getPute())));
}
public void execOptfm(double s1, double s2,
          double r, double sig1,double sig2,double time) {// after
          Fischer-//Margrabe (1978) for index -linked compensation
Probnorm p = new Probnorm();
double sigs=((sig1*sig1)-2.0*jump*sig1*sig2+(sig2*sig2));
double sigma=sqrt(sigs);
double d1=((log(s1/s2)+(sigs*time))/(sigma*sqrt(time)));
double d2=(d1-sigma*sqrt(time));
double n=p.ncDisfnc(d1);
double n2=p.ncDisfnc(d2);
double c=(s1*n-s2*n2);
setCallfm(c);
}
```

LISTING 12.3. Executive Option valuation

12.2. Time Switch Option

A time switch option provides the investor with an amount proportional to the time a stock price deviates from the strike price of the option. In a discrete call time switch, the investor receives a payoff that reflects the time at maturity for which the stock price has exceeded the strike price:

$$value = A^{*}\Delta t, forS_{i\Delta_t} > X$$

For a discrete time switch put option:

$$value = A^{*}\Delta t, forS_{i\Delta_t} < X$$

Pechtl (1995) has developed a series of formulas to price discrete and continuous time switch options for the discrete case.

$$c = Ae^{-rT}\sum_{i=1}^{n} N\left(\frac{\ln(S/X)+(b-\sigma^2/2)i\Delta t}{\sigma\sqrt{i\Delta t}}\right)\Delta t + (mAe^{-rT}\Delta t) \qquad (12.2.1)$$

$$P = Ae^{-rT}\sum_{i=1}^{n} N\left(\frac{-\ln(S/X)-(b-\sigma^2/2)i\Delta t}{\sigma\sqrt{i\Delta t}}\right)\Delta t + (mAe^{-rT}\Delta t) \qquad (12.2.2)$$

Where

$n = T/\Delta t$ and $(mAe^{-rT}\Delta t)$ is a term added to adjust for any period m of the options prior lifetime.

Class ***Timeswop***, provides time switch methods as shown in Listing 12.4.

Example 12.3

Price a call time switch option that has one year to maturity. The accumulation rate is 2 for each day of the year (365 day year) that the stock price is in excess of the strike price which is \$105.0. The initial stock price is \$95.0, the volatility of the stock is 30% per annum, the risk free rate is 7%. There is no prior period for which the option has fulfilled its accumulation condition.

S = 95.0, X = 105.0, r = 0.07, σ = 0.30, T = 1.0, A = 2, n = 365, Δt = 1/365.

$$c = Ae^{-rT}\sum_{i=1}^{n} N\left(\frac{\ln(S/X)+(b-\sigma^2/2)i\Delta t}{\sigma\sqrt{i\Delta t}}\right)\Delta t + 0.0$$

$$= 1.8647^{*}\sum_{i=1}^{n} N\left(\frac{\ln(95.0/105.0)+(0.07-0.30^2/2)i^{*}1/365}{0.30\sqrt{i^{*}1/365}}\right)1/365 + 0.0$$

$$= 0.5683$$

where Σ N*Δt = 0.3048 is the value shown in the final row (for i = 365) of Table 12.1.

```
package FinApps;
import static java.lang.Math.*;
import BaseStats.Probnorm;
public class Timeswop {
   public Timeswop(int mperiod,int dayterm,double yield) {
      m=mperiod;
      crate=yield;
      daycount=dayterm;
   } private int m;
   private double q;
   private int daycount;
   private double brate=0.0;
   private double crate=0.0;
   public double cTswitch(double s, double x,
double accumulate,double tmaturity,double rate,double volatility)
  {
      Probnorm p=new Probnorm();
      brate=crate<0.0?0.0:(brate=crate!=rate?(rate-crate):rate);
      double deltat=(1.0/daycount);
      int n=(int)(tmaturity/deltat);
                              //discards fraction..rounds down to o
      double d=0.0;
      double sum=0.0;
      double call=0.0;
      double prevalue=(deltat*accumulate*exp(-rate*tmaturity)*m);
      for(int i=1;i<n+1;i++)
      {
         d=((log(s/x)+(brate-((volatility*volatility)*0.5))*i*deltat)
                              /(volatility*sqrt(i*deltat)));
```

```
        sum+=(p.ncDisfnc(d)*deltat);
      }
      return (accumulate*exp(-rate*tmaturity)*sum+prevalue);
}
public double pTswitch(double s, double x,double accumulate,
                   double tmaturity, double rate,double volatility)
  {
        Probnorm p=new Probnorm();
      brate=crate<0.0?0.0:(brate=crate!=rate?(rate-crate):rate);
      double deltat=(1.0/daycount);
      int n=(int)(tmaturity/deltat);
                         //discards fraction..rounds down to o
      double d;
      double sum=0.0;
      double put=0.0;
      double prevalue=(deltat*accumulate*exp(-rate*tmaturity)*m);
      for(int i=1;i<n+1;i++)
      {
        d=((-log(s/x)-(brate-((volatility*volatility)*0.5))*i
                        *deltat)/(volatility*sqrt(i*deltat)));
        sum+=(p.ncDisfnc(d)*deltat);
      put=(accumulate*exp(-rate*tmaturity)*sum+prevalue);
      }
    return put;
}
```

LISTING 12.4. Valuation of Time Switch options

12.3. Chooser Option

A chooser option gives the holder a right to choose at some time after purchasing an option to make it a call or put. For a standard chooser, the right can be exercised after a pre-determined time to make the option a standard European put or call with the same strike price and remaining time to maturity. Chooser options tend to be more expensive than the equivalent standard call or put options, since the inclusion of choice represents a form of premium.

The structure of a simple chooser option is similar to that of a straddle (recall that this is a simultaneous position in a call and put).The chooser option exhibits a higher price as time to choice increases and although the option is structured as a straddle, its cost is somewhat less than a straddle.

Chooser options are generally accepted as first appearing in the OTC market by Bankers Trust, the original options were American call or put derivatives.

The payoff from a standard chooser is given as:

$$\max\left[c(S, X, T), p(S, X, T)\right]$$

where $c(S_t, X, T)$ is the plain vanilla call value and $P(S, X, T)$ is the value of a plain vanilla put option with time T being the time to option maturity and t the time at which choice is made.

TABLE 12.1. The running values from Equation 12.2.1 with parameters as in Example 12.3

i	i*Δt	d	N(d)	ΣN*Δt
1	0.0027	−6.3693	0.000	0.000
2	0.0055	−4.5007	0.000	0.000
3	0.0082	−3.6723	0.0001	0.000
4	0.011	−3.1781	0.0007	0.000
5	0.0137	−2.8406	0.0023	0.000
6	0.0164	−2.5913	0.0048	0.000
7	0.0192	−2.3975	0.0083	0.000
8	0.0219	−2.2411	0.0125	0.0001
9	0.0247	−2.1115	0.0174	0.0001
10	0.0274	−2.0017	0.0227	0.0002
11	0.0301	−1.9073	0.0282	0.0003
12	0.0329	−1.8248	0.034	0.0004
13	0.0356	−1.752	0.0399	0.0005
14	0.0384	−1.6871	0.0458	0.0006
15	0.0411	−1.6288	0.0517	0.0007
16	0.0438	−1.576	0.0575	0.0009
17	0.0466	−1.5279	0.0633	0.0011
18	0.0493	−1.4838	0.0689	0.0013
19	0.0521	−1.4432	0.745	0.0015
20	0.0548	−1.4057	0.0799	0.0017
....		....	....	....
....		....	....	....
....		....	....	....
345	0.9452	−0.2621	0.3966	0.2829
346	0.9479	−0.2615	0.3968	0.284
347	0.9507	−0.2609	0.3971	0.2851
348	0.9534	−0.2603	0.3973	0.2862
349	0.9562	−0.2597	0.3976	0.2872
350	0.9589	−0.2591	0.3978	0.2883
351	0.9616	−0.2585	0.398	0.2894
352	0.9644	−0.2579	0.3982	0.2905
353	0.9671	−0.2573	0.3985	0.2916
354	0.9699	−0.2567	0.3987	0.2927
355	0.9726	−0.2561	0.3989	0.2938
356	0.9753	−0.2555	0.3992	0.2949
357	0.9781	−0.2549	0.3994	0.296
358	0.9808	−0.2543	0.3996	0.2971
359	0.9836	−0.2537	0.3998	0.2982
360	0.9863	−0.2532	0.4001	0.2993
361	0.989	−0.2526	0.4003	0.3004
362	0.9918	−0.252	0.4005	0.3015
363	0.9945	−0.2514	0.4007	0.3026
364	0.9973	−0.2509	0.401	0.3037
365	1.000	−0.2503	0.4012	0.3048

12.3.1. Simple Chooser

Rubinstein (1991) has developed a formula based on the above to provide a valuation based on the put-call parity relationship:

$$\max(c(S_t, X, T-t), (c(S_t, X, T-t) - S_t q^{T-t} + Xr^{-(T-t)}))$$

which reduces to:

$$c(S_t, X, T) + \max(0, Xr^{-(T-t)} - S_t q^{(T-t)}; t) \tag{12.3.1}$$

where q is the dividend yield (or foreign interest rate) and r is the risk free rate.

The simple chooser option can now be priced in terms of plain vanilla calls and puts, where the positions taken are long a call and short a put.

Taking $t_2 = (T-t)$, $t_1 = t$ and the cost of carry rate is b

$$V = Se^{(b-r)t_2} N(d) - Xe^{-rt_2} N(d - \sigma\sqrt{t_2}) - Se^{(b-r)t_2} N(-y) + Xe^{-rt_2} N(-y + \sigma\sqrt{t_1}) \tag{12.3.2}$$

Where

$$d = \frac{\log(S/X) + (b + \sigma^2/2)t_2}{\sigma\sqrt{t_2}}$$

$$y = \frac{\log(S/X) + bt_2 + \sigma^2 t_1/2}{\sigma\sqrt{t_1}}$$

Figure 12.2. shows the characteristic for a simple chooser option when the time to make a choice varies. The basic parameters are:

$$\mathrm{S} = £60.0, \mathrm{X} = £60.0, \mathrm{T} = 1.0, \mathrm{q} = 3\%, \mathrm{r} = 7\%, \sigma = 24\%$$

The graph shows that as time to make a choice increases, the value of the option increases. The graph shown in Figure 12.3. shows the relative change in option value around the strike price. This clearly shows that the option value is relatively high either side of the point at which the strike price is equal to the asset price.

Although Figure 12.2. does show that time to make a choice does have an effect on the base value of the Chooser option, the relative increase is proportionally small as we reach points away from the strike /stock price equivalence price.

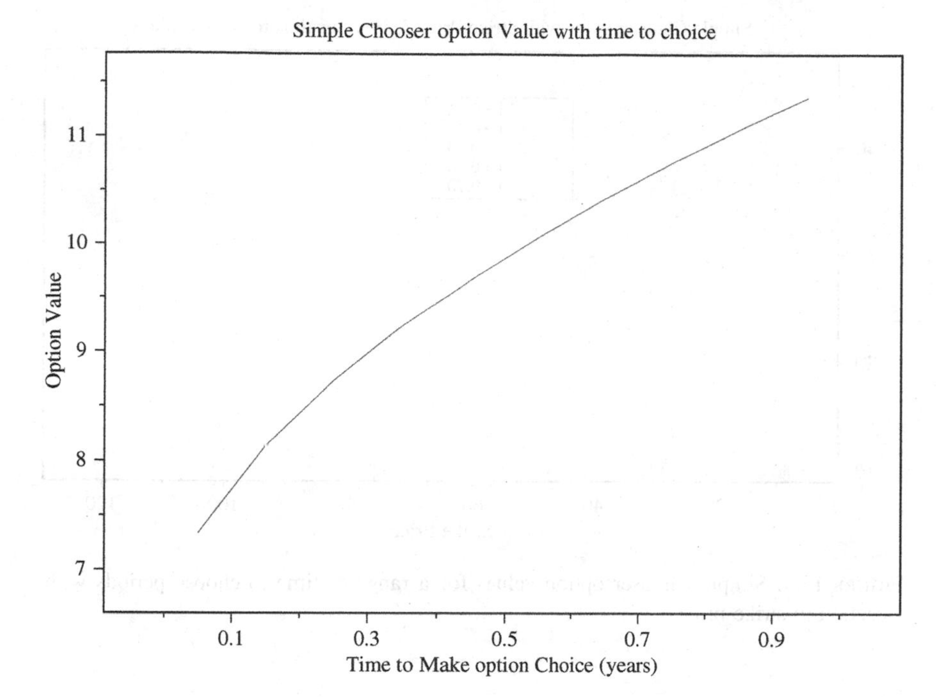

FIGURE 12.2. Simple choose value with variation in time to choose.

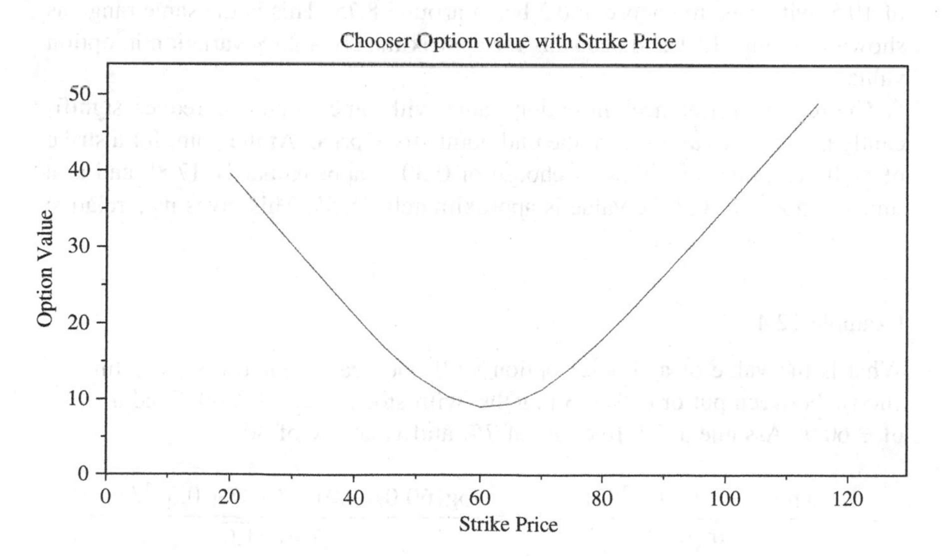

FIGURE 12.3. Simple Chooser option with variation in strike price.

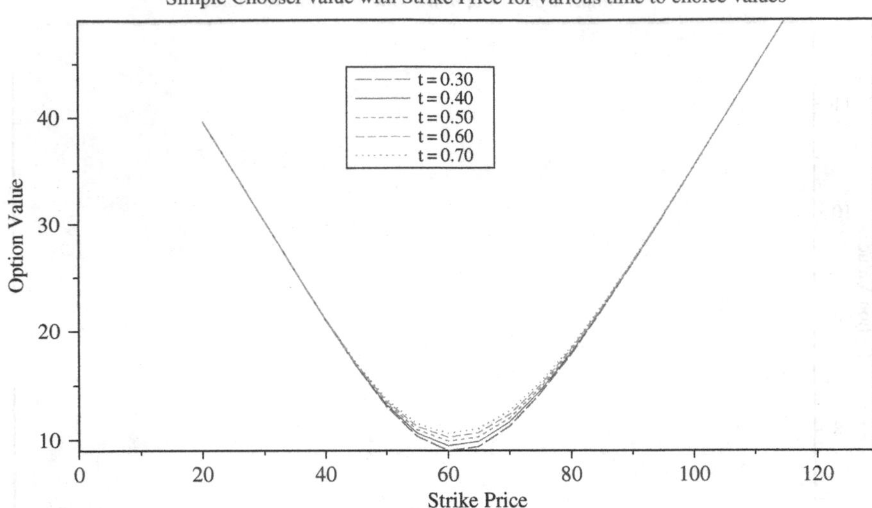

FIGURE 12.4. Simple Chooser option values for a range of time to choose periods with variation in strike price.

Figure 12.4. shows the relative effect of time to choose when superimposed on the graph of changing strike prices. Note that at the mid point the graph is reasonably symmetric at which point the different values for the range of time to choose periods can be more clearly seen. At this point the relative differences can be seen as significant with time to choose at 0.7 being approximately a value of 10.5, with time to choose at 0.3 being around 8.75. This is the same range as shown in Figure 12.1 and accounts for approximately a 25% variation in option value.

The relative differences in option value with time to choose reduce significantly as we move away from the mid point strike price. At the point for a strike of 80.0 the value for a time to choose of 0.30 is approximately 17.88 and at a time to choose of 0.7 the value is approximately 18.55. This gives us a relative change of approximately 3.5%.

Example 12.4

What is the value of a chooser option with one year to maturity and a time to choose between put or call of 3 months, with stock price of € 60.0 and a strike of € 60.0? Assume a risk free rate of 7% and volatility of 30%.

$$d = \frac{\log(S/X) + (b + \sigma^2/2)t_2}{\sigma\sqrt{t_2}} = d = \frac{\log(60.0/60.0) + (0.07 + 0.30^2/2)1.0}{0.30\sqrt{1.0}}$$

$$= 0.405$$

$$y = \frac{\log(S/X) + bt_2 + \sigma^2 t_1/2}{\sigma\sqrt{t_1}} = \frac{\log(60.0/60.0) + 0.07^*1.0 + 0.30^{2^*}0.25/2}{0.30\sqrt{1.0}}$$
$$= 0.6225$$

$$N(d) = N(0.405) = 0.657$$
$$N(d - \sigma\sqrt{t_2}) = N(0.405 - 0.30^*\sqrt{1.0}) = 0.5615$$
$$N(-y + \sigma\sqrt{t_1}) = N(-0.6225 + 0.30^*\sqrt{0.25}) = 0.3094$$
$$N(-y) = N(-0.6225) = 0.266$$

$$V = Se^{(b-r)t_2}N(d) - Xe^{-rt_2}N(d - sigma\sqrt{t_2})$$
$$- Se^{(b-r)t_2}N(-y) + Xe^{-rt_2}N(-y + \sigma\sqrt{t_1}) =$$

$$V = 60.0e^{(0.07-0.07)*1}N(d) - 60.0e^{(-0.70*1)}N(d - 0.30\sqrt{1})$$
$$- 60.0e^{(0.07-0.07)1}N(-y) + 60.0e^{0.07*1}N(-y + 0.30\sqrt{0.25})$$
$$= 9.3199$$

Listing 12.5. shows the implementation of code for valuing simple chooser options.

```
package FinApps;
import static java.lang.Math.*;
import BaseStats.Probnorm;
public class Chosersimple {
   public Chosersimple(double yield, double choicetime) {
      crate=yield;
      time=choicetime;
   }
   private double crate;
   private double time;
   private double brate=0.0;
   public double simpChose(double s, double x, double tmaturity,
                                double rate, double volatility)
   {
      Blackscholecp b=new Blackscholecp(crate);
      Probnorm p=new Probnorm();
      b.bscholEprice(s,x,volatility,tmaturity,rate);
      double call=b.getCalle();
      double put=b.getPute();
      brate=crate==0.0?0.0:(brate=crate!=rate?(rate-crate):rate);
      double d;
      double y;
      d=((log(s/x)+(brate+(volatility*volatility)*0.5)
```

```
                *tmaturity)/(volatility*sqrt(tmaturity)));
    y=((log(s/x)+(brate*tmaturity)+((volatility*volatility)
                *time)*0.5)/(volatility*sqrt(time)));
    double probd=p.ncDisfnc(d);
    double probdmv=p.ncDisfnc(d-volatility*sqrt(tmaturity));
    double probdmy=p.ncDisfnc(-y+volatility*sqrt(time));
    double proby=p.ncDisfnc(-y);
    double w=(s*exp((brate-rate)*tmaturity)*probd-x*exp
                  (-rate*tmaturity)*probdmv-s*exp
                  ((brate-rate)*tmaturity)*proby+x*
                  exp(-rate*tmaturity)*probdmy);

    return w;
}
```

LISTING 12.5. Simple Chooser Option Valuation

12.3.2. Complex Chooser Options

A complex chooser described by Rubinstein (1991) follows the same basic logic of the simple chooser, with the added complexity that both the strike price and time to maturity of the put or call may be different.

The payoff from a complex chooser is:

$$\max\left[c(S_t, X_1, T_1 - t), p(S_t, X_2, T_2 - t); t\right] \tag{12.3.3}$$

Thus, a call has strike price X_1 and time to expiry $T_1 - t$ and the put has strike of X_2 with maturity $T_2 - t$. This implies that it may not be possible to interpret the complex chooser in terms of standard options. If we take $Tc = (T_1 - t)$ as the maturity of the call and $Tp = (T_2 - t)$ as the maturity of a put, with call strike as Xc and put strike Xp. The valuation formula in the complex case becomes:

$$\begin{aligned} &Se^{-(b-r)T_c} M(d_1, y_1; \rho_1) - X_c e^{-rT_c} M(d_2, y_1 - \sigma\sqrt{T_c}; \rho_1) \\ &- Se^{(b-r)T_p} M(-d_1, -y_2; \rho_2) + X_p e^{-rT_p} M(-d_2, -y_2 + \sigma\sqrt{T_p}; \rho_2) \end{aligned} \tag{12.3.4}$$

Where

$$d_1 = \frac{\log(S/k) + (b = \sigma^2/2)t}{\sigma\sqrt{t}}, \quad d_2 = d_1 - \sigma\sqrt{t}$$

$$y_1 = \frac{\log(S/X_c) + (b + \sigma^2/2)T_c}{\sigma\sqrt{T_c}}, \quad y_2 = \frac{\log(S/X_p) + (b + \sigma^2/2)T_p}{\sigma\sqrt{T_p}}$$

$\rho_1 = \sqrt{\dfrac{t}{T_c}}, \quad \rho_2\sqrt{\dfrac{t}{T_p}}$, t is the time for the choice to be made for the option.

k is the value which solves :

$$ke^{(b-r)(T_c-t)}N(z_1) - X_c e^{-r(T_c-t)}N(z_1 - \sigma\sqrt{T_c - t}) + ke^{(b-r)(T_p-t)}N(-z_2)$$
$$- X_p e^{-r(T_p-t)}N(-z_2 + \sigma\sqrt{T_p - t}) = 0,$$

when,

$$z_1 = \frac{\log(k/X_c) + (b + \sigma^2/2)(T_c - t)}{\sigma\sqrt{T_c - t}}$$
$$z_2 = \frac{\log(k/X_p) + (b + \sigma^2/2)(T_p - t)}{\sigma\sqrt{T_p - t}}$$

Example 12.5

Consider a complex chooser with a call that has nine months maturity and a put with ten months to maturity. The time to choose is three months, with the risk free rate at 8% and stock price of $58.0 with volatility of 30%. The call strike is $60.0, the put strike is $55.0 and the underlying stock yield is 4%.

Initially solve for k using an iterative system :

{

$$ke^{(b-r)(T_c-t)}N(z_1) - X_c e^{-r(T_c-t)}N(z_1 - \sigma\sqrt{T_c - t}) + ke^{(b-r)(T_p-t)}N(-z_2)$$
$$- X_p e^{-r(T_p-t)}N(-z_2 + \sigma\sqrt{T_p - t}) = 0$$

$$z_1 = \frac{\log(k/X_c) + (b + \sigma^2/2)(T_c - t)}{\sigma\sqrt{T_c - t}}, \quad z_2 = \frac{\log(k/X_p) + (b + \sigma^2/2)(T_p - t)}{\sigma\sqrt{T_p - t}}$$

}

{

$k = 56.438$

$$z_1 = \frac{\log(k/X_c) + (b + \sigma^2/2)(T_c - t)}{\sigma\sqrt{T_c - t}} =$$
$$\frac{\log(56.438/60.0) + (0.04 + 0.30^2/2)(0.75 - 0.25)}{0.30\sqrt{0.75 - 0.25}} = -0.0881$$
$$z_2 = \frac{\log(k/X_p) + (b + \sigma^2/2)(T_p - t)}{\sigma\sqrt{T_p - t}} =$$
$$\frac{\log(56.438/55.0) + (0.04 + 0.30^2/2)(0.833 - 0.25)}{0.30\sqrt{0.833 - 0.25}} = 0.3290$$

}

$$d_1 = \frac{\log(S/k) + (b = \sigma^2/2)t}{\sigma\sqrt{t}} = \frac{\log(58.0/56.438) + (0.04 + 0.30^2/2)0.25}{0.30\sqrt{0.25}}$$
$$= 0.3236$$

$$d_2 = d_1 - \sigma\sqrt{t} = 0.3236 - 0.30\sqrt{0.25} = 0.1736$$

$$y_1 = \frac{\log(S/X_c) + (b + \sigma^2/2)T_c}{\sigma\sqrt{T_c}} = \frac{\log(58.0/60.0) + (0.04 + 0.30^2/2)0.75}{0.30\sqrt{0.75}}$$
$$= 0.11488$$

$$y2 = \frac{\log(S/X_p) + (b + \sigma^2/2)T_p}{\sigma\sqrt{T_p}} = \frac{\log(58.0/55.0) + (0.04 + 0.30^2/2)0.833}{0.30\sqrt{0.833}}$$
$$= 0.4525$$

$$\rho_1 = \sqrt{\frac{t}{T_c}} = \sqrt{\frac{0.25}{0.75}} = 0.5773, \quad \rho_2\sqrt{\frac{t}{T_p}} = \sqrt{\frac{0.25}{0.833}} = 0.5478$$

$M(a, b; \rho)$ is the Bivariate Cumulative Normal Distribution, for the variables:

$$M(d_1, y_1; \rho_1) = M(0.3236, 0.11488; 0.5773) = 0.4349$$
$$M(d_2, y_1 - \sigma\sqrt{T_c}; \rho_1) = M(0.1736, 0.11488 - 0.30^*\sqrt{0.75}; 0.5773) = 0.3459$$
$$M(-d_1, -y_2; \rho_2) = M(-0.3236, -0.4525; 0.5478) = 0.2028$$

$$M(-d_2, -y_2 + \sigma\sqrt{T_p}; \rho_2) = M(-0.1736, -0.4525 + 0.30^*\sqrt{0.833}; 0.5478)$$
$$= 0.2749$$

These variables are placed into 12.3.4 to give:

$$58.0e^{-(0.04-0.08)^*0.75} M(d_1, y_1; \rho_1) - 60.0^* e^{-0.08^*0.75} M(d_2, y_1 - \sigma\sqrt{T_c}; \rho_1)$$
$$- 58.0^* e^{(0.04-0.08)^*0.833} M(-d_1, -y_2; \rho_1)$$
$$+ 55.0^* e^{-0.08^*0.833} M(-d_2, -y_2 + \sigma\sqrt{T_p}; \rho_2) = 7.7027$$

Implementation code for complex chooser options is shown in Listing 12.6.

```
package FinApps;
import static java.lang.Math.*;
import BaseStats.Probnorm;
import BaseStats.Bivnorm;
import CoreMath.NewtonRaphson;
public class Choosercmpl extends NewtonRaphson {
  public Choosercmpl() {
  }

  public Choosercmpl(double yield, double rate, double volatility) {
  crate=yield;
  r=rate;
  sigma=volatility;
  brate=crate==0.0?0.0:(brate=crate!=r?(r-crate):r);
  accuracy(1e-6,10);
  }

  double r;
  double sigma;
  double brate;
  double crate;
  double stprice;
  double callx;
  double ival;
  double putx;
  double timec;
  double timep;
  double t;
  private double d1;
  private double d2;
  private double y1;
  private double y2;
  private double rho1;
  private double rho2;
  Probnorm p=new Probnorm();
  public double newtonroot(double rootvalue)
  {
    double solution =0.0;
    double z1=(log(rootvalue/callx)+(brate+
             (sigma*sigma)*0.5)*(timec-t))/(sigma*sqrt(timec-t));
    double z2=(log(rootvalue/putx)+(brate+
             (sigma*sigma)*0.5)*(timep-t))/(sigma*sqrt(timep-t));
    double factor1=(rootvalue*exp((brate-r)*(timec-t))
             *p.ncDisfnc(z1)-callx*exp(-r*(timec-t))
             *p.ncDisfnc(z1-sigma*sqrt(timec-t)));
    double factor2=(rootvalue*exp((brate-r)*(timep-t))
                 *p.ncDisfnc(-z2)-putx*exp(-r*(timep-t))
                 *p.ncDisfnc(-z2+sigma*sqrt(timep-t)));
    solution =(factor1+factor2);
       return solution;
  }

  public double Rubinchooser(double s, double xc,double xp,
```

```
                                        double tc, double tp,double time)
{
    Blackscholecp b=new Blackscholecp(crate);
    b.bscholEprice(s,xc,sigma,tc,r);
    double call=b.getCalle();
    b.bscholEprice(s,xp,sigma,tp,r);
    double put=b.getPute();
     stprice=s;
     putx=xp;
     callx=xc;
     timec=tc;
     timep=tp;
     t=time;
     ival= newtraph(s);
      paR();
      return ival;

}
private void paR()
{

     d1=(log(stprice/ival)+(brate+(sigma*sigma)*0.5)*t)
            /(sigma*sqrt(t));
     d2=(d1-sigma*sqrt(t));
     y1=(log(stprice/callx)+(brate+(sigma*sigma)*0.5)*timec)
            /(sigma*sqrt(timec));
     y2=(log(stprice/putx)+(brate+(sigma*sigma)*0.5)*timep)
            /(sigma*sqrt(timep));
     rho1=(sqrt(t/timec));
     rho2=(sqrt(t/timep));
    double m1=Bivnorm.bivar_params.evalArgs(d1,y1,rho1);
    double m2=Bivnorm.bivar_params.evalArgs(d2,y1-sigma
            *sqrt(timec),rho1);
    double m3=Bivnorm.bivar_params.evalArgs(-d1,-y2,rho2);
    double m4=Bivnorm.bivar_params.evalArgs(-d2,-y2+sigma
            *sqrt(timep),rho2);
    double w=((stprice*exp((brate-r)*timec)*m1)
            -(callx*(exp(-r*timec))*m2)
            -(stprice*exp((brate-r)*timep)*m3)
            +(putx*(exp(-r*timep))*m4));

}
```

LISTING 12.6. Complex Chooser Option Valuation

12.4. Options on Options

Models for valuing compound options of this type were introduced by Geske (1977,1979), Rubinstein (1991), Whaley (1981) and others.

The payoff at exercise of a compound option is dependent on the value of another option. The option on an option has two expiry dates and two strike prices, in a similar way to a complex chooser (which is another class of compound option). For the straightforward case of a call on call option, the payoff is given as:

$$\max[c(S, X_1, T_2) - X_2; 0]$$

where X_1 is the underlying option strike price and X_2 is the strike price of the option on the option. T_2 is the time to maturity of the underlying option and T_1 is the time to maturity of the option on the option. Let *Sval* be the critical price for which $(c(Sval, X_1, T_2 - T_1) - X_2) = 0.0$. The value of a call on call option depends on the joint probability of the asset price being above the critical price at the time to maturity of the option on option and greater than the strike price of the option on option at time to maturity of the underlying option.

12.4.1. Call on Call

Assuming the lognormal property of the underlying asset, the price for a call on call option can be shown as:

$$Se^{(b-r)T_2} M(z_1, y_1; \rho) - X_1 e^{-rT_2} M(z_2, y_2; \rho) - X_2 e^{-rT_1} N(y_2) \tag{12.4.1}$$

where $b - r$ is the dividend yield. The $M(a, b, \rho)$ represent the bivariate cumulative normal distribution. Since we have overlapping Brownian motion increments the correlation coefficient is given as $\rho = \sqrt{T_1/T_2}$.

The term $Se^{(b-r)T_2} M(z_1, y_1; \rho)$ represents a risk neutral expectation of the asset price $S > Sval$; T_1 and $S > X_2$; T_2. The term $X_1 e^{-rT_2} M(z_2, y_2; \rho)$ represents the expected cash value of exercising at T_2. The final term of $X_2 e^{-rT_1} N(y_2)$ is the expected cash value of exercising at T_1.

The variables are given by:

$$y_1 = \frac{\log(S/Sval) + (b + \sigma^2/2)T_1}{\sigma\sqrt{T_1}}$$

$$y_2 = y_1 - \sigma\sqrt{T_1}$$

$$z_1 = \frac{\log(S/X_1) + (b + \sigma^2/2)T_2}{\sigma\sqrt{T_2}}$$

$$z_2 = z_1 - \sigma\sqrt{T_2}$$

12.4.2. Put on Call

The put on call option has a payoff of :

$\max[X_2 - c(S, X_1, T_2); 0]$

The pricing formula is:

$$Pcall = X_1 e^{-rT_2} M(z_2, -y_2; -\rho) - Se^{(b-r)T_2} M(z_1, -y_1; -\rho) + X_2 e^{-rT_1} N(-y_2) \qquad (12.4.2)$$

Call on Put

This has a payoff of:

$$\max[p(S, X_1, T_2) - X_2; 0]$$

The pricing formula is:

$$Cput = X_1 e^{-rT_2} M(-z_2, -y_2; \rho) - Se^{(b-r)T_2} M(-z_1, -y_1; \rho) - X_2 e^{-rT_1} N(-y_2)$$

Put on Put

The payoff is:

$$\max[X_2 - p(S, X_1, T_2); 0]$$

The pricing formula is:

$$Pput = Se^{(b-r)T_2} M(-z_1, -y_1; -\rho) - X_1 e^{-rT_2} M(-z_2, -y_2; -\rho) + X_2 e^{-rT_1} N(y_2) \qquad (12.4.3)$$

Example 12.6

A put on call option, gives the holder the right to sell a call option for £60.0, in four months time. The underlying call strike price is £615.0, with time to maturity of nine months.The price of the underlying asset is £600.0, with a volatility of 30%. The risk free rate is 6% and the dividend yield from the stock is 2.5%.

$$S = 600.0,\ X_1 = 615.0,\ X_2 = 60.0,\ T_1 = 0.333,\ T_2 = 0.75,\ \sigma = 0.30,\ q = 0.025$$

$$\frac{\log(600.0/630.0599) + (0.035 + 0.30^2/2)0.333}{0.30\sqrt{0.333}} = -0.1284,$$

$$y_2 = y_1 - \sigma\sqrt{T_1} = -0.1284 - 0.30^* \sqrt{0.333} = -0.3016$$

$$z_1 = \frac{\log(S/X_1) + (b + \sigma^2/2)T_2}{\sigma\sqrt{T_2}} =$$

$$z_1 = \frac{\log(600.0/615.0) + (0.035 + 0.30^2/2)0.75}{0.30\sqrt{0.75}} = 0.1358$$

$$z_2 = z_1 - \sigma\sqrt{T_2} = -0.1239$$

$$\rho = \sqrt{T_1/T_2} = 0.6663$$

$$M_1(z_2, -y_2; -\rho) = 0.1681, \quad M_2(z_1, -y_1; -\rho) = 0.19266, \quad N(-y_2) = 0.61852$$

$$\begin{aligned} Pcall &= X_1 e^{-rT_2} M(z_2, -y_2; -\rho) - S e^{(b-r)T_2} M(z_1, -y_1; -\rho) + X_2 e^{-rT_1} N(-y_2) \\ &= 615.0^* e^{-0.06^*0.75} M_1(z_2, -y_2; -\rho) \\ &\quad - 600.0^* e^{(0.035-0.06)0.75} M_2(z_1, -y_1; -\rho) \\ &\quad + 600.0^* e^{-0.06*0.333} N(-y_2) = 21.7611 \end{aligned}$$

The code implementing option on option derivatives is shown in Listing 12.7.

```
package FinApps;
import static java.lang.Math.*;
import BaseStats.Probnorm;
import CoreMath.NewtonRaphson;
import BaseStats.Bivnorm;
public class Oponop extends NewtonRaphson{
  public Oponop(double yield,double rate, double strikeop) {
      crate=yield;
      r=rate;
      strike2=strikeop;
      brate=crate==0.0?0.0:(brate=crate!=r?(r-crate):r);
      accuracy(1e-9,10);

  }

  private double crate;
  private double brate;
  private double r;
  private double strike;
  private double sigma;
  private double time;
  private double maturity;
  private double strike2;
  private double timediff;
```

```
private double y;
private double z;
private double y2;
private double z2;
private double rho;
private double ccpayoff=0.0;
private double pcpayoff=0.0;
private double cppayoff=0.0;
private double pppayoff=0.0;
private int type;
public double getpcC() {
    return ccpayoff;
}
public double getppC() {
    return pcpayoff;
}
public double getpcP() {
    return cppayoff;
}
public double getppP() {
    return pppayoff;
}

public void parAms(double s, double x, double volatility, double t,
                   double tmaturity) {
    strike=x;
    sigma=volatility;
    time=t;
    maturity=tmaturity;
    timediff=(tmaturity-t);
    double sval=newtraph(x);
    y=(log(s/sval)+(brate+(sigma*sigma)*0.5)*(time))
                   /(sigma*sqrt(time));
    y2=(y-sigma*sqrt(time));
    z=(log(s/x)+(brate+(sigma*sigma)*0.5)*(tmaturity))
                   /(sigma*sqrt(tmaturity));
    z2=(z-sigma*sqrt(tmaturity));
    rho=sqrt(t/tmaturity);


}
public double callCall(double s, double x,double volatility,
                           double t, double tmaturity) {
    type=1;
    Blackscholecp b= new Blackscholecp(crate);
    b.bscholEprice(s,x,volatility,tmaturity,r);
    ccpayoff=max((b.getCalle()-strike2),0);
    Probnorm p=new Probnorm();
    parAms(s,x,volatility,t,tmaturity);
    double m1=(Bivnorm.bivar_params.evalArgs(z,y,rho));
    double m2=(Bivnorm.bivar_params.evalArgs(z2,y2,rho));
    double value=((s*exp((brate-r)*maturity)*m1)
                   -(strike*exp(-r*maturity)*m2)
                   -(strike2*exp(-r*time)*p.ncDisfnc(y2)));
    return value;
```

```
}

public double putCall(double s, double x,double volatility,
                       double t, double tmaturity) {
    type=1;
    Blackscholecp b= new Blackscholecp(crate);
    b.bscholEprice(s,x,volatility,tmaturity,r);
    pcpayoff=max((strike2-b.getCalle()),0);
    Probnorm p=new Probnorm();
    parAms(s,x,volatility,t,tmaturity);
    double m1=(Bivnorm.bivar_params.evalArgs(z2,-y2,-rho));
    double m2=(Bivnorm.bivar_params.evalArgs(z,-y,-rho));
    double fact1=(strike*exp(-r*maturity)*m1);
    double fact2=(s*exp((brate-r)*maturity)*m2);
    double fact3=(strike2*exp(-r*time)*p.ncDisfnc(-y2));
    double value=(fact1-fact2+fact3);
    return value;
}

public double callPut(double s, double x,double volatility,
                       double t, double tmaturity) {
    type=0;
    Blackscholecp b= new Blackscholecp(crate);
    b.bscholEprice(s,x,volatility,tmaturity,r);
    cppayoff=max((b.getPute()-strike2),0);
    Probnorm p=new Probnorm();
    parAms(s,x,volatility,t,tmaturity);
    double m1=(Bivnorm.bivar_params.evalArgs(-z2,-y2,rho));
    double m2=(Bivnorm.bivar_params.evalArgs(-z,-y,rho));
    double fact1=(strike*exp(-r*maturity)*m1);
    double fact2=(s*exp((brate-r)*maturity)*m2);
    double fact3=(strike2*exp(-r*time)*p.ncDisfnc(-y2));
    double value=(fact1-fact2-fact3);
    return value;
}

public double putPut(double s, double x,double volatility,
                       double t, double tmaturity) {
    type=0;
    Blackscholecp b= new Blackscholecp(crate);
    b.bscholEprice(s,x,volatility,tmaturity,r);
    pppayoff=max((strike2-b.getPute()),0);
    Probnorm p=new Probnorm();
    parAms(s,x,volatility,t,tmaturity);
    double m1=(Bivnorm.bivar_params.evalArgs(-z,y,-rho));
    double m2=(Bivnorm.bivar_params.evalArgs(-z2,y2,-rho));
    double value=((s*exp((brate-r)*maturity)*m1)
                 -(strike*exp(-r*maturity)*m2)
                 +(strike2*exp(-r*time)*p.ncDisfnc(y2)));
    return value;
}

public double newtonroot(double rootvalue)
```

```
{   Blackscholecp b= new Blackscholecp(crate);
    double solution =0.0;
    b.bscholEprice(rootvalue,strike,sigma,timediff,r);
    solution=type==1?(b.getCalle()-strike2):(b.getPute());
    return solution;
}
```

LISTING 12.7. Option on Option Valuation

12.5. Extendible Options

Holder Extendible options give the holder a right to exercise at maturity (T_1) or to extend the maturity date to a new date (T_2) in exchange for a fee. The strike price can also be adjusted at the extension time (T_1) from the original (X_1) to the adjusted strike (X_2).

The payoff from a holder extendible option is given as:

For a call

$$\max[S - X_1; c(S, X_2, T_2 - T_1) - A; 0]$$

For a Put

$$\max[X_1 - S; p(S, X_2, T_2 - T_1) - A; 0]$$

Where $c(s, x, t)$ is the Black Scholes formula for a call option value and $p(s, x, t)$ is the Black Scholes formula for a put option value. A is the value of the premium (fee) paid to the writer. The fundamental formulae are due to Longstaff (1990).

12.5.1. Extendible Call

An extendible call is valued by:

$$\begin{aligned} cvalue = {} & c(S, X_1, T_1) + Se^{(b-r)T_2} M(y_1, y_2, -\infty, z_1; \rho) \\ & - X_2 e^{-rT_2} M(y_1 - \sigma\sqrt{T_1}, y_2 - \sigma\sqrt{T_1}, -\infty, z_1 - \sigma\sqrt{T_2}; \rho) \\ & - Se^{(b-r)T_1} N(y_1, z_2) + X_1 e^{-rT_1} N(y_1 - \sigma\sqrt{T_1}, z_2 - \sigma\sqrt{T_1}) \\ & - Ae^{-rT_1} N(y_1 - \sigma\sqrt{T_1}, y_2 - \sigma\sqrt{T_1}) \end{aligned} \tag{12.5.1}$$

Where the parameters are given by:

$$y_1 = \frac{\log(S / Sval_2) + (b + \sigma^2 / 2)T_1}{\sigma\sqrt{T_1}}$$

$$y_2 = \frac{\log(S / Sval_1) + (b + \sigma^2 / 2)T_1}{\sigma\sqrt{T_1}}$$

$$z_1 = \frac{\log(S / X_2) + (b + \sigma^2 / 2)T_2}{\sigma\sqrt{T_2}}, \; z_2 \frac{\log(S / X_1) + (b + \sigma^2 / 2)T_1}{\sigma\sqrt{T_1}}$$

$$\rho = \sqrt{\frac{T_1}{T_2}}$$

The critical values for $Sval_1$ and $Sval_2$ are derived from a root finding process that satisfies:

$$Sval_1 = (c(Sval_1, X_1, T_2 - T_1) - A \cong 0)$$

$$Sval_2 = (c(Sval_2, X_2, T_2 - T_1) - (Sval_2 - X_1 + A) \cong 0)$$

There are conditions for which the value of $Sval_1$ can determine the viability of extension;

a) The call is always extended if $Sval_1 = 0$, $Sval_2 = \infty$
b) If $Sval_1 > 0$, $Sval_2 = \infty$, then the extendible call is a call on call option with strike price A and underlying call has strike X_2 with maturity $(T_2 - T_1)$
c) When $Sval_1 \geq X_1$, it is never optimal to extend the call

The premium payable will determine the range of values for which the critical values fall. If $A = 0$ then $Sval_1 = 0$ and if $A < (X_1 - X_2 e^{-r(T_2 - T_1)})$, $Sval_2 = \infty$. The call should be extended when $Sval_1 < S < Sval_2$.

The probability measure $M(a, b, c, d; \rho)$ can be derived from the bivariate normal distribution using:

$$M(a, b, c, d; \rho) \equiv M(b, d; \rho) - M(a, d; \rho) - M(b, c; \rho) + M(a, c; \rho),$$

this gives us:

$$\begin{aligned} Ma(y_1, y_2, -\infty, z_1; \rho) = {} & Ma(y_2, z_1, \rho) - Ma(y_1, z_1, \rho) \\ & - Ma(y_2, -\infty, \rho) + Ma(y_1, -\infty, \rho) \end{aligned}$$

$$\begin{aligned} & Mb(y_1 - \sigma\sqrt{T_1}, y_2 - \sigma\sqrt{T_1}, -\infty, z_1 - \sigma\sqrt{T_2}; \rho) \\ & = Mb(y_2 - \sigma\sqrt{T_1}, z_1 - \sigma\sqrt{T_2}, \rho) \\ & \quad - Mb(y_1 - \sigma\sqrt{T_1}, z_1 - \sigma\sqrt{T_2}, \rho) \\ & \quad - Mb(y_2 - \sigma\sqrt{T_1}, -\infty, \rho) + Mb(y_1 - \sigma\sqrt{T_1}, -\infty, \rho) \end{aligned}$$

Similarly, the probability measure $N(a, b)$ can be derived from the standard normal distribution using:

$$N(a, b) \equiv N(b) - N(a).$$

This gives us:

$$Na(y_1, z_1) = Na(z_1) - Na(y_1)$$
$$Nb(y_1 - \sigma\sqrt{T_1}, z_2 - \sigma\sqrt{T_1}) = Nb(z_2 - \sigma\sqrt{T_1}) - Nb(y_1 - \sigma\sqrt{T_1})$$
$$Nc(y_1 - \sigma\sqrt{T_1}, y_2 - \sigma\sqrt{T_1}) = Nc(y_2 - \sigma\sqrt{T_1}) - Nc(y_1 - \sigma\sqrt{T_1})$$

12.5.2. Extendible Put

An extendible put is valued by:

$$\begin{aligned} pvalue = {} & p(S, X_1, T_1) - Se^{(b-r)T_2} M(y_1, y_2, -\infty, -z_1; \rho) \\ & + X_2 e^{-rT_2} M(y_1 - \sigma\sqrt{T_1}, y_2 - \sigma\sqrt{T_1}, -\infty, -z_1 + \sigma\sqrt{T_2} : \\ & + Se^{(b-r)T_1} N(z_2, y_2) - X_1 e^{-rT_1} N(z_2 - \sigma\sqrt{T_1}, y_2 - \sigma\sqrt{T1}) \\ & - Ae^{-rT_1} N(y_1 - \sigma\sqrt{T_1}, y_2 - \sigma\sqrt{T_1}) \end{aligned} \tag{12.5.2}$$

The critical values are given by the root finding process that satisfies:

$$Sval_1 = (p(Sval_1, X_2, T_2 - T_1) - X_1 - Sval_1 + A) \cong 0$$
$$Sval_2 = (p(Sval_2, X_1, T_2 - T_1) - A) \cong 0$$

The special cases for an extendible put are:

a) The put is always extended when $A = 0$, $Sval_1 = 0$
b) When $A > 0$, $Sval_1 = 0$, the put becomes a call on put option with a strike price A. The underlying put option has a strike price of X_2 and maturity $T_2 - T_1$
c) The put will never be extended when $Sval_2 < X_1 or Sval_1 = X_1$

The put is only extended if the relationship $Sval_1 < SVal_{2,} S > Sval_1$ holds.

Example 12.7

An extendible call has an initial time to maturity of nine months. The underlying stock price is €110.0, the intial strike price is €109.0. The option is extendible by three months with an adjusted strike price of €112.0 and a premium of €1.0 is payable. The underlying stock volatility is 30% per annum and the risk free rate is 6%.

$$S = 110.0, X_1 = 109.0, X_2 = 112.0, T_1 = 0.75, T_2 = 1.0, A = 1.0, r = 0.06,$$
$$\sigma = 0.30$$

The critical values are computed using a root finding process as:

$$Sval_1 = (c(Sval_1, X_1, T_2 - T_1) - A \cong 0)\,,\; Sval_1 = 90.94$$

$$Sval_2 = (p(Sval_2, X_1, T_2 - T_1) - A) \cong 0\,,\; Sval_2 = 123.276$$

$$y_1 = \frac{\log(S/Sval_2) + (b + \sigma^2/2)T_1}{\sigma\sqrt{T_1}} = \frac{\log(110.0/123.276) + (0.06 + 0.30^2/2)0.75}{0.30\sqrt{0.75}} = -0.1354$$

$$y_2 = \frac{\log(S/Sval_1) + (b + \sigma^2/2)T_1}{\sigma\sqrt{T_1}} = \frac{\log(S/90.94) + (0.06 + 0.30^2/2)0.75}{0.30{*}\sqrt{0.75}} = 1.035$$

$$z_1 = \frac{\log(S/X_2) + (b + \sigma^2/2)T_2}{\sigma\sqrt{T_2}} = \frac{\log(110.0/112.0) + (0.06 + 0.30^2/2){*}1.0}{0.30{*}\sqrt{1.0}} = 0.289$$

$$z_2 = \frac{\log(S/X_1) + (b + \sigma^2/2)T_1}{\sigma\sqrt{T_1}} = \frac{\log(110.0/112.0) + (0.06 + 0.30^2/2){*}0.75}{0.30{*}\sqrt{0.75}} = 0.338,$$

$$\rho = \sqrt{\frac{T_1}{T_2}} = \sqrt{\frac{0.75}{1.0}} = 0.866$$

$$Ma(y_1, y_2, -\infty, z_1; \rho) = Ma(y_2, z_1, \rho) - Ma(y_1, z_1, \rho) - Ma(y_2, -10.0, \rho) + Ma(y_1, -10.0, \rho) = 0.1866$$

$$\begin{aligned} &Mb(y_1 - \sigma\sqrt{T_1}, y_2 - \sigma\sqrt{T_1}, -\infty, z_1 - \sigma\sqrt{T_2}; \rho) \\ &= Mb(y_2 - \sigma\sqrt{T_1}, z_1 - \sigma\sqrt{T_2}, \rho) - Mb(y_1 - \sigma\sqrt{T_1}, z_1 - \sigma\sqrt{T_2}, \rho) \\ &\quad - Mb(y_2 - \sigma\sqrt{T_1}, -10.0, \rho) + Mb(y_1 - \sigma\sqrt{T_1}, -10.0, \rho) = 0.1714 \end{aligned}$$

$$Na(y_1, z_1) = Na(0.289) - Na(-0.1354) = 0.1863$$

$$Nb(y_1 - \sigma\sqrt{T_1}, z_2 - \sigma\sqrt{T_1}) = Nb(0.338 - 0.30\sqrt{0.75}) - Nb(-0.1354 - 0.30\sqrt{0.75}) = 0.1849$$

$$Nc(y_1, -\sigma\sqrt{T_1}, y_2 - \sigma\sqrt{T_1}) = Nc(1.035 - 0.30\sqrt{0.75}) - Nc(-0.1354 - 0.30\sqrt{0.75}) = 0.4347$$

The standard Black Scholes call value $c(S, X_1, T_1) = 14.2059$

Using 12.4.4. The extendible call is €15.011.

The implementation code for extendible options is shown in Listing 12.8.

```
package FinApps;
import static java.lang.Math.*;
import BaseStats.Probnorm;
import CoreMath.NewtonRaphson;
import BaseStats.Bivnorm;
public class Holderextop {
    public Holderextop(double adpremium, double strikeadj,
                        double newtime, double yield, double rate) {
        ap=adpremium;
        x2=strikeadj;
        t2=newtime;
        crate=yield;
        r=rate;
        brate=crate==0.0?0.0:(brate=crate!=r?(r-crate):r);
    }
    private double ap;
    private double x2;
    private double t2;
    private double crate;
    private double brate;
    private double r;
    private double strike;
    private double sigma;
    private double time;
    private double timediff;
    private double y;
    private double z;
    private double y2;
    private double z2;
    private double rho;
    private int typeofop;
    private double sval;
    private double sval2;

    public void parAms(double s, double x, double volatility,
                        double t) {
        time=t;
        timediff=(t2-t);
        strike=x;
        sigma=volatility;
        if(typeofop==0)// If call
        {
            i1 ival=new i1();
            i2 ival2=new i2();
            sval=ival.pars(x);
            sval2=ival2.pars(x);
        } else// If put
        {
```

```
        pi1 ival=new pi1();
        pi2 ival2=new pi2();
        sval=ival.pars(x);
        sval2=ival2.pars(x);
      }

      y2=(log(s/sval)+(brate+(sigma*sigma)*0.5)*(time))
                         /(sigma*sqrt(time));
      y=(log(s/sval2)+(brate+(sigma*sigma)*0.5)*(time))
                         /(sigma*sqrt(time));
      z=(log(s/x2)+(brate+(sigma*sigma)*0.5)*(t2))
                         /(sigma*sqrt(t2));
      z2=(log(s/strike)+(brate+(sigma*sigma)*0.5)*(time))
                         /(sigma*sqrt(time));
      rho=(sqrt(time/t2));
    }

      public double extCall(double s, double x,
                                double volatility, double t) {
      typeofop=0;
      Probnorm p=new Probnorm();
      Blackscholecp b= new Blackscholecp(crate);
      parAms(s,x,volatility,t);
      double m1=(Bivnorm.bivar_params.evalArgs(y2,z,rho));
      double m2=(Bivnorm.bivar_params.evalArgs(y,z,rho));
      double m3=(Bivnorm.bivar_params.evalArgs(y2,-10.0,rho));
      double m4=(Bivnorm.bivar_params.evalArgs(y,-10.0,rho));
      double mm=(m1-m2-m3+m4);
      double n1=(Bivnorm.bivar_params.evalArgs((y2-sigma*sqrt(t)),
                         (z-sigma*sqrt(t2)),rho));
      double n2=(Bivnorm.bivar_params.evalArgs((y-sigma*sqrt(t)),
                         (z-sigma*sqrt(t2)),rho));
      double n3=(Bivnorm.bivar_params.evalArgs((y2-sigma*sqrt(t)),
                         -10.0,rho));
      double n4=(Bivnorm.bivar_params.evalArgs((y-sigma*sqrt(t)),
                         -10.0,rho));
      double nn=(n1-n2-n3+n4);
      double prob1=(p.ncDisfnc(z2)-p.ncDisfnc(y));
      double prob2=(p.ncDisfnc(z2-sigma*sqrt(time))
                         -p.ncDisfnc(y-sigma*sqrt(time)));
      double prob3=(p.ncDisfnc(y2-sigma*sqrt(time))
                         -p.ncDisfnc(y-sigma*sqrt(time)));
      b.bscholEprice(s,strike,sigma,time,r);
      double cval=b.getCalle();
        double call=(cval+s*exp((brate-r)*t2)*mm-x2*exp(-r*t2)
               *nn-s*exp((brate-r)*time)*prob1+strike*exp
               (-r*time)*prob2-ap*exp(-r*time)*prob3);
    return call;
  }

public double extPut(double s, double x, double volatility,
                         double t) {
    Blackscholecp b= new Blackscholecp(crate);
    typeofop=1;
    Probnorm p=new Probnorm();
```

```
    parAms(s,x,volatility,t);
    double m1=(Bivnorm.bivar_params.evalArgs(y2,z,rho));
    double m2=(Bivnorm.bivar_params.evalArgs(y,z,rho));
    double m3=(Bivnorm.bivar_params.evalArgs(y2,-10.0,rho));
    double m4=(Bivnorm.bivar_params.evalArgs(y,-10.0,rho));
    double mm=(m1-m2-m3+m4);
    double n1=(Bivnorm.bivar_params.evalArgs((y2-sigma*sqrt(t)),
                       (z+sigma*sqrt(t2)),rho));
    double n2=(Bivnorm.bivar_params.evalArgs((y-sigma*sqrt(t)),
                       (z+sigma*sqrt(t2)),rho));
    double n3=(Bivnorm.bivar_params.evalArgs((y2-sigma*sqrt(t))
                       ,-10.0,rho));
    double n4=(Bivnorm.bivar_params.evalArgs((y-sigma*sqrt(t))
                       -10.0,rho));
    double nn=(n1-n2-n3+n4);
    double prob1=(p.ncDisfnc(y2)-p.ncDisfnc(z2));
    double prob2=(p.ncDisfnc(y2-sigma*sqrt(time))
                       -p.ncDisfnc(z2-sigma*sqrt(time)));
    double prob3=(p.ncDisfnc(y2-sigma*sqrt(time))
                       -p.ncDisfnc(y-sigma*sqrt(time)));
     b.bscholEprice(s,strike,sigma,time,r);
    double pval=b.getPute();
    double put=(pval-s*exp((brate-r)*t2)*mm+x2*exp(-r*t2)
                       *nn+s*exp((brate-r)*time)*prob1-strike*exp
                       (-r*time)*prob2-ap*exp(-r*time)*prob3);

    return put;
}
private class i1 extends NewtonRaphson {
    i1() {
        accuracy(1e-6,10);
    }
    public double pars(double seedvalue) {
        double sval2=newtraph(seedvalue);

        return sval2;
    }
    Blackscholecp b= new Blackscholecp(crate);
    public double newtonroot(double rootvalue) {
        double solution =0.0;
        b.bscholEprice(rootvalue,strike,sigma,timediff,r);
        solution=(b.getCalle()-(ap));

        return solution;
    }
}
private class pi1 extends NewtonRaphson {
    pi1() {
        accuracy(1e-6,10);
    }
    public double pars(double seedvalue) {
        double sval2=newtraph(seedvalue);
        return sval2;
    }
```

```
        Blackscholecp b= new Blackscholecp(crate);
        public double newtonroot(double rootvalue) {
            double solution =0.0;
            b.bscholEprice(rootvalue,x2,sigma,timediff,r);
            solution=(b.getPute()-(strike-rootvalue+ap));
            return solution;
        }
    }
    private class pi2 extends NewtonRaphson {
        pi2() {
            accuracy(1e-6,10);
        }
        public double pars(double seedvalue) {
            double sval2=newtraph(seedvalue);
            return sval2;
        }
        Blackscholecp b= new Blackscholecp(crate);
        public double newtonroot(double rootvalue) {
            double solution =0.0;
            b.bscholEprice(rootvalue,strike,sigma,timediff,r);
            solution=(b.getPute()-(ap));
            return solution;
        }
    }
    private class i2 extends NewtonRaphson {
        i2() {
            accuracy(1e-6,10);
        }
        public double pars(double seedvalue) {
            double sval2=newtraph(seedvalue);
            return sval2;
        }
        Blackscholecp b= new Blackscholecp(crate);
        public double newtonroot(double rootvalue) {
            double solution =0.0;
            b.bscholEprice(rootvalue,x2,sigma,timediff,r);
            solution=(b.getCalle()-(rootvalue-strike+ap));
            return solution;
        }
    }
```

LISTING 12.8. Holder Extendible Option valuation

12.6. Writer Extendible

Writer extendible options can be exercised at the initial maturity date or can be extended to another maturity, if the option is out-of-the-money at the initial maturity date.

For a writer extendible call the payoff at initial maturity T_1 is given as:

For $S \geq X_1$; $(S - X_1)$ otherwise $c(S, X_2, T_2 - T_1)$.Where $T_2$is the extended maturity date. For a put the payoff is:

For $S < X_1$; $(X - S)$ otherwise $p(S, X_2, T_2 - T_1)$.

The value of a call is given by:

$$cext = c(S, X, T_1) + Se^{(b-r)T_2} M(z_1, -z_2; \rho) - X_2 e^{-rT_2} M(z_1 - \sigma\sqrt{T_2}, -z_2 + \sigma\sqrt{T_1}; -\rho) \tag{12.6.1}$$

The value of the writer extendible put is given by:

$$pext = p(S, X_1, T_1) + X_2 e^{-rT_2} M(-z_1 + \sigma\sqrt{T_2}, z_2 - \sigma\sqrt{T_1}; -\rho) - Se^{(b-r)T_2} M(-z_1, z_2; -\rho) \tag{12.6.2}$$

The basic parameters are calculated using the same process as in the general case of extendible maturities.

Example 12.8

A writer extendible call with six months to maturity has the possibility of extension for a further six months, if the option is out-of-the-money at the end of initial maturity. The stock price is £100.0, the strike price is £110.0, the extended strike is £103.0. Stock price volatility is 25% per annum and the risk free rate is 8%.

$T_1 = 0.5$, $T_2 = 1.0$, $S_1 = 100.0$, $X_1 = 110.0$, $X_2 = 103.0$, $\sigma = 0.25$, $r = 0.08$, $b = 0.08$

The Black Scholes call price is:

$$c(s, X_1, T_1) = c(100, 110, 0.5) = 4.7507$$

The parameters are:

$$z_1 = \frac{\log(100.0/103.0) + (0.08 + 0.25^2/2)^*1.0}{0.25^*\sqrt{1.0}} = 0.3267$$

$$z_2 = \frac{\log(100.0/110.0) + 90.08 + 0.25^*\sqrt{0.5}}{0.25^*\sqrt{0.5}} = -0.2244$$

$$\rho = \sqrt{\frac{0.5}{1.0}} = 0.707$$

$$M1(z_1, -z_2; -\rho) = M1(0.3267, -(-0.2244); -0.707) = 0.26088$$

$$M2(z_1 - \sigma\sqrt{T_2}, -z_2 + \sigma\sqrt{T_2}; -\rho) = M2(0.3267 - 0.25^*\sqrt{1.0},$$
$$-(-0.2244 + 0.25^*\sqrt{0.5}); -0.707) = 0.237$$

$$\begin{aligned} cext &= c(S, X, T_1) + Se^{(b-r)T_2}M(z_1, -z_2; \rho) \\ &\quad - X_2e^{-rT_2}M(z_1 - \sigma\sqrt{T_2}, -z_2 + \sigma\sqrt{T_1}; -\rho) \\ &= 4.7507 + 100e^{(0.08-0.08)^*1.0}M(z_1, -z_2; \rho) \\ &\quad - 103e^{-0.08^*1.0}M(z_1 - \sigma\sqrt{T_2}, -z_2 + \sigma\sqrt{T_1}; -\rho) \\ &= 8.2706 \end{aligned}$$

Writer extendible options are valued using the class shown in Listing 12.9.

```
package FinApps;
import static java.lang.Math.*;
import BaseStats.Bivnorm;
public class Writerexop {
   public Writerexop(double exttime, double newstrike,
                              double yield, double rate) {
      t2=exttime;
      x2=newstrike;
      r=rate;
      crate=yield;
      brate=crate==0.0?0.0:(brate=crate!=r?(r-crate):r);
   }
   double t2;
   double x2;
   double z1;
   double z2;
   double rho;
   double brate;
   double crate;
   double r;
   Blackscholecp b= new Blackscholecp(crate);
   private void params(double s,double x,double volatility,
                              double time) {
      z1=((log(s/x2)+(brate+(volatility*volatility)*0.5)*t2)
                           /(volatility*sqrt(t2)));
      z2=((log(s/x)+(brate+(volatility*volatility)*0.5)*time)
                           /(volatility*sqrt(time)));
      rho=(sqrt(time/t2));
   }
   public double writeExtcall(double s, double x, double volatility,
                              double time) {
      params(s,x,volatility,time);
      double m1=(Bivnorm.bivar_params.evalArgs(z1,-z2,-rho));
      double m2=(Bivnorm.bivar_params.evalArgs((z1-volatility
               *sqrt(t2)), (-z2+volatility*sqrt(time)), -rho));
      b.bscholEprice(s,x,volatility,time,r);
      double c=(b.getCalle()+s*exp((brate-r)*t2)*m1-x2*exp(-r*t2)
```

```
                        *m2);
BlackSholes=="+(b.getCalle())+"rho=="+rho);
    return c;
  }
  public double writeExtput(double s, double x, double volatility,
                                  double time) {
    params(s,x,volatility,time);
    double m2=(Bivnorm.bivar_params.evalArgs(-z1,z2,-rho));
    double m1=(Bivnorm.bivar_params.evalArgs((-z1+volatility
             *sqrt(t2)), (z2-volatility*sqrt(time)), -rho));
    b.bscholEprice(s,x,volatility,time,r);
    double p=(b.getPute()+x2*exp(-r*t2)*m1-s*exp((brate-r)*t2)
             *m2);
    return p;
  }
```

LISTING 12.9. Writer Extendible Option Valuation

12.7. Rainbow Options

Rainbow options cover a generic class of options that depend on having two or more underlying assets.

12.7.1. Two Asset Correlated

A fundamental rainbow option is the two asset correlated option, first proposed by Zhang (1995). The two asset correlation model has two underlying assets, with an associated correlation between their returns. The correlations are derived from historical data. The two asset model is also refered to as a digital option. This type of option differs from the 'norm' by having a predefined payoff at expiration, as opposed to the differences between strike and stock price at expiry.

In a cash-or-nothing two asset correlated option, the payoff is a cash amount or nothing at expiry. In the asset-or-nothing option the payoff is a number of units of the stock or nothing.

If we have two assets S_1, S_2. With corresponding strike prices X_1, X_2.

The payoff from a call is:

When $S_1 > X_1$ call = max $(S_2 - X_2; 0)$ Otherwise 0.

The payoff from a put is:

When $S_1 < X_1$ put = max $(X_2 - S_2; 0)$ Otherwise 0.

The formula for valuing a call is :

$$S_2 e^{(b_2-r)T} M1(y_2 + \sigma_2\sqrt{T}, y_1 + \rho\sigma_2\sqrt{T}; \rho) - X_2 e^{-rT} M2(y_2, y_1; \rho) \qquad (12.7.1)$$

Where

$$y_1 = \frac{\log(S_1/X_1) + (b_1 - \sigma_1^2/2)T}{\sigma_1\sqrt{T}}$$

$$y_2 = \frac{\log(S_2/X_2) + (b_2 - \sigma_2^2/2)T}{\sigma_2\sqrt{T}}$$

The value for ρ is obtained empirically.

The formula for valuing a put is:

$$X_2 e^{-rT} M1(-y_2, -y_1; \rho) - S_2 e^{(b_2-r)T} M2(-y_2 - \sigma_2\sqrt{T}, -y_1 - \rho\sigma_2\sqrt{T}; \rho) \tag{12.7.2}$$

Example 12.9

A cash-or-nothing call with three months to expiry has stock 1 with a price of \$65.0 and strike of \$ 60.0. Stock 2 has a price of \$70.0. The payout is \$72.0. The volatility of stock 1 is 25% and thc volatility of stock 2 is 30%. The correlation between both stocks is seen as 0.80, the risk free rate is 6%.

$$y_1 = \frac{\log(S_1/X_1) + (b_1 - \sigma_1^2/2)T}{\sigma_1\sqrt{T}} = \frac{\log(65.0/60.0) + (0.06 - 0.25^2/2)0.25}{0.25\sqrt{0.25}}$$

$$= 0.697$$

$$y_2 = \frac{\log(S_2/X_2) + (b_2 - \sigma_2^2/2)T}{\sigma_2\sqrt{T}} = \frac{\log(70.0/72.0) + (0.06 - 0.30^2/2)0.25}{0.30\sqrt{0.25}}$$

$$= -0.162$$

$$\begin{aligned}
M1(y_2 + \sigma_2\sqrt{T}, y_1 + \rho\sigma_2\sqrt{T}; \rho) &= M1(-0.162 + 0.30\sqrt{0.25}, 0.697 \\
&\quad + 0.80 * 0.30\sqrt{0.25}; 0.80) \\
&= 0.484 M2(y_2, y_1; \rho) \\
&= M2(-0.162, 0.697; 0.80) = 0.425 \\
c &= S_2 e^{(b_2-r)T} M1(y_2 + \sigma_2\sqrt{T}, y_1 + \rho\sigma_2\sqrt{T}; \rho) \\
&\quad - X_2 e^{-rT} M2(y_2, y_1; \rho) \\
&= 70.0 e^{(0.06-0.06)*0.25} M1(y_2 + \sigma_2\sqrt{T}, y_1 \\
&\quad + \rho\sigma_2\sqrt{T}; \rho) - 72.0 e^{-0.06*0.25} M2(y_2, y_1; \rho) \\
&= 3.735
\end{aligned}$$

The cash-or-nothing option is valued using code shown in Listing 12.10

```
package FinApps;
import static java.lang.Math.*;
import BaseStats.Probnorm;
public class Cashornonop {
    public Cashornonop(double rate, double yield, double time) {
```

```
        r=rate;
        crate=yield;
        t=time;
        b=crate==0.0?0.0:(b=crate!=r?(r-crate):r);
    }
    double b;
    double crate;
    double r;
    double t;
    public double cashornoCall(double s, double x, double sigma,
                        double payout) {
        return payout*exp(-r*t)*N(d(s,x,sigma));
    }
    public double cashornoPut(double s, double x, double sigma,
                        double payout) {
        return payout*exp(-r*t)*N(-d(s,x,sigma));
    }
    private double d(double s, double x, double sigma) {
        double sig=(sigma*sigma);
        double f= (log(s/x)+((b-sig*0.5)*t))/(sigma*sqrt(t));
        return f;
    }
    private double N(double x) {
        Probnorm p=new Probnorm();
        double ret=x>(6.95)?1.0:x<(-6.95)?0.0:p.ncDisfnc(x);
                    //restrict the range of cdf values to stable values
        return ret;
    }
```

LISTING 12.10. Cash-or-Nothing Option Valuation

12.7.2. Exchange Assets Option

An option to exchange one asset for another on expiration first described by Margrabe (1978). Gives the holder a right to exchange asset S_1 for asset S_2. Exchange options are involved in a range of applications including bonds and futures.

The payoff from an exchange option is given by:

$$\max\left[\lambda(S_1(T) - S_2(T)), 0\right], \tag{12.7.3}$$

where $\lambda = 1$; *call*, $\lambda = -1$; *put*.

There are closed from formulae proposed by Margrabe to provide valuations for these options as follows:

$$c = S_1(T)\varphi(d_1) - S_2(T)\varphi(d_2) \tag{12.7.4}$$

where $\varphi(.)$ is the standard Gaussian distribution function and;

$$d_1 = \frac{\log\left(\dfrac{S_1(T)}{S_2(T)}\right)}{\sqrt{\sigma_1^2 + \sigma_2^2 - 2\rho_{1,2}\sigma_1\sigma_2 T}} + \frac{1}{2}\sqrt{(\sigma_1^2 + \sigma_2^2 - 2\rho_{1,2})T} \tag{12.7.5}$$

$$d_2 = d_1 - \sqrt{(\sigma_1^2 + \sigma_2^2 - 2\rho_{1,2}\sigma_1\sigma_2)T} \tag{12.7.6}$$

This model assumes that interest rates are given a constant value for the option lifetime and the assets have no yield. To incorporate the effects of yield and the prevailing risk free rates the following formulae are used:

$$call = S_1 e^{(b_1 - r)T} N(d_1) - S_2 e^{(b_2 - r)T} N(d_2) \tag{12.7.7}$$

For assets that have proportional quantities the general formula becomes:

$$c = Q_1 S_1 e^{(b_1 - r)T} N(d_1) - Q_2 S_2 e^{(b_2 - r)T} N(d_2)$$

where Q is the relative quantity of the asset.

$$d_1 = \frac{\log(Q_1 S_1 / Q_2 S_2) + (b_1 - b_2 + \varsigma^2/2)T}{\varsigma\sqrt{T}} \tag{12.7.8}$$

where $\varsigma = \sqrt{(\sigma_1^2 + \sigma_2^2 - 2\rho_{1,2}\sigma_1\sigma_2}$, in this case

$$d_2 = d_1 - \varsigma\sqrt{T} \tag{12.7.9}$$

Exchange options are sensitive to the correlation between assets. The call option values for various degrees of correlation are shown in Figure 12.5. The higher degree of correlation tends to decrease the option value

Figure 12.6 shows the sensitivity of exchange options to changes in volatility of one asset. As volatility increases so does the option value. The degree of correlation effect is maintained.

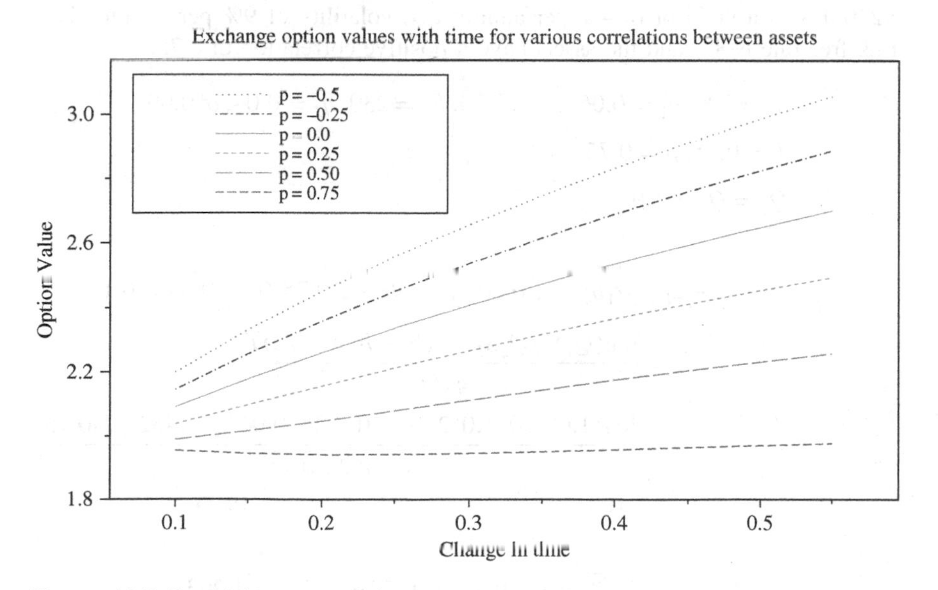

FIGURE 12.5. Sensitivity to correlation between assets.

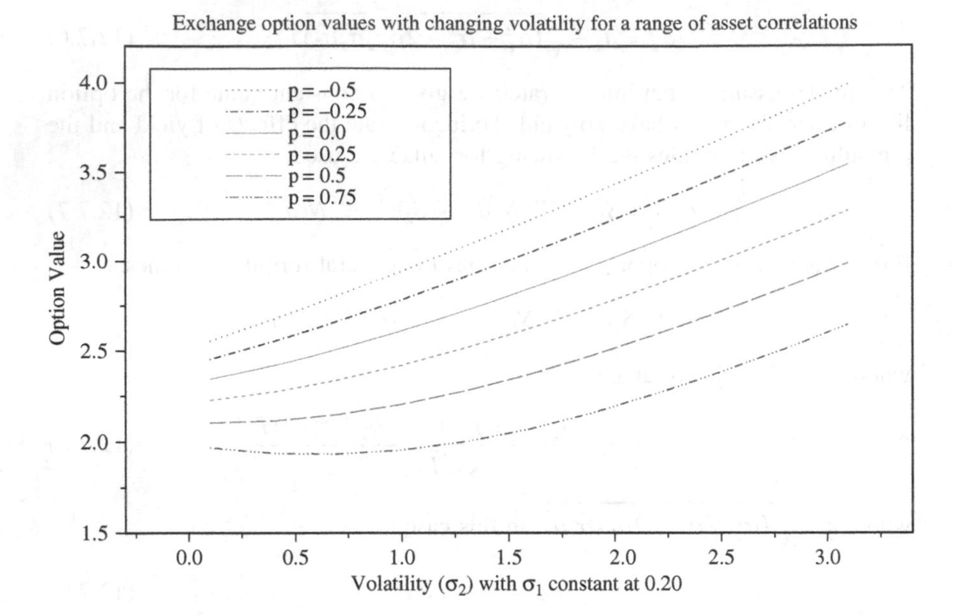

FIGURE 12.6. Exchange option sensitivity to asset volatility.

Example 12.10

Value an exchange option to exchange asset 1 for asset 2, that has nine months to maturity. The price of asset 1 is \$200.0, which pays a dividend yield of of 6% per annum and has a volatility of 20% per annum. Asset 2 has a price of \$230.0 with a yield of of 4% per annum and volatility of 9% per annum. The risk free rate is 8% and the assets have a positive correlation of 0.75.

$$S_1 = 200,\ q_1 = 0.06,\ \sigma_1 = 0.20,\ S_2 = 230,\ q_2 = 0.04,\ \sigma_2 0.09$$

$$T = 0.75,\ \rho = 0.75$$

$$Q_1 = Q_2 = 1.0$$

$$\varsigma = \sqrt{(\sigma_1^2 + \sigma_2^2 - 2\rho_{1,2}\sigma_1\sigma_2} = \sqrt{0.20^2 + 0.09^2 - 2{*}0.75{*}0.20{*}0.09} = 0.1452$$

$$\begin{aligned} d_1 &= \frac{\log(Q_1 S_1/Q_2 S_2) + (b_1 - b_2 + \varsigma^2/2)T}{\varsigma\sqrt{T}} \\ &= \frac{\log(1.0{*}200/1.0{*}230) + (0.02 - 0.04 + 0.1452^2/2)0.75}{0.1452\sqrt{0.75}} \\ &= -1.167 \end{aligned}$$

$$d_2 = d_1 - \varsigma\sqrt{T} = -1.167 - 0.1452\sqrt{0.75} = -1.293$$

$$N(d_1) = N(-1.167) = 0.121, N(d_2) = N(-1.293) = 0.097$$

$$\text{Using, } c = Q_1 S_1 e^{(b_1 - r)T} N(d_1) - Q_2 S_2 e^{(b_2 - r)T} N(d_2) = 1.0^{*}200$$
$$e^{(0.02-0.08)T} N(d_1) - 1.0^{*}230 e^{(0.04-0.08)T} N(d_2) = 1.367$$

The code for valuing an exchange-one-asset for another option is shown in Listing 12.11.

12.7.3. American Exchange Option

Bjerksund & Stensland (1993) have provided a closed form analytical solution for pricing an American exchange one asset for another option.The process to exchange asset S_2 for asset S_1 involves using a standard (plain vanilla) American call with the input parameters:

S_1 for asset price, S_2 the strike price, $\varsigma = \sqrt{\sigma_1^2 + \sigma_2^2 - 2\rho_{1,2}\sigma_1\sigma_2}$ the asset volatility, the adjusted risk free rate $(r - b_2)$ and a risk adjusted cost of carry $(b_1 - b_2)$. The call price is given by:

$$c = camer(Q_1 S_1, Q_1 S_2, T, (r - b_2), \varsigma, (b_1 - b_2)) \tag{12.7.10}$$

Figure 12.7 shows the value of an American exchange option. In general the behaviour is similar to the European version, with the exception that the value of an American option is higher than the equivalent European.

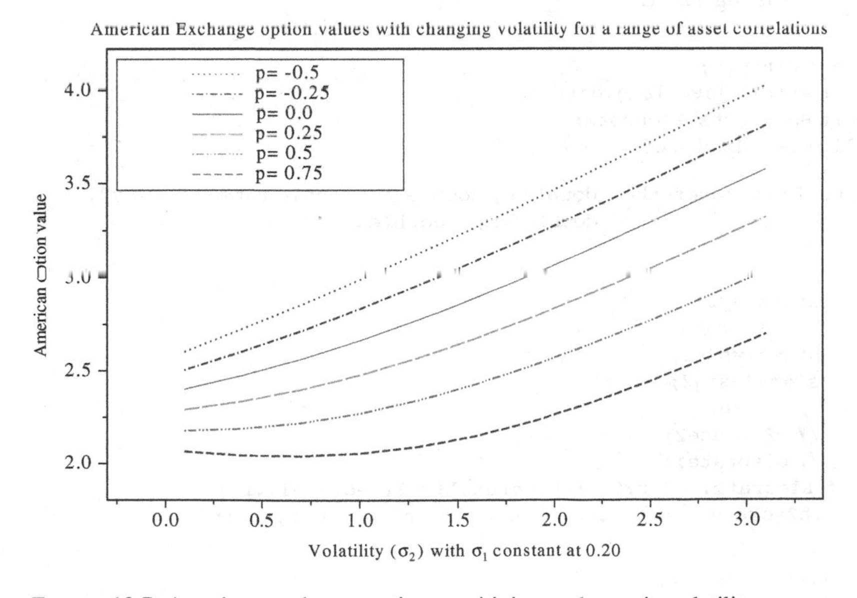

FIGURE 12.7. American exchange option sensitivity to change in volatility.

Example 12.11

An American exchange one asset for another option has nine months to maturity. Asset 1 has a price of \$100.0 and asset 2 has a price of \$90.0. The exchange option is to exchange asset 2 for asset 1. Asset 1 has a yield of 5% and price volatility of 25%, asset 2 has a yield of 5% and price volatility of 25%. The correlation between assets is 0.75 and the risk free rate is 8%. Assume that the quantity for exchange is 1:1 (exchange ratio 1.0).

$S_1 = 100, y_1 = 0.05, \sigma_1 = 0.25, S_2 = 90, y_2 = 0.03, \sigma_2 = 0.25, r = 0.08, Q_1 = Q_2 = 1, \rho = 0.75$

$$\varsigma = \sqrt{(\sigma_1^2 + \sigma_2^2 - 2\rho_{1,2}\sigma_1\sigma_2)} = 0.17677.$$

$$c = camer(Q_1 S_1, Q_1 S_2, T, (r - b_2), \varsigma, (b_1 - b_2))$$

From the standard American call parameters:

$$\beta = 2.934, B_\infty = 136.527, B_0 = 90.0, h(T) = -0.563$$

$$I^* = 110.036, \alpha = 2.047E - 5$$

The call price is \$11.1761. The equivalent European call price is \$10.7333.

The code for valuing an American exchange one asset for another option is shown in Listing 12.11.

```
package FinApps;
import static java.lang.Math.*;
import BaseStats.Probnorm;
public class Exchonefother {

    public Exchonefother(double p, double y1, double rate,double y2,
                         double sig1, double sig2)
    {
        cratel=y1;
        crate2=y2;
              r=rate;
        sigmal=sig1;
        sigma2=sig2;
               =p;
        // b2=brate2;
        // b1=bratel;
         b1=cratel==0.0?0.0:(b1=cratel!=r?(r-cratel):r);
         b2=crate2==0.0?0.0:(b2=crate2!=r?(r-crate2):r);
    }
    private double r;
    private double sigmal;
```

```
    private double sigma2;
    private double b2;
    private double b1;
    private double rho;
    private double d1;
    private double d2;
    private double sigma;
    private double crate1;
    private double crate2;

private void params(double s1,double q1,double s2,
                        double q2, double time)
{
    sigma=(sqrt((sigma1*sigma1)+(sigma2*sigma2)
                        -2.0*rho*sigma1*sigma2));
      d1=((log((q1*s1)/(q2*s2))+(b1-b2+(sigma*sigma)*0.5)
                        *time)/(sigma*sqrt(time)));
      d2=(d1-sigma*sqrt(time));
        }
public double exoneFother(double s1, double q1, double s2,
                        double q2, double time)
{
    Probnorm p = new Probnorm();
    params(s1,q1,s2,q2,time);
    double n=p.ncDisfnc(d1);
    double n2=p.ncDisfnc(d2);
    double c=(q1*s1*exp((b1-r)*time)*n-q2*s2*exp((b2-r)*time)*n2);
    return c;

}
public double amexoneFother(double s1, double q1, double s2,
                              double q2, double time)
{
    params(s1,q1,s2,q2,time);
    Americbs a=new Americbs(b2);
    double c= a.amerBs(s1,s2,sigma,time,r-b2,1);
    return c;

}
```

LISTING 12.11. Exchange-One-Asset-For-Another Valuation

12.8. Sequential Exchange Options

Sequential exchange options are valued on a model of compound options (an option written on another option). Sequential exchange refers to the process where the possibility exists to exchange assets and those exchanges have the potential to generate further exchange possibilities.

At its simplist an exchange option is the right to exchange one asset for another (see above). The straightforward exchange option becomes a simple call or put if either of the assets maintains a constant value. Carr (1988) has developed valuation formulae to price a compound exchange option, where the delivery of an asset is made and an exchange option received. The received exchange option can be again exchanged at a future time.

Sequential exchange options have many possible applications, some quite complex. Carr provides example applications in valuing sinking fund debt, capital budgeting and various other scenarios outside of simple finance.

The fundamental pricing equation developed by Carr provides a generic solution to the sequential valuation equation. This basic formula exhibits different behaviours dependent on the parameter values applied and will decompose into the valuation formulas of Margrabe (which values the simple exchange case) if the exchange ratios (quantity of asset 1 to asset 2) become 0. If the exchange ratio is non zero and positive but the volatility of the delivered asset is zero, the formula becomes the Geske (1979) formula for a compound option with an assumed constant risk free rate. In the final special case, if both volatility is zero and the exchange ratio is 0, the formula becomes the Black-Scholes formula for pricing an option.

The basic valuation formula described by Carr is:

$$VN_2(d_1(\frac{P}{P*},t_1),d_1(P,t_2)-DN_2(d_2(\frac{P}{P*},t_1),d_2(P,t_2))-qDN_1(d_2(\frac{P}{P*},t_1)). \tag{12.8.1}$$

Where

$N(d_1)$ is the standard univariate normal distribution function evaluated at d, $N_2(a,b)$ is the standard bivariate normal distribution with upper and lower integrals a,b and correlation coefficient

$$\rho=\sqrt{\frac{t_1}{t_2}} \tag{12.8.2}$$

V is the optioned asset and D is the delivery asset.

$$d_1(y,t)\equiv\frac{\log y+\sigma^2 t}{\sigma\sqrt{t}}, \tag{12.8.3}$$

$$d_2(y,t)\equiv d_1(y,t)-\sigma\sqrt{t}, \tag{12.8.4}$$

$$P\equiv\frac{V}{D}, \tag{12.8.5}$$

$$\sigma^2\equiv\sigma_\mu^2+\sigma_d^2-2\rho_{1,2}\sigma_u\sigma_d\,. \tag{12.8.6}$$

σ^2 is the instantaneous variance of the percentage change in the ratio $\left(\text{var}\left(\frac{dP}{p}\right)=\sigma^2 dt\right)$.

$\rho_{1,2}$ is the correlation between asset prices.

The compound exchange option is exercised at maturity if the simple option value is in excess of the exercise value, where q is the exchange ratio for assets.

$$VN_1(d_1(P))-DN_1(d_2(P))\geq q$$

The simple option value is dependent on the random values taken by the two assets (ratio $P = V/D$). Taking the delivery asset as numeraire (dividing by D) gives:

$$PN_1(d_1(P)) - N_1(d_2(P)) \geq q \tag{12.8.7}$$

$PN_1(d_1(P))$ is the Black-Scholes formula for a call on the price ratio, with risk free rate of zero and exercise price of unity. The option price of 12.6.1 will increase with P, there is therefore a unique price at which

$$P^* N_1(d_1(P^*)) - N_1(d_2(P^*)) = q \tag{12.8.8}$$

will hold. This is the critical price P^*.

For continuity we will continue with the development of valuation formulae using our familiar terminology, including the term for cost of carry.

S_1 is the optioned asset, S_2 the delivery asset , T_1, T_2 are respectively the time to maturity of the original option and time to maturity of the underlying. b_1, b_2 are the cost of carry for asset 1 and asset 2 respectively and Q the quantity delivered at expiration (the exchange ratio, usually a constant value)

The formula for the value of an option to exchange a quantity Q of asset S_2 for the option to exchange S_2 for S_1 is given by:

$$c = S_1 e^{(b_1 - r)T_2} M_1(d_1, y_1; \rho) - s_2 e^{(b_2 - r)T_2} M_2(d_2, y_2; \rho) - QS_2 e^{(b_2 - r)T_1} N(d_2) \tag{12.8.9}$$

The formula for the value of an option to exchange S_2 for S_1 in return for a quantity Q of S_2 is given by:

$$c = S_2 e^{(b_2 - r)T_2} M_1(d_3, y_2; -\rho) - S_1 e^{(b_1 - r)T_2} M_2(d_4, y_1; -\rho) + QS_2 e^{(b_2 - r)T_1} N(d_2). \tag{12.8.10}$$

Where

ρ is the identity from 12.6.2,

$$d_1 = \frac{\log(S_1/P^* S_2) + (b_1 - b_2 + \sigma^2/2)T_1}{\sigma\sqrt{T_1}} \tag{12.8.11}$$

$$d_2 = d_1 - \sigma\sqrt{T_1} \tag{12.8.12}$$

$$d_3 = \frac{\log(P^* S_2/S_1) + (b_2 - b_1 + \sigma^2/2)T_1}{\sigma\sqrt{T_1}} \tag{12.8.13}$$

$$d_4 = d_3 - \sigma\sqrt{T_1} \tag{12.8.14}$$

$$y_1 = \frac{\log(S_1/S_2) + (b_1 - b_2 + \sigma^2/2)T_2}{\sigma\sqrt{T_2}} \tag{12.8.15}$$

$$y_2 = y_1 - \sigma\sqrt{T_2} \tag{12.8.16}$$

$$y_3 = \frac{\log(S_2/S_1) + (b_2 - b_1 + \sigma^2/2)T_2}{\sigma\sqrt{T_2}} \tag{12.8.17}$$

$$y_4 = y_3 - \sigma\sqrt{T_2} \tag{12.8.18}$$

and σ^2 is given by 12.6.6 giving,

$$\sigma \equiv \sqrt{\sigma^2} \tag{12.8.19}$$

P^* the critical price ratio which is solved by a root finding process where the initial ratio $P_1^* = \dfrac{S_1 e^{(b_1-r)(T_2-T_1)}}{S_2 e^{(b_2-r)(T_2-T_1)}}$ is used in 12.6.8 to give:

$$P_1^* N_1(d_1(P_1^*)) - N_1(d_2(P_1^*)) = q \tag{12.8.20}$$

Simplifying 12.6.20

$$P^* N(z_1) - N(z_2) = Q \tag{12.8.21}$$

$$z_1 = \frac{\log(P^*) + (T_2 - T_1)\sigma^2/2}{\sigma\sqrt{T_2 - T_1}} \tag{12.8.22}$$

$$z_2 = z_1 - \sigma\sqrt{T_2 - T_1} \tag{12.8.23}$$

Table 12.2. shows a range of exchange options on exchange option values for various asset and volatility values with three different asset correlations, for 12.6.9 and 12.6.10.

The value of an option to exchange Q of an asset S_2 for the option to exchange $S_1^{'}$ for S_2 is given by:

$$c = S_2 e^{(b_2-r)T_2} M_1(d_3, y_3; \rho) - S_1 e^{(b_1-r)T_2} M_2(d_4, y_4; \rho) - Q S_2 e^{(b_2-r)T_1} N(d_3) \tag{12.8.24}$$

The value of an option to exchange the option S_1 for S_2 in return for Q of S_2 is given by:

$$c = S_1 e^{(b_1-r)T_2} M_1(d_1, y_4; -\rho) - S_2 e^{(b_2-r)T_2} M_2(d_2, y_3; -\rho) + Q S_2 e^{(b_2-r)T_1} N(d_2) \tag{12.8.25}$$

The critical price ratio is given by:

$$P_2^* = \frac{S_2 e^{(b_2-r)(T_2-T_1)}}{S_1 e^{(b_1-r)(T_2-T_1)}} \tag{12.8.26}$$

used in a root finding process which solves;

$$N(z_1) - P_2^* N(z_2) = Q \tag{12.8.27}$$

TABLE 12.2. Exchange option values satisfying $P^* N(z_1) - N(z_2) = Q$

		Exchange options on Exchange Options									
		P = −0.50		P = −0.25		P = 0.0		P = 0.25		P = 0.50	
σ1	σ2	S1	S2	S1	S2	S1	S2	S1	S2	S1	S2
		Option to Exchange Quantity qs2 for the option to exchange S2 for S1									
		$c = S_1 e^{(b_1-r)T_2} M_1(d_1, y_1; \rho) - S_2 e^{(b_2-r)T_2} M_2(d_2, y_2; \rho) - QS_2 e^{(b_2-r)T_1} N(d_2)$									
0.100	0.100	1.7867	4.2204	1.4133	3.703	1.0362	3.1435	0.6638	2.5275	0.3175	1.8296
0.100	0.150	3.031	5.8045	2.5109	5.1612	1.9709	4.4661	1.4133	3.703	0.8485	2.8438
0.200	0.150	5.7867	8.997	4.929	8.0291	4.0139	6.9747	3.031	5.8045	1.9709	4.4661
0.200	0.200	7.2376	10.604	6.1968	9.4545	5.0757	8.1958	3.8551	6.7887	2.5109	5.1612
0.200	0.250	8.7928	12.298	7.6103	11.0121	6.331	9.6036	4.929	8.0291	3.3669	6.2099
0.250	0.250	10.3192	13.9416	8.963	12.4821	7.4871	10.8773	5.8559	9.0744	4.0139	6.9747
0.250	0.300	11.9176	15.6487	10.4233	14.0532	8.7928	12.298	6.9841	10.3254	4.929	8.0291
0.300	0.300	13.481	17.3083	11.8216	15.5465	10.0032	13.6026	7.9743	11.4093	5.6474	8.8408
0.350	0.300	15.0967	19.0157	13.3033	17.1201	11.3342	15.027	9.1315	12.6641	6.5957	9.8969
0.350	0.350	16.6757	20.6783	14.7251	18.6236	12.5764	16.3491	10.1619	13.7729	7.3628	10.7413

TABLE 12.2. (Continued)

Exchange options on Exchange Options											
		P = −0.50		P = −0.25		P = 0.0		P = 0.25		P = 0.50	
σ1	σ2	S1	S2	S1	S2	S1	S2	S1	S2	S1	S2
Option to Exchange Quantity qs2 for the option to exchange S2 for S1											
$c = S_2 e^{(b_2-r)T_2} M_1(d_3, y_2; -\rho) - S_1 e^{(b_1-r)T_2} M_2(d_4, y_1; -\rho) + QS_2 e^{(b_2-r)T_1} N(d_2)$											
0.100	0.100	1.3957	0.4986	1.0815	0.330	0.7705	0.187	0.4722	0.0793	0.2082	0.0173
0.100	0.150	2.4709	1.1993	2.0172	0.8846	1.5524	0.590	1.0815	0.330	0.6188	0.1281
0.200	0.150	4.9373	3.1874	4.161	2.525	3.3406	1.857	2.4709	1.1993	1.5524	0.590
0.200	0.200	6.2631	4.3673	5.3106	3.5141	4.2934	2.636	3.1991	1.7461	2.0172	0.8846
0.200	0.250	7.6976	5.6927	6.6057	4.6799	5.433	3.6223	4.161	2.525	2.7665	1.4158
0.250	0.250	9.1162	7.0375	7.8554	5.8407	6.4923	4.5762	5.0002	3.2421	3.3406	1.857
0.250	0.300	10.6107	8.4804	9.2133	7.1305	7.6976	5.6927	6.0304	4.1566	4.161	2.525
0.300	0.300	12.0797	9.9179	10.5207	8.3929	8.8218	6.7562	6.941	4.9885	4.8108	3.0777
0.350	0.300	13.6041	11.4249	11.9124	9.7535	10.0643	7.9503	8.0116	5.9878	5.6748	3.8373
0.350	0.350	15.0989	12.9141	13.253	11.0767	11.2289	9.0834	8.9696	6.8973	6.3781	4.472

S1 = 107.0, S2 = 100.0. t1 = 0.50, t2 = 1.0. r= 0.08, b1 = 0.08, b2 = 0.08. Q = 0.10.

$$z_1 = \frac{-\log(P_2^*)(T_2 - T_1)\sigma^2/2}{\sigma\sqrt{T_2 - T_1}} \qquad (12.8.28)$$

$$z_2 = z_1 - \sigma\sqrt{T_2 - T_1} \qquad (12.8.29)$$

Table 12.3. shows a range of exchange options on exchange option values for various asset and volatility values with three different asset correlations, for 12.6.24 and 12.6.25.

Example 12.12

Derive the value of an exchange option on exchange option to exchange quantity 0.10 of asset S_2, for the option to exchange asset S_1 for asset S_2. The correlation between assets is 0.5. The price of asset 1 is £110.0 and asset 2 is £110.0. The time to expiration of the original option is six months, the time to maturity of the underlying option is one year. The risk free rate is 8%, $b_1 = b_2 = 8\%$. The volatility of both assets is 25%.

$N(z_1) - P_2^* N(z_2) = Q$ gives a critical price ratio of 0.908

$$d_1 = \frac{\log(S_1/P^*S_2) + (b_1 - b_2 + \sigma^2/2)T_1}{\sigma\sqrt{T_1}}$$

$$= \frac{\log(110/0.908^*110) + (0.10 - 0.10 + 0.20^2/2)0.75}{0.25\sqrt{0.75}} = 0.640$$

$$d_2 = d_1 - \sigma\sqrt{T_1} = 0.640 - 0.20^*\sqrt{0.75} = 0.4669$$

$$d_3 = \frac{\log(P^*S_2/S_1) + (b_2 - b_1 + \sigma^2/2)T_1}{\sigma\sqrt{T_1}}$$

$$= \frac{\log(0.908^*110/110) + (0.10 - 0.10 + 0.20^2/2)0.75}{0.20\sqrt{0.75}} = -0.4669$$

$$d_4 = d_3 - \sigma\sqrt{T_1} = -0.4669 - 0.20\sqrt{0.75} = -0.640$$

$$y_1 = \frac{\log(S_1/S_2) + (b_1 - b_2 + \sigma^2/2)T_2}{\sigma\sqrt{T_2}}$$

$$= \frac{\log(110/110) + (0.10 - 0.10 + 0.20^2/2)^*1.0}{0.20\sqrt{1.0}} = 0.10$$

$$y_2 = y_1 - \sigma\sqrt{T_2} = 0.10 - 0.20\sqrt{1.0} = -0.099$$

TABLE 12.3. Exchange option values satisfying $N(z_1) - P_2^* N(z_2) = Q$

Exchange options on Exchange Options											
		P = −0.50		P = −0.25		P = 0.0		P = 0.25		P = 0.50	
σ1	σ2	S1	S2	S1	S2	S1	S2	S1	S2	S1	S2
Option to Exchange Quantity S1 for S2 the option to exchange S2 for S1											
$c = S_1 e^{(b_1-r)T_2} M_1(d_1, y_4; -\rho) - S_2 e^{(b_2-r)T_2} M_2(d_2, y_3; -\rho) + Q S_2 e^{(b_2-r)T_1} N(d_2)$											
0.100	0.100	4.4945	6.3192	4.7803	6.7215	5.1333	7.1986	5.5892	7.7746	6.2204	8.4809
0.100	0.150	3.7934	5.2906	4.0516	5.6747	4.3701	6.1405	4.7803	6.7215	5.3451	7.4724
0.200	0.150	2.8506	3.8718	3.0907	4.2328	3.3929	4.6886	3.7934	5.2906	4.3701	6.1405
0.200	0.200	2.5121	3.3673	2.7473	3.7172	3.0471	4.1671	3.4515	4.7769	4.0516	5.6747
0.200	0.250	2.2169	2.9333	2.4359	3.2546	2.7149	3.6689	3.0907	4.2328	3.6455	5.0689
0.250	0.250	1.9755	2.5842	2.1879	2.8911	2.4607	3.2912	2.8327	3.845	3.3929	4.6886
0.250	0.300	1.7612	2.279	1.9605	2.5627	2.2169	2.9333	2.5662	3.4475	3.0907	4.2328
0.300	0.300	1.5809	2.0262	1.7731	2.2959	2.0222	2.6513	2.365	3.1502	2.8873	3.9268
0.350	0.300	1.4187	1.8021	1.6001	2.053	1.8355	2.3843	2.1597	2.8501	2.6529	3.5764
0.350	0.350	1.2796	1.6125	1.4541	1.8507	1.6821	2.1676	1.9986	2.6173	2.4861	3.3287

0.100	0.100	4.9619	3.0577	5.1944	3.1093	5.4897	3.1671	5.8837	3.2304	6.453	3.2925
0.100	0.150	4.4172	2.9159	4.6134	2.9703	4.825	3.034	5.1944	3.1093	5.671	3.1982
0.200	0.150	3.7511	2.7047	3.9126	2.7595	4.1241	2.8279	4.4172	2.9159	4.8625	3.034
0.200	0.200	3.5343	2.6281	3.6835	2.6812	3.8828	2.7495	4.1661	2.841	4.6134	2.9703
0.200	0.250	3.3572	2.5636	3.4875	2.6112	3.6626	2.6739	3.9126	2.7595	4.0374	2.8838
0.250	0.250	3.2219	2.514	3.3404	2.5575	3.5026	2.6167	3.7393	2.7006	4.1241	2.8279
0..250	0.300	3.1102	2.4736	3.2138	2.511	3.3572	2.5636	3.568	2.6403	3.9126	2.7595
0..300	0.300	3.0235	2.4436	3.1162	2.4758	3.2474	2.5233	3.4446	2.5956	3.7754	2.7131
0..350	0.300	2.9521	2.4207	3.0324	2.4466	3.148	2.4871	3.3243	2.5515	3.6229	2.6598
0..350	0.350	2.8969	2.4051	2.9671	2.4253	3.0713	2.4599	3.2344	2.5186	3.5182	2.6223

S1 = 107.0, S2 = 100.0. t1 = 0.50, t2 = 1.0. r = 0.07, b1 = 0.07, b2 = 0.07. Q = 0.10.

$$y_3 = \frac{\log(S_2/S_1) + (b_2 - b_1 + \sigma^2/2)T_2}{\sigma\sqrt{T_2}}$$
$$= \frac{\log(110/110) + (0.10 - 0.10 + 0.20^2/2)^*1.0}{0.20\sqrt{1.0}} = 0.10$$

$$y_4 = y_3 - \sigma\sqrt{T_2} = 0.10 - 0.20\sqrt{1.0} = -0.099$$

$$M_1(d_3, y_3; \rho) = M_1(-0.4669, 0.10; 0.866) = 0.3059$$

$$M_2(d_4, y_4; \rho) = M_2(-0.640, -0.099; 0.866) = 0.2459$$

$$N(d_3) = N(-0.4669) = 0.3202$$

$$c = S_2 e^{(b_2-r)T_2} M_1(d_3, y_3; \rho) - S_1 e^{(b_1-r)T_2} M_2(d_4, y_4; \rho) - QS_2 e^{(b_2-r)T_1} N(d_3)$$
$$= 110e^{(0.10-0.10)1.0} M_1(d_3, y_3; \rho) - 110e^{(0.10-0.10)1.0} M_2(d_4, y_4; \rho)$$
$$-0.10^*110e^{(0.10-0.10)0.75} N(d_3) = 3.068$$

The value is £3.07.

The code for valuing exchange options on exchange options is shown in Listing 12.12.

```
package FinApps;
import static java.lang.Math.*;
import BaseStats.Probnorm;
import BaseStats.Bivnorm;
import CoreMath.NewtonRaphson;
public class Exonexop {
   public Exonexop(double rate, double yield1, double yield2,double
                  sig1,double sig2,double p) {
      r=rate;
      crate1=yield1;
      crate2=yield2;
      sigma1=sig1;
      sigma2=sig2;
      rho=p;
      b1=crate1==0.0?0.0:(b1=crate1!=r?(r-crate1):r);
      b2=crate2==0.0?0.0:(b2=crate2!=r?(r-crate2):r);
   }

   private double crate1;
   private double crate2;
   private double r;
```

```
private double sigma;
private double sigma1;
private double sigma2;
private double b1;
private double b2;
private double rho;
private double y1;
private double y2;
private double y3;
private double y4;
private double qval;
private double d1;
private double d2;
private double d3;
private double d4;
private double t1;
private double t2;
private int typeop=0;
private double sv1;
private double sv2;

Probnorm p=new Probnorm();
private void params(double s1, double s2,double time1,
                    double time2,double q) {
  double pratio;
  t1=time1;
  t2=time2;
  qval=q;
  sigma=(sqrt((sigma1*sigma1)+(sigma2*sigma2)-2*rho*sigma1;
                  *sigma2))
  y1=((log(s1/s2)+(b1-b2+(sigma*sigma)*0.5)*t2)(sigma*sqrt(t2)));
  y3=((log(s2/s1)+(b2-b1+(sigma*sigma)*0.5)*t2) /(sigma*sqrt(t2)));
  y2=(y1-sigma*sqrt(t2));
  y4=(y3-sigma*sqrt(t2));

  if(typeop==1) {
    i1 ival=new i1();
    pratio=ival.pars();
  } else {
    i2 ival=new i2();
    pratio=ival.pars();
  }
  d1=((log(s1/(pratio*s2))+(b1-b2+(sigma*sigma)*0.5)*t1)
    /(sigma*sqrt(t1)));
  d2=(d1-sigma*sqrt(t1));
  d3=((log((pratio*s2)/s1)+(b2-b1+(sigma*sigma)*0.5)*t1)
    /(sigma*sqrt(t1)));
  d4=(d3-sigma*sqrt(t1));

}


public double exchqfs(double s1, double s2,double time1,
                    double time2,double q) {
```

```
    typeop=1;
    sv1=s1;
    sv2=s2;
    params(s1,s2,time1,time2,q);
    double m1=Bivnorm.bivar_params.evalArgs(d1,y1,sqrt(t1/t2));
    double m2=Bivnorm.bivar_params.evalArgs(d2,y2,sqrt(t1/t2));
    double n=p.ncDisfnc(d2);
    double c=(s1*exp((b1-r)*t2)*m1-s2*exp((b2-r)*t2)*m2-qval*s2
           *exp((b2-r)*t1)*n);
    return c;

  }

  public double exchsfq(double s1, double s2,double time1,
                        double time2,double q) {
    typeop=1;
    sv1=s1;
    sv2=s2;
    params(s1,s2,time1,time2,q);
    double m1=Bivnorm.bivar_params.evalArgs(d3,y2,-sqrt(t1/t2));
    double m2=Bivnorm.bivar_params.evalArgs(d4,y1,-sqrt(t1/t2));
    double n=p.ncDisfnc(d3);
    double c=(s2*exp((b2-r)*t2)*m1-s1*exp((b1-r)*t2)*m2+qval*s2
           *exp((b2-r)*t1)*n);
    return c;

  }
  public double exchqfs2(double s1, double s2,double time1,
                        double time2,double q) {
    typeop=0;
    sv1=s1;
    sv2=s2;
    params(s1,s2,time1,time2,q);
    double m1=Bivnorm.bivar_params.evalArgs(d3,y3,sqrt(t1/t2));
    double m2=Bivnorm.bivar_params.evalArgs(d4,y4,sqrt(t1/t2));
    double n=p.ncDisfnc(d3);
    double c=(s2*exp((b2-r)*t2)*m1-s1*exp((b1-r)*t2)*m2-qval*s2
           *exp((b2-r)*t1)*n);
    return c;

  }
  public double exchs1fs2(double s1, double s2,double time1,
                          double time2,double q) {
    typeop=0;
    sv1=s1;
    sv2=s2;
    params(s1,s2,time1,time2,q);
    double m1=Bivnorm.bivar_params.evalArgs(d1,y4,-sqrt(t1/t2));
    double m2=Bivnorm.bivar_params.evalArgs(d2,y3,-sqrt(t1/t2));
    double n=p.ncDisfnc(d2);
    double c=(s1*exp((b1-r)*t2)*m1-s2*exp((b2-r)*t2)*m2+qval*s2
           *exp((b2-r)*t1)*n);
    return c;

  }
```

```
private class i1 extends NewtonRaphson {
   i1() {
      accuracy(1e-6,10);
   }
   public double pars() {
      double seedval=(sv1*exp((b1-r)*(t2-t1))/sv2
                         *exp((b2-r)*(t2-t1)));
      double sval2=newtraph(seedval);
      return sval2;
   }
   public double newtonroot(double rootvalue) {
      double solution =0.0;
      double z1=((log(rootvalue)+((t2-t1)*(sigma*sigma)*0.5))
              /(sigma*sqrt(t2-t1)));
      double z2=(z1-sigma*sqrt(t2-t1));
      double factor1=p.ncDisfnc(z1);
      double factor2=p.ncDisfnc(z2);
      solution =(((rootvalue*factor1)-factor2)-qval);

      return solution;
   }
}

private class i2 extends NewtonRaphson {
   i2() {
      accuracy(1e-6,10);
   }
   public double pars() {
      double seedval=(sv2*exp((b2-r)*(t2-t1))/sv1
                         *exp((b1-r)*(t2-t1)));
      double sval2=newtraph(seedval);
      return sval2;
   }
   public double newtonroot(double rootvalue) {
      double solution =0.0;
      double z1=((-log(rootvalue)+((t2-t1)*(sigma*sigma)*0.5))
              /(sigma*sqrt(t2-t1)));
      double z2=(z1-sigma*sqrt(t2-t1));
      double factor1=p.ncDisfnc(z1);
      double factor2=p.ncDisfnc(z2);
      solution=((factor1-(rootvalue*factor2))-qval);

      return solution;
   }
      }
```

LISTING 12.12. Listing Exchange Options on Exchange Option Valuation

References

Bjerksund, P. and G. Stensland (1993). "American Exchange Options and a Put-Call Transformation: A Note," *Journal of Business Finance and Accounting*, 20(5), 761–764.

Carr, P. P. (1988). "The Valuation of Sequential Exchange Opportunities," *Journal of Finance*, 43, 1235–1256.

Fischer, S. (1978). "Call option pricing when the exercise price is uncertain and the valuation of index bonds." *Journal of Finance*, 33 (March), 169–176.

Geske, R. (1977). "The Valuation of Corporate Liabilities as Compound Options," *Journal of Financial and Quantitative Analysis*, 541–552.

Geske, R. (1979). "The Valuation of Compound Options," *Journal of Financial Economics*, 7, 63–81.

Haug, E. G. (1998). "*The complete guide to option pricing formulas*," McGraw-Hill.

Jennergren, L. P. and B. Naslund (1993). "A Comment on Valuation of Executive Stock Options and the FASB Proposal," *The Accounting Review*, 68(1), 179–183.

Longstaff, F. A. (1990). "Pricing Options and Extendible Maturities: Analysis and Applications," *Journal of Finance*, 45(3), 935–957.

Margrabe, W. (1978). "The Value of an Option to Exchange One Asset for Another," *Journal of Finance*, 33(1), 177–186.

Pechtl, A. (1995). "Classified Information," *Risk Magazine*, p. 8.

Rubinstein, M. (1990). "Forward Start Options" published as "Pay now, choose later" in *Risk Magazine* 4(1991), 13.

Rubinstein, M. (1991). "One for Another," *Risk Magazine*, 4(7).

Whaley, R.E. (1981). "On the Valuation of American Call Options on Stocks with Known Dividends," *Journal of Financial Economics*, 9, 207–211.

Zhang, P.G. (1995). "Correlation Digital Options," *Journal of Financial Engineering*, 4, 75–96.

Schnusenberg, O. and W. R. McDaniel (2000). " How to Value Indexed Executive Stock Options," *Journal of Financial and Strategic Decisions*, 13(3), 45–48.

13 Complex Conditional Options

Complex conditional options are those which involve an added degree of complexity in monitoring the underlying behaviour. In some cases the assumption of path independency is abandoned and the payoff is linked to the previous behaviour of the underlying or another asset.

13.1. Fixed Look Back Options

The payoff from a lookback option is dependent on the maximum or minimum price that the underlying asset achieves during the lifetime of the option. For a European lookback call the payoff is the amount by which the asset price at expiry has exceeded the minimum asset price achieved during the option life. For a European put option the payoff is the amount by which the maximum asset price has exceeded the minimum during the lifetime of the option.

Lookback options are sensitive to the frequency of price measurement; prices can be assumed to be constantly or discretely monitored. The following formulae assume continuous measurement.

13.1.1. Fixed Strike Lookback Call Option

The payoff from a lookback call is dependent on the difference between the strike price at expiry and the minimum achieved in the option life. The call effectively presents the holder an opportunity to buy an asset at the cheapest price it has been at during the option life.

If $S(t)$ is the price of a stock at time $t, t \geq 0$ and we assume that:

$$S(t) = S(0)e^{x(t)} \tag{13.1.1}$$

where $x(t)$ denotes a Wiener process with drift μ and volatility σ. A European lookback call with maturity at time T and strike price K has a payoff at time T given by:

$$W = (\max(L, \max(S(t) - K, 0))) \tag{13.1.2}$$

The variable L is a positive value $L \geq S(0)$ that represents the maximum level of the stock's past price. The possibility exists that $K \geq L$ so that the strike is greater than a previous stock price high or $K < L$, where the strike is less than any previous stock maximum. For the former a payoff at expiry is proportional to the maximum value of the stock in the option lifetime minus the ratio of the strike to stock price at time 0.

The price of a fixed price lookback call in the Black Scholes framework for $L < K$ is:

$$S(0)e^{-\xi T}\Phi(d_{1,k}) - Ke^{-rT}\Phi(d_{2,K})$$
$$+\frac{S(0)\sigma^2}{2(r-\xi)}\left\{e^{-\xi T}\Phi(d_{1,K}) - \left[\frac{K}{S(0)}\right]^{2(r-\xi)/\sigma^2} e^{-rT}\Phi(d_{3,K})\right\} \tag{13.1.3}$$

The first two terms equate to the Black Scholes formula for a plain vanilla option on a dividend paying stock. 13.1.3 is due to Conze and Viswanathan (1991). The parameters are:

$d_{n,K} = d_n(\log(\frac{K}{S(0)}), \delta)$, where the identities can be written as, $\mu = r - \xi - \frac{\sigma^2}{2}$ and $\delta = -r + \xi$.

δ is the 'force of interest', ξ is the dividend yield, r is the risk free rate and μ is the drift rate.

In general, the terms for d_n are ;

$$d_1(\mu, \delta) = \frac{-\mu + (-\delta + \sigma^2/2)T}{\sigma\sqrt{T}} \tag{13.1.4}$$

$$d_2(\mu, \delta) = \frac{(-1 - 2\delta/\sigma^2)\sigma\sqrt{T}}{2} = d_1(\mu, \delta) - \sigma\sqrt{T} \tag{13.1.5}$$

$$d_3(\mu, \delta) = \frac{(2\delta/\sigma^2 + 1)\sigma\sqrt{T}}{2} = d_1(\mu, \delta) - \frac{2\sigma\sqrt{T}}{\sigma} \tag{13.1.6}$$

$$d_{1,K} = d_1(\log(K/S(0)), -r + \xi),\ \Phi(d_{1,K}) = \Phi(\log(K/S(0)), -r + \xi)$$

where Φ is the univariate normal distribution function.

Given that the leading terms $S(0)e^{-\xi T}\Phi(d_{1,k}) - Ke^{-rT}\Phi(d_{2,k})$ are the Black Scholes plain vanilla value, the remaining term $\frac{S(0)\sigma^2}{2(r-\xi)}\left\{e^{-\xi T}\Phi(d_{1,k}) - \left[\frac{K}{S(0)}\right]^{2(r-\xi)/\sigma^2} e^{-rT}\Phi(d_3, k)\right\}$ represents the excess price for being able to exercise the plain vanilla at the maximum price that the stock obtains.

For the case where the strike price is less than or equal to the maximum stock price obtained. The valuation formula is:

$$S(0)e^{-\xi T}\Phi(d_{1,L}) + Le^{-rT}\Phi(-d_{2,L}) - Ke^{-rT} + \frac{S(0)\sigma^2}{2(r-\xi)}\left\{e^{-\xi T}\Phi(d_{1,L}) - \left[\frac{L}{S(0)}\right]^{2(r-\xi)/\sigma^2} e^{-rT}\Phi(d_3, L)\right\} \tag{13.1.7}$$

13.1.2. Fixed Strike Lookback Put

For the case of a fixed strike lookback put the with past minimum higher than the fixed strike:

$$p = Ke^{-rT}\Phi(-d_{2,K}) - S(0)e^{-\xi T}\Phi(-d_{1,K}) + \frac{S(0)\sigma^2}{2(r-\xi)}\left\{\left[\frac{K}{S(0)}\right]^{2(r-\xi)/\sigma^2} e^{-rT}\Phi(-d_{3,K}) - e^{-\xi T}\Phi(-d_{1,K})\right\} \tag{13.1.8}$$

The two terms $Ke^{-rT}\Phi(-d_{2,K}) - S(0)e^{-\xi T}\Phi(-d_{1,K})$, represent a Black-Scholes formula for a plain vanilla put option on a dividend paying asset.

The final term, $\frac{S(0)\sigma^2}{2(r-\xi)}\left\{\left[\frac{K}{S(0)}\right]^{2(r-\xi)/\sigma^2} e^{-rT}\Phi(-d_{3,K}) - e^{-\xi T}\Phi(-d_{1,K})\right\}$ represents a premium to be paid for the right to exercise at the minimum stock price.

The case where the strike is greater or equal to the past minimum:

$$Ke^{-rT} - Le^{-rT}\Phi(d_{2,L}) - S(0)e^{-\xi T}\Phi(-d_{1,L}) + \frac{S(0)\sigma^2}{2(r-\xi)}\left\{\left[\frac{L}{S(0)}\right]^{2(r-\xi)/\sigma^2} e^{-rT}\Phi(-d_{3,L}) - e^{-\xi T}\Phi(-d_{1,L})\right\} \tag{13.1.9}$$

If we incorporate our usual terminology to include the cost of carry term, then transpose $L = S_{\max}$, *call* : $L = S_{\min}$, *put* The usual cost of carry term is b and assume $\xi = 0$, so that $b = r$.

Then, 13.1.3 becomes

$$S(0)e^{(b-r)T}\Phi(d_1) - Ke^{-rT}\Phi(d_2) + \frac{S(0)e^{-rT}\sigma^2}{2b}\left\{-\left[\frac{S}{X}\right]^{2b/\sigma^2}\Phi(d_1 - \frac{2b}{\sigma}\sqrt{T}) + e^{bT}\Phi(d_1)\right\},$$

further if we define

$$d_1 = \frac{\log(S/X) + (b + \sigma^2/2)T}{\sigma\sqrt{T}}$$

and

$$d_3 = d_1(u,\delta) + \frac{2\delta\sqrt{T}}{\sigma}$$

then for computational ease we can represent 13.1.3 as:

$$S(0)\Phi(d_{1,k}) - Ke^{-rT}\Phi(d_{2,K}) + \frac{S(0)e^{-rT}\sigma^2}{2b}\left\{-\left[\frac{K}{S(0)}\right]^{2b/\sigma^2}\Phi(d_{1,K}) + e^{-rT}\Phi(d_{3,K})\right\}$$

We can do this for the remaining three cases and dropping 0 from $S(0)$ and the $_{K,L}$ terms from Φ, we get: 13.1.4 as:

$$e^{-rT}(S_{\max} - K) + Se^{(b-r)T}\Phi(-d_4) - S_{\max}e^{-rT}\Phi(d_5) + \frac{Se^{-rT}\sigma^2}{2b}\left\{-\left[\frac{S}{S_{\max}}\right]^{-2b/\sigma^2} e^{-rT}\Phi(d_4 - \frac{2b}{\sigma}\sqrt{T}) + e^{bT}\Phi(d_5)\right\}$$

where

$$d_4 = \frac{\log(S/S_{\max}) + (b+\sigma^2/2)T}{\sigma\sqrt{T}}$$

$$d_5 = d_4 - \sigma\sqrt{T}$$

13.1.5 becomes:

$$p = Ke^{-rT}\Phi(-d_2) - Se^{(b-r)T}\Phi(-d_1) + \frac{Se^{-rT}\sigma^2}{2b}\left\{\left[\frac{S}{K}\right]^{-2b/\sigma^2}\Phi(-d_1 + \frac{2b}{\sigma}\sqrt{T}) - e^{-bT}\Phi(-d_1)\right\}$$

13.1.6 becomes:

$$e^{-rT}(K - S_{\min}) - Se^{(b-r)T}\Phi(-d_6) + S_{\min}e^{-rT}\Phi(-d_7) + \frac{Se^{-rT}\sigma^2}{2b}\left\{\left[\frac{S}{S_{\min}}\right]^{-2b/\sigma^2} e^{-rT}\Phi(-d_6 + \frac{2b}{\sigma}\sqrt{T}) - e^{bT}\Phi(-d_6)\right\}$$

where

$$d_6 = \frac{\log(S/S_{\min}) + (b+\sigma^2/2)T}{\sigma\sqrt{T}}$$

$$d_7 = d_6 - \sigma\sqrt{T}$$

Table 13.1(a). shows a range of fixed strike lookback call options for a range of strike prices with different volatilities. Table 13.1(b). shows a similar range for fixed strike Lookback put options.

Fixed strike lookback options are valued using code implemented in Listing 13.1.

```
package FinApps;
import static java.lang.Math.*;
import BaseStats.Probnorm;
import java.text.*;
import java.io.*;
public class FixedXlook {

   public FixedXlook(double rate, double yield) {
      crate=yield;
      r=rate;
      brate=crate==0.0?0.0:(brate=crate!=r?(r-crate):r);

   }
   private double r;
   private double crate;
   private double brate;

   public double callFixlook(double s,double smax, double k,
                             double t, double sigma) {
      Probnorm p=new Probnorm();
      double sig=(sigma*sigma);
      double callvalue;
      if(k>smax) {
         double d1=((log(s/k)+(brate+(sig*0.5))*t)/
                  (sigma*sqrt(t)));
         double d2=(d1-sigma*sqrt(t));
         double n1=(p.ncDisfnc(d1));
         double n2=(p.ncDisfnc(d2));
         double n3=(p.ncDisfnc(d1-((2.0*brate/sigma)*sqrt(t))));
         double term1=(-pow((s/k),(-2.0*brate/(sigma*sigma)))
                        *n3+exp(brate*t)*n1);
         callvalue=(s*exp((brate-r)*t)*n1-k*exp(-r*t)
                  *n2+s*exp(-r*t)*((sigma*sigma)/
                  (2.0*brate))*term1);
      } else {
         double e1=((log(s/smax)+(brate+(sig*0.5))*t)/
                  (sigma*sqrt(t)));
         double e2=(e1-sigma*sqrt(t));
         double na=(p.ncDisfnc(e1));
         double nb=(p.ncDisfnc(e2));
         double nc=(p.ncDisfnc(e1-((2.0*brate/sigma)*sqrt(t))));
         double term=(-pow((s/smax),(-2.0*brate/(sig)))
                        *nc+exp(brate*t)*na);
         callvalue=(exp(-r*t)*(smax-k)+s*exp((brate-r)*t)
                  *na-smax*exp(-r*t)*nb+s*
exp(-r*t)*((sigma*sigma)/(2.0*brate))*term);
      }
      return callvalue;

   }
```

```
public double putFixlook(double s,double smin, double k,
                          double t, double sigma) {
   Probnorm p=new Probnorm();
   double sig=(sigma*sigma);
   double putvalue;

   if(k<smin) {
      double d1=((log(s/k)+(brate+(sig*0.5))*t)/(sigma*sqrt(t)));
      double d2=(d1-sigma*sqrt(t));
      double n1=(p.ncDisfnc(-d1));
      double n2=(p.ncDisfnc(-d2));
      double n3=(p.ncDisfnc(-d1+((2.0*brate/sigma)*sqrt(t))));
      double term1=(pow((s/k),(-2.0*brate/(sig)))
                  *n3-exp(brate*t)*n1);
      putvalue=(k*exp(-r*t)*n2-s*exp((brate-r)*t)*n1+s*exp(-r*t)
               *((sigma*sigma)/(2.0*brate))*term1);
   } else {
      double d5=((log(s/smin)+(brate+(sig*0.5))*t)/
               (sigma*sqrt(t)));
      double d6=(d5-sigma*sqrt(t));
      double na=(p.ncDisfnc(-d5));
      double nb=(p.ncDisfnc(-d6));
      double nc=(p.ncDisfnc(-d5+((2.0*brate/sigma)*sqrt(t))));
      double term=(pow((s/smin),(-2.0*brate/(sig)))
                 *nc-exp(brate*t)*na);
      putvalue=(exp(-r*t)*(k-smin)-s*exp((brate-r)*t)
               *na+smin*exp(-r*t)*nb+s*exp(-r*t)*((sig)/
               (2.0*brate))*term);
   }
   return putvalue;
}
```

LISTING 13.1. Fixed Strike Lookback Option Valuation.

TABLE 13.1(a). Fixed Strike Lookback call values.

Fixed Strike Lookback Call option values							
Time	S = 100.0, S max = 100.0, r = 0.10, b = 0.10						
	Strike (X)	σ = 0.10	σ = 0.15	σ = 0.20	σ = 0.25	σ = 0.30	σ = 0.35
	90.000	18.0249	20.7866	23.6825	26.6721	29.7419	32.8864
	95.000	13.2687	16.0304	18.9263	21.916	24.9858	28.1302
T= 0.5	100.000	8.5126	11.2743	14.1702	17.1598	20.2296	23.3741
	105.000	4.3908	7.0666	9.8905	12.8246	15.8512	18.9615
	110.000	1.8083	4.0374	6.5832	9.3255	12.2115	15.2143
	90.000	22.8483	26.5695	30.5972	34.8311	39.2358	43.7969
	95.000	18.3242	22.0454	26.0731	30.3069	34.7116	39.2728
T =1.0	100.000	13.800	17.5212	21.5489	25.7827	30.1874	34.7486
	105.000	9.5445	13.2816	17.2965	21.5125	25.9002	30.4463
	110.000	6.0419	9.6924	13.6173	17.7574	22.0828	26.5778

TABLE 13.1(b). Fixed Strike Lookback put values.

Time	Fixed Strike Lookback Put option values						
	S = 100.0, S min = 100.0, r = 0.10, b = 0.10						
	Strike (X)	$\sigma = 0.10$	$\sigma = 0.15$	$\sigma = 0.20$	$\sigma = 0.25$	$\sigma = 0.30$	$\sigma = 0.35$
	90.000	0.0877	0.7709	2.0931	3.7928	5.6898	7.6849
	95.000	0.6899	2.3658	4.4448	6.6631	8.9213	11.1771
T= 0.5	100.000	3.3917	5.8486	8.3177	10.7587	13.1579	15.5098
	105.000	8.1478	10.6047	13.0739	15.5148	17.914	20.266
	110.000	12.904	15.3609	17.830	20.271	22.6702	25.0221
	90.000	0.230	1.5127	3.6295	6.1418	8.8174	11.5434
	95.000	1.0534	3.4333	6.2813	9.2563	12.2375	15.176
T =1.0	100.000	3.8079	6.9343	10.1294	13.2926	16.3889	19.4036
	105.000	8.3321	11.4585	14.6535	17.8168	20.913	23.9278
	110.000	12.8563	15.9827	19.1777	22.341	25.4372	28.452

13.2. Floating Strike Look Back Options

European floating strike call option gives the holder a right to purchase the underlying at the lowest observed price during the option lifetime. A European floating strike lookback put gives the holder a right to sell the underlying at the highest price obtained by the asset during the option lifetime. The floating strike option does not have a payoff dependent on the previously fixed strike price, rather the payoff is dependent on the final price achieved by the underlying stock at expiry. The floating strike put and call formula (with the dividend yield assumed zero) is due to Goldman, Sosin & Gatto (1979).

13.2.1. Floating Strike Lookback Put

European Floating Strike Lookback Put

The payoff at exercise is:

$$\max(L, S(t)_{\max}) - S(T) \tag{13.2.1}$$

which is effectively $S_{\max} - S$.

The put value at time zero is the same as the Black Scholes value of a fixed strike call with the discounted strike Ke^{-rT} replaced by the time zero price of the T payoff $S(T)$.

The value of a put is therefore:

$$Le^{-rT}\Phi(-d_{2,L}) - S(0)e^{-\xi T}\Phi(d_{1,L})$$
$$+\frac{S(0)\sigma^2}{2(r-\xi)}\left\{e^{-\xi T}\Phi(d_1, L) - \left[\frac{L}{S(0)}\right]^{2(r-\xi)/\sigma^2} e^{-rT}\Phi(d_3, L)\right\} \tag{13.2.2}$$

The first two terms are the Black-Scholes valuation for a plain vanilla put on a non dividend paying stock.

If we rearrange the formula of 13.2.2 into the more easily computed form including a cost of carry term and including the parameter transposition as in the fixed strike case:

$$S_{\max}e^{-rT}\Phi(-b_2) - Se^{(b-r)T}\Phi(-b_1) + \frac{Se^{-rT}\sigma^2}{2b}\left\{-\left[\frac{S}{S_{\max}}\right]^{-2b/\sigma^2} e^{-rT}\Phi(b_1 - \frac{2b}{\sigma}\sqrt{T}) + e^{bT}\Phi(b_1)\right\} \tag{13.2.3}$$

Where

$$b_1 = \frac{\log(S/S_{\max}) + (b+\sigma^2/2)T}{\sigma\sqrt{T}}$$

$$b_2 = b_1 - \sigma\sqrt{T}$$

13.2.2. Floating Strike Lookback Call

The payoff is given by:

$$S(T) - \min(L, S(t)_{\min}) \tag{13.2.4}$$

which is effectively, $S - S_{\min}$. The value of a floating strike lookback call is given by:

$$S(0)e^{-\xi T}\Phi(d_1, L) - Le^{-rT}\Phi(d_{2,L}) + \frac{S(0)\sigma^2}{2(r-\xi)}\left\{\left[\frac{L}{S(0)}\right]^{2(r-\xi)/\sigma^2} e^{-rT}\Phi(-d_{3,L}) - e^{-\xi T}\Phi(-d_{1,L})\right\} \tag{13.2.5}$$

where d_1, d_2, d_3 are computed as in 13.1.4–13.1.6. Rearranging this formula in terms of cost of carry and transposing the parameters we get:

$$Se^{(b-r)T}\Phi(a_1) + S_{\min}e^{-rT}\Phi(a_2) + \frac{Se^{-rT}\sigma^2}{2b}\left\{\left[\frac{S}{S_{\min}}\right]^{-2b/\sigma^2} \Phi(-a_1 + \frac{2b}{\sigma}\sqrt{T}) - e^{bT}\Phi(-a_1)\right\} \tag{13.2.6}$$

Where

$$a_1 = \frac{\log(S/S_{\min}) + (b+\sigma^2/2)T}{\sigma\sqrt{T}}$$

$$a_2 = a_1 - \sigma\sqrt{T}$$

Example 13.0

A floating strike lookback call option has six months to expiry gives the holder a right to purchase an underlying asset at the lowest price observed during the option lifetime. The minimum price so far is \$75.0 and the current stock price is \$100.0. The risk free rate is 8%, the stock yield pays 5% and the stock volatility is 30%. What is the option value?

$$S = 100.0,\ S_{\min} = 75.0,\ T = 0.5,\ r = 0.08,\ q = 0.08,\ q = 0.05,\ \sigma = 0.30$$

$$\begin{aligned} a_1 &= \frac{\log(S/S_{\min}) + (b + \sigma^2/2)T}{\sigma\sqrt{T}} \\ &= \frac{\log(100/75) + (0.03 + 0.30^2/2)0.5}{0.30\sqrt{0.5}} = 1.532 \end{aligned}$$

$$a_2 = a_1 - \sigma\sqrt{T} = 1.532 - 0.30\sqrt{0.5} = 1.3207$$

$$\Phi(a_1) = \Phi(1.532) = 0.937,\ \ \Phi(-a_1) = \Phi(-1.532) = 0.0626,$$

$$\Phi(a_2) = \Phi(1.3207) = 0.9067.$$

$$\Phi(-a_1 + \frac{2b}{\sigma}\sqrt{T}) = \Phi(-1.532 + \frac{2*0.03}{0.30}\sqrt{0.5}) = 0.0626$$

$$\begin{aligned} &Se^{(b-r)T}\Phi(a_1) + S_{\min}e^{-rT}\Phi(a_2) \\ &+ \frac{Se^{-rT}\sigma^2}{2b}\left\{\left[\frac{S}{S_{\min}}\right]^{-2b/\sigma^2}\Phi(-a_1 + \frac{2b}{\sigma}\sqrt{T}) - e^{bT}\Phi(-a_1)\right\} \\ &= 100e^{(0.03-0.08)0.5}\Phi(a_1) + 75e^{-0.08*0.5}\Phi(a_2) \\ &+ \frac{100e^{\ 0.088*0.5}*0.30^2}{2*0.03}\left\{\left[\frac{100}{75}\right]^{-2*0.03/0.30^2}\right. \\ &\left.\times\,\Phi(-a_1 + \frac{2*0.03}{0.30}\sqrt{0.5}) - e^{0.03*0.5}\Phi(-a_1)\right\} = 26.678 \end{aligned}$$

The value of the floating strike call option is \$26.68

Example 13.1

A newly issued lookback put on a non dividend paying stock with underlying asset price of £120.0 that has an annual volatility of 30%, has 6 months to expiry. The risk free rate is 8%. What is the put option value?

$$b_1 = \frac{\log(S/S_{\max}) + (b + \sigma^2/2)T}{\sigma\sqrt{T}} = \frac{\log(120/120) + (0.08 + 0.30^2/2)0.5}{0.30\sqrt{0.5}}$$

$$= 0.294$$

$$b_2 = b_1 - \sigma\sqrt{T} = 0.294 - 0.30\sqrt{0.5} = 0.0824$$

$$\Phi(b_1) = \Phi(0.294) = 0.6158,\ \Phi(-b_1) = \Phi(-0.294) = 0.384,$$
$$\Phi(-b_2) - \Phi(-0.0824) = 0.467.$$

$$\Phi(b_1 - \frac{2b}{\sigma}\sqrt{T}) = \Phi(0.294 - \frac{2*0.08}{0.30}\sqrt{0.5}) = 0.467$$

$$\begin{aligned}
&S_{\max} e^{-rT}\Phi(-b_2) - Se^{(b-r)T}\Phi(-b_1) \\
&+ \frac{Se^{-rT}\sigma^2}{2b}\left\{-\left[\frac{S}{S_{\max}}\right]^{-2b/\sigma^2} e^{-rT}\Phi(b_1 - \frac{2b}{\sigma}\sqrt{T}) + e^{bT}\Phi(b_1)\right\} \\
&= 120e^{-0.08*0.5}\Phi(-b_2) - 120e^{(0.08-0.08)*0.5}\Phi(-b_1) \\
&+ \frac{120e^{-0.08*0.5}*0.30^2}{2*0.08}\left\{-\left[\frac{120}{120}\right]^{-2*0.08/0.30^2}\right. \\
&\left.\times\, e^{-0.08*0.5}\Phi(b_1 - \frac{2*0.08}{0.30}\sqrt{0.5}) + e^{0.08*0.5}\Phi(b_1)\right\} = 19.036
\end{aligned}$$

The put option value is £19.04. An equivalent call would be £21.09.

Floating strike Look back options are valued using code shown in Listing 13.2.

```
package FinApps;
import static java.lang.Math.*;
import BaseStats.Probnorm;
public class Floatxlook {
   public Floatxlook(double rate, double yield) {
      crate=yield;
      r=rate;

      brate=crate==0.0?0.0:(brate=crate!=r?(r-crate):r);
   }
   private double r;
   private double crate;
   private double brate;

   public double callFxlook(double s1, double s2,
                                    double t, double sigma) {
      Probnorm p=new Probnorm();

      double a1=((log(s1/s2)+(brate+(sigma*sigma)*0.5)*t)/
                 (sigma*sqrt(t)));
      double a2=(a1-(sigma*sqrt(t)));
      double n1=(p.ncDisfnc(a1));
      double n2=(p.ncDisfnc(a2));
```

```
    double n3=(p.ncDisfnc(-a1+((2.0*brate/sigma)*sqrt(t))));
    double n4=(p.ncDisfnc(-a1));
    double term1=(pow((s1/s2),(-2.0*brate/(sigma*sigma)))
                *n3-exp(brate*t)*n4);
    double c=(s1*exp((brate-r)*t)*n1-s2*exp(-r*t)*n2+s1*exp(-r*t)*
             ((sigma*sigma)/(2.0*brate))*term1);
    return c;
}
public double putFxlook(double s1, double s2, double t, double sigma)
             {//s2 = smax
    Probnorm p=new Probnorm();

    double b1=((log(s1/s2)+(brate+(sigma*sigma)*0.5)*t)/
              (sigma*sqrt(t)));
    double b2=(b1-(sigma*sqrt(t)));
    double n1=(p.ncDisfnc(b1));
    double n2=(p.ncDisfnc(-b2));
    double n4=(p.ncDisfnc(b1-((2.0*brate/sigma)*sqrt(t))));
    double n3=(p.ncDisfnc(-b1));
    double term1=(-(pow((s1/s2),(-2.0*brate/(sigma*sigma))))
                *n4+exp(brate*t)*n1);
    double put=(s2*exp(-r*t)*n2-s1*exp((brate-r)*t)*n3+s1*
               exp(-r*t)*((sigma*sigma)/(2.0*brate))*term1);
    return put;
}
```

LISTING 13.2. Floating Strike Lookback Option Valuation.

13.3. Partial Time Fixed Strike Lookback

Lookback options in general tend to be expensive. To reduce the cost a partial or fractional solution offers an attractive alternative. In the partial lookback option, it is possible to have a predetermined start date t_1 which is somewhat later than the initial start of the option. The payoff is given as:

$$put = max(K - S_{min'} 0), \; call = max(S_{max} - K, 0) \qquad (13.3.1)$$

within the time frame, $(t_2 - t_1)$ where t_2 is the option expiry date.

13.3.1. Partial Time Fixed Strike Call

Partial time lookback option formula have been developed by Heynen & Kat (1997). The partial time call is given by:

$$c = Se^{(b-r)t_2}\Phi(d_1) - Ke^{-rt_2}\Phi(d_2)$$

$$+ \frac{Se^{-rt_2}\sigma^2}{2b}\left[-\left[\frac{S}{K}\right]^{-2b/\sigma^2} M(d_1 - \frac{2b\sqrt{t_2}}{\sigma}, -f_1 + \frac{2b\sqrt{t_1}}{\sigma};\right.$$

$$-\sqrt{\frac{t_1}{t_2}}) + e^{bt_2} M_2(e_1, d_1; \sqrt{1 - \frac{t_1}{t_2}})\Bigg]$$

$$- Se^{(b-r)t_2} M_2(e_1, d_1; \sqrt{1 - \frac{t_1}{t_2}}) - Ke^{-rt_2} M_3(f_2, -d_2;$$

$$-\sqrt{\frac{t_1}{t_2}}) + e^{-b(t_2-t_1)}(1 - \frac{\sigma^2}{2b}) Se^{(b-r)t_2} \Phi(f_1)\Phi(-e_2) \qquad (13.3.2)$$

where

$$d_1 = \frac{log(S/S_{\min}) + (b + \sigma^2/2)t_2}{\sigma\sqrt{t_2}}, \quad d_2 = d_1 - \sigma\sqrt{t_2}$$

$$e_1 = \frac{(b + \sigma^2/2)(t_2 - t_1)}{\sigma\sqrt{t_2 - t_1}}, \quad e_2 = e_1 - \sigma\sqrt{t_2 - t_1}$$

$$f_1 = \frac{\log(S/S_{\min}) + (b + \sigma^2/2)t_1}{\sigma\sqrt{t_1}}$$

13.3.2. Partial Time Fixed Strike Put

For the partial time fixed strike lookback put:

$$p = Ke^{-rt_2}\Phi(-d_2) - Se^{(b-r)t_2}\Phi(-d_1)$$

$$+ \frac{Se^{-rt_2}\sigma^2}{2b}\left[\left(\frac{S}{K}\right)\right]^{-2b/\sigma^2} M(-d_1 + \frac{2b\sqrt{t_2}}{\sigma}, f_1 - \frac{2b\sqrt{t_1}}{\sigma};$$

$$-\sqrt{\frac{t_1}{t_2}}) - e^{bt_2} M_2(-e_1, -d_1; -\sqrt{1 - \frac{t_1}{t_2}})$$

$$+ Se^{(b-r)t_2} M_3(e_1, -d_1; -\sqrt{1 - \frac{t_1}{t_2}})$$

$$- e^{-b(t_2-t_1)}(1 - \frac{\sigma^2}{2b}) Se^{(b-r)t_2} \Phi(-f_1)\Phi(e_2) \qquad (13.3.3)$$

Tables 13.2 and 13.3 give a range of option values for a range of strike prices with changing volatilities and time to start the partial time. From Table 13.2, call values; it can be observed that as the time from the lookback option beginning increases (the time difference period $t_2 - t_1$ decreasing) the value of the option decreases.

As an example from Table 13.2 with a strike of 115 and volatility of 15%. The value of an option ranges from 14.3579 with a time to lookback period of

TABLE 13.2. Partial Time Fixed Strike European Call values.

Partial Time Fixed Strike Lookback Call Option Values for a range of strike prices & volatilities												
	X 95.0	X 105.0	X 115.0	X95.0	X105.0	X115.0	X95.0	X105.0	X115.0	X95.0	X105.0	X115.0
T1	σ	σ	σ	σ	σ	σ	σ	σ	σ	σ	σ	σ
	0.150	0.150	0.150	0.200	0.200	0.200	0.250	0.250	0.250	0.300	0.300	0.300
0.15	14.8372	14.6145	1[illegible].3579	18.6519	18.3058	17.914	22.6148	22.1358	21.5992	26.7102	26.0891	25.3981
0.25	7.4492	7.4096	[illegible].3353	11.1653	11.0586	10.8933	15.0791	14.881	14.6026	19.1538	18.8459	18.4377
0.35	3.098	3.096	[illegible].0869	6.0558	6.0394	5.9951	9.4815	9.4281	9.3182	13.2289	13.114	12.9123

T2 = 1.0, S = 95.0, r = 0.07, b = 0.07.

TABLE 13.3. Partial Time Fixed Strike European Put values.

Partial Time Fixed Strike Lookback Put Option Values for a range of strike prices & volatilities												
	X 95.0	X 105.0	X 115.0	X95.0	X105.0	X115.0	X95.0	X105.0	X115.0	X95.0	X105.0	X115.0
T1	σ	σ	σ	σ	σ	σ	σ	σ	σ	σ	σ	σ
	0.150	0.150	0.150	0.200	0.200	0.200	0.250	0.250	0.250	0.300	0.300	0.300
0.15	7.0898	6.7251	6.3618	10.1668	9.727	9.2807	13.1839	12.6761	12.1553	16.1246	15.5551	14.9669
0.25	15.7259	14.9195	14.1407	18.7608	17.9013	17.0722	21.7547	20.8491	19.9709	24.6856	23.7385	22.8134
0.35	25.025	24.128	23.201	28.0118	26.9942	25.9607	30.947	29.8367	28.7253	33.8208	32.641	31.470

T2 = 1.0, S = 95.0, r = 0.07, b = 0.07.

1.8 months. With a time to lookback period increasing to 3 months the value of the option is 7.2255 and with a lookback period of 4.2 months the value decreases to 3.0869.

For a put option the behaviour of the option value is the reverse of a call option with changing time periods, as can be observed from Table 13.3.

Partial time fixed strike lookback options are valued using code shown in Listing 13.3.

```
package FinApps;
import static java.lang.Math.*;
import BaseStats.Probnorm;
import BaseStats.Bivnorm;
public class Parttfxlook {
   public Parttfxlook(double rate, double yield) {
      crate=yield;
      r=rate;
      brate=crate==0.0?0.0:(brate=crate!=r?(r-crate):r);
   }
   private double r;
   private double crate;
   private double brate;
   private double d1;
   private double d2;
   private double e1;
   private double e2;
   private double f1;
   private double f2;
   private double sig;

   private void params(double s,double x, double t1, double t2,
                        double sigma) {

      sig=(sigma*sigma);
      d1=((log(s/x)+(brate+sig*0.5)*t2)/(sigma*sqrt(t2)));
      e1=(((brate+sig*0.5)*(t2-t1))/(sigma*sqrt(t2-t1)));
      f1=((log(s/x)+(brate+sig*0.5)*t1)/(sigma*sqrt(t1)));
      d2=(d1-(sigma*sqrt(t2)));
      e2=(e1-(sigma*sqrt(t2-t1)));
      f2=(f1-(sigma*sqrt(t1)));
   }

   public double partFxCall(double s,double x, double t1,
                              double t2,double sigma) {

      Probnorm p=new Probnorm();
      params(s,x,t1,t2,sigma);
      double n1=p.ncDisfnc(d1);
      double n2=p.ncDisfnc(d2);
      double n3=p.ncDisfnc(f1);
      double n4=p.ncDisfnc(-e2);
      double fac1=(d1-2.0*brate*sqrt(t2)/sigma);
```

```
    double fac2=(-f1+2.0*brate*sqrt(t1)/sigma);
    double fac3=(-pow((s/x),(-2.0*brate/(sigma*sigma))));
    double m1=(Bivnorm.bivar_params.evalArgs(fac1,fac2,
             (-sqrt(t1/t2))));
    double m2=(Bivnorm.bivar_params.evalArgs(e1,d1,
             (sqrt(1.0-t1/t2))));
    double m3=(Bivnorm.bivar_params.evalArgs(-e1,d1,
             (-sqrt(1.0-t1/t2))));
    double m4=(Bivnorm.bivar_params.evalArgs(f2,-d2,
             (-sqrt(t1/t2))));
    double term=fac3*m1+exp(brate*t2)*m2;
    double factor1=s*exp((brate-r)*t2)*n1-exp(-r*t2)*x*
                   n2+s*exp(-r*t2)*sig
                     /(2.0*brate)*term;
    double factor2=-s*exp((brate-r)*t2)*m3-x*exp(-r*t2)*m4+exp
                   (-brate*(t2-t1))*(1.0-sig/(2.0*brate))*s*exp
                   ((brate-r)*t2)*n3*n4;
    double c=(factor1+factor2);
    return c;
  }

  public double partFxPut(double s,double x, double t1,
                          double t2,double sigma) {

    Probnorm p=new Probnorm();
    params(s,x,t1,t2,sigma);
    double n1=p.ncDisfnc(-d1);
    double n2=p.ncDisfnc(-d2);
    double n3=p.ncDisfnc(-f1);
    double n4=p.ncDisfnc(e2);
    double fac1=(-d1+2.0*brate*sqrt(t2)/sigma);
    double fac2=(f1-2.0*brate*sqrt(t1)/sigma);
    double fac3=(pow((s/x),(-2.0*brate
                /(sigma*sigma))));
    double m1=(Bivnorm.bivar_params.evalArgs(fac1,fac2,
             (-sqrt(t1/t2))));
    double m2=(Bivnorm.bivar_params.evalArgs(-e1,-d1,
             (sqrt(1.0-t1/t2))));
    double m3=(Bivnorm.bivar_params.evalArgs(e1,-d1,
             (-sqrt(1.0-t1/t2))));
    double m4=(Bivnorm.bivar_params.evalArgs(-f2,d2,
             (-sqrt(t1/t2))));
    double term=fac3*m1-exp(brate*t2)*m2;
    double factor1=x*exp(-r*t2)*n2-s*exp((brate-r)*t2)*n1+s*exp
                   (-r*t2)*sig/(2.0*brate)*term;
    double factor2=s*exp((brate-r)*t2)*m3+x*exp(-r*t2)*m4-exp
                   (-brate*(t2-t1))*(1.0-sig/(2.0*brate))*s*exp
                   ((brate-r)*t2)*n3*n4;
    double c=(factor1+factor2);
    return c;
  }
```

LISTING 13.3. Partial Time Fixed Strike Look back Option Valuation.

13.4. Partial Time Floating Strike Lookback

In a partial time floating strike lookback the lookback period is taken from the initiation of the option life. The lookback period ends at some time before the expiry date. If time to expiry is t_2 and time to end of lookback period is t_1. The value is determined over a time zero to t_1 with $(t_1 \leq t_2)$. The introduction of a fractional time for the lookback period, reduces the lifetime and thus reduces the comparative cost of these options.

The fractional time period λ is set at some percentage of the limit prices, greater than the min for calls and less than the max for puts. The assumption is that the percentage term is log normal.

Formula for partial time floating strike lookback options have been developed by Heynen & Kat (1997).

13.4.1. Partial Time Floating Strike Call

A closed form formula for the call is given by:

$$
\begin{aligned}
c = {} & Se^{(b-r)t_2}\Phi(d_1-\gamma_1) - \lambda S_{\min}e^{-rt_2}\Phi(d_2-\gamma_2)\frac{e^{-rt_2}\sigma^2}{2b}\lambda S^* \\
& \times \left[\left(\frac{S}{S_{\min}}\right)^{-2b/\sigma^2} M(-f_1+\frac{2b\sqrt{t_1}}{\sigma}, -d_1+\frac{2b\sqrt{t_2}}{\sigma}-\gamma_1;\sqrt{\frac{t_1}{t_2}}) \right. \\
& \left. -e^{bt_2}\lambda^{2b/\sigma^2}M_1(-d_1-\gamma_1, \varepsilon_1+\gamma_2; -\sqrt{1-\frac{t_1}{t_2}}\right] \\
& + Se^{(b-r)t_2}M_2(-d_1+\gamma_1, \varepsilon_1-\gamma_2; -\sqrt{1-\frac{t_1}{t_2}}) + \lambda S_{\min}e^{-rt_2}M_3(-f_2, d_2-\gamma_1; \\
& -\sqrt{\frac{t_1}{t_2}}) - e^{-b(t_2-t_1)}(1+\frac{\sigma^2}{2b})\lambda Se^{(b-r)t_2}\Phi(\varepsilon_2-\gamma_2)\Phi(-f_1)
\end{aligned}
\tag{13.4.1}
$$

Where

$$
d_1 = \frac{\log(S/S_{\min}) + (b+\sigma^2/2)t_2}{\sigma\sqrt{t_2}}, \quad d_2 = d_1 - \sigma\sqrt{t_2}
$$

$$
\varepsilon_1 = \frac{(b^2+\sigma^2/2)+(t_2-t_1)}{\sigma\sqrt{t_2-t_1}}, \quad \varepsilon_2 = \varepsilon_1 - \sigma\sqrt{t_2-t_1}
$$

$$
f_1 = \frac{\log(S/S_{\min}) + (b+\sigma^2/2)t_1}{\sigma\sqrt{t_1}}, \quad f_2 = f_1 - \sigma\sqrt{t_1}
$$

$$
\gamma_1 = \frac{\log(\lambda)}{\sigma\sqrt{t_2}}, \quad \gamma_2 = \frac{\log(\lambda)}{\sigma\sqrt{t_2-t_1}}
$$

TABLE 13.4. Partial time floating strike lookback call values for a range of predetermined dates (T_1).

	Partial Time Floating Strike Lookback Call Option Values											
	S = 85.0	S = 90.0	S = 95.0	S = 85.0	S = 90.0	S = 95.0	S = 85.0	S = 90.0	S = 95.0	S = 85.0	S = 90.0	S = 95.0
T1	σ 0.100	σ 0.100	σ 0.100	σ 0.200	σ 0.200	σ 0.200	σ 0.250	σ 0.250	σ 0.250	σ 0.300	σ 0.300	σ 0.300
0.25	11.2876	8.6524	11.0732	13.6057	13.3402	14.9512	15.5391	15.6775	17.2069	17.5526	17.9831	19.5041
0.35	11.3332	8.9163	11.2193	14.0204	13.8763	15.4374	16.0863	16.3312	17.8336	18.2153	18.7445	20.2559
0.45	11.3933	9.1239	11.3511	14.3773	14.3171	15.8511	16.5526	16.8731	18.3645	18.7777	19.3787	20.8917
0.55	11.4561	9.2937	11.4682	14.6903	14.6934	16.2116	16.9605	17.3394	18.8274	19.2694	19.927	21.4465

T2 = 1.0, Smin = 90.0, λ = 1.0, r = 0.06, b = 0.06.

TABLE 13.5. Partial time floating strike lookback put values for a range of predetermined dates (T_1).

	Partial Time Floating Strike Lookback Put Option Values											
	S = 80.0	S = 85.0	S = 90.0	S = 80.0	S = 85.0	S = 90.0	S = 80.0	S = 85.0	S = 90.0	S = 80.0	S = 85.0	S = 90.0
T1	σ 0.100	σ 0.100	σ 0.100	σ 0.200	σ 0.200	σ 0.200	σ 0.250	σ 0.250	σ 0.250	σ 0.300	σ 0.300	σ 0.300
0.25	3.8726	2.9213	3.3364	3.8364	7.9381	9.0591	10.5859	10.5644	11.7676	13.0326	13.2451	14.5464
0.35	4.0473	3.2224	4.0805	4.0805	8.6172	9.7103	11.3502	11.4463	12.6416	14.0066	14.341	15.6537
0.45	4.2369	3.5058	4.3284	4.3284	9.2328	10.3174	12.0547	12.2406	13.4435	14.8946	15.3242	16.6603
0.55	4.4363	3.7801	4.5793	4.5793	9.8074	10.8933	12.7169	12.9772	14.1952	15.7226	16.2324	17.5972

T2 = 1.0, Smax = 85.0, λ = 1.0, r = 0.05, b = 0.05.

Table 13.4 shows a range of values for a partial time floating strike lookback call. The table covers pre determined times from 3 months to 6.6 months, keeping the expiry date constant at 1 year. The asset prices cover a range slightly below and slightly above the minimum. The change in value is slight with extension of the pre determined time period. For increasing volatility the increase is marked.

13.4.2. Partial Time Floating Strike Put

The closed formula for a partial time floating strike lookback put is:

$$p = \lambda S_{\max} e^{-rt_2}\Phi(-d_2+\gamma_1) - Se^{(b-r)t_2}\Phi(-d_1+\gamma_1) - \frac{e^{-rt_2}\sigma^2}{2b}\lambda S^*$$

$$\left[\left(\frac{S}{S_{\max}}\right)^{-\frac{2b}{\sigma^2}} M(f_1 - \frac{2b\sqrt{t_1}}{\sigma}, d_1 - \frac{2b\sqrt{t_2}}{\sigma} + \gamma_1; \sqrt{\frac{t_1}{t_2}}) - e^{bt_2}\lambda^{\frac{2b}{\sigma^2}} M_2(d_1\right.$$

$$\left. +\gamma_1, -\varepsilon_1 - \gamma_2; -\sqrt{1-\frac{t_1}{t_2}})\right]$$

$$- Se^{(b-r)t_2} M_2(d_1-\gamma_1, -\varepsilon_1+\gamma_2; -\sqrt{1-\frac{t_1}{t_2}}) - \lambda S_{\max} e^{-rt_2} M_3(f_2, -d_2+\gamma_1;$$

$$-\sqrt{\frac{t_1}{t_2}}) + e^{-b(t_2-t_1)}(1+\frac{\sigma^2}{2b})\lambda Se^{(b-r)t_2}\Phi(-\varepsilon_2+\gamma_2)\Phi(f_1) \quad (13.4.2)$$

Table 13.5 shows values for the put option. As the period length increases the value of an option decreases. As the volatility increases (risk) so does the option value.

Implementation of partial time floating strike lookback options is shown in Listing 13.4.

```
package FinApps;
import static java.lang.Math.*;
import BaseStats.Probnorm;
import BaseStats.Bivnorm;
public class PartfloatX {
   public PartfloatX(double rate, double yield, double l) {
      crate=yield;
      r=rate;
      lambda=l;
      brate=crate==0.0?0.0:(brate=crate!=r?(r-crate):r);
   }
   private double r;
   private double crate;
   private double brate;
   private double lambda;
   private double d1;
   private double d2;
   private double e1;
   private double e2;
   private double f1;
```

```
private double f2;
Private double sig;
private double g1;
private double g2;


private void params(double s,double sminax, double t1, double t2,
                double sigma) {

   sig=(sigma*sigma);
   d1=((log(s/sminax)+(brate+sig*0.5)*t2)/(sigma*sqrt(t2)));
   e1=(((brate+sig*0.5)*(t2-t1))/(sigma*sqrt(t2-t1)));
   f1=((log(s/sminax)+(brate+sig*0.5)*t1)/(sigma*sqrt(t1)));
   d2=(d1-(sigma*sqrt(t2)));
   e2=(e1-(sigma*sqrt(t2-t1)));
   f2=(f1-(sigma*sqrt(t1)));
   // System.out.println("d1=="+d1+"e1=="+e1+"e2=="+e2+"d2
                         =="+d2+"f1=="+f1+"f2=="+f2);
   g1=(log(lambda)/(sigma*sqrt(t2)));
   g2=(log(lambda)/(sigma*sqrt(t2-t1)));

}

public double partxCall(double s,double smin, double t1,
                     double t2,double sigma) {


   Probnorm p=new Probnorm();
   params(s,smin,t1,t2,sigma);
   double n1=p.ncDisfnc(d1-g1);
   double n2=p.ncDisfnc(d2-g1);
   double n3=p.ncDisfnc(e2-g2);
   double n4=p.ncDisfnc(-f1);
   double m1=(Bivnorm.bivar_params.evalArgs(-f1+2.0*brate*
            sqrt(t1)/sigma,-d1+2.0*brate*sqrt(t2)/sigma-g1,
            sqrt(t1/t2)));
   double m2=(Bivnorm.bivar_params.evalArgs((-d1-g1),(e1+g2),
            (-sqrt(1.0-t1/t2))));
   double m3=(Bivnorm.bivar_params.evalArgs(-d1+g1,e1-g2,-sqrt
            (1.0-t1/t2)));
   double m4=(Bivnorm.bivar_params.evalArgs(-f2,d2-g1,-sqrt
            (t1/t2)));
   double term=(pow((s/smin),((-2.0*brate/sig)))*m1-exp(brate*t1)*
             pow(lambda,(2.0*brate/sig))*m2);
   double fact1=s*exp((brate-r)*t2)*n1-lambda*smin*exp(-r*t2)*n2;
   double fact2=exp(-r*t2)*((sigma*sigma)/(2.0*brate))*lambda*s*
              term+s*exp((brate-r)*t2)*m3;
   double fact3=exp(-r*t2)*lambda*smin*m4-exp(-brate*(t2-t1))*exp
            ((brate-r)*t2)*(1.0+sig/(2.0*brate))*lambda*s*n3*n4;
   double op=(fact1+fact2+fact3);
   return op;
}

public double partxPut(double s,double smax, double t1, double t2,
                    double sigma) {
```

```
    Probnorm p=new Probnorm();
    params(s,smax,t1,t2,sigma);
    double n1=p.ncDisfnc(-d2+g1);
    double n2=p.ncDisfnc(-d1+g1);
    double n3=p.ncDisfnc(-e2+g2);
    double n4=p.ncDisfnc(f1);
    double m1=(Bivnorm.bivar_params.evalArgs(f1-2.0*brate*sqrt(t1)/
               sigma,d1-2.0*brate*sqrt(t2) /sigma+g1,sqrt(t1/t2)));
    double m2=(Bivnorm.bivar_params.evalArgs(d1+g1,-e1-g2,-sqrt
               (1.0-t1/t2)));
    double m3=(Bivnorm.bivar_params.evalArgs(d1-g1,-e1+g2,-sqrt
               (1.0-t1/t2)));
    double m4=(Bivnorm.bivar_params.evalArgs(f2,-d2+g1,-sqrt
               (t1/t2)));
    double term=(pow((s/smax),((-2.0*brate/sig)))*m1-exp(brate*t2)*
                 pow(lambda, (2.0*brate/sig))*m2);
    double fact1=lambda*smax*exp(-r*t2)*n1-s*exp
                 ((brate-r)*t2)*n2;
    double fact2=-exp(-r*t2)*((sigma*sigma)/(2.0*brate))*lambda*s*
                  term-s*exp((brate-r)*t2)*m3;
    double fact3=-exp(-r*t2)*lambda*smax*m4+exp(-brate*(t2-t1))*
                  exp((brate-r)*t2)*(1.0+sig/(2.0*brate))*lambda*
                  s*n3*n4;
    double op=(fact1+fact2+fact3);
    return op;
}
```

LISTING 13.4. Partial Time Floating Strike Lookback Option Valuation.

13.5. Min or Max of Two Risky Assets

The original work on providing analytical formulas for European put and call options on the minimum or maximum of two risky assets is due to Stulz[3] (1982). The original work has been further extended by several others including Rich & Chance, (1993). Options of this kind are also sometimes referred to as rainbow or outperformance options. These options have found a wide range of application including contingent claims in foreign currency, option-bonds and risk sharing contracts.

The various pricing formulas are as follows.

13.5.1. Minimum of Two Risky Assets

Call on the minimum of two risky assets:

$$\begin{aligned} &S_1 e^{(b_1-r)T} M(\gamma_1, -d; -\rho_1) + S_2 e^{(b_2-r)T} M_1(y_2, d - \sigma\sqrt{T}; \rho_2) \\ &- K e^{-rT} M_2(y_1 - \sigma_1\sqrt{T}, y_2 - \sigma_2\sqrt{T}; \rho) \end{aligned} \tag{13.5.1}$$

Where

$$d = \frac{\log(S_1/S_2) + (b_1 - b_2 + \sigma^2/2)T}{\sigma\sqrt{T}}, \gamma_1 = \frac{\log(S_1/K) + (b_1 + \sigma_1^2/2)T}{\sigma_1\sqrt{T}}$$

$$\gamma_2 = \frac{\log(S_2/K) + (b_2 + \sigma_2^2/2)T}{\sigma_2\sqrt{T}}, \sigma = \sqrt{\sigma_1^2 + \sigma_2^2 - 2\rho\sigma_1\sigma_2}$$

$$\rho_1 = \frac{\sigma_1 - \rho\sigma_2}{\sigma}, \rho_2 = \frac{\sigma_2 - \rho\sigma_1}{\sigma}$$

Put on the minimum of two assets:

$$Ke^{-rT} - call_{\min}(S_1, S_2, 0, T) + call_{\min}(S_1, S_2K, T) \tag{13.5.2}$$

Where

$$call_{\min}(S_1, S_2, 0, T) = S_1e^{(b-r)T} - S_1e^{(b_1-r)T}\Phi(d) + S_2e^{(b_2-r)T}\Phi(d - \sigma\sqrt{T}) \tag{13.5.3}$$

13.5.2. *Maximum of Two Risky Assets*

Call on the maximum of two assets:

$$\begin{aligned} &S_1e^{(b_1-r)T}M(y_1, d; \rho_1) + S_2e^{(b_2-r)T}M_1(y_2, -d + \sigma\sqrt{T}; \rho_2) \\ &- Ke^{-rT}\left(1 - M_2(-y_1 + \sigma_1\sqrt{T}, -y_2 + \sigma_2\sqrt{T}; \rho)\right) \end{aligned} \tag{13.5.4}$$

Put on the maximum of two assets:

$$Ke^{-rT} - call_{\max}(S_1, S_2, 0, T) + call_{\max}(S_1, S_2K, T) \tag{13.5.5}$$

Where

$$call_{\max}(S_1, S_2, 0, T) = S_2e^{(b_2-r)T} + S_1e^{(b_1-r)T}\Phi(d) - S_2e^{(b_2-r)T}\Phi(d - \sigma\sqrt{T}) \tag{13.5.6}$$

Example 13.2

A put on the maximum of two assets gives the holder a right to sell stock index A and index B at a strike of £105. The A yield is 8% and the B yield is 11%. The A index price is £110 and the B index price is £118. Index A has a volatility of 16% and B has volatility of 20%. The correlation of return between both indices is 0.59, the risk free rate is 6% and time to expiry is 9 months. What is the option value?

$$\sigma = \sqrt{\sigma_1^2 + \sigma_2^2 - 2\rho\sigma_1\sigma_2} = \sqrt{0.16^2 + 0.20^2 - 2*0.59*0.16*0.20} = 0.1668$$

$$\rho_1 = \frac{\sigma_1 - \rho\sigma_2}{\sigma} = \frac{0.16 - 0.59*0.20}{0.1668} = 0.2517,$$

$$\rho_2 = \frac{\sigma_2 - \rho\sigma_1}{\sigma} = \frac{0.20 - 0.59*0.16}{0.1668} = 0.632$$

$$d = \frac{\log(S_1/S_2) + (b_1 - b_2 + \sigma^2/2)T}{\sigma\sqrt{T}}$$
$$= \frac{\log(110/118) + (-0.02 - (-0.05) + 0.1668^2/2)0.75}{0.1668\sqrt{0.75}} = -0.2578$$

$$y_1 = \frac{\log(S_1/K) + (b_1 + \sigma_1^2/2)T}{\sigma_1\sqrt{T}}$$
$$= \frac{\log(110/105) + (-0.02 + 0.16^2/2)0.75}{0.16\sqrt{0.75}} = 0.2967$$

$$y_2 = \frac{\log(S_2/K) + (b_2 + \sigma_2^2/2)T}{\sigma_2\sqrt{T}}$$
$$= \frac{\log(118/105) + (-0.05 + 0.20^2/2)0.75}{0.20\sqrt{0.75}} = 0.544$$

$$Ke^{-rT} - call_{\max}(S_1, S_2, 0, T) + call_{\max}(S_1, S_2K, T)$$
$$= 105e^{-0.06*0.75} - call_{\max}(110, 118, 0, 0.75)$$
$$+ call_{\max}(110, 118, 105, 0.75)$$

$$call_{\max}(110, 118, 0, 0.75) = S_2e^{(b_2-r)T} + S_1e^{(b_1-r)T}\Phi(d)$$
$$- S_2e^{(b_2-r)T}\Phi(d - \sigma\sqrt{T}) = 118e^{(-0.05-0.06)*0.75} + 110e^{(-0.02-0.06)T}\Phi(d)$$
$$- 118e^{(-0.05-0.06)T}\Phi(d - 0.1668\sqrt{0.75}) = 112.5668$$
$$call_{\max}(110, 118, 105, 0.75) = S_1e^{(b_1-r)T}M(y_1, d; \rho_1)$$
$$+ S_2e^{(b_2-r)T}M_1(y_2, -d + \sigma\sqrt{T}; \rho_2)$$
$$- Ke^{-rT}\left(1 - M_2(-y_1 + \sigma_1\sqrt{T}, -y_2 + \sigma_2\sqrt{T}; \rho)\right).$$

$$= 110e^{(-0.02-0.06)T}M(0.2967, -0.2578; 0.2517) + 118e^{(-0.05-0.06)*0.75}$$

$$\times M_1(0.544, -(-0.2578) + 0.1668\sqrt{0.75}; 0.6$$

$$105e^{-0.06*0.75}\left(1 - M_2(-0.2967 + 0.16\sqrt{0.75}, -0.544\right.$$

$$\left.+0.20\sqrt{0.75}; 0.59)\right). = 14.1799$$

$$105e^{-0.06*0.75} - 112.5668 + 14.1799 = 1.9928$$

The option value is £1.99.

Listing 13.5 shows code for implementing options on two risky assets.

```
package FinApps;
import static java.lang.Math.*;
import BaseStats.Bivnorm;
public class Optwoassets {
  public Optwoassets(double p, double yield, double rate,double
                    brate2, double sig1, double sig2) {

    r=rate;
    crate=yield;
    brate=crate==0.0?0.0:(brate=crate!=r?(r-crate):r);
    sigma1=sig1;
    sigma2=sig2;
    rho=p;
    b2=brate2;

  }
  private double crate;
  private double brate;
  private double r;
  private double sigma1;
  private double sigma2;
  private double b2;
  private double y1;
  private double y2;
  private double rho;


  private void params(double s1, double s2, double x1, double x2,
                      double time){

    y1=((log(s1/x1)+(brate-(sigma1*sigma1)*0.5)*time)/(sigma1*
                      sqrt(time)));
    y2=((log(s2/x2)+(b2-(sigma2*sigma2)*0.5)*time)/(sigma2*
                      sqrt(time)));
  }

  public double twoAssetcall(double s1, double s2, double x1,
                              double x2, double time) {
    params(s1,s2,x1,x2,time);
```

```
    double m1=(Bivnorm.bivar_params.evalArgs((y2+sigma2*
            sqrt(time)),(y1+rho*sigma2*sqrt(time)),rho));
    double m2=(Bivnorm.bivar_params.evalArgs(y2,y1,rho));
    double c=(s2*exp((b2-r)*time)*m1-x2*exp(-r*time)*m2);
    return c;
}
public double twoAssetput(double s1, double s2, double x1, double x2,
                         double time) {
    NumberFormat formatter=NumberFormat.getNumberInstance();
    formatter.setMaximumFractionDigits(4);
    formatter.setMinimumFractionDigits(3);
    params(s1,s2,x1,x2,time);
    double m2=(Bivnorm.bivar_params.evalArgs((-y2-sigma2*sqrt
            (time)),(y1-rho*sigma2*sqrt(time)),rho));
    double m1=(Bivnorm.bivar_params.evalArgs(-y2,-y1,rho));
    double p=(x2*exp(-r*time)*m1-s2*exp(b2-r*time)*m2);
    return p;
}
```

LISTING 13.5. Options on Two Risky Assets.

13.6. Spread Option Approximation

Spread options are written on the differences between two assets such as the difference or spread between two stock indices or the differences between two types of bond etc. Spreads are widely used in commodity markets where they can be used to cover price differences of commodities over geographical regions (location spread) or in time (calendar spreads).

13.6.1. Analytical Spread Approximation

There are several models proposed for analytical spread option approximations. Bachelier's model, assumes a Gaussian distribution of the variables at maturity, this approach has been developed by Shimko (1994). The Kirk (1995) model offers a very good approximation and is in its general form:

$$\sigma\Phi\left(\frac{\log(\frac{\alpha}{\gamma+\kappa})}{\sigma^K}+\frac{\sigma^K}{2}\right)-(\gamma+\kappa)\Phi\left(\frac{\log(\frac{\alpha}{\gamma+\kappa})}{\sigma^K}-\frac{\sigma^K}{2}\right) \quad (13.6.1)$$

Where

$$\sigma^K=\sqrt{\beta^2-2\rho\beta\delta\frac{\gamma}{\gamma+\kappa}+\delta^2(\frac{\gamma}{\gamma+\kappa})^2} \quad (13.6.2)$$

In situations where variables γ and κ are zero, the approximation is exact. If the variable $\alpha = 0$ or $\rho \neq \pm 1$ the approximation is never exact.

Applying the Kirk model to approximating the spread on two futures contracts.

$$\alpha = F_1, \gamma = F_2, \kappa = X, \beta = \sigma_1, \delta = \sigma_2$$

Arranging into the standard Black Scholes model and rearranging terms for pricing futures:

$$call = (F_2 + X)(e^{-rT}(F\Phi(d_1) - \Phi(d_2)) \tag{13.6.3}$$

$$put = (F_2 + X)(e^{-rT}(\Phi(-d_2) - F\Phi(-d_1)) \tag{13.6.4}$$

Where

$$F = (\frac{F_1}{F_2 + X}),\ d_1 = \frac{\log(F) + (\sigma^2/2)T}{\sigma\sqrt{T}},$$

$d_2 = d_1 - \sigma\sqrt{T}$ and the volatility of F is:

$$\sigma = \sqrt{\sigma_1^2 \left(\sigma_2 \frac{F_2}{(F_2 + X)}\right)^2 - 2\rho\sigma_1\sigma_2 \frac{F_2}{(F_2 + X)}}$$

Table 13.6. gives a range of values for call spread options using the above formulae. The better the correlation the lower is the relative value for the option. Higher volatilities result in higher option values.

Example 13.3

A put option on the spread between two futures contracts with 6 months to expiry has a strike price of \$9.0. The price of the first futures contract is \$35.0, the second contract price is \$30.0. The first contract has volatility of 30% per annum, the second futures contract has 40% volatility per annum. Value this option when the risk free rate is 4% and the correlation between the futures is 0.5.

$$F_1 = 35.0, F_2 = 30.0, X = 9.0, \sigma_1 = 30\%, \sigma_2 = 40\%, r = 4\%, t = 0.5, \rho = 0.5$$

$$F = (\frac{F_1}{F_2 + X}) = (\frac{35}{30 + 9}) = 0.897,$$

$$\sigma = \sqrt{\sigma_1^2 + \left(\sigma_2 \frac{F_2}{(F_2 + X)}\right)^2 - 2\rho\sigma_1\sigma_2 \frac{F_2}{(F_2 + X)}}$$

$$= \sqrt{0.30 + \left(0.40 \frac{30}{(30 + 9)}\right)^2 - 2{*}0.5{*}0.30{*}0.40 \frac{30}{(30 + 9)}} = 0.3039$$

TABLE 13.6. Call Spread options with changing correlation and times with volatility.

			Call Spread Options on Futures Options											
		t	0.1	0.5	0.1	0.5	0.1	0.5	0.1	0.5	0.1	0.5	0.1	0.5
σ_1	σ_2	ρ	−0.50	−0.50	−0.25	−0.25	0.0	0.0	0.25	0.25	0.5	0.5	0.75	0.75
0.100	0.100		2.8751	5.5294	2.6738	5.0922	2.4519	4.6088	2.2015	4.0606	1.9074	3.4114	1.5338	2.5695
0.100	0.150		3.4671	6.8106	3.2258	6.2892	2.961	5.7159	2.6641	5.0712	2.3195	4.3193	1.8924	3.3781
0.150	0.100		3.4814	6.8413	3.2414	6.3228	2.9782	5.7531	2.6836	5.1136	2.3426	4.3699	1.9225	3.4449
0.200	0.150		4.641	9.337	4.2949	8.5935	3.9138	7.7736	3.4844	6.8478	2.9818	5.7609	2.3475	4.3805
0.150	0.200		4.6267	9.3062	4.2792	8.5598	3.8964	7.736	3.4644	6.8047	2.9578	5.7088	2.3151	4.3096
0.200	0.200		5.2037	10.5433	4.7968	9.6713	4.3471	8.7057	3.8373	7.6087	3.2339	6.3066	2.4524	4.6099
0.200	0.250		5.7912	11.8004	5.3396	10.8344	4.8411	9.7662	4.2771	8.5553	3.6126	7.1243	2.7604	5.2805
0.250	0.200		5.8056	11.8312	5.3553	10.8681	4.8586	9.8038	4.2973	8.5986	3.637	7.177	2.7939	5.3533
0.250	0.300		6.9575	14.2874	6.4027	13.1058	5.7898	11.7973	5.0954	10.3115	4.2751	8.5509	3.2172	6.2705
0.300	0.250		6.9719	14.3181	6.4185	13.1394	5.8074	11.8349	5.1157	10.3548	4.2997	8.6039	3.2513	6.3443
0.300	0.300		7.5398	15.5251	6.9293	14.2275	6.2539	12.7883	5.4873	11.1505	4.5786	9.2029	3.3971	6.6593
0.350	0.350		8.7073	17.9971	7.9954	16.4913	7.2077	14.8196	6.3133	12.9152	5.2526	10.6481	3.8722	7.6839

F1 = 110.0, F2 = 107.0, X = 2.0, r = 0.10.

$$d_1 = \frac{\log(F) + (\sigma^2/2)T}{\sigma\sqrt{T}} = \frac{\log(0.897) + (0.3039^2/2)0.5}{0.3039\sqrt{0.5}} = -0.39609,$$

$$d_2 = d_1 - \sigma\sqrt{T} = -0.39609 - 0.3039\sqrt{0.5} = -0.6109,$$

$$\Phi(-d_1) = \Phi(-(-0.39609)) = 0.7293,$$

$$\Phi(-d_2) = \Phi(-(-0.6109)) = 0.6539.$$

$$\begin{aligned} put &= (F_2 + X)(e^{-rT}(\Phi(-d_2) - F\Phi(-d_1)) \\ &= (30+9)(e^{-0.04*0.5}(\Phi(-(-0.6109)) - 0.8974*\Phi(-(-0.39609))) \\ &= 2.587 \end{aligned}$$

The put option value is \$2.59. The equivalent call option value is \$1.53.

Listing 13.6 shows implementation code for valuing spread options.

```
package FinApps;
import java.util.*;
import static java.lang.Math.*;
import static FinApps.PresentValue.*;
import CoreMath.NewtonRaphson;
import static FinApps.Intr.*;
import FinApps.Tyield.*;

public class Spread extends NewtonRaphson {

   private double precision=1e-5;
   private int iterations=20;

   public Spread() //default constructor//
   {
      this.terms=2.0;// default twice annual coupon payments
      this.dataperiod=1;//period of spot rate data
      this.facevalue=100.0;//default par
   }
   public Spread(double frequency,int dataterms,double parvalue ) {
      this.terms=frequency;
      this.dataperiod=dataterms;
      this.facevalue=parvalue;
   }
   double terms=0.0;
   int dataperiod=0;
   double facevalue=0.0;
   double periodyield=0.0;
   double nperiods=0.0;
   double periodcoupon=0.0;
   double price=0.0;
   double[]spots;
   double coupon=0.0;

   /** Method computes spread for the annual period rates
    *
```

```
 * provides the static spread from the corporate bond price.
            The amount by which each spot needs to be adjusted
 * Assumes annual coupon rate and annual yield estimate
 */
 public double spreadsT(double[]spotrates,double pcorp,double
            maturity,double couponrate,double estimate) {
   accuracy(precision,iterations);
   spots=spotrates;
   price=pcorp;
   coupon=couponrate;

   return newtraph(estimate);
 }
 /**
  *Calculates the yield spread for YTM of corretly priced risk free and
arbitrary priced corporatecorporate
  */
 public double spreadrateS(double[]spotrates,double priceval,double
couponrate,double maturity,double yieldapprox ) {
   double baseyield=0.0;
   double curveyield=0.0;
   double spotapprox=0.0;
   spots=spotrates;
   coupon=couponrate;
   nperiods=maturity*terms;// number of compounding periods
   if(((int)maturity*dataperiod)!=spots.length) {
     System.out.println("error: spots data is not == to the
             maturity*dataperiods");
     return 0.0;
   }
   periodyield=((yieldapprox/100.0)/terms);
   periodcoupon=(coupon/(terms));
   spotapprox=(((spots[0]+spots[(spots.length-1)])/2.0)/
            100.0);//first guess
   spotapprox=spotapprox/terms;

   price=dataperiod==1?spotPvannual(spots,coupon):
               spotPvperiod(spots,coupon);

   Tyield c=new Tyield();// create a yield object

   curveyield=c.yieldEstimate(facevalue,6.0,coupon,price,
            maturity,spotapprox);
   price=priceval;
   Tyield t=new Tyield();
   baseyield=t.yieldEstimate(facevalue,6.0,coupon,price,
            maturity,periodyield);
   return (abs(baseyield-curveyield));// returns annualised rates
 }
 /**
  *Assumes annual coupon and annual yield with years to maturity
            as input parameters
  *assumes coupon and yield is entered as percentage value
  */
 public double[] spreadrateS(double[]spotrates,
```

```
                double prices[],double couponrate,double maturity,
                double yieldapprox ) {
        double curveyield=0.0;
        double spotapprox=0.0;
        spots=spotrates;
        coupon=couponrate;
        double curvest=yieldapprox;
        int index=0;
        double spreads[]=new double[prices.length];
        Tyield t=new Tyield();
        nperiods=maturity*terms;
        periodcoupon=(coupon/(terms));
        price=dataperiod==1?spotPvannual(spots,coupon):
                spotPvperiod(spots,coupon);

        curveyield=t.yieldEstimate(facevalue,6.0,coupon,price,
                maturity,(curvest/100.0));

        for(double p:prices){
            Tyield yld=new Tyield();
            price=p;
            double y=yld.yieldEstimate(facevalue,6.0,coupon,price,
                maturity,(yieldapprox/100.0));
            spreads[index]=(y-curveyield);
            index++;

        }

        return spreads;
    }
    public double[] spreadsT(double[]spotrates,double prices[],double
couponrate,double maturity,double yieldapprox )
{
        spots=spotrates;
        coupon=couponrate;
        int index=0;
        double spreads[]=new double[prices.length];
        accuracy(precision,iterations);
        periodyield=yieldapprox;
        nperiods=maturity*terms;
        periodcoupon=(coupon/(terms));

        for(double p:prices) {
            price=p;
            periodyield=((periodyield/100.0)/terms);
            periodyield=newtraph(periodyield);
            spreads[index]=periodyield;
            index++;
        }

        return spreads;
    }

    /**
     *Method computes the PV for an array of period spots and the
     annual coupon
     *periods are user defined
```

```
 **/
private double spotPvperiod(double[]periodspot,double coupon) {
  double pv=0.0;
  double par=0.0;
  double periodcoupon=0.0;
  double couponadjust=coupon/terms;
  int size=0;
  size=periodspot.length*(int)terms;
  pv= pVonecash(periodspot,couponadjust);
  par=(100.0*exp(-(double)size*log(1.0+(periodspot[(periodspot.
            length-1)]/100.0))));
  return(pv+par);
}
/** Method to compute the PV of an array of annual spots and
 annual coupon with given annual frequency of compounding
 **
 **/
private double spotPvannual(double[]periodspot,double coupon ) {
  double pv=0.0;
  double par=0.0;
  if(terms>1.0) {
    int size=0;
    int compfreq=0;
    int index=0;
    compfreq=(int)terms;
    size=periodspot.length*(int)terms;
    double[]periodspotadj=new double[(size)];
    for(double d:periodspot) {
      for (int i=0;i<compfreq;i++) {
        periodspotadj[index]=d/terms;
        index++;
      }
    }
    double couponadjust=(coupon/terms); //from an annual
            coupon to the period rate
    pv=pVonecash(periodspotadj,couponadjust);
    par=(100.0*exp(-(double)size*log(1.0+(periodspotadj
            [(periodspotadj.length-1)]/100.0))));
    return(pv+par);
  } else {
    double couponadjust=(coupon/terms); //from an annual
            coupon to the period rate
    pv=pVonecash(periodspot,couponadjust);
    par=(100.0*exp(-(double)periodspot.length*log(1.0+
            (periodspot[(periodspot.length-1)]/100.0))));
    return(pv+par);
  }
}
public double spotPvannualT(double[]periodspot,double
            coupon,double terms ) {
  double pv=0.0;
  double par=0.0;
```

```
   if(terms>1.0) {
      int size=0;
      int compfreq=0;
      int index=0;
      compfreq=(int)terms;
      size=periodspot.length*(int)terms;
      double[]periodspotadj=new double[(size)];
      for(double d:periodspot) {
         for (int i=0;i<compfreq;i++) {
            periodspotadj[index]=d/terms;
            index++;
         }
      }
      double couponadjust=(coupon/terms); //from an annual
               coupon to the period rate
      pv=pVonecash(periodspotadj,couponadjust);
      par=(100.0*exp(-(double)size*log(1.0+(periodspotadj
               [(periodspotadj.length-1)]/100.0))));
      return(pv+par);
   } else {
      double couponadjust=(coupon/terms); //from an annual
               coupon to the period rate
      pv=pVonecash(periodspot,couponadjust);
      par=(100.0*exp(-(double)periodspot.length*log(1.0+
               periodspot[(periodspot.length-1)]/100.0))));
      return(pv+par);
   }
}

/**
 *Assumes period spot and the yield approximation are
               period rates with period percentages
 *Assumes flat rate spot for entire maturity period
 *assumes the coupon is an annual coupon percent of par rate.
 */
public double spreadrate(double periodspot,double priceval,double
 coupon,double maturity,double yieldapprox ) {
   accuracy(precision,iterations);
   double baseyield=0.0;
   double frequency=0.0;
   price=priceval;
   nperiods=maturity*terms;
   periodcoupon=coupon;
   periodyield=yieldapprox/100.0;
   periodspot=periodspot/100.0;
   baseyield=newtraph(periodyield);
   return (abs(baseyield-periodspot)*terms*100.0);//returns
               annualised spread
}
/**
 *credit spread computes probability of default and forward
               prob of default
```

```
    *assumes corporate bond zero and treasury zero (riskless)
    */
   public double[] creditS(double[]riskless,double[]risky) {
      int size=riskless.length;
      double[]fdefault=new double[size];
      double[]pdefault=new double[size];
      pdefault[0]=(1.0-(risky[0]/riskless[0]));
      fdefault[0]=pdefault[0];
      for(int i=1;i<size;i++) {
         fdefault[i]=(1.0-(exp(-log(1.0-pdefault[i-1])))*(risky[i]/
                  riskless[i]));
         pdefault[i]=(pdefault[i-1]*fdefault[i]);
      }
      return fdefault;
   }
   public double newtonroot(double spread) {
      int indx=0;
      double[] spotspreads=new double[spots.length];
      for(double d:spots) {
         spotspreads[indx]=(d+spread);
         indx++;
      }
      spread=(spotPvannual(spotspreads,coupon)-price);
      return spread;
   }
}
```

LISTING 13.6. Spread-Option Valuation.

13.7. Extreme Spreads

An extreme spread option is split into two sequential periods; initial to t_1 and t_1 to maturity. The payoff from an extreme spread is :

$$\text{Call, } C_{pay} = abs(S_{\max}(t_2 - t_1) - S_{\max}(t_1 - t_0)),$$

$$\text{put, } P_{pay} = abs(S_{\min}(t_1 - t_0) - S_{\min}(t_2 - t_1))$$

Where the maximum observed price for the asset in the time period is $S_{\max}$ and the minimum observed asset price within a time period is $S_{\min}$.

The payoff for a reverse spread is given by:

$$\text{Call, } Cr_{pay} = abs(S_{\min}(t_2 - t_1) - S_{\min}(t_1 - t_0)),$$

$$\text{put, } Pr_{pay} = abs(S_{\max}(t_1 - t_0) - S_{\max}(t_2 - t_1))$$

Extreme spread option formulae are due initially to the work of Bermin (1996). There are valuation formulae for extreme and reverse extreme spread options. For the extreme spread option the European valuation formula is:

13.7.1. Extreme Spread

$$\omega = \eta \begin{pmatrix} Se^{(b-r)t_2}(1+\frac{\sigma^2}{2b})\Phi\left(\eta\frac{-m+\mu_2 t_2}{\sigma\sqrt{t_2}}\right) - e^{-b(t_2-t_1)}Se^{(b-r)t_2}(1+\frac{\sigma^2}{2b})\Phi\left(\eta\frac{-m+\mu_2 t_1}{\sigma\sqrt{t_1}}\right) \\ +e^{-rt_2}S_m\Phi\left(\eta\frac{m-\mu_1 t_2}{\sigma\sqrt{t_2}}\right) - e^{-rt_2}S_m\frac{\sigma^2}{2b}e^{\frac{2\mu m}{\sigma^2}}\Phi\left(\eta\frac{-m-\mu_1 t_2}{\sigma\sqrt{t_2}}\right) - e^{-rt_2}S_m\Phi\left(\eta\frac{m-\mu_1 t_1}{\sigma\sqrt{t_1}}\right) \\ +e^{-rt_2}S_m\frac{\sigma^2}{2b}e^{\frac{2\mu m}{\sigma^2}}\Phi\left(\eta\frac{-m-\mu_1 t_1}{\sigma\sqrt{t_1}}\right) \end{pmatrix} \tag{13.7.1}$$

Where $m = \log(\frac{S_m}{S})$, $\mu_1 = b - \frac{\sigma^2}{2}$, $\mu_2 = b + \frac{\sigma^2}{2}$. For $\eta = 1$, the calculation is for a call option with $S_m = S_{\max}$. For $\eta = -1$, the calculation is for a put option with $S_m = S_{\min}$.

13.7.2. Reverse Extreme Spread

For the reverse extreme spread option the valuation formula is given by:

$$\omega = -\eta \begin{pmatrix} Se^{(b-r)t_2}(1+\frac{\sigma^2}{2b})\Phi\left(\eta\frac{m-\mu_2 t_2}{\sigma\sqrt{t_2}}\right) + e^{-rt_2}S_m\Phi\left(\eta\frac{-m+\mu_1 t_2}{\sigma\sqrt{t_2}}\right) \\ -e^{-rt_2}S_m\frac{\sigma^2}{2b}e^{\frac{2\mu m}{\sigma^2}}\Phi\left(\eta\frac{m+\mu_1 t_2}{\sigma\sqrt{t_2}}\right) - Se^{(b-r)t_2}(1+\frac{\sigma^2}{2b})\Phi\left(\eta\frac{-\mu_2(t_2-t_1)}{\sigma\sqrt{(t_2-t_1)}}\right) - \\ e^{-b(t_2-t_1)}Se^{(b-r)t_2}(1-\frac{\sigma^2}{2b})\Phi\left(\eta\frac{\mu_1(t_2-t_1)}{\sigma\sqrt{(t_2-t_1)}}\right) \end{pmatrix} \tag{13.7.2}$$

Table 13.7 shows the values of an extreme call option, for a range of time periods and volatilities. As the time period $(t_2 - t_1)$ decreases, so does the option value. As the volatility increases so does the option value.

Table 13.8 shows the values of the reverse extreme call option with the same parameters of Table 13.7, with the excretion that $S_m = S_{\min}$ is decreasing in value. Table 13.9 shows the reverse extreme put values where the differences between the observed maximum values of the asset are used (in the extreme put we would use the observed differences between the minimum asset values for each period).

The code for implementing extreme spread options is shown in Listing 13.7.

```
package FinApps;
import static java.lang.Math.*;
import BaseStats.Probnorm;
import BaseStats.Bivnorm;
import CoreMath.Csmallnumber;
public class Extremesop {

   public Extremesop(double rate, double yield, double eta) {
      //eta=1=call, -1 = put
      e=eta;
      r=rate;
```

```
        crate=yield;
        b=crate==0.0?0.0:(b=crate!=r?(r-crate):r);
    }
    private double e;
    private double r;
    private double b;
    private double crate;
    private double m;
    private double mu;
    private double mu1;
    public double extremeSp(double s, double s2, double t1, double t2,
        double vol)
    {
        Probnorm p=new Probnorm();

        double sig=(vol*vol);
        mu1=(b-sig*0.5);
        mu=(mu1+sig);
        m=(log(s2/s));//where s2=Mo

        t1=(t1==0.0&m==0.0)?floorvalue(t1):t1==0.0?(sig*m*0.50):t1;//
An approximation for zero values, which works for s2<50% of s1
        // for situations where t==0 and log s1/s2=0
        double term1=(1.0+sig/(2.0*b));
        double term2=p.ncDisfnc(e*(-m+mu*t2)/(vol*sqrt(t2)));
        double term3= p.ncDisfnc(e*(-m+mu*t1)/(vol*sqrt(t1)));
        double term4= p.ncDisfnc(e*(m-mu1*t2)/(vol*sqrt(t2)));
        double term5= p.ncDisfnc(e*(-m-mu1*t2)/(vol*sqrt(t2)));
        double term6= p.ncDisfnc(e*(m-mu1*t1)/(vol*sqrt(t1)));
        double term7= p.ncDisfnc(e*(-m-mu1*t1)/(vol*sqrt(t1)));

        double w=e*(s*exp((b-r)*t2)*term1*term2-exp(-r*(t2-t1))
                *s*exp((b-r)*t2)*term1*term3+exp(-r*t2)*s2*term4
                -exp(-r*t2)*s2*sig/(2.0*b)*exp(2.0*mu1*m/sig)
                *term5-exp(-r*t2)*s2*term6+exp(-r*t2)
                *s2*sig/(2.0*b)*exp(2.0*mu1*m/sig)*term7);
        return w;
    }
    public double extremeSprev(double s, double s2, double t1, double t2,
        double vol)
    {
        Probnorm p=new Probnorm();
        double sig=(vol*vol);
        mu1=(b-sig*0.5);
        mu=(mu1+sig);
        m=(log(s2/s));//where s2=Mo
        double term1=(1.0+sig/(2.0*b));
        double term2= p.ncDisfnc(e*(m-mu*t2)/(vol*sqrt(t2)));
        double term3=p.ncDisfnc(e*(-m+mu1*t2)/(vol*sqrt(t2)));
        double term4=p.ncDisfnc(e*(m+mu1*t2)/(vol*sqrt(t2)));
        double term5=p.ncDisfnc(e*(-mu*(t2-t1))/(vol*sqrt(t2-t1)));
        double term6=p.ncDisfnc(e*(mu1*(t2-t1))/(vol*sqrt(t2-t1)));

        double w=-e*(s*exp((b-r)*t2)*term1*term2+exp(-r*t2)*s2
```

```
                    *term3-exp(-r*t2)*s2*sig/(2.0*b)*exp(2.0*mu1*m
                    /sig)*term4-s*exp((b-r)*t2)*term1*term5
                    -exp(-r*(t2-t1))*s*exp((b-r)*t2)
                    *(1.0-sig/(2.0*b))*term6);

    return w;

}
public double floorvalue(double x) {
    return abs(x)<Csmallnumber.getSmallnumber()?Csmallnumber.
    getSmallnumber():x;
}
```

LISTING 13.7. Extreme Spread Option Valuation.

13.8. Value or Nothing Options

Value or nothing options are simple binary choices that payoff the underlying or a cash value. The performance of various models is discussed in Peng & Han (2004).

13.8.1. Cash-or-Nothing Option

The cash-or-nothing option has a discontinuous payoff. For a cash-or-nothing call the payoff is a fixed amount, if the asset price exceeds the strike price at expiration, the option pays an amount Q. If the asset price is below the strike price, the option pays nothing. Pricing formulae have been developed by Reiner & Rubinstein (1991).

The pricing of this option is based on a Black-Scholes economy; it assumes lognormal distribution of the asset price, assumes no transaction costs and assumes continuous risk free rates over the option lifetime. The payoff for a call is given by:

$$Q : iff, S_T > X : else, 0$$

The payoff from a put cash-or-nothing option is:

$$Q : iff, S_T < X : else, 0.$$

The value of this option is given by:

$$C = Qe^{-r(T-t)}P(S_T > X)$$

showing that the path of the fixed amount, follows the probability distribution of the asset price being greater than the strike at expiry.

TABLE 13.7. Extreme option values.

	Extreme Call Option Values											
	S Max = 100.0				S Max = 115.0				S Max = 130.0			
T1	σ 0.150	σ 0.200	σ 0.250	σ 0.300	σ 0.150	σ 0.200	σ 0.250	σ 0.300	σ 0.150	σ 0.200	σ 0.250	σ 0.300
0.00	15.3931	19.629	24.026	28.5715	5.1212	8.742	12.7179	16.9575	1.2164	3.325	6.1986	9.6177
0.15	10.4844	13.200	16.0567	19.0427	5.0733	8.4542	11.9298	15.4337	1.2163	3.3227	6.170	9.4893
0.30	8.0965	10.1412	12.3048	14.5769	4.6415	7.3106	9.9411	12.5612	1.2092	3.2223	5.7685	8.5568
0.45	6.1038	7.6154	9.2224	10.9163	3.8735	5.8445	7.768	9.6879	1.1492	2.8789	4.9111	7.0466
0.60	4.3068	5.356	6.4759	7.6599	2.924	4.2769	5.5973	6.9218	0.9873	2.3013	3.7628	5.2628

T2 = 1.0, S = 100.0, r = 0.06, b = 0.06.

TABLE 13.8. Reverse extreme call values for various time periods and observed minima.

	Reverse Extreme Call Option Values											
	S Min = 100.0				S Min = 85.0				S Min = 70.0			
T1	σ 0.150	σ 0.200	σ 0.250	σ 0.300	σ 0.150	σ 0.200	σ 0.250	σ 0.300	σ 0.150	σ 0.200	σ 0.250	σ 0.300
0.00	0.000	0.000	0.000	0.000	6.6028	4.9078	3.797	3.0372	19.7975	16.6249	13.9226	11.7287
0.15	1.2724	1.4838	1.[illegible]58	1.8759	7.8751	6.3916	5.4828	4.913	21.0698	18.1087	15.6084	13.6045
0.30	2.6354	3.0896	3.[illegible]241	3.9342	9.2382	7.9973	7.3211	6.9713	22.4329	19.7144	17.4468	15.6629
0.45	4.1192	4.8576	5.[illegible]55	6.2357	10.7219	9.7654	9.3625	9.2728	23.9166	21.4825	19.4881	17.9643
0.60	5.7761	6.8585	7.[illegible]83	8.8864	12.3788	11.7662	11.6953	11.9236	25.5735	23.4834	21.8209	20.6151

T2 = 1.0, S = 100.0, r = 0.06, b = 0.06.

TABLE 13.9. Reverse extreme put values for various time periods and observed maxima.

	Reverse Extreme Put Option Values											
	S Max = 100.0				S Max = 115.0				S Max = 130.0			
T1	σ 0.150	σ 0.200	σ 0.250	σ 0.300	σ 0.150	σ 0.200	σ 0.250	σ 0.300	σ 0.150	σ 0.200	σ 0.250	σ 0.300
0.00	0.000	0.000	0.000	0.000	3.8546	3.2394	2.8184	2.5124	14.0762	11.9489	10.4256	9.2991
0.15	0.5806	0.9162	1.2778	1.663	4.4352	4.1556	4.0962	4.1754	14.6568	12.8651	11.7034	10.9621
0.30	1.2456	1.9492	2.7045	3.5065	5.1002	5.1886	5.5229	6.019	15.3218	13.8981	13.1301	12.8057
0.45	2.025	3.1394	4.3305	5.5913	5.8797	6.3788	7.1489	8.1037	16.1013	15.0883	14.7561	14.8904
0.60	2.9712	4.5571	6.2442	8.0234	6.8258	7.7965	9.0626	10.5358	17.0474	16.506	16.6697	17.3225

T2 = 1.0, S = 100.0, r = 0.06, b = 0.06.

Given that we assume stock prices follow a lognormal distribution;

$$\log S_T : \phi\left[\log S + \left(r - \frac{\sigma^2}{2}\right), \sigma\sqrt{T-t}\right].$$

The probability $P(S_T > X)$ can be rewritten $P(\log S_T > \log X)$. By rearrangement. The distribution $d =$

$$P\left[\left(\frac{\log S_T - \log S - (r - \frac{1}{2}\sigma^2)(T-t)}{\sigma\sqrt{T-t}}\right) > \left(\frac{\log X - \log S - (r - \frac{1}{2}\sigma^2)(T-t)}{\sigma\sqrt{T-t}}\right)\right];$$

$$d = 1 - \phi\left[\frac{\log(X/S) - (r - \frac{1}{2}\sigma^2)(T-t)}{\sigma\sqrt{T-t}}\right] = 1 - \phi(-d) = \phi(d)$$

The valuation is therefore the familiar second term of a standard Black-Scholes formula:

$$C = Qe^{-r(T-t)}\phi(d) \qquad (13.8.1)$$

Where,

$$d = \frac{\left[\log(S/X) + (r - \frac{1}{2}\sigma^2)(T-t)\right]}{\sigma\sqrt{T-t}}$$

Using our familiar notation;

$$d = \left[\frac{\log(S/X) + (b - \frac{1}{2}\sigma^2)T}{\sigma\sqrt{T}}\right].$$

Where $b = r - q = 0$. The value of a cash-or-nothing put is given by:

$$P = Qe^{-rT}\phi(-d) \qquad (13.8.2)$$

Example 13.4

A cash-or-nothing call has six months to expiry. The asset price is set at €85.0. The strike price is set at €76.0 and the cash payout is €4.50. If the risk-free rate is 5.25% and the asset volatility is 30% per annum. What is the option value?

$$T = 0.5,\ S = €85.0,\ X = €76.0,\ Q = €4.50,\ r = 0.0525,\ \sigma = 0.30$$

$$d = \frac{\log(S/X) + (b - \frac{1}{2}\sigma^2)T}{\sigma\sqrt{T}} = \frac{\log(85.0/76.0) + (0 - \frac{1}{2}0.30^2)0.5}{0.30\sqrt{0.5}} = 0.42152$$

$$\phi(0.42152) = 0.6633$$

$$C = Qe^{-r(T-t)}\phi(d) = C = 4.50e^{-0.0525*0.5*}0.6633 = 2.9075.$$

The option value is therefore €2.91.

Cash or nothing options are valued using code shown in Listing 13.8

```
package FinApps;
import static java.lang.Math.*;
import BaseStats.Probnorm;
public class Cashornonop {
   public Cashornonop(double rate, double yield, double time) {
      r=rate;
      crate=yield;
      t=time;
      b=crate==0.0?0.0:(b=crate!=r?(r-crate):r);
   }
   double b;
   double crate;
   double r;
   double t;
   public double cashornoCall(double s, double x, double sigma,
      double payout){return payout*exp(-r*t)*N(d(s,x,sigma));
   }
   public double cashornoPut(double s, double x,
      double sigma, double payout) {
      return payout*exp(-r*t)*N(-d(s,x,sigma));
   }
   private double d(double s, double x, double sigma) {
      double sig=(sigma*sigma);
      double f= (log(s/x)+((b-sig*0.5)*t))/(sigma*sqrt(t));
      return f;
   }
   private double N(double x) {
      Probnorm p=new Probnorm();

      double ret=x>(6.95)?1.0:x<(-6.95)?0.0:p.ncDisfnc(x);//
      restrict the range of cdf values to stable values

      return ret;
   }
```

LISTING 13.8. Cash-or-Nothing Option.

13.8.2. Asset-or-Nothing Option

The asset-or-nothing option has a payoff of zero if the asset price is below or at the strike at expiry or the asset value, if the asset price is above the strike price. The pricing formulae have been developed by Cox&Rubinstein (1985).

For a call the payoff is:

$$S_T : iff, S_T > X : else, 0$$

For a put:

$$S_T : iff, S_T < X : else, 0.$$

It can be shown that the value of a call follows the lognormal distribution of the asset path. The probability distribution is given by:

$$V_{S_T} call = \frac{1}{\sqrt{2\pi(T-t)\sigma S_T}} \exp\left\{-\frac{(\log S_T - u)^2}{2(T-t)\sigma^2}\right\} \tag{13.8.3}$$

Where $u = \log S + (r - \frac{1}{2}\sigma^2)(T-t)$. Given that,

$$S_T = S.\exp(\log S_T - \log S) = S.\exp(\log S_T - u + \left(r - \frac{1}{2}\sigma^2\right)(T-t)).$$

Using the transformation;

$$y = \frac{1}{\sigma\sqrt{T-t}}\left[(\log S_T - u) - (T-t)\sigma^2\right]$$

for $S_T = X$. (boundary point of zero)

$$y = \frac{-1}{\sqrt{T-t\sigma}}\left[\log(S/X) + \left(r + \frac{1}{2}\sigma^2\right)(T-t)\right] = -d$$

The value of a call can then be given by:

$$C = S(1 - \phi(-d)) = S\phi(d) \tag{13.8.4}$$

Including a cost of carry term and allowing for the asset to make a return 13.5.4 is rearranged to give

$$C = Se^{(b-r)T}\phi(d) \tag{13.8.5}$$

where

$$d = \left[\frac{\log(S/X) + (b + \frac{1}{2}\sigma^2)T}{\sigma\sqrt{T}}\right]$$

The value of a put is given by:

$$P = Se^{(b-r)T}\phi(-d) \tag{13.8.6}$$

Example 13.5

An asset-or-nothing put has 9 months to expiry. The asset price is \$110.0 with a yield of 3% per annum. The strike price is \$100.0. The asset has an annual volatility of 25%, the risk free rate is 5.25%. What is the option valuation?

$$d = \frac{\log(S/X) + (b + \frac{1}{2}\sigma^2)}{\sigma\sqrt{T}} = \frac{\log(110.0/100.0) + (0.0225 + \frac{1}{2}0.0625)0.5}{0.25\sqrt{0.75}} = 0.6264$$

$$\phi(-0.6264) = 0.2655$$

$$P = Se^{(b-r)T}\phi(-d) = 110.0e^{(0.0225-0.0525)0.5} * 0.2655 = 28.5575.$$

The put valuation is $28.56.

Asset-or-nothing options are valued using code shown in Listing 13.9.

```
package FinApps;
import static java.lang.Math.*;
import BaseStats.Probnorm;
public class Assetornop {
   public Assetornop(double rate, double yield, double time) {
      r=rate;
      crate=yield;
      t=time;
      b=crate==0.0?0.0:(b=crate!=r?(r-crate):r);
   }
   double b;
   double crate;
   double r;
   double t;
   public double assetornoPut(double s, double x, double sigma) {
      return s*exp((b-r)*t)*N(-d(s,x,sigma));
   }
   public double assetornoCall(double s, double x, double sigma) {
      return s*exp((b-r)*t)*N(d(s,x,sigma));
   }

   private double d(double s, double x, double sigma) {
      double sig=(sigma*sigma);
      double f= (log(s/x)+((b+sig*0.5)*t))/(sigma*sqrt(t));
      return f;
   }
   private double N(double x) {
      Probnorm p=new Probnorm();

      double ret=x>(6.95)?1.0:x<(-6.95)?0.0:p.ncDisfnc(x);
                    //restrict the range of cdf values to stable values

      return ret;
   }
```

LISTING 13.9. Asset-or-nothing Option.

References

Bermin, H. P. (1996). "Exotic Lookback Options: The Case of Extreme Spread Options," Working Paper, Department of Economics, Lund University, Sweden.

Conze, A. and Viswanathan (1991). "Path Dependent Options: The Case of Lookback Options," *Journal of Finance*, 46, 1893–1907.

Cox, J. C. and M. Rubinstein (1985). "Innovations in Options Markets," *Options Markets*, Chapter 8. Prentice-Hall, New Jersey.

Goldman, B. M., H. B. Sosin and M. A. Gatto (1979). "Path Dependent Options: Buy at the Low, Sell at the High," *Journal of Finance*, 34(5), 1111–1127.

Heynen, R. C. and H. M. Kat (1997). "Lookback Options-Pricing and Applications" in *Exotic Options: The State-of-the Art* 99–123. Eds: Clewlow, L. and Strickland, C. London, International Thompson.

Kirk, E. (1995). "Correlation in the Energy Markets," *Managing Energy Price Risk*. London: Risk Publications and Enron, 71–78.

Peng, B. and Y. Han (2004). "A Study on the Binary Option Model and its Pricing," *Proceedings of the Academy of Accounting and Financial Studies* (New Orleans 2004) 9(1), 71–78.

Reiner, E. and M. Rubinstein (1991). "Breaking down the Barriers," *Risk Magazine*, 4(8).

Rich, D. R. and D. M. Chance (1993). "An Alternative Approach to the Pricing Options on Multiple Assets," *Journal of Financial Engineering*, 2(3), 271–285.

Shimko, D. (1994). "Options on Futures Spreads: Hedging, Speculation, and Valuation," *The Journal of Futures Markets*, 14(2), 183–213.

Stulz, R. M. (1982). "Options on the Minimum or the Maximum of Two Risky Assets," *Journal of Financial Economics*, 10, 161–185.

14 Barrier Type Options

Barrier options have a payoff dependent on the underlying asset reaching a given level in a specified period. Barrier options are classified as either knock-in or knock-out. A knock-out option ceases to exist when the underlying reaches an appropriate level (the barrier). A knock-in option comes into existence when the asset reaches the barrier.

Barrier options can be priced using the Black-Scholes economy, a barrier call or put is similar to a plain vanilla call or put. Barrier options unlike their plain vanilla counterparts however are path dependent, as the history of the underlying asset process determines the payoff at expiration. The path dependency is described as weak since the valuation is dependent on the current asset price and time only. The weak path dependency introduces a degree of complexity to the pricing formulas used.

The knock-in barrier option has no value until the option hits the barrier. Within the knock-in option there are two possibilities; up and-in option and down and-in option. For the up and-in option there is no value until the asset prices rises up to the barrier level, for the down and-in option there is no value until the asset declines sufficiently to hit the barrier. For the basic barrier option it is only necessary for the asset to hit the barrier at some time. The asset can move freely around the barrier level as often as conditions dictate, throughout the option lifetime. The single event of hitting the barrier is the only condition for instantiating an option.

A knock-out option is the converse of a knock-in option. If the asset hits a barrier level the option ceases to exist and will expire with no value. For an up and-out barrier option the asset has to remain below a barrier level for the option to retain value. For the down and-out option the asset has to remain above the barrier level for the option to have value.

There is a comprehensive range of barrier option types with increasingly complex conditions attached to achieving the barrier level; we will examine more complex barrier options later.

14.1. In Barrier Valuation

Valuation formulae for standard barrier options have been developed by Merton (1973) and Rubinstein & Reiner (1991). More recent work on the basic models, showing the effects of model risk suggest that calibration of risk factors in any barrier option has to be given particular focus. The relative merits of using plain vanilla calibration and applying this to a barrier type is discussed in Hirsa et al (2003).

The usual assumptions concerning risk free rate, risk-neutral valuation and the log normal characteristic of an underlying asset are made. For the standard barrier option, when an option comes into existence, the value is a standard put or call (in the Black Scholes sense). This infers that a knock-in option takes on the value of a plain vanilla if the barrier is hit in the option lifetime. For a knock-out option the plain vanilla option value is extinguished when the barrier is hit.

With the assumption of a Black Scholes economy the valuation can be closed form (the stock follows geometric Brownian motion). If we intuitively consider the mechanics of a down and-out call barrier option. A standard call option will evolve towards maturity according to geometric Brownian motion, with a probability density of the asset achieving the strike price at some time in the future. This path towards being in the money has a price attached to it based on the likelihood of the event (normal probability density). For a barrier option there is an additional probability associated with the barrier being hit at some earlier time prior to the evolution to maturity; extinguishing the call option. Based on there being a joint probability density it is reasonable to assume that the price of a barrier call option is somewhat less than the standard vanilla counterpart.

The valuation formula for a barrier option takes account of the standard vanilla density and the density value for hitting the barrier. If we have a portfolio of a down and-in call option with a down and-out call option we are left with the payoff from a standard vanilla call option, therefore the price of a down and-out barrier call will be related to the price of a standard call option minus the down and-in call option. It is common to incorporate a rebate into a standard barrier option; this is essentially a form of compensation to the holder. The rebate is payable if the option has not been knocked in (or if knocked out) before expiration. If we ignore the role of a rebate (simplifying the explanation) Merton shows that the valuation of a down and-out call with strike price greater than barrier price is given by an equation which incorporates the joint probability of the barrier being hit from above and the standard vanilla call. If we denote the joint event probability as:

$$\{S_\tau = H, S_\tau > X\}$$

Where S_τ is the asset price at some time τ after the initial start date, H the barrier level and X, the option strike price. The valuation is:

$$C_{do} = S_0 \left\{ N(x_1) - \frac{H}{S_0}^{2(r+\frac{1}{2}\sigma^2)/\sigma^2} N(y_1) \right\}$$

$$- Xe^{-rT} \left\{ N(x_2) - \frac{H}{S_0}^{2(r+\frac{1}{2}\sigma^2)/\sigma^2-2} N(y_2) \right\} \quad (14.1.1)$$

Where,

$$x_1 = \frac{\log(\frac{S_0}{X}) + (r + \frac{1}{2}\sigma^2)T}{\sigma\sqrt{T}}, \tag{14.1.2}$$

$$y_1 = \frac{\log(\frac{H^2}{S_0 X}) + (r + \frac{1}{2}\sigma^2)T}{\sigma\sqrt{T}} \tag{14.1.3}$$

$$x_2 = x_1 - \sigma\sqrt{T} \quad (14.1.4), \quad y_2 = y_1 - \sigma\sqrt{T} \tag{14.1.5}$$

The joint probability $\{S_\tau = H, S_\tau > X\}$ is represented by the densities $N(y_1), N(y_2)$. When the densities are zero, the call is a standard plain vanilla. $N(x_1), N(x_2)$, represent a probability of the option being in the money with $S_T > X$ (at maturity). Subtracting the joint probability density from the first term of 17.1 reflects a potential loss of value if the asset price at any time within the option lifetime hits the barrier. The term;

$\frac{H}{S_0}^{2(r+\frac{1}{2}\sigma^2)/\sigma^2} N(y_1)$, is often referred to as the absorption probability.

The complete term;

$N(x_1) - \frac{H}{S_0}^{2(r+\frac{1}{2}\sigma^2)/\sigma^2} N(y_1)$, is often referred to as the survival probability.

14.1.1. Valuation with a Rebate

Rubinstein has developed a range of valuation formulas that cover barrier options including a rebate, which can be knock-out, knock-in and combinations of down and-out, down and-in, up and-out and up and-in.

For a down and-in call, the call option does not exist until the asset price drops down to hit a barrier. If at time τ, the asset price hits the knock-in boundary, the holder receives a standard European call with strike X and time to maturity $(T - \tau)$. If the knock-in boundary is not hit, a rebate K is paid at expiration. The payoff is:

$$\max[0, S_\tau - X] if S_\tau \leq H; else, K.$$

The assumption is that the initial asset price is greater than the barrier. In terms of probabilities of the asset price we have three possibilities given that $S_0 > H$;

The strike can be greater than the barrier. $P = prob(S_\tau > X : S_\tau \leq H)$.
Strike less than barrier. $P = prob(H \geq S_T > X) + prob(S_T > H : S_\tau \leq H)$.
Barrier not hit. $P = prob(S_T > H) - prob(S_T > H : S_\tau \leq H)$.

In a situation where the strike is greater than the barrier, there are three payoff possibilities:

$S_T - X$, when the barrier has been hit and at maturity, $S_T > X$.
0, when the barrier has been hit and at maturity, $S_T \leq X$.
K (Rebate), when the barrier has never been hit and at maturity, $S_T > H$.

For the situation where the strike is less than the barrier, there are four payoff possibilities:

$S_T - X$, when the barrier has been hit and at maturity, $S_T > H$.
$S_T - X$, when the barrier has been hit and at maturity, $S_T > X$.
0, when the barrier has been hit and at maturity, $S_T \leq X$.
K (Rebate), when the barrier has never been hit and at maturity, $S_T > H$.

If we consider the case where, $X > H$. The valuation is a sum of the call payoff and a term to cover the rebate payoff:

$$P = prob(S_T > X : S_\tau \leq H) + prob(S_T > H) - prob(S_T > H : S_\tau \leq H).$$

To take account of situations where the initial asset price is above the barrier (down and-in) or starts below the barrier (up and-in) and the density is given by the same expression. Rubinstein suggests the use of binary variables η, ϕ, to allow the same expressions to be used in either case. The sum can be expressed as:

$$\begin{aligned} C_{di} &= \phi S d^{-T} (H/S)^{2\lambda} N(\eta y) - \phi X r^{-T} (H/S)^{2\lambda-2} N(\eta y - \eta\sigma\sqrt{T}) \\ &\quad + X r^{-T} \left[N(\eta x_1 - \eta\sigma\sqrt{T}) - (H/S)^{2\lambda-2} N(\eta y_1 - \eta\sigma\sqrt{T} \right] \end{aligned}$$

Where,

$$\begin{aligned} x_1 &\equiv \left[\frac{\log(S/H)}{\sigma\sqrt{T}} \right] + \lambda\sigma\sqrt{T} \\ y &\equiv \left[\frac{\log(H^2/SX)}{\sigma\sqrt{T}} \right] + \lambda\sigma\sqrt{T} \\ y_1 &= \left[\frac{\log(H/S)}{\sigma\sqrt{T}} \right] + \lambda\sigma\sqrt{T} \\ \lambda &\equiv 1 + (\mu/\sigma^2) \end{aligned}$$

By using terms from Equations 14.1 to 14.5, we can re-write the equivalent expressions as:

14.1.2. Down and In Call Valuation

$$\begin{aligned} C_{di} &= S e^{-rT} (H/S)^{2(\mu+1)} N(y_1) - X e^{-rT} (H/S)^{2\mu} N(y_1 - \sigma\sqrt{T}) \\ &\quad + X e^{-rT} [N(x_2 - \sigma\sqrt{T}) - (H/S)^{2\mu} N(y_2 - \sigma\sqrt{T})] \end{aligned} \tag{14.1.6}$$

For convenience and allowing for a cost of carry term (b). We will re-write 14.1.6 as:

$$C_{di} = \phi S e^{(b-r)T}(H/S)^{2(\mu+1)}N(\eta y_1) - \phi X e^{-rT}(H/S)^{2\mu}N(\eta y_1 - \eta\sigma\sqrt{T}) + Xe^{-rT}[N(\eta x_2 - \eta\sigma\sqrt{T}) - (H/S)^{2\mu}N(\eta y_2 - \eta\sigma\sqrt{T})] \quad (14.1.7)$$

$$\eta = 1, \phi = 1$$

Where parameters 14.2–14.5 are re-written as:

$$x_1 = \frac{\log(S/X)}{\sigma\sqrt{T}} + (1+\mu)\sigma\sqrt{T} \quad (14.1.8)$$

$$x_2 = \frac{\log(S/H)}{\sigma\sqrt{T}} + (1+\mu)\sigma\sqrt{T} \quad (14.1.9)$$

$$y_1 = \frac{\log(H^2/SX)}{\sigma\sqrt{T}} + (1+\mu)\sigma\sqrt{T} \quad (14.1.10)$$

$$y_2 = \frac{\log(H/S)}{\sigma\sqrt{T}} + (1+\mu)\sigma\sqrt{T} \quad (14.1.11)$$

Using the same argument for the down and-in call, when the strike price is less than the barrier, we have the following probabilities:

$P = prob(H \geq S_T > X) + prob(S_T > H : S_\tau \leq H)$ Plus a rebate term.

Since, $prob(H \geq S_T > X) = prob(S_T > X) - prob(S_T > H)$, we have four probability terms:

$$prob(S_T > X) - prob(S_T > H) + prob(S_T > H : S_\tau \leq H) + (prob(S_T > H) - prob(S_T > H : S_\tau \leq H))$$

This gives us the valuation of a down and-in call with $X < H$

$$\begin{aligned} C_{di} &= \phi S e^{(b-r)T}N(\phi x_1) - \phi X e^{-rT}N(\phi x_1 - \phi\sigma\sqrt{T}) \\ &- \phi S e^{(b-r)T}N(\phi x_2) - \phi X e^{-rT}N(\phi x_2 - \phi\sigma\sqrt{T}) \\ &+ \phi S e^{(b-r)T}(H/S)^{2(\mu+1)}N(\eta y_2) - \phi X e^{-rT}(H/S)^{2\mu}N(\eta y_2 - \eta\sigma\sqrt{T}) \\ &+ Xe^{-rT}[N(\eta x_2 - \eta\sigma\sqrt{T}) - (H/S)^{2\mu}N(\eta y_2 - \eta\sigma\sqrt{T})] \end{aligned} \quad (14.1.12)$$

$$\eta = 1, \phi = 1$$

Table 14.1. shows a range of valuations for a down and-in barrier option. The table shows that in general the value of the option increases with volatility when the strike price is less than the barrier level. When the strike price exceeds the barrier level, values decrease with increased volatility. All values reflect an increase which is proportionate to expiry time.

Table 14.2. shows a range of valuations for up and-in options. Figures 14.1 and 14.2 show characteristics with changes to time and volatility.

TABLE 14.1. Down and-in Barrier Call Option Values

Valuations for Down and In Standard Barrier Call with S = 115.0, K = 5.0.								
X	H	$\sigma = 0.10$	$\sigma = 0.15$	$\sigma = 0.20$	$\sigma = 0.25$	$\sigma = 0.30$	$\sigma = 0.35$	T
80.0	90.0	4.8766	4.8801	4.937	5.1201	5.4312	5.8408	
90.0	90.0	4.8765	4.8738	4.8372	4.7536	4.6816	4.6822	3
100.0	90.0	4.8765	4.8731	4.8194	4.660	4.4351	4.2252	months
110.0	90.0	4.8765	4.8731	4.8189	4.6513	4.3919	4.1061	
X	H	$\sigma = 0.10$	$\sigma = 0.15$	$\sigma = 0.20$	$\sigma = 0.25$	$\sigma = 0.30$	$\sigma = 0.35$	T
80.0	115.0	35.5467	35.5467	35.5468	35.5511	35.5747	35.6403	
90.0	115.0	25.7936	25.7941	25.8093	25.8833	26.0554	26.3376	3
100.0	115.0	16.0422	16.1022	16.3331	16.7506	17.3161	17.9894	months
110.0	115.0	6.6377	7.3554	8.2419	9.2075	10.2148	11.2464	
X	H	$\sigma = 0.10$	$\sigma = 0.15$	$\sigma = 0.20$	$\sigma = 0.25$	$\sigma = 0.30$	$\sigma = 0.35$	T
80.0	90.0	4.6462	4.8514	5.4451	6.322	7.3765	8.5483	
90.0	90.0	4.634	4.560	4.5424	4.7918	5.3376	6.1274	9
100.0	90.0	4.632	4.4737	4.169	4.0193	4.1663	4.6072	months
110.0	90.0	4.632	4.4631	4.0667	3.7075	3.5776	3.7299	
X	H	$\sigma = 0.10$	$\sigma = 0.15$	$\sigma = 0.20$	$\sigma = 0.25$	$\sigma = 0.30$	$\sigma = 0.35$	T
80.0	115.0	36.5479	36.5512	36.6033	36.794	37.1722	37.7348	
90.0	115.0	27.2717	27.3358	27.6286	28.2038	29.017	30.0086	9
100.0	115.0	18.0586	18.5141	19.3884	20.5364	21.852	23.2725	months
110.0	115.0	9.5639	10.9279	12.5058	14.1773	15.8952	17.6379	

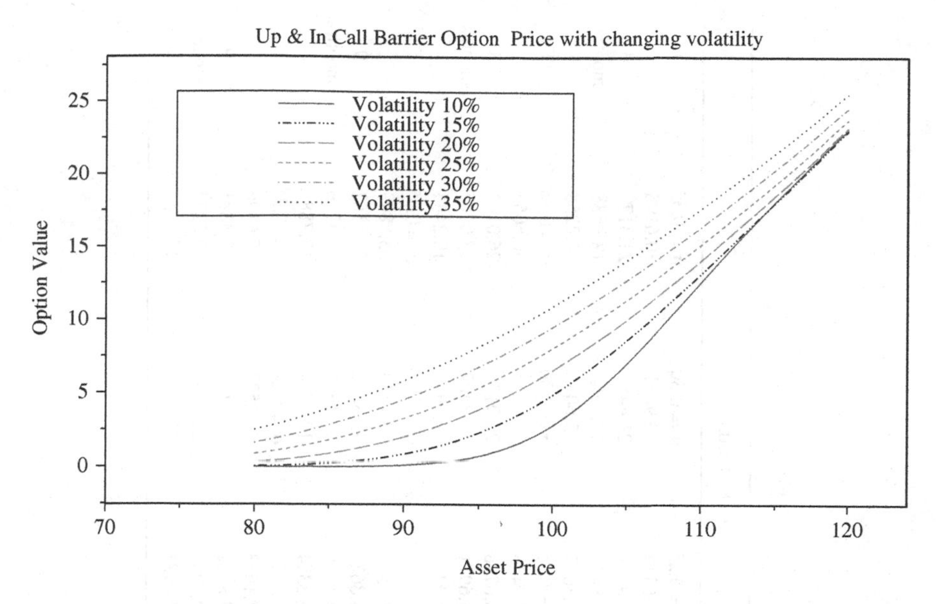

FIGURE 14.1. Up and-in call barrier values for varying volatilities of underlying.

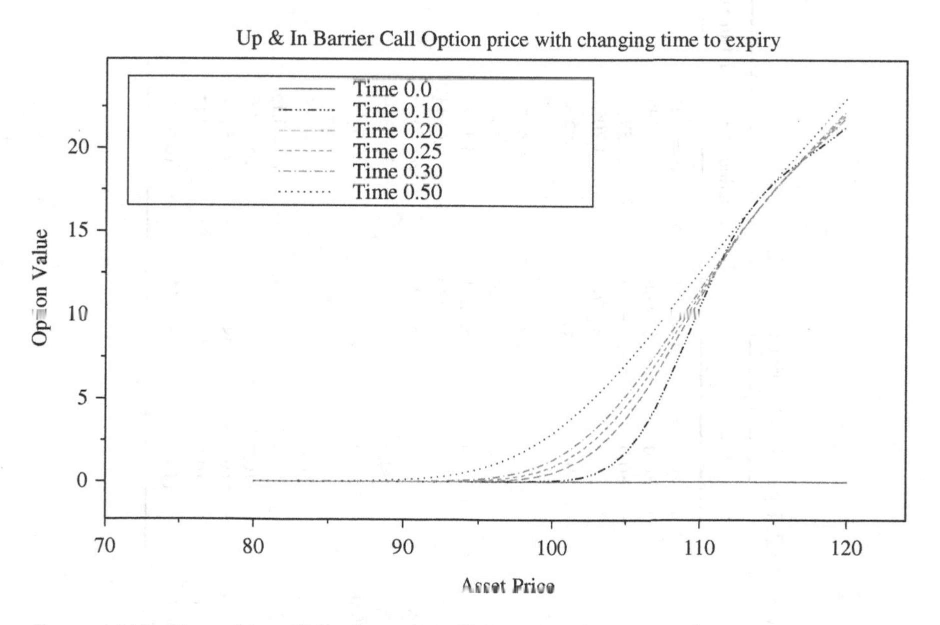

FIGURE 14.2. Up and-in call barrier values for varying times to expiry.

Table 14.2. Up and-in Barrier Call Values

Valuations for Up and In Standard Barrier Call with S = 115.0, K = 5.0.								
X	H	$\sigma = 0.10$	$\sigma = 0.15$	$\sigma = 0.20$	$\sigma = 0.25$	$\sigma = 0.30$	$\sigma = 0.35$	T
80.0	80.0	35.3606	34.1495	32.7185	31.6449	30.9651	30.6198	
90.0	80.0	25.6075	24.397	22.9809	21.9771	21.4459	21.317	3
100.0	80.0	15.8561	14.7051	13.5047	12.8444	12.7066	12.9688	months
110.0	80.0	6.4516	5.9582	5.4135	5.3013	5.6053	6.2258	
X	H	$\sigma = 0.10$	$\sigma = 0.15$	$\sigma = 0.20$	$\sigma = 0.25$	$\sigma = 0.30$	$\sigma = 0.35$	T
80.0	110.0	42.0448	40.6754	39.126	37.9448	37.1244	36.6055	
90.0	110.0	28.7045	27.564	26.6017	26.0778	25.913	26.0103	3
100.0	110.0	15.5281	15.1195	15.2482	15.6971	16.3307	17.0767	months
110.0	110.0	4.8439	5.6633	6.7747	7.9335	9.0932	10.2444	
X	H	$\sigma = 0.10$	$\sigma = 0.15$	$\sigma = 0.20$	$\sigma = 0.25$	$\sigma = 0.30$	$\sigma = 0.35$	T
80.0	80.0	36.3709	35.2349	34.0387	33.4522	33.4553	33.8849	
90.0	80.0	27.0947	26.0194	25.064	24.862	25.3002	26.1587	9
100.0	80.0	17.8817	17.1977	16.8239	17.1946	18.1351	19.4226	months
110.0	80.0	9.387	9.6116	9.9413	10.8354	12.1784	13.788	
X	H	$\sigma = 0.10$	$\sigma = 0.15$	$\sigma = 0.20$	$\sigma = 0.25$	$\sigma = 0.30$	$\sigma = 0.35$	T
80.0	110.0	39.0845	38.2196	37.4729	37.1633	37.2523	37.6496	
90.0	110.0	28.1156	27.5171	27.444	27.8622	28.6147	29.5881	9
100.0	110.0	17.5507	17.8237	18.7016	19.8955	21.2614	22.7274	months
110.0	110.0	8.7065	10.0755	11.7387	13.4921	15.278	17.0756	

14.1.3. Up and In Call Valuation

The up and-in call barrier (with asset price less than barrier) option has a payoff:

$$\max[0, S_T - X], S_{0,T} \geq H; \; else, K$$

The three probability conditions for this barrier option are:

The strike can be greater than the barrier. $P = prob(S_T > X)$.
Strike less than barrier. $P = prob(H > S_T > X : S_\tau \geq H) + prob(S_T > H)$.
Barrier not hit. $P = prob(S_T < H) - prob(S_T < H : S_\tau \geq H)$.

The valuation for an up and-in call with strike less than barrier is:

$$\begin{aligned}
C_{ui} = {} & \phi S e^{(b-r)T} N(\phi x_2) - \phi X e^{-rT} N(\phi x_2 - \phi\sigma\sqrt{T}) \\
& - \phi S e^{(b-r)T} (H/S)^{2(\mu+1)} N(\eta y_1) - \phi X e^{-rT} (H/S)^{2\mu} N(\eta y_1 - \eta\sigma\sqrt{T}) \\
& + \phi S e^{(b-r)T} (H/S)^{2(\mu+1)} N(\eta y_2) - \phi X e^{-rT} (H/S)^{2\mu} N(\eta y_2 - \eta\sigma\sqrt{T}) \\
& + X e^{-rT} [N(\eta x_2 - \eta\sigma\sqrt{T}) - (H/S)^{2\mu} N(\eta y_2 - \eta\sigma\sqrt{T})] \qquad (14.1.13)
\end{aligned}$$

$$\eta = -1, \phi = 1$$

An up and-in call with strike greater than barrier is:

$$\begin{aligned}
& \phi S e^{(b-r)T} N(\phi x_1) - \phi X e^{-rT} N(\phi x_1 - \phi\sigma\sqrt{T}) + X e^{-rT} [N(\eta x_2 - \eta\sigma\sqrt{T}) \\
& - (H/S)^{2\mu} N(\eta y_2 - \eta\sigma\sqrt{T})] \qquad (14.1.14)
\end{aligned}$$

$$\eta = -1, \phi = 1$$

14.1.4. Down and In Put Valuation

For the down and-in barrier put option, where the asset price is greater than the barrier. The payoff is:

$$\max[0, X - S_T], S_{0,T} \leq H; \; else, K$$

The three probability conditions for a down and-in put option are:

Strike greater than barrier.

$$P = prob[S_T \leq H] + prob[H < S_T < X : S_\tau \leq H]$$

Strike less than barrier.

$$P = prob[S_T < X]$$

TABLE 14.3. Down and-in Barrier Put Values

Valuations for Down and In Standard Barrier Put with S = 115.0, K = 5.0.								
X	H	$\sigma = 0.10$	$\sigma = 0.15$	$\sigma = 0.20$	$\sigma = 0.25$	$\sigma = 0.30$	$\sigma = 0.35$	T
80.0	80.0	35.3606	34.1495	32.7185	31.6449	30.9651	30.6198	
90.0	80.0	25.6075	24.397	22.9809	21.9771	21.4459	21.317	3
100.0	80.0	15.8561	14.7051	13.5047	12.8444	12.7066	12.9688	months
110.0	80.0	6.4516	5.9582	5.4135	5.3013	5.6053	6.2258	
X	H	$\sigma = 0.10$	$\sigma = 0.15$	$\sigma = 0.20$	$\sigma = 0.25$	$\sigma = 0.30$	$\sigma = 0.35$	T
80.0	110.0	42.0448	40.6754	39.126	37.9448	37.1244	36.6055	
90.0	110.0	28.7045	27.564	26.6017	26.0778	25.913	26.0103	3
100.0	110.0	15.5281	15.1195	15.2482	15.6971	16.3307	17.0767	months
110.0	110.0	4.8439	5.6633	6.7747	7.9335	9.0932	10.2444	
X	H	$\sigma = 0.10$	$\sigma = 0.15$	$\sigma = 0.20$	$\sigma = 0.25$	$\sigma = 0.30$	$\sigma = 0.35$	T
80.0	80.0	36.3709	35.2349	34.0387	33.4522	33.4553	33.8849	
90.0	80.0	27.0947	26.0194	25.064	24.862	25.3002	26.1587	9
100.0	80.0	17.8817	17.1977	16.8239	17.1946	18.1351	19.4226	months
110.0	80.0	9.387	9.6116	9.9413	10.8354	12.1784	13.788	
X	H	$\sigma = 0.10$	$\sigma = 0.15$	$\sigma = 0.20$	$\sigma = 0.25$	$\sigma = 0.30$	$\sigma = 0.35$	T
80.0	110.0	39.0845	38.2196	37.4729	37.1633	37.2523	37.6496	
90.0	110.0	28.1156	27.5171	27.444	27.8622	28.6147	29.5881	9
100.0	110.0	17.5507	17.8237	18.7016	19.8955	21.2614	22.7274	months
110.0	110.0	8.7065	10.0755	11.7387	13.4921	15.278	17.0756	

Barrier never hit.

$$P = prob[S_T > H] - prob[S_T > H : S_\tau \leq H]$$

The valuation of a down and-in put with strike greater than barrier is:

$$\begin{aligned} P_{di} &= \phi S e^{(b-r)T} N(\phi x_2) - \phi X e^{-rT} N(\phi x_2 - \phi\sigma\sqrt{T}) \\ &- \phi S e^{(b-r)T} (H/S)^{2(\mu+1)} N(\eta y_1) - \phi X e^{-rT} (H/S)^{2\mu} N(\eta y_1 - \eta\sigma\sqrt{T}) \\ &+ \phi S e^{(b-r)T} (H/S)^{2(\mu+1)} N(\eta y_2) - \phi X e^{-rT} (H/S)^{2\mu} N(\eta y_2 - \eta\sigma\sqrt{T}) \\ &+ X e^{-rT} [N(\eta x_2 - \eta\sigma\sqrt{T}) - (II/S)^{2\mu} N(\eta y_2 - \eta\sigma\sqrt{T})] \end{aligned} \quad (14.1.15)$$

$$\eta = 1, \phi = -1$$

A down and-in put with strike less than barrier is:

$$\begin{aligned} &\phi S e^{(b-r)T} N(\phi x_1) - \phi X e^{-rT} N(\phi x_1 - \phi\sigma\sqrt{T}) + X e^{-rT} [N(\eta x_2 - \eta\sigma\sqrt{T}) \\ &- (H/S)^{2\mu} N(\eta y_2 - \eta\sigma\sqrt{T})] \end{aligned} \quad (14.1.16)$$

$$\eta = 1, \phi = -1$$

Table 14.3 shows a range of valuations for down and-in options.

14.1.5. Up and In Put Valuation

For an up and-in put (with asset price less than barrier) the payoff is:

$$\max[0, X - S_T], S_{0,T} \geqslant H; \; else, K$$

For the up and-in put the three probabilities are:

Strike greater than barrier.

$$P = prob[H \leqslant S_T < X] + prob[S_T < H : S_\tau \geqslant H]$$

Strike less than barrier.

$$P = prob[S_T < X : S_\tau \geqslant H]$$

Barrier never hit.

$$P = prob[S_T < H] - prob[S_T < H : S_\tau \geqslant H]$$

The up and-in valuation for strike less than barrier is:

$$\phi S e^{(b-r)T} N(\phi x_1) - \phi X e^{-rT} N(\phi x_1 - \phi\sigma\sqrt{T}) + X e^{-rT} [N(\eta x_2 - \eta\sigma\sqrt{T}) - (H/S)^{2\mu} N(\eta y_2 - \eta\sigma\sqrt{T})] \tag{14.1.17}$$

$$\eta = -1, \phi = -1$$

The up and-in put valuation for strike greater than barrier:

$$\begin{aligned} P_{ui} = {} & \phi S e^{(b-r)T} N(\phi x_1) - \phi X e^{-rT} N(\phi x_1 - \phi\sigma\sqrt{T}) - \phi S e^{(b-r)T} N(\phi x_2) \\ & - \phi X e^{-rT} N(\phi x_2 - \phi\sigma\sqrt{T}) \\ & + \phi S e^{(b-r)T} (H/S)^{2(\mu+1)} N(\eta y_2) - \phi X e^{-rT} (H/S)^{2\mu} N(\eta y_2 - \eta\sigma\sqrt{T}) \\ & + X e^{-rT} [N(\eta x_2 - \eta\sigma\sqrt{T}) - (H/S)^{2\mu} N(\eta y_2 - \eta\sigma\sqrt{T})] \end{aligned} \tag{14.1.18}$$

$$\eta = -1, \phi = -1$$

Table 14.4 shows a range of valuations for up and-in options.

14.2. Out Barrier Valuation

A European "out" barrier option has a corresponding set of possibilities to the in options. In the case of a call down and-out, the standard call exists when the option contract is made. If the barrier is hit from above, the call is extinguished prior to expiration and a rebate (if part of the contract) is paid. If the barrier is never hit during the option period, the payoff is that of a standard call at expiration. In the case of an up and-out barrier, the option is extinguished if the barrier is hit from below. The payoff from a down and-out call can be priced from the parity relationship;

$$P_{d,o} = P_{s,o} + P_{d,i},$$

This is the payoff from a standard call minus the payoff from a down and in call. This parity relationship holds if no rebate is payable. If we have a portfolio of one up and-out and one up and-in barrier option without rebate, the payoff at expiration, when the barrier is never hit, will be a standard call

TABLE 14.4. Up and-in Barrier Put Values

Valuations for Up and In Standard Barrier Put with S = 105.0, K = 5.0, r = 0.10, b = 0.05.

X	H	$\sigma = 0.10$	$\sigma = 0.15$	$\sigma = 0.20$	$\sigma = 0.25$	$\sigma = 0.30$	$\sigma = 0.35$	T maturity
80.0	90.0	0.000	0.000	0.000	0.972	2.1696	3.3832	
90.0	90.0	1.683	4.1388	6.0936	7.7335	9.1868	10.5176	3
100.0	90.0	4.3758	10.4242	14.2357	16.6171	18.3016	19.6471	months
110.0	90.0	10.9014	20.2529	25.489	28.2068	29.7836	30.8675	
X	H	$\sigma = 0.10$	$\sigma = 0.15$	$\sigma = 0.20$	$\sigma = 0.25$	$\sigma = 0.30$	$\sigma = 0.35$	T maturity
80.0	110.0	2.786	2.0584	1.6387	1.3716	1.2051	1.1287	
90.0	110.0	2.786	2.059	1.6543	1.4555	1.4329	1.5645	3
100.0	110.0	2.7882	2.1258	1.9379	2.0737	2.4282	2.932	months
110.0	110.0	3.2763	3.2544	3.6676	4.2977	5.0498	5.8769	
X	H	$\sigma = 0.10$	$\sigma = 0.15$	$\sigma = 0.20$	$\sigma = 0.25$	$\sigma = 0.30$	$\sigma = 0.35$	T maturity
80.0	90.0	0.000	0.3343	1.8895	3.6775	5.487	7.262	
90.0	90.0	1.3724	4.1092	6.7655	9.1451	11.3073	13.3175	9
100.0	90.0	4.0254	9.4519	13.2997	16.170	18.5423	20.6487	months
110.0	90.0	9.2953	17.0823	21.7454	24.8049	27.1563	29.1802	
X	H	$\sigma = 0.10$	$\sigma = 0.15$	$\sigma = 0.20$	$\sigma = 0.25$	$\sigma = 0.30$	$\sigma = 0.35$	T maturity
80.0	110.0	1.3433	1.0446	0.9159	0.9836	1.2656	1.7443	
90.0	110.0	1.345	1.1143	1.2321	1.6778	2.379	3.2667	9
100.0	110.0	1.4369	1.6322	2.3223	3.3103	4.4802	5.7657	months
110.0	110.0	2.5776	3.5729	4.8524	6.2792	7.7905	9.3535	

option value. However due to the difference in rebate payouts for in and out options (for an in option the rebate is not paid until expiration, the out options pays arbitrarily on hitting the barrier) we need to take account of the randomness associated with an "out" rebate.

The additional probability densities are given by distributions of the following parameters:

$$\lambda = \sqrt{\mu^2 + \frac{2r}{\sigma^2}} \tag{14.2.1}$$

$$z = \frac{\log(H/S)}{\sigma\sqrt{T}} + \lambda\sigma\sqrt{T} \tag{14.2.2}$$

14.2.1. Down and Out Call Valuation

For a down and-out call with the initial underlying asset price greater than the barrier, the payoff is:

$$\max[0, S_T - X], S_{0,T} > H; \textit{else}, K$$

The down and-out call valuation with strike greater than barrier is:

$$C_{do} = \phi S e^{(b-r)T} N(\phi x_1) - \phi X e^{-rT} N(\phi x_1 - \phi\sigma\sqrt{T})$$
$$- \phi S e^{(b-r)T} (H/S)^{2(\mu+1)} N(\eta y_1) - \phi X e^{-rT} (H/S)^{2\mu} N(\eta y_1 - \eta\sigma\sqrt{T})$$
$$+ K[(H/S)^{\mu+\lambda} N(\eta z) + (H/S)^{\mu-\lambda} N(\eta z - 2\eta\lambda\sigma\sqrt{T})] \quad (14.2.3)$$

$$\eta = 1, \phi = 1$$

With strike less than barrier, the call down and-out valuation is given by:

$$C_{do} = \phi S e^{(b-r)T} N(\phi x_2) - \phi X e^{-rT} N(\phi x_2 - \phi\sigma\sqrt{T})$$
$$- \phi S e^{(b-r)T} (H/S)^{2(\mu+1)} N(\eta y_2) - \phi X e^{-rT} (H/S)^{2\mu} N(\eta y_2 - \eta\sigma\sqrt{T})$$
$$+ K[(H/S)^{\mu+\lambda} N(\eta z) + (H/S)^{\mu-\lambda} N(\eta z - 2\eta\lambda\sigma\sqrt{T})] \quad (14.2.4)$$

$$\eta = 1, \phi = 1$$

Table 14.5 shows a range of valuations for down and-out options.

TABLE 14.5. Down and-out Barrier Call Values

Valuations for Down and Out Standard Barrier Call with S = 115.0, K = 5.0, r = 0.10, b = 0.05.								
X	H	σ = 0.10	σ = 0.15	σ = 0.20	σ = 0.25	σ = 0.30	σ = 0.35	T maturity
80.0	90.0	35.5467	35.5431	35.4867	35.309	35.0239	34.6835	
90.0	90.0	25.7936	25.7969	25.8489	26.0077	26.2543	26.5393	3
100.0	90.0	16.0422	16.1057	16.3905	16.9687	17.7614	18.648	months
110.0	90.0	6.6377	7.3588	8.2998	9.4342	10.7034	12.0242	
X	H	σ = 0.10	σ = 0.15	σ = 0.20	σ = 0.25	σ = 0.30	σ = 0.35	T maturity
80.0	105.0	34.6562	31.7185	28.7436	26.3863	24.5897	23.2092	
90.0	105.0	25.341	23.8337	22.2926	21.0616	20.1174	19.388	3
100.0	105.0	16.0257	15.9488	15.8415	15.7369	15.645	15.5668	months
110.0	105.0	6.8537	8.1897	9.4695	10.461	11.204	11.7666	
X	H	σ = 0.10	σ = 0.15	σ = 0.20	σ = 0.25	σ = 0.30	σ = 0.35	T maturity
80.0	90.0	36.5405	36.3422	35.8135	35.1476	34.4942	33.9081	
90.0	90.0	27.2765	27.4182	27.7415	28.0876	28.3781	28.6027	9
100.0	90.0	18.0654	18.6827	19.8748	21.1927	22.3843	23.3869	months
110.0	90.0	9.5707	11.1072	13.0944	15.1453	17.0163	18.6295	
X	H	σ = 0.10	σ = 0.15	σ = 0.20	σ = 0.25	σ = 0.30	σ = 0.35	T maturity
80.0	105.0	32.7873	28.1477	24.9411	22.7942	21.2988	20.2108	
90.0	105.0	25.2456	22.6679	20.8434	19.6011	18.7247	18.081	9
100.0	105.0	17.7039	17.188	16.7458	16.4079	16.1506	15.9511	months
110.0	105.0	10.2346	11.7457	12.6673	13.2255	13.5831	13.8255	

14.2.2. *Up and Out Call Valuation*

For an up and-out call with the initial asset price being less than the barrier, the payoff is given by:

$$\max[0, S_T - X], S_{0,T} < H; \textit{else}, K$$

An up and-out call with strike price greater than the barrier has a value given by:

$$K[(H/S)^{\mu+\lambda}N(\eta z) + (H/S)^{\mu-\lambda}N(\eta z - 2\eta\lambda\sigma\sqrt{T})] \tag{14.2.5}$$
$$\eta = -1, \phi = 1$$

We can see from 14.2.3 that the single contribution to the option value is from a rebate proportion. This is because the asset price is below the barrier, which in turn is below the strike. For the asset to exceed the strike, it would need to breach the barrier.

An up and-out call with strike less than the barrier has a value of:

$$\begin{aligned}
&\phi Se^{(b-r)T}N(\phi x_1) - \phi Xe^{-rT}N(\phi x_1 - \phi\sigma\sqrt{T})\\
&-\phi Se^{(b-r)T}N(\phi x_2) - \phi Xe^{-rT}N(\phi x_2 - \phi\sigma\sqrt{T})\\
&+\phi Se^{(b-r)T}(H/S)^{2(\mu+1)}N(\eta y_1) - \phi Xe^{-rT}(H/S)^{2\mu}N(\eta y_1 - \eta\sigma\sqrt{T})\\
&-\phi Se^{(b-r)T}(H/S)^{2(\mu+1)}N(\eta y_2) - \phi Xe^{-rT}(H/S)^{2\mu}N(\eta y_2 - \eta\sigma\sqrt{T})\\
&+K[(H/S)^{\mu+\lambda}N(\eta z) + (H/S)^{\mu-\lambda}N(\eta z - 2\eta\lambda\sigma\sqrt{T})]
\end{aligned} \tag{14.2.6}$$

$$\eta = -1, \phi = 1$$

Figure 14.3 shows the characteristics for a range of volatilities. Table 14.6 gives a range of valuations.

14.2.3. *Down and Out Put Valuation*

For a down and-out put with asset price greater than barrier, the payoff is given by:

$$\max[0, X - S_T], S_{0,T} > H; \textit{else}, K$$

A down and-out put with strike less than barrier has a value given by:

$$K\left[(H/S)^{\mu+\lambda}N(\eta z) + (H/S)^{\mu-\lambda}N(\eta z - 2\eta\lambda\sigma\sqrt{T})\right] \tag{14.2.7}$$
$$\eta = 1, \phi = -1$$

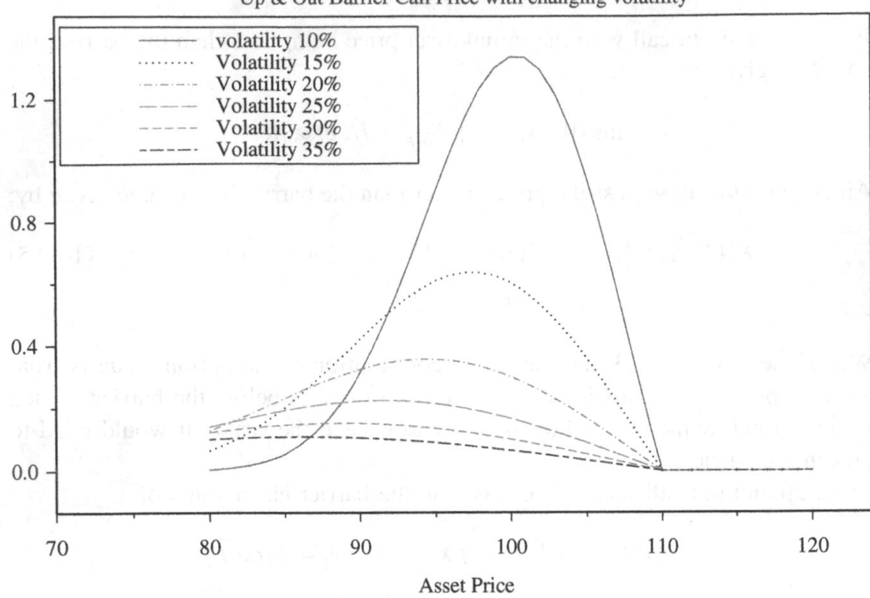

FIGURE 14.3. Up and-out call barrier valuations for a range of volatilities.

TABLE 14.6. Up and-out Barrier call values

Valuations for Up and Out Standard Barrier Call with S = 95.0, K = 5.0, r = 01.0, b = 0.05								
X	H	$\sigma = 0.10$	$\sigma = 0.15$	$\sigma = 0.20$	$\sigma = 0.25$	$\sigma = 0.30$	$\sigma = 0.35$	T maturity
80.0	96.0	5.6716	5.2476	5.0349	4.9318	4.8836	4.8623	
90.0	96.0	4.5252	4.5775	4.6396	4.6894	4.7277	4.7574	3
100.0	96.0	4.3387	4.510	4.609	4.6732	4.7181	4.7513	months
110.0	96.0	4.3387	4.510	4.609	4.6732	4.7181	4.7513	
X	H	$\sigma = 0.10$	$\sigma = 0.15$	$\sigma = 0.20$	$\sigma = 0.25$	$\sigma = 0.30$	$\sigma = 0.35$	T maturity
80.0	126.0	15.7952	15.799	15.6692	15.0802	14.0221	12.7184	
90.0	126.0	6.2351	6.7328	7.2366	7.3787	7.116	6.6159	3
100.0	126.0	0.5997	1.3752	2.1219	2.600	2.7869	2.7966	months
110.0	126.0	0.0056	0.1077	0.3538	0.6186	0.8387	1.023	
X	H	$\sigma = 0.10$	$\sigma = 0.15$	$\sigma = 0.20$	$\sigma = 0.25$	$\sigma = 0.30$	$\sigma = 0.35$	T maturity
80.0	96.0	5.0871	4.9324	4.8791	4.8653	4.8654	4.8702	
90.0	96.0	4.7112	4.748	4.7838	4.8115	4.8325	4.8489	9
100.0	96.0	4.6733	4.7349	4.778	4.8084	4.8307	4.8477	months
110.0	96.0	4.6733	4.7349	4.778	4.8084	4.8307	4.8477	
X	H	$\sigma = 0.10$	$\sigma = 0.15$	$\sigma = 0.20$	$\sigma = 0.25$	$\sigma = 0.30$	$\sigma = 0.35$	T maturity
80.0	126.0	17.1594	15.6785	13.0917	10.5869	8.5895	7.1132	
90.0	126.0	8.4624	8.2061	7.0607	5.8625	4.9358	4.2953	9
100.0	126.0	2.513	3.1952	3.1086	2.8712	2.7066	2.6328	months
110.0	126.0	0.3746	0.8735	1.1839	1.4313	1.660	1.8709	

For the case where the strike price is greater than barrier, a down and-out put option has a value given by:

$$\begin{aligned}
P_{do} &= \phi S e^{(b-r)T} N(\phi x_1) - \phi X e^{-rT} N(\phi x_1 - \phi\sigma\sqrt{T}) \\
&\quad - \phi S e^{(b-r)T} N(\phi x_2) - \phi X e^{-rT} N(\phi x_2 - \phi\sigma\sqrt{T}) \\
&\quad + \phi S e^{(b-r)T} (H/S)^{2(\mu+1)} N(\eta y_1) \\
&\quad - \phi X e^{-rT} (H/S)^{2\mu} N(\eta y_1 - \eta\sigma\sqrt{T}) \\
&\quad - \phi S e^{(b-r)T} (H/S)^{2(\mu+1)} N(\eta y_2) - \phi X e^{-rT} (H/S)^{2\mu} N(\eta y_2 - \eta\sigma\sqrt{T}) \\
&\quad + K\left[(H/S)^{\mu+\lambda} N(\eta z) + (H/S)^{\mu-\lambda} N(\eta z - 2\eta\lambda\sigma\sqrt{T})\right]
\end{aligned} \tag{14.2.8}$$

$$\eta = 1, \phi = -1$$

Table 14.7 gives a range of down and-out option valuations.

For the situation where the initial asset price is less than the barrier, the payoff is:

$$\max[0, X - S_T], S_{0,T} < H; \; else, K$$

TABLE 14.7. Down and-out Barrier put values

Valuations for Down and Out Standard Barrier Put with S = 115.0, K = 5.0, r = 0.10, b = 0.05.

X	H	$\sigma = 0.10$	$\sigma = 0.15$	$\sigma = 0.20$	$\sigma = 0.25$	$\sigma = 0.30$	$\sigma = 0.35$	T maturity
80.0	90.0	4.8765	4.8731	4.8191	4.6551	4.4128	4.1658	
90.0	90.0	4.8765	4.8737	4.8347	4.7404	4.6467	4.6161	3
100.0	90.0	4.8766	4.8799	4.9321	5.0985	5.3837	5.768	months
110.0	90.0	4.8766	4.8867	5.0469	5.5416	6.324	7.2578	
X	H	$\sigma = 0.10$	$\sigma = 0.15$	$\sigma = 0.20$	$\sigma = 0.25$	$\sigma = 0.30$	$\sigma = 0.35$	T maturity
80.0	115.0	0.000	0.000	0.0002	0.0044	0.028	0.0937	
90.0	115.0	0.000	0.0006	0.0157	0.0897	0.2619	0.544	3
100.0	115.0	0.0017	0.0618	0.2926	0.7102	1.2757	1.9489	months
110.0	115.0	0.3503	1.068	1.9546	2.9201	3.9275	4.959	
X	H	$\sigma = 0.10$	$\sigma = 0.15$	$\sigma = 0.20$	$\sigma = 0.25$	$\sigma = 0.30$	$\sigma = 0.35$	T maturity
80.0	90.0	4.632	4.4654	4.0914	3.7761	3.6824	3.8396	
90.0	90.0	4.6332	4.5273	4.3941	4.4634	4.8047	5.3908	9
100.0	90.0	4.6447	4.7945	5.2262	5.9083	6.7947	7.8427	months
110.0	90.0	4.6581	5.1372	6.3294	7.814	9.3672	10.9375	
X	H	$\sigma = 0.10$	$\sigma = 0.15$	$\sigma = 0.20$	$\sigma = 0.25$	$\sigma = 0.30$	$\sigma = 0.35$	T maturity
80.0	115.0	0.000	0.0033	0.0554	0.2461	0.6243	1.1869	
90.0	115.0	0.0012	0.0653	0.3581	0.9334	1.7466	2.7381	9
100.0	115.0	0.0656	0.521	1.3954	2.5434	3.859	5.2795	months
110.0	115.0	0.8484	2.2123	3.7902	5.4617	7.1797	8.9223	

14.2.4. Up and Out Put Valuation

For a put up and-out option with strike greater than barrier, the valuation is given by:

$$P_{uo} = \phi S e^{(b-r)T} N(\phi x_2) - \phi X e^{-rT} N(\phi x_2 - \phi\sigma\sqrt{T}) \\ - \phi S e^{(b-r)T} (H/S)^{2(\mu+1)} N(\eta y_2) - \phi X e^{-rT} (H/S)^{2\mu} N(\eta y_2 - \eta\sigma\sqrt{T}) \tag{14.2.9}$$

$$+ K\left[(H/S)^{\mu+\lambda} N(\eta z) + (H/S)^{\mu-\lambda} N(\eta z - 2\eta\lambda\sigma\sqrt{T})\right] \tag{14.2.10}$$

$$\eta = -1, \phi = -1$$

An up and-out put option with strike less than barrier is valued by:

$$P_{uo} = \phi S e^{(b-r)T} N(\phi x_1) - \phi X e^{-rT} N(\phi x_1 - \phi\sigma\sqrt{T}) \\ - \phi S e^{(b-r)T} (H/S)^{2(\mu+1)} N(\eta y_1) - \phi X e^{-rT} (H/S)^{2\mu} N(\eta y_1 - \eta\sigma\sqrt{T}) \\ + K\left[(H/S)^{\mu+\lambda} N(\eta z) + (H/S)^{\mu-\lambda} N(\eta z - 2\eta\lambda\sigma\sqrt{T})\right] \tag{14.2.11}$$

$$\eta = -1, \phi = -1$$

Table 14.8 provides example data for up and-out put options over a range of strike and barrier prices.

TABLE 14.8. Up and-out Barrier put option values

Valuations for Up and Out Standard Barrier Put with S = 95.0, K = 5.0, r = 01.0, b = 0.05								
X	H	σ = 0.10	σ = 0.15	σ = 0.20	σ = 0.25	σ = 0.30	σ = 0.35	T maturity
80.0	96.0	4.3388	4.5183	4.6537	4.7763	4.8861	4.9814	
90.0	96.0	4.4504	4.7811	5.0034	5.1568	5.2676	5.3509	3
100.0	96.0	5.522	5.6465	5.7176	5.7635	5.7955	5.8191	months
110.0	96.0	6.780	6.5794	6.4625	6.3864	6.3329	6.2934	
X	H	σ = 0.10	σ = 0.15	σ = 0.20	σ = 0.25	σ = 0.30	σ = 0.35	T maturity
80.0	126.0	0.0001	0.0159	0.1379	0.4587	0.9749	1.6333	
90.0	126.0	0.1931	0.7003	1.4013	2.2571	3.230	4.271	3
100.0	126.0	4.3108	5.0932	5.9827	6.9783	8.0622	9.192	months
110.0	126.0	13.4698	13.5761	13.9108	14.4967	15.2753	16.1586	
X	H	σ = 0.10	σ = 0.15	σ = 0.20	σ = 0.25	σ = 0.30	σ = 0.35	T maturity
80.0	96.0	4.6822	4.8055	4.9381	5.0531	5.1468	5.2225	
90.0	96.0	4.8547	5.0728	5.2239	5.3298	5.407	5.4656	9
100.0	96.0	5.3652	5.5114	5.5992	5.6572	5.6983	5.7289	months
110.0	96.0	5.9136	5.9632	5.9803	5.9878	5.9914	5.9933	
X	H	σ = 0.10	σ = 0.15	σ = 0.20	σ = 0.25	σ = 0.30	σ = 0.35	T maturity
80.0	126.0	0.0358	0.4654	1.3859	2.5728	3.8731	5.1987	
90.0	126.0	0.5818	1.8308	3.458	5.1885	6.886	8.4814	9
100.0	126.0	3.8754	5.6575	7.6091	9.5374	11.3235	12.9195	months
110.0	126.0	10.9802	12.1735	13.7876	15.4375	16.9436	18.2582	

Single Barrier option methods are in the class ***Sbarrierop*** shown in Listing 14.1

```
package FinApps;
import static java.lang.Math.*;
import BaseStats.Probnorm;
public class Sbarrierop {//Alows for negative returns,
                          //i.e a cost to the seller to sell an
option
   //the user program should trap negative
                          //values and return zero for no value
   /** Creates a new instance of Sbarrierop */
   public Sbarrierop(double rate,double q,double time,
                    int period ) {
      crate=q;
      r=rate;
      t=time;
      b=crate<0.0?0.0:(b=crate!=r?(r-crate):r);
       tau=period==1?(1.0/(24*365)):period==2?(1.0/(365.0)):
       period==3?(1.0/(52.0)) :period==4?(1.0/(12.0)):0.0;
   }
   double b;
   double tau;
   double crate;
   double t;
   double r;
   double x1;
   double x2;
   double y1;
   double y2;
   double mu;
   double z
   double lambda;
   double s;
   double x;
   double sigma;
   double k;
   double h;
   public double upencall(double sprice, double strike,
                          double volatility, double barrier,
                          double rebate) {
      s=sprice;
      x=strike;
      sigma=volatility;
      k=rebate;
      h=(barrier*exp(sqrt(tau)*sigma*0.5826));
                         // barrier above asset price

      inPars(s,x,sigma,h);
      return x> h?(f1(1.0,-1.0)+f5(1.0,-1.0)):
                  (f2(1.0,-1.0)-f3(1.0,-1.0)+f4(1.0,-1.0)
                  +f5 (1.0,-1.0));

   }
```

```
public double downIcall(double sprice, double strike,
                double volatility, double barrier, double rebate)
{
   s=sprice;
   x=strike;
   sigma=volatility;
   k=rebate;
   h=(barrier*exp(-sqrt(tau)*sigma*0.5826));
                       // barrier below asset price

   inPars(s,x,sigma,h);
   return x>h?(f3(1.0,1.0)+f5(1.0,1.0)):
              (f1(1.0,1.0)-f2(1.0,1.0)+f4(1.0,1.0)
              +f5 (1.0,1.0));

}

public double downIput(double sprice, double strike,
                double volatility, double
barrier, double rebate)
{
   s=sprice;
   x=strike;
   sigma=volatility;
   k=rebate;
   h= (barrier*exp(-sqrt(tau)*sigma*0.5826));

   inPars(s,x,sigma,h);
   return x<h?(f1(-1.0,1.0)+f5(-1.0,1.0)):
              (f2(-1.0,1.0)-f3(-1.0,1.0)+f4(-1.0,1.0)
              +f5 (-1.0,1.0));


}
public double upIput(double sprice, double strike,
                double volatility, double barrier, double rebate)
{
   s=sprice;
   x=strike;
   sigma=volatility;
   k=rebate;
   h=(barrier*exp(sqrt(tau)*sigma*0.5826));

   inPars(s,x,sigma,h);
   return x<h?(f3(-1.0,-1.0)+f5(-1.0,-1.0)):(f1(-1.0,-1.0)
              -f2(-1.0,-1.0)+f4 (-1.0,-1.0) +f5 (-1.0,-1.0));


}
public double downOcall(double sprice, double strike,
                double volatility, double barrier, double rebate)
{
   s=sprice;
   x=strike;
   sigma=volatility;
   k=rebate;
```

```
    h=(barrier*exp(-sqrt(tau)*sigma*0.5826));
    outPars(s,x,sigma,h);
    return ((s<=h)&&(t!=0.0))?k: x<h?(f2(1.0,1.0)-f4(1.0,1.0)
                +f6(1.0,1.0)):
    (f1(1.0,1.0)-f3(1.0,1.0)+f6 (1.0,1.0));

 }

 public double upOput(double sprice, double strike, double volatility,
                       double barrier, double rebate)
 {
    s=sprice;
    x=strike;
    sigma=volatility;
    k=rebate;
     h=(barrier*exp(-sqrt(tau)*sigma*0.5826));
    outPars(s,x,sigma,h);
    return ((s>=h)&&(t!=0.0))?k:x<h?(f1(-1.0,-1.0)-f3(-1.0,-1.0)
    +f6(-1.0,-1.0)):(f2(-1.0,-1.0) -f4(-1.0,-1.0)+f6 (-1.0,-1.0));
 }
 public double upOcall(double sprice, double strike,
 double volatility, double barrier, double rebate)
 {
    s=sprice;
    x=strike;
    sigma=volatility;
    k=rebate;
     h=(barrier*exp(sqrt(tau)*sigma*0.5826));

     outPars(s,x,sigma,h);
    return ((s>=h)&&(t!=0.0))?k: x<h?(f1(1.0,-1.0)-f2(1.0,-1.0)
    +f3(1.0,-1.0)-f4(1.0,-1.0)+f6(1.0,-1.0)):f6 (1.0,-1.0);

 }
 public double downOput(double sprice, double strike,
 double volatility, double barrier, double rebate)
 {
    s=sprice;
    x=strike;
    sigma=volatility;
    k=rebate;
    h=(barrier*exp(-sqrt(tau)*sigma*0.5826));
    outPars(s,x,sigma,h);
    return ((s<=h)&&(t!=0.0))?k: x>h?(f1(-1.0,1.0)
                -f2(-1.0,1.0)+f3(-1.0,1.0)-f4
                (-1.0,1.0) +f6(-1.0,1.0)):f6 (-1.0,1.0);

 }
 private void inPars(double s,double x,double sigma,double h) {

    mu=((b-(sigma*sigma)/2.0)/(sigma*sigma));
    x1=(((log(s)-log(x))/(sigma*sqrt(t)))+((1.0+mu)
                 *sigma*sqrt(t)));
    x2=(((log(s)-log(h))/(sigma*sqrt(t)))+((1.0+mu)
                 *sigma*sqrt(t)));
```

```
      y1=(((log(h*h)-log(s*x))/(sigma*sqrt(t)))+((1.0+mu)
                   *sigma*sqrt(t)));
      y2=(((log(h)-log(s))/(sigma*sqrt(t)))+((1.0+mu)
                   *sigma*sqrt(t)));

   }

   private void outPars(double s, double x, double sigma, double h) {

      mu=((b-(sigma*sigma)/2.0)/(sigma*sigma));
      x1=(((log(s)-log(x))/(sigma*sqrt(t)))+((1.0+mu)
                   *sigma*sqrt(t)));
      x2=(((log(s)-log(h))/(sigma*sqrt(t)))+((1.0+mu)
                   *sigma*sqrt(t)));
      y1=(((log(h*h)-log(s*x))/(sigma*sqrt(t)))+((1.0+mu)
                   *sigma*sqrt(t)));
      y2=(((log(h)-log(s))/(sigma*sqrt(t)))+((1.0+mu)
                   *sigma*sqrt(t)));
      lambda=(sqrt((mu*mu)+((2.0*r)/(sigma*sigma))));
      z=(((log(h)-log(s))/(sigma*sqrt(t)))+lambda*sigma*sqrt(t));
   }
   private double f1(double phi,double eta)
   {
      Probnorm p=new Probnorm();
      return (phi*s*exp((b-r)*t)*p.ncDisfnc(phi*x1)-phi*x*exp(-r*t)
      *p.ncDisfnc (phi*x1-phi*sigma*sqrt(t)));
   }

   private double f2 (double phi,double eta)
   {
      Probnorm p=new Probnorm();
      return (phi*s*exp((b-r)*t)*p.ncDisfnc(phi*x2)-phi*x*exp(-r*t)
      *p.ncDisfnc (phi*x2-(phi*sigma*sqrt(t))));
   }
   private double f3(double phi,double eta)
   {
      Probnorm p=new Probnorm();
      return (phi*s*exp((b-r)*t)*pow((h/s),(2.0*(mu+1.0)))
                   *p.ncDisfnc(eta*y1)
      -phi*x*exp(-
r*t) *pow((h/s),(2.0*mu))*p.ncDisfnc(eta*y1-eta*sigma*sqrt(t)));
   }

   private double f4(double phi, double eta)
   {
      Probnorm p=new Probnorm();
       return (phi*s*exp((b-r)*t)*pow((h/s),(2.0*(mu+1.0)))
      *p.ncDisfnc(eta*y2)-phi*x*exp(-
r*t)*pow((h/s),(2.0*mu))*p.ncDisfnc(eta*y2-eta*sigma*sqrt(t)));
   }

   private double f5(double phi,double eta)
   { Probnorm p=new Probnorm();
```

```
    return(k*exp(-r*t)*(p.ncDisfnc(eta*x2-eta*sigma*sqrt(t))
                 -pow((h/s),(2*mu))
    *p.ncDisfnc(eta*y2-eta*sigma*sqrt(t))));
}
private double f6(double phi,double eta)
{
    Probnorm p=new Probnorm();
    return(k*(pow((h/s),(mu+lambda))*p.ncDisfnc(eta*z)
                 +pow((h/s),
    (mu-lambda))*p.ncDisfnc(eta*z
                 -2.0*eta*lambda*sigma*sqrt(t))));
}
```

LISTING 14.1. Computation of Single Barrier Methods

References

Merton, R. C. (1973). "Theory of Rational Option Pricing," *Bell Journal of Economics and Management Science*, 4, 141–183.

Hirsa et al. (2003) "The Effect of Model Risk on the Valuation of Barrier Options", *Journal of Risk Finance*, Winter 2003, 1–9.

Rubinstein, M. and E. Reiner (1991). "Breaking Down the Barriers," *Risk, 4*, 28–35.

15
Double Barrier Options

A double barrier option is knocked out or in if the asset price touches a lower (L) or upper (U) barrier level within the option lifetime. Double barrier option valuation has been described by Geman & Yor (1996), Kunitomo & Ikeda (1992) and others. The approach taken by Geman & Yor is a probabilistic model which uses the Laplace transform of the option price. The pricing formulae of Kunitomo & Ikeda take an approach based on the pricing being constrained by curved boundaries, this approach therefore has the advantage of covering barriers which are flat or have exponential growth/decay or are concave. We will develop classes for the Kunitomo & Ikeda formulae and a simpler set of methods based on the work of Haug (1999), which gives a set of methods for pricing barrier options with flat boundaries. The methods based on Haug have the advantage of using standard (plain vanilla) barrier options to construct a double barrier pricing mechanism.

15.1. Double Knock In/Out

15.1.1. Double Knock Out Call

The price of a double knock-out call has payoff:max$[S_T - K; 0]$; $L < S_T < U$: *else*, 0. Thus, if the underlying asset remains above the lower barrier and below the upper barrier until maturity, the payoff is the standard Black Scholes call payoff. If the asset price touches either the upper or lower boundary, then the option is knocked out worthless (zero payoff).

The formula for the double knock-out call is:

$$c = Se^{(b-r)T} \sum_{n=-\infty}^{\infty} \left\{ \left(\frac{U^n}{L^n}\right)^{\mu_1} \left(\frac{L}{S}\right)^{\mu_2} [\phi(d_1) - \phi(d_2)] - \left(\frac{L^{n+1}}{U^n S}\right)^{\mu_3} [\phi(d_3) - \phi(d_4)] \right\}$$
$$-Xe^{-rT} \sum_{n=-\infty}^{\infty} \left\{ \left(\frac{U^n}{L^n}\right)^{\mu_1 - 2} \left(\frac{L}{S}\right)^{\mu_2} [\phi(d_1 - \sigma\sqrt{T}) - \phi(d_2 - \sigma\sqrt{T})] - \left(\frac{L^{n+1}}{U^n S}\right)^{\mu_3 - 2} [\phi(d_3 - \sigma\sqrt{T}) - \phi(d_4 - \sigma\sqrt{T})] \right\} \quad (15.1.1)$$

Where

$$d_1 = \frac{\log(SU^{2n} / XL^{2n}) + (b + \sigma^2 / 2)T}{\sigma\sqrt{T}}$$

$$d_2 = \frac{\log(SU^{2n} / FL^{2n}) + (b + \sigma^2 2)T}{\sigma\sqrt{T}}$$

$$d_3 = \frac{\log(L^{2n+2} / XSU^{2n}) + (b + \sigma^2 / 2)T}{\sigma\sqrt{T}}$$

$$d_4 = \frac{\log(L^{2n+2}/FSU^{2n}) + (b + \sigma^2 / 2)T}{\sigma\sqrt{T}}$$

$$\mu_1 = \frac{2[b - \delta_2 - n(\delta_1 - \delta_2)]}{\sigma^2} + 1$$

$$\mu_2 = 2n\frac{(\delta_1 - \delta_2)}{\sigma^2}$$

$$\mu_3 = \frac{2[b - \delta_2 + n(\delta_1 - \delta_2)]}{\sigma^2} + 1, F = Ue^{\delta_1 T}$$

The variables δ_1, δ_2 represent the boundary curvature. For $\delta_1 = \delta_2 = 0$, the boundaries are flat.

For $\delta_1 < 0 < \delta_2$, the lower boundary exhibits exponential growth and the upper boundary exhibits exponential decay. The case of $\delta_1 > 0 > \delta_2$ is a convex lower boundary and convex upper boundary.

The double barrier knock-out option can be represented as a portfolio of a standard option and a double barrier knock-in. The resultant characteristics are shown in Figure 15.1. At each point on the graph we can see that the value of an up and-out down and-out call option is given by the Black Scholes standard call value minus the value of the up and-in down and-in call option.

With

$$S = 100.0, X = 100.0, \delta_1 = \delta_2 = 0.0, L = 80.0, U = 120.0, T = 0.25.$$

Table 15.1 shows a range of valuations for up and-out down and-out calls, with various curve boundary parameters.

Table 15.2 shows an example range of valuations for call up and-in down and-in barrier calls.

15.1.2. Double Knock Out Put

The price of a double knock-out put has payoff:$\max[X - S; 0], L < S < U; else, 0$. Thus, if the underlying asset remains above the lower barrier and below the upper barrier until maturity, the payoff is the standard Black Scholes put payoff. If the asset price touches either the upper or lower boundary, then the option is knocked out.

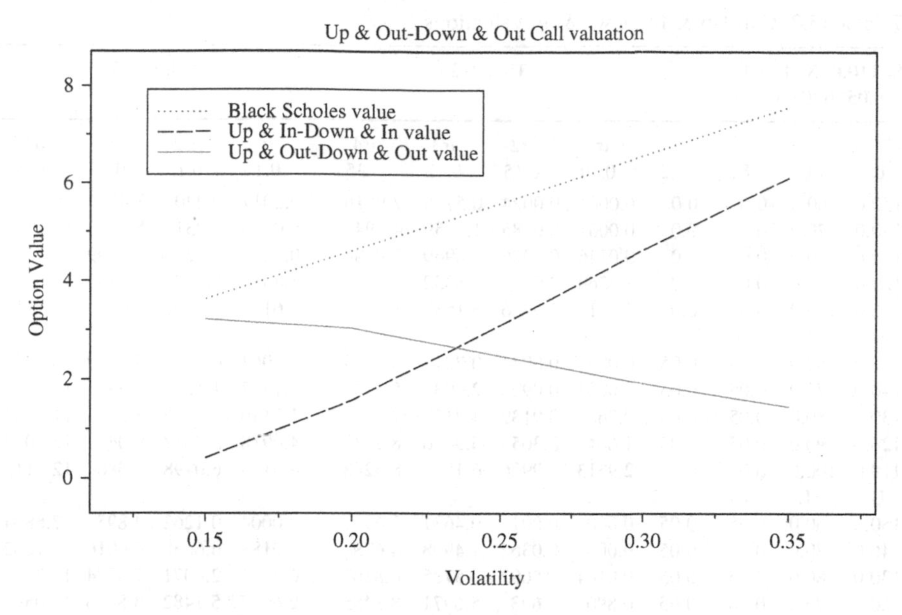

FIGURE 15.1. Up & Out-Down & Out Call barrier option as a portfolio of a long vanilla call and short an Up & In-Down & In barrier call.

TABLE 15.1. Call Up & Out-Down & Out valuations

S=110.0, X=110.0 r=0.05, b=0.05				Time 0.25				Time 0.5			
				$\sigma 1$	$\sigma 2$	$\sigma 3$	$\sigma 4$	$\sigma 1$	$\sigma 2$	$\sigma 3$	$\sigma 4$
U	L	$\delta 1$	$\delta 2$	0.10	0.15	0.25	0.35	0.10	0.15	0.25	0.35
150.0	60.0	0.0	0.0	2.9313	3.9962	5.5857	5.3532	4.6097	5.8498	5.5975	3.795
140.0	70.0	0.0	0.0	2.9312	3.9401	4.4224	3.3301	4.557	5.0486	3.4107	1.8949
130.0	80.0	0.0	0.0	2.8968	3.308	2.2613	1.2561	3.8655	2.955	1.2562	0.5492
120.0	90.0	0.0	0.0	1.7046	0.9845	0.323	0.1164	1.1608	0.4927	0.1125	0.0184
110.0	100.0	0.00	0.00	0.000	0.000	0.000	0.000	0.000	0.000	0.000	0.000
U	L	$-\delta 1$	$\delta 2$								
150.0	60.0	0.05	0.05	2.9313	3.9942	5.4568	5.0659	4.6051	5.6815	4.9531	3.1875
140.0	70.0	0.05	0.05	2.931	3.9065	4.1546	3.0139	4.4708	4.5591	2.7475	1.4537
130.0	80.0	0.05	0.05	2.8626	3.0847	1.9411	1.0408	3.2969	2.2022	0.8387	0.3451
120.0	90.0	0.05	0.05	1.3268	0.6932	0.2116	0.0711	0.5197	0.1981	0.0372	0.0042
110.0	100.0	0.05	0.05	0.000	0.000	0.000	0.000	0.000	0.000	0.000	0.000
U	L	$\delta 1$	$-\delta 2$								
150.0	60.0	0.05	0.05	2.9313	3.9973	5.6952	5.6271	4.6111	5.9536	6.191	4.4221
140.0	70.0	0.05	0.05	2.9313	3.9624	4.6674	3.6468	4.5927	5.414	4.0844	2.3852
130.0	80.0	0.05	0.05	2.9149	3.4894	2.5857	1.4887	4.2324	3.6727	1.7486	0.8071
120.0	90.0	0.05	0.05	2.041	1.3056	0.4612	0.1763	1.9468	0.9316	0.2454	0.0521
110.0	100.0	0.05	0.05	0.000	0.000	0.000	0.000	0.000	0.000	0.000	0.000

TABLE 15.2. Call Up & In-Down & In valuations

S=110.0, X=110.0 r=0.05, b=0.05				Time 0.25				Time 0.5			
				σ1	σ2	σ3	σ4	σ1	σ2	σ3	σ4
U	L	δ1	δ2	0.10	0.15	0.25	0.35	0.10	0.15	0.25	0.35
150.0	60.0	0.0	0.0	0.000	0.0024	0.5725	2.9716	0.0017	0.230	3.4885	8.3165
140.0	70.0	0.0	0.0	0.0001	0.0585	1.7359	4.9947	0.0545	1.0313	5.6753	10.2166
130.0	80.0	0.0	0.0	0.0346	0.6906	3.8969	7.0688	0.746	3.1248	7.8299	11.5623
120.0	90.0	0.0	0.0	1.2268	3.014	5.8352	8.2084	3.4507	5.5871	8.9735	12.0931
110.0	100.0	0.0	0.0	2.9313	3.9986	6.1582	8.3248	4.6115	6.0798	9.086	12.1115
U	L	−δ1	δ2								
150.0	60.0	0.05	0.05	0.000	0.0044	0.7014	3.2589	0.0064	0.3983	4.1329	8.924
140.0	70.0	0.05	0.05	0.0004	0.0921	2.0036	5.3109	0.1407	1.5207	6.3385	10.6578
130.0	80.0	0.05	0.05	0.0687	0.9139	4.2172	7.284	1.3146	3.8776	8.2473	11.7663
120.0	90.0	0.05	0.05	1.6045	3.3053	5.9466	8.2537	4.0918	5.8817	9.0489	12.1073
110.0	100.0	0.05	0.05	2.9313	3.9986	6.1582	8.3248	4.6115	6.0798	9.086	12.1115
U	L	δ1	−δ2								
150.0	60.0	0.05	0.05	0.000	0.0013	0.4631	2.6977	0.0004	0.1262	2.895	7.6894
140.0	70.0	0.05	0.05	0.000	0.0362	1.4908	4.678	0.0188	0.6658	5.0016	9.7262
130.0	80.0	0.05	0.05	0.0164	0.5092	3.5725	6.8362	0.3791	2.4071	7.3374	11.3044
120.0	90.0	0.05	0.05	0.8903	2.693	5.6971	8.1485	2.6647	5.1482	8.8406	12.0594
110.0	100.0	0.05	0.05	2.9313	3.9986	6.1582	8.3248	4.6115	6.0798	9.086	12.1115

The Kunitomo & Ikeda formula for an up and-out down and-out put is:

$$p = Xe^{-rT} \sum_{n=-\infty}^{n=\infty} \left\{ \left(\frac{U^n}{L^n}\right)^{\mu_1-2} \left(\frac{L}{S}\right)^{\mu_2} [\phi(y_1-\sigma\sqrt{T}) - \phi(y_2-\sigma\sqrt{T})] - \left(\frac{L^{n+1}}{U^n S}\right)^{\mu_3-2} [\phi(y_3-\sigma\sqrt{T}) - \phi(y_4-\sigma\sqrt{T})] \right\}$$
$$- Se^{(b-r)T} \sum_{n=-\infty}^{n=\infty} \left\{ \left(\frac{U^n}{L^n}\right)^{\mu_1} \left(\frac{L}{S}\right)^{\mu_2} [\phi(y_1) - \phi(y_2)] - \left(\frac{L^{n+1}}{U^n S}\right)^{\mu_3} [\phi(y_3) - \phi(y_4)] \right\} \tag{15.1.2}$$

Where

$$y_1 = \frac{\log(SU^{2n} \;/\; EL^{2n}) + (b+\sigma^2 \;/\; 2)T}{\sigma\sqrt{T}}$$

$$y_2 = \frac{\log(SU^{2n} \;/\; XL^{2n}) + (b+\sigma^2 \;/\; 2)T}{\sigma\sqrt{T}}$$

$$y_3 = \frac{\log(L^{2n+2} \;/\; ESU^{2n}) + (b+\sigma^2 \;/\; 2)T}{\sigma\sqrt{T}}$$

$$y_3 = \frac{\log(L^{2n+2} \;/\; XSU^{2n}) + (b+\sigma^2/2)T}{\sigma\sqrt{T}}, E = Le^{\delta_1 T}$$

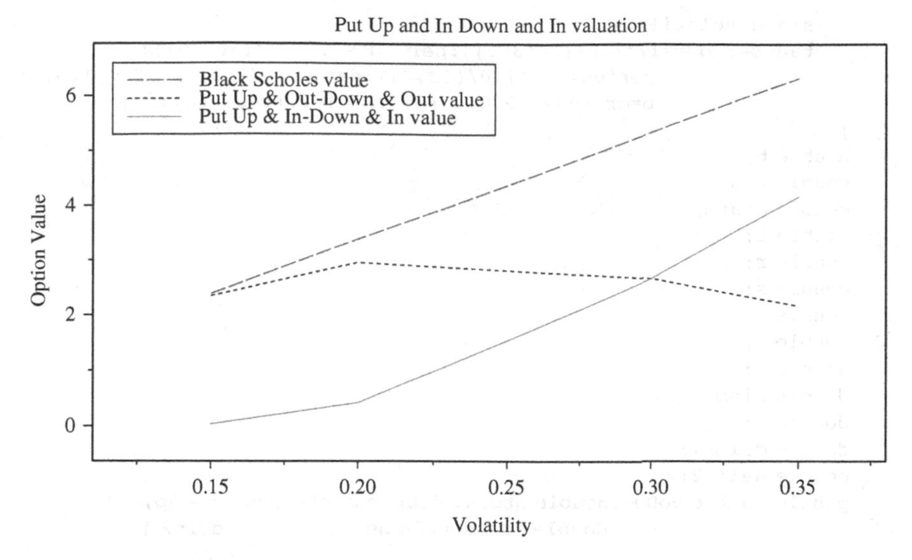

FIGURE 15.2. Up & In-Down & In barrier option as a portfolio of a long vanilla put and short an Up & Out-Down & Out barrier put.

15.1.3. Double Knock In Put

The up and-in down and-in barrier put option can be represented by a portfolio of long a standard Black-Scholes put option and a short an up and-out down and-out put. The resultant characteristic is shown in Figure 15.2.

The parameters for Figure 15.2. are:

$$S = 100.0, X = 100.0, \delta_1 = \delta_2 = 0.0, L = 80.0, U = 120.0, T = 0.25.$$

Tables 15.3. and 15.4. show example values for the remaining two double barrier options, the put up and-in down and-in, together with the put up and-out down and-out.

Double Barrier pricing methods are in the class ***Dbarrierop*** shown in Listing 15.1

```
package FinApps;
import static java.lang.Math.*;
import BaseStats.Probnorm;
public class Dbarrierop {

    /** Creates a new instance of Dbarrierop */
    public Dbarrierop(double rate,double q,double time, double volatility,
        int period ) {
        crate=q;
        r=rate;
        t=time;
        //0 continuos,1 hourly,2 daily,3 weekly, 4 monthly
```

```
    sigma=volatility;
    tau=period==1?(1.0/(24*365)):period==2?(1.0/(365.0)):
                period==3?(1.0/(52.0)):period==4?(1.0/(12.0)):0.0;
                b=crate==0.0?0.0:(b=crate!=r?(r-crate):r);
  }
  double b;
  double tau;
  double crate;
  double t;
  double r;
  double s;
  double u;
  double x;
  double l;
  double sigma;
  double f;
  double delta1;
  double delta2;
  public double uoDoc(double stock, double strike, double up,
                      double low,double delt1, double delt2 )
  {
    s=stock;
    x=strike;
    u= up>s? (up*exp(sqrt(tau)*sigma*0.5826)):up;
    l= low<s? (low*exp(-sqrt(tau)*sigma*0.5826)):low;
      if(s>=u|s<=l)
      return 0.0;// no need to continue
    delta1=delt1;
    delta2=delt2;
    double mu2;
    int n=0;
    double sum=0.0;
    double sum1=0.0;
    double c;
    f=(u*exp(delta1*t));
    for(n=-5;n<6;n++)
    {
      mu2=2.0*n*((delta1-delta2)/(sigma*sigma));
      sum+=pow((pow(u,n)/pow(l,n)),mu1(n)) *pow((l/s),mu2)
              *(N(d1(n))-N(d2(n)))-pow((pow(l,(n+1.0))
              /(pow(u,n)*s)),mu3(n))*(N(d3(n))-N(d4(n)));
      sum1+=pow((pow(u,n)/pow(l,n)),(mu1(n)-2.0))*pow((l/s),mu2)
              *(N(d1(n)-sigma*sqrt(t))-N(d2(n)-sigma*sqrt(t)))
              -pow(pow(l,(n+1.0))/(pow(u,n)*s),mu3(n)-2.0)
              *(N(d3(n)-sigma*sqrt(t))-N(d4(n)-sigma*sqrt(t)));
    }
    c=s*exp((b-r)*t)*sum-x*exp(-r*t)*sum1;
      return c;
    public double uoDop(double stock, double strike, double up,
                        double low,double delt1, double delt2 )
    {
    s=stock;
    x=strike;
    delta1=delt1;
```

```
        delta2=delt2;
           u= up>s? (up*exp(sqrt(tau)*sigma*0.5826)):up;
           l= low<s? (low*exp(-sqrt(tau)*sigma*0.5826)):low;
           if(s>=u|s<=l)
           return 0.0;// no need to continue
           double mu2;
           int n=0;
        double sum=0.0;
        double sum1=0.0;
        double p;
        f=(l*exp(delta1*t));
        for(n=-5;n<6;n++)
           {
        mu2=2.0*n*((delta1-delta2)/(sigma*sigma));
        sum1+=pow((pow(u,n)/pow(l,n)),mu1(n))*pow((l/s),mu2)*(N(d2(n))
             -N(d1(n)))-pow((pow(l,(n+1.0))/(pow(u,n)*s)),mu3(n))
             *(N(d4(n))-N(d3(n)));
        sum+=pow((pow(u,n)/pow(l,n)),(mu1(n)-2.0))*pow((l/s),mu2)
             *(N(d2(n)-sigma*sqrt(t))-N(d1(n)-sigma*sqrt(t)))-pow
             (pow(l,(n+1.0))/(pow(u,n)*s),mu3(n)-2.0)*(N(d4(n)
             -sigma*sqrt(t))-N(d3(n)-sigma*sqrt(t)));
    }
             p=-s*exp((b-r)*t)*sum1+x*exp(-r*t)*sum;
             return p;
}
public double uiDinc(double stock, double strike, double up,
                       double low,double delt1, double delt2 )
{
   Blackscholecp b=new Blackscholecp(crate);
   b.bscholEprice(stock,strike,sigma,t,r);
   return (b.getCalle()-uoDoc(stock,strike,up,low,delt1,delt2));
}
public double uiDinp(double stock, double strike, double up,
                      double low,double delt1, double delt2 )
{
   Blackscholecp b=new Blackscholecp(crate);
   b.bscholEprice(stock,strike,sigma,t,r);
   double outvalue=uoDop(stock,strike,up,low,delt1,delt2);
   return b.getPute()-outvalue;
}
private double d1(double n)
{
   double dvalue=(log((s*pow(u,2.0*n))/(x*pow(l,2.0*n)))
                +((b+((sigma*sigma)*0.5))*t))/(sigma*sqrt(t));
                return dvalue;
}
private double d2(double n)
{
      double dvalue=(log((s*pow(u,2.0*n))/(f*pow(l,2.0*n)))
                   +((b+((sigma*sigma)*0.5))*t))/(sigma*sqrt(t));
                   return dvalue;
}
private double d3(double n)
{
```

```
double dvalue=(log((pow(l,2.0*n+2.0))/(x*s*pow(u,2.0*n)))
              +((b+((sigma*sigma)*0.5))*t))/(sigma*sqrt(t));
              return dvalue;
}
private double d4(double n)
{
   double dvalue=(log((pow(l,2.0*n+2.0))/(f*s*pow(u,2.0*n)))
                 +((b+((sigma*sigma)*0.5))*t))/(sigma*sqrt(t));
                 return dvalue;
}
private double mu1(double n)
{
   double mvalue= 2.0*(b-delta2-n*(delta1-delta2))/(sigma*sigma)+1.0;
   return mvalue;
}
   private double mu3(double n)
{
   double mvalue=2.0*(b-delta2+n*(delta1-delta2))/(sigma*sigma)+1.0;
   return mvalue;
}
private double N(double d)
{
   Probnorm p=new Probnorm();
      double probvalue=d>(6.95)?1.0:d<(-6.95)?0.0:p.ncDisfnc(d);
                 //restrict the range of cdf values
   return probvalue;
}
```

LISTING 15.1. Double Barrier Options

15.2. Valuing With a Single Put/Call Model

The Haug pricing formulae make use of single barrier pricing using a mechanism of put-call transformations. Following the observations of Bjerksund & Stensland (1993) for American options that:

$$C(S, X, T, r, b, \sigma) = P(X, S, T, r - b, -b, \sigma)$$

Where S is the asset price, X the strike price, b the cost of carry, r is the risk free rate and T the time to maturity. The call put transformation is therefore a call in terms of a put with the asset price being equal to the strike price and the strike price being substituted by the asset price, with the risk free rate replaced by $r - b$ and cost of carry $-b$. If we re-write the payoff from a call as $\max(S - X; 0) \equiv X/S\max(S^2/X - S; 0)$ Then we can express the parity equation as:

$$C(S, X, T, r, b, \sigma) = X/S^{*}P(S, S^2/X, r - b, -b, \sigma)$$

TABLE 15.3. Put Up & In-Down & In valuations

S=110.0, X=110.0 r=0.05, b=0.05				Time 0.25				Time 0.5			
				$\sigma 1$	$\sigma 2$	$\sigma 3$	$\sigma 4$	$\sigma 1$	$\sigma 2$	$\sigma 3$	$\sigma 4$
U	L	$\delta 1$	$\delta 2$	0.10	0.15	0.25	0.35	0.10	0.15	0.25	0.35
150.0	60.0	0.0	0.0	0.000	0.000	0.0001	0.0287	0.000	0.000	0.0255	0.7828
140.0	70.0	0.0	0.0	0.000	0.000	0.0105	0.418	0.000	0.0004	0.3734	2.9587
130.0	80.0	0.0	0.0	0.000	0.0004	0.3049	2.2692	0.000	0.0464	2.0568	6.6114
120.0	90.0	0.0	0.0	0.0008	0.1277	2.3991	5.8693	0.0471	0.9673	5.3775	9.2216
110.0	100.0	0.0	0.0	1.5649	2.6321	4.7918	6.9584	1.8956	3.3639	6.3701	9.3956
U	L	$-\delta 1$	$\delta 2$								
150.0	60.0	0.05	0.05	0.000	0.000	0.0001	0.0409	0.000	0.000	0.0622	1.1095
140.0	70.0	0.05	0.05	0.000	0.000	0.0191	0.5305	0.000	0.0038	0.6768	3.7134
130.0	80.0	0.05	0.05	0.000	0.0017	0.4396	2.6186	0.0026	0.2067	2.9248	7.4285
120.0	90.0	0.05	0.05	0.0056	0.2676	2.8394	6.1623	0.3972	2.0002	5.9912	9.3507
110.0	100.0	0.05	0.05	1.5649	2.6321	4.7918	6.9584	1.8956	3.3639	6.3701	9.3956
U	L	$\delta 1$	$-\delta 2$								
150.0	60.0	0.05	0.05	0.000	0.000	0.000	0.0255	0.000	0.000	0.0241	0.7041
140.0	70.0	0.05	0.05	0.000	0.000	0.0094	0.3787	0.000	0.000	0.3406	2.7138
130.0	80.0	0.05	0.05	0.000	0.0004	0.274	2.1094	0.0001	0.0001	1.8764	6.2528
120.0	90.0	0.05	0.05	0.0008	0.1146	2.2198	5.6811	0.0548	0.0548	5.1141	9.1073
110.0	100.0	0.05	0.05	1.5649	2.6321	4.7918	6.9584	1.8956	1.8956	6.3701	9.3956

TABLE 15.4. Put Up & Out-Down & Out valuations

S=110.0, X=110.0 r=0.05, b=0.05				Time 0.25				Time 0.5			
				$\sigma 1$	$\sigma 2$	$\sigma 3$	$\sigma 4$	$\sigma 1$	$\sigma 2$	$\sigma 3$	$\sigma 4$
U	L	$\delta 1$	$\delta 2$	0.10	0.15	0.25	0.35	0.10	0.15	0.25	0.35
150.0	60.0	0.0	0.0	1.5649	2.6321	4.7917	6.9297	1.8956	3.3639	6.3446	8.6127
140.0	70.0	0.0	0.0	1.5649	2.6321	4.7813	6.5404	1.8956	3.3636	5.9967	6.4368
130.0	80.0	0.0	0.0	1.5649	2.6318	4.4869	4.6892	1.8955	3.3175	4.3133	2.7841
120.0	90.0	0.0	0.0	1.5641	2.5044	2.3927	1.0891	1.8485	2.3966	0.9926	0.174
110.0	100.0	0.0	0.0	0.000	0.000	0.000	0.000	0.000	0.000	0.000	0.000
U	L	$-\delta 1$	$\delta 2$								
150.0	60.0	0.05	0.05	1.5649	1.5649	4.7917	6.9174	1.8956	3.3639	6.3079	8.2861
140.0	70.0	0.05	0.05	1.5649	1.5649	4.7727	6.4279	1.8956	3.3601	5.6933	5.6822
130.0	80.0	0.05	0.05	1.5649	2.5649	4.3522	4.3398	1.893	3.1572	3.4453	1.9671
120.0	90.0	0.05	0.05	1.5592	1.5592	1.9524	0.796	1.4984	1.3637	0.3789	0.0449
110.0	100.0	0.05	0.05	0.000	0.000	0.000	0.000	0.000	0.000	0.000	0.000
U	L	$\delta 1$	$-\delta 2$								
150.0	60.0	0.05	0.05	1.5649	2.6321	4.7917	6.9328	1.8956	3.3639	6.3461	8.6915
140.0	70.0	0.05	0.05	1.5649	2.6321	4.7824	6.5797	1.8956	3.3634	6.0295	6.6817
130.0	80.0	0.05	0.05	1.5649	2.6317	4.5178	4.849	1.8955	3.3134	4.4937	3.1428
120.0	90.0	0.05	0.05	1.5641	2.5175	2.572	1.2773	1.8408	2.4726	1.256	0.2882
110.0	100.0	0.05	0.05	0.000	0.000	0.000	0.000	0.000	0.000	0.000	0.000

For a standard barrier 'in' option, the call put transformation can be written as:

$$\begin{aligned} C_{di}(S, X, H, r, b) &= P_{ui}(X, S, SX/H, r-b, -b) \\ &= X/S^* P_{ui}(S, S^2/X, S^2/H, r-b, -b) \end{aligned} \quad (15.2.1)$$

$$\begin{aligned} C_{ui}(S, X, H, r, b) &= P_{di}(X, S, SX/H, r-b, -b) \\ &= X/S^* P_{di}(S, S^2/X, S^2/H, r-b, -b) \end{aligned} \quad (15.2.2)$$

It can be argued that constructing a series of positions in up and in and down and in put options which protect the upper and lower barriers; we can value a double barrier call from a series of single barrier puts. The formula uses four such terms.

15.2.1. Valuing Double Calls

For the double barrier call:

$$\begin{aligned} &C_{ui,di}(S, X, L, U) \approx \min[C(S, X, T); C_{di}(S, X, L) + C_{ui}(S, X, U) \\ &-\frac{X}{U}P_{ui}\left(S, \frac{U^2}{X}, \frac{U^2}{L}\right) - \frac{X}{L}P_{di}\left(S, \frac{L^2}{X}, \frac{L^2}{U}\right) + \frac{U}{L}C_{di}\left(S, \frac{L^2X}{U^2}, \frac{L^3}{U^2}\right) \\ &+\frac{L}{U}C_{ui}\left(S, \frac{U^2X}{L^2}, \frac{U^3}{L^2}\right) - \frac{LX}{U^2}P_{ui}\left(S, \frac{U^4}{L^2X}, \frac{U^4}{L^3}\right) + \frac{UX}{L^2}P_{di}\left(S, \frac{L^4}{U^2X}, \frac{L^4}{U^3}\right) \\ &+\frac{U^3}{L^2}C_{di}\left(S, \frac{L^4X}{U^4}, \frac{L^5}{U^4}\right) + \frac{L^2}{U^2}C_{ui}\left(S, \frac{U^4X}{L^4}, \frac{U^5}{L^4}\right)] \end{aligned} \quad (15.2.3)$$

15.2.2. Valuing Double Put's

For the double barrier put:

$$\begin{aligned} &P_{ui,di}(S, X, L, U) \approx \min[P(S, X, T); P_{di}(S, X, L) + P_{ui}(S, X, U) \\ &-\frac{X}{U}C_{ui}\left(S, \frac{U^2}{X}, \frac{U^2}{L}\right) - \frac{X}{L}C_{di}\left(S, \frac{L^2}{X}, \frac{L^2}{U}\right) + \frac{U}{L}P_{di}\left(S, \frac{L^2X}{U^2}, \frac{L^3}{U^2}\right) \\ &+\frac{L}{U}P_{ui}\left(S, \frac{U^2X}{L^2}, \frac{U^3}{L^2}\right) - \frac{LX}{U^2}C_{ui}\left(S, \frac{U^4}{L^2X}, \frac{U^4}{L^3}\right) + \frac{UX}{L^2}C_{di}\left(S, \frac{L^4}{U^2X}, \frac{L^4}{U^3}\right) \\ &+\frac{U^3}{L^2}P_{di}\left(S, \frac{L^4X}{U^4}, \frac{L^5}{U^4}\right) + \frac{L^2}{U^2}P_{ui}\left(S, \frac{U^4X}{L^4}, \frac{U^5}{L^4}\right)] \end{aligned} \quad (15.2.4)$$

Table 15.5 shows example values using this approximation for a call up and-in down and-in option.

TABLE 15.5. Haug's approximation for a double barrier call option derived from a series of single barrier options

S=110.0, X=110.0 r=0.05, b=0.05				Time 0.25				Time 0.5			
				σ1	σ2	σ3	σ4	σ1	σ2	σ3	σ4
U	L	δ1	δ2	0.10	0.15	0.25	0.35	0.10	0.15	0.25	0.35
150.0	60.0	0.0	0.0	0.000	0.0024	0.5725	2.9716	0.000	0.230	3.4885	8.3165
140.0	70.0	0.0	0.0	0.000	0.0585	1.7359	4.9947	0.000	1.0313	5.6753	10.2166
130.0	80.0	0.0	0.0	0.0344	0.6906	3.8969	7.0688	0.7508	3.1248	7.8299	11.5704
120.0	90.0	0.0	0.0	1.2268	3.014	5.8356	8.222	3.4507	5.5872	9.0008	12.1115
110.0	100.0	0.0	0.0	2.9313	3.9986	6.1582	8.3248	4.6115	6.0798	9.086	12.1115

TABLE 15.6. Kunitomo & Ikeda direct calculation for a double barrier call option with flat boundaries

S=110.0, X=110.0 r=0.05, b=0.05				Time 0.25				Time 0.5			
				σ1	σ2	σ3	σ4	σ1	σ2	σ3	σ4
U	L	δ1	δ2	0.10	0.15	0.25	0.35	0.10	0.15	0.25	0.35
150.0	60.0	0.0	0.0	0.000	0.0024	0.5725	2.9716	0.000	0.230	3.4885	8.3165
140.0	70.0	0.0	0.0	0.000	0.0585	1.7359	4.9947	0.000	1.0313	5.6753	10.2166
130.0	80.0	0.0	0.0	0.0344	0.6906	3.8969	7.0688	0.7508	3.1248	7.8299	11.5704
120.0	90.0	0.0	0.0	1.2268	3.014	5.8356	8.222	3.4507	5.5872	9.0008	12.1115
110.0	100.0	0.0	0.0	2.9313	3.9986	6.1582	8.3248	4.6115	6.0798	9.086	12.1115

TABLE 15.7. Differences between Haug's approximation and Kunitomo & Ikeda formula for option valuation (Haug-Kunitomo & Ikeda)

Time = 0.25				Time = 0.5			
σ1	σ2	σ3	σ4	σ1	σ2	σ3	σ4
0.10	0.15	0.25	0.35	0.10	0.15	0.25	0.35
−0.0000001	−0.0000285	0.000000	0.000000	−0.001748	−0.000025	0.000000	0.000000
−0.0001215	−0.0000007	0.000000	0.000000	−0.054508	0.000000	0.000000	0.0000472
−0.0001492	0.0000001	0.000000	0.0000285	0.0048285	0.000000	0.0000619	0.0081228
−0.0000012	0.000000	0.0003737	0.0136462	0.0000001	0.0000347	0.0273155	0.0183606
0.000000	0.000000	0.000000	0.000000	0.000000	0.000000	0.000000	0.000000

Table 15.6 shows the valuations for direct calculation and Table 15.7 shows comparative differences.

For lower volatilities the errors show that the Haug approximation is lower. Whereas at higher volatilities, the approximation is higher

Haug's double barrier approximation methods are shown in Listing 15.2.

```
package FinApps;
import static java.lang.Math.*;
import java.text.*;
import java.io.*;

public class Dbarrierh {

   public Dbarrierh(double rate,double q,double time,
                          double volatility, int period ) {
      crate=q;
      r=rate;
      t=time;
      //0 continuos,1 hourly,2 daily,3 weekly, 4 monthly
      sigma=volatility;
      tau=period==1?(1.0/(24*365)):period==2?(1.0/(365.0)):
      period==3?(1.0/(52.0)):period==4?(1.0/(12.0)):0.0;
      b=crate==0.0?0.0:(b=crate!=r?(r-crate):r);
}
double b;
double tau;
double crate;
double t;
double r;
double s;
double u;
double x;
double l;
double sigma;
double f;
double delta1;
double delta2;
double xu1;
double xu2;
double xu3;
double xu4;
double up1;
double up2;
double up3;
double up4;
double xl1;
double xl2;
double xl3;
double xl4;
double low1;
double low2;
double low3;
double low4;

public double uiDinc(double stock, double strike, double up, double low ) {
   s=stock;
   x=strike;
   paraMs(up,low,x);
   Blackscholecp b=new Blackscholecp(crate);
   b.bscholEprice(stock,strike,sigma,t,r);
   double bscall=b.getCalle();
```

```
    Sbarrierop sb=new Sbarrierop(r,crate,t,0);
    double cui=sb.uplncall(s,x,sigma,up,0.0);
    double cdi=sb.downlcall(s,x,sigma,low,0.0);
    double puil=sb.uplput(s,xul,sigma,upl,0.0);
    double pdil=sb.downlput(s,xll,sigma,lowl,0.0);
    double cdil=sb.downlcall(s,xl2,sigma,low2,0.0);
    double cuil=sb.uplncall(s,xu2,sigma,up2,0.0);
    double pui2=sb.upIput(s,xu3,sigma,up3,0.0);
    double pdi2=sb.downIput(s,xl3,sigma,low3,0.0);
    double cdi2=sb.downIcall(s,xl4,sigma,low4,0.0);
    double cui2=sb.upIncall(s,xu4,sigma,up4,0.0);
    double term=(cui+cdi-(x/up)*puil-(x/low)*pdil+(up/low)*cdil
               +(low/up)*cuil-(low*x/(up*up))*pui2-(up*x/(low*low))
               *pdi2+(up*up/(low*low))*cdi2+(low*low/(up*up))*cui2);
    return(min(bscall,term));
}
public double uiDinp(double stock, double strike, double up, double low ){
    s=stock;
    x=strike;
    paraMs(up,low,x);
    Blackscholecp b=new Blackscholecp(crate);
    b.bscholEprice(stock,strike,sigma,t,r);
    double bscall=b.getPute();
    Sbarrierop sb=new Sbarrierop(r,crate,t,0);
    double pui=sb.uplput(s,x,sigma,up,0.0);
    double pdi=sb.downlput(s,x,sigma,low,0.0);
    double cuil=sb.uplncall(s,xul,sigma,upl,0.0);
    double cdil=sb.downlcall(s,xll,sigma,lowl,0.0);
    double pdil=sb.downlput(s,xl2,sigma,low2,0.0);
    double puil=sb.uplput(s,xu2,sigma,up2,0.0);
    double cui2=sb.uplncall(s,xu3,sigma,up3,0.0);
    double cdi2=sb.downlcall(s,xl3,sigma,low3,0.0);
    double pdi2=sb.downlput(s,xl4,sigma,low4,0.0);
    double pui2=sb.uplput(s,xu4,sigma,up4,0.0);
    double term=(pui+pdi-(x/up)*puil-(x/low)*pdil+(up/low)*cdil+(low/up)
               *cuil-(low*x/(up*up))*pui2-(up*x/(low*low))
               *pdi2+(up*up/(low*low))*cdi2+(low*low/(up*up))*cui2);
    return(min(bscall,term));
}
public double uoDoc(double stock, double strike, double up, double low ) {
    if(stock>=up|stock <=low)
        return 0.0;// no need to continue
    Blackscholecp b=new Blackscholecp(crate);
    b.bscholEprice(stock,strike,sigma,t,r);
    double bscall=b.getCalle();
    double vall=(bscall-uiDinc(stock,strike,up,low));
    return vall;
}
public double uoDop(double stock, double strike, double up, double low ) {
    if(stock>=up|stock<=low)
        return 0.0;// no need to continue
    Blackscholecp b=new Blackscholecp(crate);
    b.bscholEprice(stock,strike,sigma,t,r);
    double bscall=b.getPute();
```

```
    double val1=(bscall-uiDinp(stock,strike,up,low));
    System.out.println(" r=="+r+" b=="+crate);
    return val1;
}
private void paraMs(double up, double low, double x) {
    xu1=((up*up)/x);
    xu2=((up*up*x)/(low*low));
    xu3=(pow(up,4)/(low*low*x));
    xu4=((pow(up,4)*x)/(pow(low,4)));
    up1=((up*up)/low);
    up2=(pow(up,3)/(low*low));
    up3=(pow(up,4)/(pow(low,3)));
    up4=(pow(up,5)/(pow(low,4)));
    xl1=((low*low)/x);
    xl2=((low*low*x)/(up*up));
    xl3=(pow(low,4)/(up*up*x));
    xl4=((pow(low,4)*x)/(pow(up,4)));
    low1= ((low*low)/up);
    low2=(pow(low,3)/(up*up));
    low3=(pow(low,4)/(pow(up,3)));
    low4=(pow(low,5)/(pow(up,4)));
}
```

LISTING 15.2. Double Barrier Approximation

References

Bjerksund and Stensland (1993). "American Exchange Options and a put call Transformation: A Note," *Journal of Business Finance and Accountancy*, 20(5), 61–64.

Geman, H. and M. Yor (1996). "Pricing and Hedging Double-Barrier Options: A Probabilistic Approach," *Mathematical Finance*, 6(4), 365–378.

Haug, E. H. (1999). "Barrier Put-Call Transformations," *Tempus Financial Engineering*.

Kunitomo, N. and M. Ikeda (1992). "Pricing Options with Curved Boundaries," *Mathematical Finance*, 2(4), 275–2.

16
Digital Options

Digital (also known as Binary) options are characterized by discontinuous payoffs. The option payoff will be either 'on' or 'off' dependent on the relative position of the underlying asset. The payoff from a barrier option is path dependent; it follows the asset price path in relation to a barrier level.

Reiner & Rubinstein (1991) developed pricing formulae for digital/binary options (They preferred to use the term binary). Their formulae are based on two categories of path independent options. Firstly, a cash or nothing barrier option, that pays a predetermined amount, if the asset price has gone below the barrier prior to expiry (put) and nothing otherwise. Next, an asset or nothing option, that pays off the value of an underlying asset at expiry, if the price has been at or below the barrier prior to expiry (put), otherwise nothing. In some cases the option will have an immediate payoff when the barrier is touched; others will have a payoff at expiry, dependent on touching the barrier. There are extended variations which include further threshold conditions. See Bermin (1996).

16.1. General (Rubinstein & Reiner) Method

The binary barrier options are constructed from parts of the general barrier option formulae (see section above). There are four probability densities: The probability associated with the asset staring above the barrier, falling through the barrier, and then ending above the barrier at expiry. There is a probability for the asset starting below the barrier, rising through the barrier and ending below the barrier at expiry. There is a probability density of the underlying risk neutral asset return. Finally, there is a probability for the first passage time density (the first time to hit the barrier).

The basic formulae use the same expressions for each of the first two probability densities. To distinguish the individual cases a binary coefficient of $+1$ is applied to the initial above barrier probability and -1 for the below barrier probability. For the risk neutral asset return the binary variable is $\varphi \pm 1$. For the barrier hit direction of the underlying asset returns, the binary variable is, $\eta \pm 1$.

The basic 'pieces' from the fundamental barrier formulae are broken down into nine terms, whic be conveniently re-arranged to price the range of digital barrier options. The nine formulae terms are:

$$t1 = Se^{(b-r)T}\ \phi(\varphi x_1),$$
$$t2 = Xe^{-rT}\ \phi(\varphi x_1 - \varphi\sigma\sqrt{T}),$$
$$t3 = Se^{(b-r)T}\ \phi(\varphi x_2),$$
$$t4 = Xe^{-rT}\ \phi(\varphi x_2 - \varphi\sigma\sqrt{T}),$$
$$t5 = Se^{(b-r)T}(H\ /\ S)^{2(\mu+1)}\ \phi(\eta y_1),$$
$$t6 = Xe^{-rT}\ (H\ /\ S)^{2\mu}\phi(\eta y_1 - \eta\sigma\sqrt{T}),$$
$$t7 = Se^{(b-r)T}\ (H\ /\ S)^{2(\mu+1)}\phi(\eta y_2),$$
$$t8 = Xe^{-rT}\ (H\ /\ S)^{2\mu}\phi(\eta y_2 - \eta\sigma\sqrt{T}),$$
$$t9 = X\left[(H\ /\ S)^{\mu+\lambda}\phi(\eta z) + (H\ /\ S)^{\mu+\lambda}\phi(\eta z - 2\eta\lambda\sigma\sqrt{T})\right].$$

Where X is the specified cash amount and the remaining parameters are:

$$x_1 = \frac{\log(S\ /\ K)}{\sigma\sqrt{T}} + (\mu+1)\sigma\sqrt{T},$$
$$x_2 = \frac{\log(S\ /\ K)}{\sigma\sqrt{(T)}} + (\mu+1)\sigma\sqrt{T},$$
$$y_1 = \frac{\log(H^2\ /\ SK)}{\sigma\sqrt{(T)}} + (\mu+1)\sigma\sqrt{T},$$
$$y_2 = \frac{\log(H\ /\ S)}{\sigma\sqrt{(T)}} + (\mu+1)\sigma\sqrt{T},$$
$$z = \frac{\log(H\ /\ S)}{\sigma\sqrt{(T)}} + \lambda\sigma\sqrt{T},$$
$$\mu = \frac{b - \sigma^2\ /\ 2}{\sigma^2},\ \lambda = \sqrt{\mu^2 + \frac{2r}{\sigma^2}}.$$

Digital barrier options are implemented in the class **Digitalbarrier** , this class is shown in Listing 16.1. This class is responsible for calculating each of the terms in the above nine formulas. The class makes available the public method ***barops()***, that accepts the input parameters for the option. The constructor is used to initialize the class with interest rates and the barrier level. The bulk of computation is carried out within an enumerated type, *barV*. All twenty eight digital barriers are computed within this enumerated type. Each type of digital barrier is selected via a switch statement. The selected type is implemented by methods that provide the fundamental calculations. Methods are used to implement all of the basic parameters.

```
package FinApps;
import static java.lang.Math.*;
import BaseStats.Probnorm;
import BaseStats.Bivnorm;
public class Digitalbarrier {
    public Digitalbarrier(double rate, double yield, double barrier)
  {
      r=rate;
      crate=yield;
      h=barrier;
      b=crate==0.0?0.0:(b=crate!=r?(r-crate):r);
  }
  static double crate;
  static double r;
  static double t;
  static double b;
  static double k;
  static double s;
  static double x;
  static double h;
  static double sigma;
  static int phi;
  static int eta;
  public double barops(double asset , double strike, double cash,
                      double volatility,double time, int tp) {
      s=asset;
      x=strike;
      k=cash;
      sigma=volatility;
      t=time;
      return barV.funC.Args(tp);
  }
  private static double N(double x) {
      Probnorm p=new Probnorm();
      double ret=x>(6.95)?1.0:x<(-6.95)?0.0:p.ncDisfnc(x);//restrict
      the range of cdf values to stable values
      return ret;
  }
private enum barV {
   funC{
       public double Args( int tp) {
          int sw=tp;
          double puterm=0.0;
          switch(sw) {
               case 1: eta=1; puterm=t9.val(); // type 1
               break;
               case 2:eta=-1; puterm=t9.val(); // type 2
               break;
               case 3: eta=1;puterm=t9.val();// type 3
               break;
```

```
case 4: eta=-1;puterm=t9.val();// type 4
break;
case 5: eta=1;phi=-1;puterm=t4.val()+t8.val();
break;
case 6:eta=-1;phi=1;puterm=t4.val()+t8.val();
break;
case 7:eta=1;phi=-1;puterm=t3.val()+t7.val();
break;
case 8:eta=-1;phi=1;puterm=t3.val()+t7.val();
break;
case 9:eta=1;phi=1;puterm=t4.val()-t8.val();
break;
case 10:eta=-1;phi=-1;puterm=t4.val()-t8.val();
break;
case 11:eta=1;phi=1;puterm=t3.val()-t7.val();
break;
case 12:eta=-1;phi=-1;puterm=t3.val()-t7.val();
break;
case 13:phi=x<h?1:0;eta=1;puterm=x<h?(t2.val()
        -t4.val()+t8.val()):t6.val();
break;
case 14:eta=-1;phi=1;puterm=x<h?(t4.val()-t6.val()
        +t8.val()):t6.val();
break;
case 15:eta=1;phi=x<h?1:0;puterm=x<h?(t1.val()
        -t3.val()+t7.val()):t5.val();
break;
case 16:phi=1;eta=x<h?-1:0;puterm=x<h?(t3.val()
        -t5.val()+t7.val()):t1.val();
break;
case 17:phi=-1;eta=x>h?1:0;puterm=x>h?(t4.val()
        -t6.val()+t8.val()):t2.val();
break;
case 18:phi=-1;eta=-1;puterm=x>h?(t2.val()
        -t4.val()+t8.val()):t6.val();
break;
case 19:phi=-1;eta=x>h?1:0;puterm=x>h?(t3.val()
        -t5.val()+t7.val()):t1.val();
break;
case 20:phi=-1;eta=-1;puterm=x>h?(t1.val()-t3.val()
        +t5.val()):t5.val();
break;
case 21:eta=1;phi=1;puterm=x>h?(t2.val()-t6.val()):
        (t4.val()-t8.val());
break;
case 22: eta=-1;phi=1;puterm=x<h?(t2.val()-t4.val()
        +t6.val()-t8.val()):0.0;
break;
case 23:eta=1;phi=1;puterm=x<h?(t3.val()-t7.val()):
        (t1.val()-t5.val());
break;
case 24:eta=-1;phi=1;puterm=x<h?(t1.val()-t3.val()
        +t5.val()-t7.val()):0.0;
break;
```

```
            case 25:eta=1;phi=-1;puterm=x>h?(t2.val()-t4.val()
                    +t6.val()-t8.val()):0.0;
            break;
            case 26:eta=-1;phi=-1;puterm=x>h?(t4.val()-t8.val()):
                    (t2.val()-t6.val());
            break;
            case 27:eta=1;phi=-1;puterm=x>h?(t1.val()-t3.val()
                    +t5.val()-t7.val()):0.0;
            break;
            case 28: eta=-1;phi=-1;puterm=x>h?(t3.val()-t7.val()):
                    (t1.val()-t5.val());
            break;

        }

        return puterm;
    }
},
t1{
    public double val() {

        return s*exp((b-r)*t)*N(phi*par.x1());
    }
},
t2{
    public double val() {
        return k*exp(-r*t)*N(phi*par.x1()-phi*sigma*sqrt(t));
    }
},
t3{
    public double val() {
        return s*exp((b-r)*t)*N(phi*par.x2());
    }
},

t4{
    public double val(){
        return k*exp(-r*t)*N(phi*par.x2()-phi*sigma*sqrt(t));
    }
},

t5{
    public double val() {

        return s*exp((b-r)*t)*pow((h/s),(2.0*(par.mu()+1.0)))
                *N(eta*par.y1());


    }
},
t6{
    public double val(){


        return k*exp(-r*t)*pow((h/s),(2.0*par.mu()))*N(eta*par.y1()-eta
                *sigma*sqrt(t));

    }
},
```

```
t7{
  public double val() {
    return s*exp((b-r)*t)*pow((h/s),(2.0*(par.mu()+1.0)))
            *N(eta*par.y2());
  }
},
t8{
  public double val() {
    return k*exp(-r*t)*pow((h/s),(2.0*par.mu()))*N(eta*par.y2()
            -eta*sigma*sqrt(t));
  }
},
t9{
  public double val() {
    return k*(pow((h/s),(par.mu()+par.lambda()))*N(eta*par.zeta())
            +pow((h/s),(par.mu()-par.lambda()))*N(eta*par.zeta()
            -2.0*eta*par.lambda()*sigma*sqrt(t)));
  }
},
par{


  public double x1() {
    double m=mu();
    return (log(s/x)/(sigma*sqrt(t)))+(m+1.0)*sigma*sqrt(t);
  }
  public double x2() {
    double m=mu();
    return log(s/h)/(sigma*sqrt(t))+(m+1.0)*sigma*sqrt(t);
  }
  public double y1() {
    double m=mu();
    return log(h*h/(s*x))/(sigma*sqrt(t))+(m+1.0)*sigma*sqrt(t);
  }
  public double y2() {
    double m=mu();
    return (log(h/s)/(sigma*sqrt(t)))+(m+1.0)*sigma*sqrt(t);
  }
  public double zeta() {
    double m=mu();
    return (log(h/s)/(sigma*sqrt(t)))+par.lambda()*sigma*sqrt(t);
  }
  public double lambda() {
    double m=mu();
    return sqrt((m*m)+(2.0*r/(sigma*sigma)));
  }
  public double mu() {
    return (b-(sigma*sigma)*0.5)/(sigma*sigma);


  }


};
```

```
    barV() {

    }
    public double x1() {

        throw new UnsupportedOperationException("Not yet implemented");
    }
    public double x2() {

        throw new UnsupportedOperationException("Not yet implemented");
    }
    public double y1() {

        throw new UnsupportedOperationException("Not yet implemented");
    }
    public double y2() {

        throw new UnsupportedOperationException("Not yet implemented");
    }
    public double zeta() {

        throw new UnsupportedOperationException("Not yet implemented");
    }
    public double mu() {

        throw new UnsupportedOperationException("Not yet implemented");
    }
    public double lambda() {

        throw new UnsupportedOperationException("Not yet implemented");
    }
    public double val() {
        throw new UnsupportedOperationException("Not yet implemented");
    }
    public double Args( int tp) {

        throw new UnsupportedOperationException("Not yet implemented");
    }
}
```

LISTING 16.1. Class **Digitalbarrier**

16.2. Valuation

The barrier options are valued in the following ways.

16.2.1. In Valuation

Down and-in-cash at-hit or nothing. Valued by:

$$V = X\left[(H\ /\ S)^{\mu+\lambda}\ \phi + (\eta z) + (H\ /\ S)^{\mu-\lambda}\ \phi(\eta z - 2\eta\lambda\sigma\sqrt{T})\right] \qquad (16.2.1)$$

$$\eta = 1.$$

This option has a payoff immediately the barrier is hit. The payoff is: $X : iff, \forall, \tau \leq T : S(\tau) \leq H.$ Otherwise, 0.

TABLE 16.1. Down and-In Cash at-hit or nothing option valuations

Asset Price	Time = 0.25			Time = 0.5		
	$\sigma = 0.15$	$\sigma = 0.20$	$\sigma = 0.30$	$\sigma = 0.15$	$\sigma = 0.20$	$\sigma = 0.30$
105.0	7.0361	9.5118	12.2125	8.6918	11.0232	13.3701
104.0	8.5565	10.7942	13.1029	10.0168	12.0716	14.0518
103.0	10.3044	12.1859	14.0286	11.4986	13.1918	14.7560
102.0	12.2905	13.6861	14.9882	13.1486	14.3855	15.4822
101.0	14.5215	15.2922	15.9794	14.9785	15.6545	16.2303
100.0	17.0000	17.0000	17.0000	17.0000	17.0000	17.0000

$r = 0.10, b = 0.10 : X = 17 : H = 100.0$

Table 16.1 shows valuations for changing asset price with various volatilities and time to maturity. When the asset price touches the barrier, the payoff is immediately payable for the cash value (X).

Up and-in-cash at-hit or nothing. Is valued using:

$$V = X\left[(H / S)^{\mu+\lambda}\phi(\eta z) + (H / S)^{\mu-\lambda}\phi(\eta z - 2\eta\lambda\sigma\sqrt{T})\right]$$

$$\eta = -1.$$

This option has a payoff immediately the barrier is hit. The payoff is:

$X : iff, \forall, \tau \leq T : S(\tau) \leq H.$ Otherwise, 0.

Table 16.2 gives the range of option valuations.

The down and-in-asset at-hit or nothing barrier option has a value given by:

$$V = X\left[(H / S)^{\mu+\lambda}\phi(\eta z) + (H / S)^{\mu-\lambda}\phi(\eta z - 2\eta\lambda\sigma\sqrt{T})\right]$$

$$\eta = 1, X = H$$

This option has a payoff immediately the barrier is hit. The payoff is given by:

$$0 : iff, \forall\tau \leq T : S(\tau) > H;\, else,\, S(\tau)$$

TABLE 16.2. Up and-In Cash at-hit or nothing option valuations

Asset Price	Time = 0.25			Time = 0.5		
	$\sigma = 0.15$	$\sigma = 0.20$	$\sigma = 0.30$	$\sigma = 0.15$	$\sigma = 0.20$	$\sigma = 0.30$
95.0	9.9999	11.2860	12.7596	12.5413	13.2093	14.0481
96.0	11.4364	12.4482	13.6146	13.5032	14.0013	14.6496
97.0	12.8845	13.6138	14.4701	14.4382	14.7801	15.2469
98.0	14.3134	14.7694	15.3221	15.3374	15.5419	15.8384
99.0	15.6940	15.9020	16.1666	16.1934	16.2829	16.4232
100.0	17.0000	17.0000	17.0000	17.0000	17.0000	17.0000

$r = 0.10, b = 0.10 : X = 17 : H = 100.0$

Table 16.3 gives a range of the option values for various asset prices with changing volatilities at different maturities.

An up and-in-asset at-hit or nothing barrier option is valued by:

$$V = X\left[(H/S)^{\mu+\lambda}\phi(\eta z) + (H/S)^{\mu-\lambda}\phi(\eta z - 2\eta\lambda\sigma\sqrt{T})\right] \tag{16.2.2}$$

$$\eta = -1, X = H$$

This option also has a payoff immediately the barrier is hit. The payoff is given by:

$$0 : iff, \forall\tau \leq T : S(\tau) < H; \, else, S(\tau)$$

Table 16.4 shows a range of option values for various asset prices and volatilities with two maturity times.

The options valued so far are based on an immediate payout on a given condition being satisfied. The following options have to satisfy a given condition but the payout is delayed until expiration.

The down-and-in cash at expiry, or nothing option is valued by:

$$V = Xe^{-rT}\phi(\varphi x_2 - \varphi\sigma\sqrt{T}) + Xe^{-rT}(H/S)^{2\mu}\phi(\eta y_2 - \eta\sigma\sqrt{T}) \tag{16.2.3}$$

$$\eta = 1, \varphi = -1$$

The option has a payoff at expiration, if the asset price is touched at some time prior to expiry.

The payoff is:

$$0 : iff \forall\tau \leq T; \, S(\tau) > H; \, else, X$$

Table 16.5 shows a range of valuations for the down-and-in cash at expiry option.

TABLE 16.3. Down and-in asset at-hit or nothing option valuations

Asset Price	Time = 0.25			Time = 0.5		
	$\sigma = 0.15$	$\sigma = 0.20$	$\sigma = 0.30$	$\sigma = 0.15$	$\sigma = 0.20$	$\sigma = 0.30$
105.0	41.3886	55.9518	71.8382	51.1280	64.8426	78.6476
104.0	50.3321	63.4951	77.0756	58.9222	71.0094	82.6579
103.0	60.6142	71.6820	82.5215	67.6386	77.5986	86.7997
102.0	72.2973	80.5066	88.1657	77.3447	84.6208	91.0718
101.0	85.4209	89.9539	93.9965	88.1087	92.0853	95.4726
100.0	100.0000	100.0000	100.0000	100.0000	100.0000	100.0000

$r = 0.10, b = 0.10 : H = 100.0$

TABLE 16.4. Up and-in asset-at-hit or nothing option valuations

Asset Price	Time = 0.25			Time = 0.5		
	$\sigma = 0.15$	$\sigma = 0.20$	$\sigma = 0.30$	$\sigma = 0.15$	$\sigma = 0.20$	$\sigma = 0.30$
95.0	58.8232	66.3881	75.0566	73.7722	77.7017	82.6357
96.0	67.2728	73.2248	80.0857	79.4306	82.3605	86.1744
97.0	75.7909	80.0813	85.1182	84.9304	86.9420	89.6874
98.0	84.1968	86.8786	90.1298	90.2200	91.4229	93.1673
99.0	92.3179	93.5414	95.0977	95.2553	95.7818	96.6070
100.0	100.0000	100.0000	100.0000	100.0000	100.0000	100.0000

$r = 0.10, b = 0.10 : H = 100.0$

The up-and-in cash at expiry option has a value given by:

$$V = Xe^{-rT}\phi(\varphi x_2 - \varphi\sigma\sqrt{T}) + Xe^{-rT}(H\ /\ S)^{2\mu}\phi(\eta y_2 - \eta\sigma\sqrt{T})$$

$$\eta = -1, \varphi = 1$$

The payoff at expiration is:

$$0 : iff \forall\tau \le T;\ S(\tau) < H;\ else,\ X$$

Table 16.6 gives a representative set of valuations for up-and-in-cash at expiry options.

For the case of an 'in' asset payoff at expiry the valuation is the same as the standard up-and-in or down-and-in barrier option with strike=rebate value=zero. The valuation formula is therefore:

$$V = Se^{(b-r)T}\phi(\varphi x_2) + Se^{(b-r)T}(H/S)^{2(\mu+1)}\phi(\eta y_2) \qquad (16.2.4)$$

$$\eta = 1, \varphi = -1$$

The payoff at expiry is given by:

$$0 : iff \forall\tau \le T;\ S(\tau) > H;\ else,\ S(T).$$

TABLE 16.5. Down-and-in cash at expiry

Asset Price	Time = 0.25			Time = 0.5		
	$\sigma = 0.15$	$\sigma = 0.20$	$\sigma = 0.30$	$\sigma = 0.15$	$\sigma = 0.20$	$\sigma = 0.30$
105.0	6.9293	9.3497	11.9778	8.3900	10.6084	12.8228
104.0	8.4140	10.5967	12.8386	9.6454	11.5939	13.4568
103.0	10.1158	11.9462	13.7314	11.0428	12.6424	14.1091
102.0	12.0426	13.3961	14.6543	12.5907	13.7546	14.7793
101.0	14.1978	14.9427	15.6048	14.2975	14.9308	15.4668
100.0	16.5803	16.5803	16.5803	16.1709	16.1709	16.1709

$r = 0.10, b = 0.10 : X = 17 : H = 100.0$

TABLE 16.6. Up-and-out cash at expiry

Asset Price	Time = 0.25			Time = 0.5		
	$\sigma = 0.15$	$\sigma = 0.20$	$\sigma = 0.30$	$\sigma = 0.15$	$\sigma = 0.20$	$\sigma = 0.30$
95.0	9.8518	11.0971	12.5175	12.1132	12.7186	13.4782
96.0	11.2490	12.2231	13.3422	13.0081	13.4519	14.0328
97.0	12.6507	13.3478	14.1649	13.8694	14.1676	14.5806
98.0	14.0258	14.4574	14.9814	14.6885	14.8617	15.1204
99.0	15.3445	15.5389	15.7878	15.4577	15.5305	15.6508
100.0	16.5803	16.5803	16.5803	16.1709	16.1709	16.1709

$r = 0.10, b = 0.10 : X = 17 : H = 100.0$

Table 16.7A gives a range of option valuations for an asset, which goes in at the barrier. Note that the values are the same as in the down-and-in asset-at-hit model. The values are effectively the same since the asset is assumed to yield at the same rate as the risk-free rate. Note however, that this option does not pay the barrier value of the asset, but rather, the asset price at expiration.

The effect of cost of carry can be seen from the graph in Figure 16.1. This displays the data from column one of Tables 16.7A and 16.7B (Time 0.25 and volatility 0.15). The difference in valuation reflects the waiting time to expiration.

The up-and-in asset-at-expiry option is valued by:

$$V = Se^{(b-r)T}\phi(\varphi x_2) + Se^{(b-r)T}(H/S)^{2(\mu+1)}\phi(\eta y_2)$$
$$\eta = -1, \varphi = 1$$

The payoff at expiry is:

$$0 : iff \forall\tau \le T;\ S(\tau) < H;\ else,\ S(T)$$

Table 16.8 shows a range of valuations for this option. This option has exactly the same value as an up-and-in standard barrier call with strike and rebate equal to zero.

TABLE 16.7A. Down and in asset at expiry or nothing valuation

Asset Price	Time = 0.25			Time = 0.5		
	$\sigma = 0.15$	$\sigma = 0.20$	$\sigma = 0.30$	$\sigma = 0.15$	$\sigma = 0.20$	$\sigma = 0.30$
105.0	41.3886	55.9518	71.8382	51.1280	64.8426	78.6476
104.0	50.3321	63.4951	77.0756	58.9222	71.0094	82.6579
103.0	60.6142	71.6820	82.5215	67.6386	77.5986	86.7997
102.0	72.2973	80.5066	88.1657	77.3447	84.6208	91.0718
101.0	85.4209	89.9539	93.9965	88.1087	92.0853	95.4726
100.0	100.0000	100.0000	100.0000	100.0000	100.0000	100.0000

$r = 0.10 : b = 0.10 : H = 100.0$

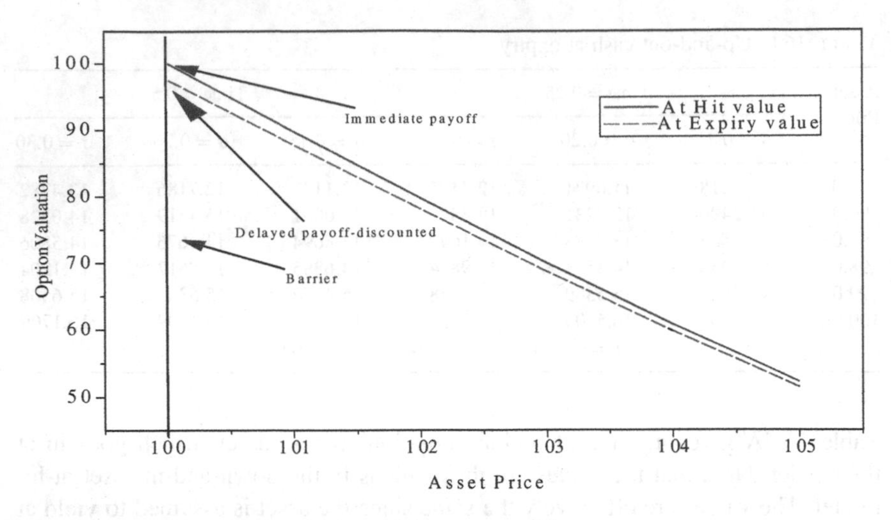

FIGURE 16.1. Effect of Asset yield with time to payoff.

16.2.2. Out Options

There are a complementary set of 'out' barrier options, which match the four 'in' options above.

The down-and-out cash-or-nothing option has a value given by:

$$V = Xe^{-rT}\phi(\varphi x_2 - \varphi\sigma\sqrt{T}) - Xe^{-rT}(H/S)^{2\mu}\phi(\eta y_2 - \eta\sigma\sqrt{T}) \qquad (16.2.5)$$

$$\eta = 1, \varphi = 1$$

This is essentially the complement of the formula for a down-and-in cash or nothing option. The down-and-out payoff is:

$$X : \forall\tau \leq T;\ S(\tau) > H;\ else,\ 0$$

Table 16.9 provides example valuations for down-and-out cash-or-nothing options.

TABLE 16.7B. Down-and-in asset-at-expiry or nothing valuation with zero cost of carry

Asset Price	Time = 0.25			Time = 0.5		
	$\sigma = 0.15$	$\sigma = 0.20$	$\sigma = 0.30$	$\sigma = 0.15$	$\sigma = 0.20$	$\sigma = 0.30$
105.0	51.4893	62.4996	74.4058	62.8934	71.1219	79.6669
104.0	59.7644	69.0943	78.9051	69.0008	75.7762	82.7143
103.0	68.6315	75.9544	83.4864	75.3269	80.5255	85.7904
102.0	77.9786	83.0242	88.1313	81.8253	85.3488	88.8886
101.0	87.6659	90.2399	92.8200	88.4433	90.2228	92.0019
100.0	97.5310	97.5310	97.5310	95.1229	95.1229	95.1229

$r = 0.10 : b = 0 : H = 100.0$

TABLE 16.8. Up-and-in asset-at-expiry option valuations

Asset Price	Time = 0.25			Time = 0.5		
	$\sigma = 0.15$	$\sigma = 0.20$	$\sigma = 0.30$	$\sigma = 0.15$	$\sigma = 0.20$	$\sigma = 0.30$
95.0	58.8232	66.3881	75.0566	73.7722	77.7017	82.6357
96.0	67.2728	73.2248	80.0857	79.4306	82.3605	86.1744
97.0	75.7909	80.0813	85.1182	84.9304	86.9420	89.6874
98.0	84.1968	86.8786	90.1298	90.2200	91.4229	93.1673
99.0	92.3179	93.5414	95.0977	95.2553	95.7818	96.6070
100.0	100.0000	100.0000	100.0000	100.0000	100.0000	100.0000

$r = 0.10 : b = 0.10 : H = 100.0$

A down-and-out cash-or-nothing option can be replicated from the standard barrier option by $(-\frac{X}{H})$ times the cash value from a down-and-out call payoff, with a strike price equivalent to H.

The up-and-out cash-or-nothing option is valued by:

$$V = Xe^{-rT}\phi(\varphi x_2 - \varphi\sigma\sqrt{T}) - Xe^{-rT}(H/S)^{2\mu}\phi(\eta y_2 - \eta\sigma\sqrt{T})$$
$$\eta = -1, \varphi = -1$$

The up-and-out payoff is given by:

$$X : \forall\tau \leq T;\ S(\tau) < H;\ else, 0$$

An up-and-out cash-or-nothing option payoff can be represented by the cash value from the payoff of an up-and-out put with strike price equivalent to H by $\frac{X}{H}$. Table 16.10 provides valuations for up and out options.

The remaining two 'out' options are asset based. An out-asset payoff can be derived from the payoff of a down-and-out (or up-and-out) call with the strike

TABLE 16.9. Down-and-out cash-or-nothing option valuations

Asset Price	Time = 0.25			Time = 0.5		
	$\sigma = 0.15$	$\sigma = 0.20$	$\sigma = 0.30$	$\sigma = 0.15$	$\sigma = 0.20$	$\sigma = 0.30$
105.0	9.6510	7.2306	4.6025	7.7809	5.5625	3.3481
104.0	8.1662	5.9836	3.7417	6.5255	4.5770	2.7141
103.0	6.4645	4.6341	2.8489	5.1281	3.5285	2.0618
102.0	4.5377	3.1841	1.9260	3.5802	2.4163	1.3916
101.0	2.3825	1.6376	0.9755	1.8734	1.2401	0.7041
100.0	0.0000	0.0000	0.0000	0.0000	0.0000	0.0000

$r = 0.10, b = 0 : X = 17 : H = 100$

TABLE 16.10. Up-and-out cash-or-nothing option valuations

Asset Price	Time = 0.25			Time = 0.5		
	$\sigma = 0.15$	$\sigma = 0.20$	$\sigma = 0.30$	$\sigma = 0.15$	$\sigma = 0.20$	$\sigma = 0.30$
95.0	6.7285	5.4831	4.0628	4.0577	3.4523	2.6927
96.0	5.3313	4.3571	3.2381	3.1628	2.7190	2.1381
97.0	3.9295	3.2325	2.4154	2.3015	2.0033	1.5903
98.0	2.5545	2.1229	1.5989	1.4824	1.3092	1.0505
99.0	1.2358	1.0414	0.7925	0.7132	0.6404	0.5201
100.0	0.0000	0.0000	0.0000	0.0000	0.0000	0.0000

$r = 0.10, b = 0.10 : X = 17 : H = 100.0$

price equal to the rebate equal to zero. The down-and-out asset or-nothing has a payoff at expiry of:

$$S(T) : \forall \tau \leq T; S(\tau) > H; else, 0$$

The option is valued by:

$$V = Se^{(b-r)T}\phi(\varphi x_2) - Se^{(b-r)T}(H/S)^{2(\mu+1)}\phi(\eta Y_2) \tag{16.2.6}$$

$$\eta = 1, \varphi = 1$$

Table 16.11 gives example valuations for the down-and-out asset or-nothing option.

The up-and-out asset-or-nothing option is valued using the following:

$$V = Se^{(b-r)T}\phi(\varphi x_2) - Se^{(b-r)T}(H/S)^{2(\mu+1)}\phi(\eta y_2)$$

$$\eta = -1, \varphi = -1$$

The payoff is:

$$S(T) : \forall \tau \leq T; S(\tau) < H; else, 0.$$

Example valuations are shown in Table 16.12.

TABLE 16.11. Down-and-out asset-or-nothing option valuations

Asset Price	Time = 0.25			Time = 0.5		
	$\sigma = 0.15$	$\sigma = 0.20$	$\sigma = 0.30$	$\sigma = 0.15$	$\sigma = 0.20$	$\sigma = 0.30$
105.0	63.6114	49.0482	33.1618	53.8720	40.1574	26.3524
104.0	53.6679	40.5049	26.9244	45.0778	32.9906	21.3421
103.0	42.3858	31.3180	20.4785	35.3614	25.4014	16.2003
102.0	29.7027	21.4934	13.8343	24.6553	17.3792	10.9282
101.0	15.5791	11.0461	7.0035	12.8913	8.9147	5.5274
100.0	0.0000	0.0000	0.0000	0.0000	0.0000	0.0000

$r = 0.10 : b = 0.10 : H = 100.0$

TABLE 16.12. Up-and-out asset-or-nothing valuations

Asset Price	Time = 0.25			Time = 0.5		
	$\sigma = 0.15$	$\sigma = 0.20$	$\sigma = 0.30$	$\sigma = 0.15$	$\sigma = 0.20$	$\sigma = 0.30$
95.0	36.1768	28.6119	19.9434	21.2278	17.2983	12.3643
96.0	28.7272	22.7752	15.9143	16.5694	13.6395	9.8256
97.0	21.2091	16.9187	11.8818	12.0696	10.0580	7.3126
98.0	13.8032	11.1214	7.8702	7.7800	6.5771	4.8327
99.0	6.6821	5.4586	3.9023	3.7447	3.2182	2.3930
100.0	0.0000	0.0000	0.0000	0.0000	0.0000	0.0000

$r = 0.10 : b = 0.10 : H = 100.0$

16.3. Valuation as a Portfolio

It is possible to deduce parity relationships between any two of the binary options and the underlying. If we consider the portfolio of a down-and-in cash-at-expiry option with a cash sum equivalent to the payoff:

$$D_{out}Xcall = \prod(Xr^{-t} - D_{in}Xcall)$$

This can be re-arranged as:

$$\prod(D_{out}Xcall + D_{in}Xcall) = Xr^{-T}$$

Thus, the combination of a down-and-out call with a down-and-in call is guaranteed to deliver the cash value, discounted to expiry (PV). The same argument is seen to hold for asset based options. In the latter case the combination of both options, gives us the current price of the asset (not the PV) at expiry

Figure 16.2 shows the relationship for valuation of an option as a function of the portfolio values of an asset and an option. The barrier is set at 100.0.

The horizontal dotted lines reflect the valuations of both options, which sum to the asset value. Similarly, taking the valuation of any option minus the asset price at that point will provide the valuation for the other option. There are two curves reflecting two times to expiry. The valuation for a down and out option is lower at all points to the barrier, and the corresponding valuation of the down and in for all points is higher, with longer time to expiry.

The following set of options have positive payoffs that depend on the barrier being hit and the asset being above a given level at expiry (call options).

16.3.1. *In Cash or Nothing Valuations*

Down-and-in cash-or-nothing.

This digital option has a payoff given by:

$$0 : \forall\tau \leq T, S(\tau) > H \vee S(T) < K; \, else, X$$

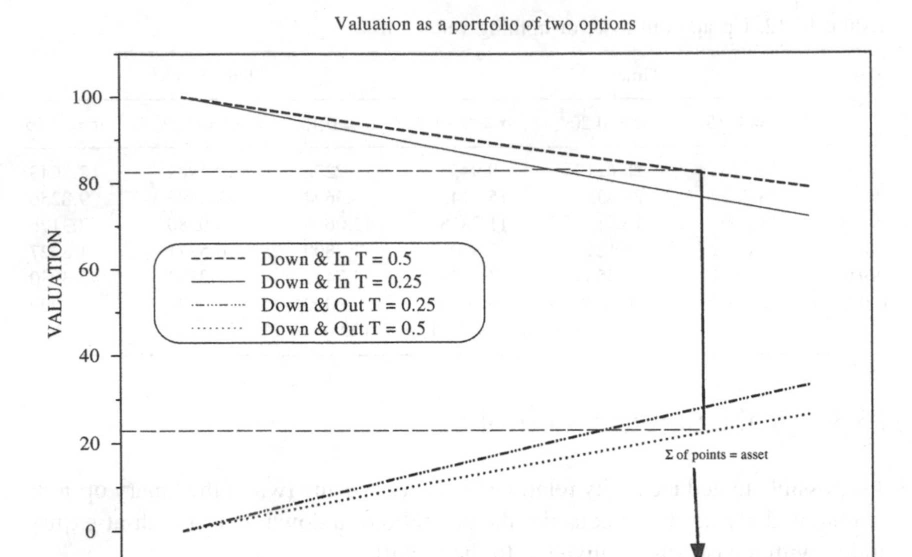

FIGURE 16.2. Portfolio valuations for binary barrier options.

The valuation has two factors; a situation where the level K , is above the barrier level and the situation where the level is below the barrier. The formulae are therefore distinct to a particular situation.

Case 1: $K > H$;

$$V = Xe^{-rT}(H\ /\ S)^{2\mu}\phi(\eta y_1 - \eta\sigma\sqrt{T})$$

$$\eta = 1. \tag{16.3.1}$$

Case 2: $K < H$;

$$V = Xe^{-rT}\phi(\varphi X_1 - \varphi\sigma\sqrt{T}) - Xe^{-rT}\phi(\varphi x_2 - \varphi\sigma\sqrt{T}) + Xe^{-rT}(H/S)^{2\mu}\phi(\eta y_2 - \eta\sigma\sqrt{T}) \tag{16.3.2}$$

$$\eta = 1, \varphi = 1$$

Table 16.13 gives a range of example valuations for this option.

Up-and-in cash-or-nothing.

This digital option has a payoff given by:

$$0 : \forall\ \tau \leq T, S(\tau) < H \vee S(T) < K;\ else,\ X$$

TABLE 16.13. Down-and-in cash-or-nothing call valuations

Asset Price	Time = 0.25			Time = 0.5		
	$\sigma = 0.15$	$\sigma = 0.20$	$\sigma = 0.30$	$\sigma = 0.15$	$\sigma = 0.20$	$\sigma = 0.30$
106.0	3.8809	5.0096	6.0790	5.3642	6.2347	6.8355
105.0	4.7927	5.7316	6.5396	6.2377	6.8590	7.1914
104.0 : K	5.8632	6.5246	7.0204	7.2260	7.5303	7.5593
103.0	7.1063	7.3901	7.5209	8.3396	8.2505	7.9393
102.0 : H	8.5343	8.3286	8.0402	9.5895	9.0210	8.3312
101.0	10.1577	9.3400	8.5773	10.9868	9.8435	8.7348
100.0	11.9845	10.4230	9.1308	12.5433	10.7193	9.1498
99.0	14.0211	11.5755	9.6995	14.2710	11.6495	9.5761

$r = 0.10, b = 0.10 : X = 17 : H = 102.0 : K = 104.0$

Case 1: $K > H$;

$$V = Xe^{rT}\phi(\varphi x_1 - \varphi\sigma\sqrt{T}) \qquad (16.3.3)$$

$$\varphi = 1$$

Case 2: $K < H$;

$$V = Xe^{-rT}\phi(\varphi x_2 - \varphi\sigma\sqrt{T}) - Xe^{-rT}(H / S)^{2\mu}\phi(\eta y_1 - \eta\sigma\sqrt{T})$$

$$+Xe^{-rT}(H / S)^{2\mu}\phi(\eta y_2 - \eta\sigma\sqrt{T}) \qquad (16.3.4)$$

$$\eta = -1, \varphi = 1.$$

Table 16.14 provides example valuations for up-and-in cash-or-nothing options.

TABLE 16.14. Up-and-in cash-or-nothing call option valuations

Asset Price	Time = 0.25			Time = 0.5		
	$\sigma = 0.15$	$\sigma = 0.20$	$\sigma = 0.30$	$\sigma = 0.15$	$\sigma = 0.20$	$\sigma = 0.30$
99.0	6.9621	7.1077	7.4969	6.1941	6.5724	7.1957
100.0	7.3992	7.5241	7.8556	6.3618	6.7846	7.4172
101.0	7.7627	7.9067	8.2038	6.4910	6.9774	7.6320
102.0 : H	8.0459	8.2516	8.5400	6.5814	7.1499	7.8397
103.0	8.2457	8.5556	8.8628	6.6334	7.3015	8.0399
104.0 : K	8.3622	8.8166	9.1710	6.6482	7.4321	8.2323
105.0	8.3983	9.0335	9.4636	6.6276	7.5415	8.4167
106.0	8.3597	9.2061	9.7398	6.5742	7.6300	8.5927

$r = 0.10, b = 0.10 : X = 17 : H = 102.0 : K = 104.0$

16.3.2. In Asset or Nothing Valuations

Down-and-in asset-or-nothing call

This option has the following payoff:

$$0 : \forall\, \tau \leq T, S(\tau) > H \vee S(T) < K;\ else,\ X$$

Case 1: $K > H$;

$$V = Se^{(b-r)T}\phi(\varphi x_2) \tag{16.3.5}$$

$$\eta = 1$$

Case 2: $K < H$;

$$V = Se^{(b-r)T}\phi(\varphi x_1) - Se^{(b-r)T}\phi(\varphi x_2) + Se^{(b-r)T}(H\ /\ S)^{2(\mu+1)}\phi(\eta y_2) \tag{16.3.6}$$

$$\eta = 1, \varphi = 1$$

Down-and-in asset or-nothing call valuation examples are shown in Table 16.15.

An up-and-in asset-or-nothing call option has a payoff given by:

$$0 : \forall\, \tau \leq T, S(\tau) < H \vee S(T) < K;\ else,\ X$$

Case 1: $K > H$;

$$V = Se^{(b-r)T}\phi(\varphi x_1) \tag{16.3.7}$$

$$\varphi = 1$$

Case 2: $K < H$;

$$\begin{aligned} V =\, & Se^{(b-r)T}\phi(\varphi x_2) - Se^{(b-r)T}(H\ /\ S)^{2(\mu+1)}\phi(\eta y_1) \\ & + Se^{(b-r)T}(H\ /\ S)^{2(\mu+1)}\phi(\eta y_2) \\ & \eta = -1, \varphi = 1 \end{aligned} \tag{16.3.8}$$

TABLE 16.15. Down-and-in asset-or-nothing call option valuations

Asset Price	Time = 0.25			Time = 0.5		
	$\sigma = 0.15$	$\sigma = 0.20$	$\sigma = 0.30$	$\sigma = 0.15$	$\sigma = 0.20$	$\sigma = 0.30$
106.0	24.9430	32.8352	41.4521	35.6148	42.5789	49.3935
105.0	30.8877	37.6762	44.7298	41.5544	47.0017	52.1426
104.0 : K	37.9002	43.0225	48.1742	48.3123	51.7868	55.0050
103.0	46.0861	8.8919	51.7838	55.9728	56.9529	57.9827
102.0 : H	55.5450	55.2980	55.5566	64.6256	62.5186	61.0771
101.0	66.3682	62.2498	59.4892	74.3662	68.5027	64.2898
100.0	78.6380	69.7519	63.5774	85.2968	74.9241	67.6220
99.0	92.4284	77.8045	67.8161	97.5270	81.8016	71.0749

$r = 0.10, b = 0.10 : K = 104.0 : H = 102.0$

TABLE 16.16. Up-and-in asset-or-nothing call option valuations

Asset Price	Time = 0.25			Time = 0.5		
	$\sigma = 0.15$	$\sigma = 0.20$	$\sigma = 0.30$	$\sigma = 0.15$	$\sigma = 0.20$	$\sigma = 0.30$
99.0	38.3521	41.9357	46.0758	51.8646	52.4934	53.8162
100.0	43.9550	46.3267	49.2100	56.1456	55.8407	56.2330
101.0	49.7168	50.7940	52.3742	60.4097	59.1869	58.6547
102.0 : H	55.5450	55.2980	55.5566	64.6256	62.5186	61.0771
103.0	61.3481	59.7999	58.7456	68.7646	65.8231	63.4964
104.0 : K	67.0404	64.2628	61.9300	72.8013	69.0886	65.9087
105.0	72.5454	68.6527	65.0993	76.7141	72.3046	68.3106
106.0	77.7989	72.9390	68.2436	80.4849	75.4618	70.6988

$r = 0.10,\ b = 0.10 : K = 104.0 : H = 102.0.$

Up-and-in asset-or-nothing call option examples are shown in Table 16.16.

The following four barrier options have a positive payoff if the barrier is breached and the asset finishes below a predetermined level.

16.3.3. In Asset Dependent Valuations

The down-and-in cash-or-nothing put has a payoff given by:

$$0 : \forall\ \tau \le T, S(\tau) > H \vee S(T) > K;\ else,\ X$$

Case 1: $K > H$;

$$V = Xe^{-rT}\phi(\varphi x_2 - \varphi\sigma\sqrt{T}) - Xe^{-rT}(H\ /\ S)^{2\mu}\phi(\eta y_1 - \eta\sigma\sqrt{T})$$
$$+Xe^{-rT}(H/S)^{2\mu}\phi(\eta y_2 - \eta\sigma\sqrt{T}) \qquad (16.3.9)$$
$$\eta = 1, \varphi = -1$$

Case 2: $K < H$

$$V = Xe^{rT}\phi(\varphi x_1 - \varphi\sigma\sqrt{T}) \qquad (16.3.10)$$
$$\varphi = -1$$

Table 16.17 gives example valuations for the down-and-in cash-or-nothing put option.

Up-and-in cash-or-nothing put has a payoff given by:

$$0 : \forall\ \tau \le T, S(\tau) < H \vee S(T) > K;\ else,\ X$$

Case 1: $K > H$;

$$V = Xe^{-rT}\phi(\varphi x_1 - \varphi\sigma\sqrt{T}) - Xe^{-rT}\phi(\varphi x_2 - \varphi\sigma\sqrt{T})$$
$$+Xe^{-rT}(H\ /\ S)^{2\mu}\phi(\eta y_2 - \eta\sigma\sqrt{T}) \qquad (16.3.11)$$
$$\eta = -1, \varphi = -1$$

TABLE 16.17. Down-and-in cash-or-nothing put option valuations

Asset Price	Time = 0.25			Time = 0.5		
	$\sigma = 0.15$	$\sigma = 0.20$	$\sigma = 0.30$	$\sigma = 0.15$	$\sigma = 0.20$	$\sigma = 0.30$
106.0	4.6586	5.6892	6.8284	4.3856	5.4391	6.6717
105.0	5.4302	6.2972	7.2453	4.8919	5.8471	6.9567
104.0 : K	6.2596	6.9304	7.6706	5.4287	6.2692	7.2466
103.0	7.1360	7.5839	8.1027	5.9930	6.7040	7.5411
102.0 : H	8.0459	8.2516	8.5400	6.5814	7.1499	7.8397
101.0	8.9734	8.9273	8.9806	7.1897	7.6049	8.1418
100.0	9.9007	9.6039	9.4226	7.8131	8.0671	8.4468
99.0	10.8097	10.2743	9.8640	8.4461	8.5342	8.7540

$r = 0.10, b = 0.10 : X = 17 : H = 102.0 : K = 104.0$

Case 2: $K < H$;

$$V = Xe^{-rT}(H\ /\ S)^{2\mu}\phi(\eta y_1 - \eta\sigma\sqrt{T}) \tag{16.3.12}$$

$$\eta = -1,$$

Table 16.18 gives a set of valuation examples for the up-and-in cash-or-nothing put.

Down-and-in asset-or-nothing put.

The down-and-in asset-or-nothing put has a payoff given by:

$$0 : \forall\ \tau \leq T, S(\tau) > H \vee S(T) > K; \ else, X$$

Case 1: $K > H$;

$$V = Se^{(b-r)T}\phi(\varphi x_2) - Se^{(b-r)T}(H/S)^{2(\mu+1)}\phi(\eta y_1) + Se^{(b-r)T}(H/S)^{2(\mu+1)}\phi(\eta y_2) \tag{16.3.13}$$

$$\eta = 1, \varphi = -1$$

TABLE 16.18. Up-and-in cash-or-nothing put option valuations

Asset Price	Time = 0.25			Time = 0.5		
	$\sigma = 0.15$	$\sigma = 0.20$	$\sigma = 0.30$	$\sigma = 0.15$	$\sigma = 0.20$	$\sigma = 0.30$
99.0	6.7802	7.0323	7.4748	6.1311	6.5462	7.1881
100.0	7.2781	7.4738	7.8409	6.3208	6.7674	7.4122
101.0	7.7032	7.8819	8.1965	6.4712	6.9689	7.6295
102.0 : H	8.0459	8.2516	8.5400	6.5814	7.1499	7.8397
103.0	8.3003	8.5793	8.8699	6.6516	7.3096	8.0424
104.0 : K	8.4641	8.8622	9.1850	6.6826	7.4477	8.2372
105.0	8.5390	9.0987	9.4842	6.6762	7.5641	8.4238
106.0	8.5296	9.2883	9.7666	6.6345	7.6590	8.6021

$r = 0.10, b = 0.10 : X = 17 : H = 102.0 : K = 104.0$

Case 2: $K < H$;

$$V = Se^{(b-r)T}\phi(\varphi x_1) \tag{16.3.14}$$
$$\varphi = -1$$

Table 16.19 gives the valuations for a set of down-and-in asset-or-nothing put options.

An up-and-in asset-or-nothing put has a payoff given by:

$$0 : \forall\, \tau \leq T, S(\tau) < H \vee S(T) > K;\ else,\ X$$

Case 1: $K > H$;

$$V = Se^{(b-r)T}\phi(\varphi x_1) - Se^{(b-r)T}\phi(\varphi x_2) + Se^{(b-r)T}(H/S)^{2(\mu+1)}\phi(\eta y_1) \tag{16.3.15}$$
$$\eta = -1, \varphi = -1$$

Case 2: $K < H$;

$$V = Se^{(b-r)T}(H/S)^{2(\mu+1)}\phi(\eta y_1) \tag{16.3.16}$$
$$\varphi = -1$$

Table 16.20 gives a set of example valuations for an up-and-in asset-or-nothing put.

16.3.4. Out Asset Dependent Valuations

The following four options require that the asset price does not hit the barrier and it finishes above the stated level.

A down-and-out cash-or-nothing call has a payoff give by:

$$X : \forall\, \tau \leq T, S(\tau) > H \wedge S(T) > K;\ else,\ 0$$

TABLE 16.19. Down-and-in asset-or-nothing put option valuations

Asset Price	Time = 0.25			Time = 0.5		
	σ = 0.15	σ = 0.20	σ = 0.30	σ = 0.15	σ = 0.20	σ = 0.30
106.0	27.1680	32.5611	37.5932	25.1482	30.3621	35.2441
105.0	31.5997	35.9510	39.7752	27.9907	32.5579	36.6458
104.0 : K	36.3402	39.4603	41.9846	30.9891	34.8165	38.0618
103.0	41.3205	43.0563	44.2110	34.1246	37.1280	39.4887
102.0 : H	46.4550	46.7020	46.4434	37.3744	39.4814	40.9229
101.0	51.6447	50.3571	48.6702	40.7111	41.8645	42.3606
100.0	56.7814	53.9786	50.8793	44.1038	44.2640	43.7978
99.0	61.7533	57.5223	53.0580	47.5180	46.6656	45.2303

$r = 0.10$: $b = 0.10$: $K = 104.0$: $H = 102.0$.

TABLE 16.20. Up-and-in asset-or-nothing put option valuations

Asset Price	Time = 0.25			Time = 0.5		
	$\sigma = 0.15$	$\sigma = 0.20$	$\sigma = 0.30$	$\sigma = 0.15$	$\sigma = 0.20$	$\sigma = 0.30$
99.0	50.6428	48.2379	46.2881	42.6092	41.9809	41.4921
100.0	53.2910	50.5988	48.1670	43.4111	43.0401	42.5432
101.0	55.3081	52.6702	49.9490	43.9231	43.9548	43.5421
102.0 : H	56.6612	54.4310	51.6258	44.1501	44.7227	44.4869
103.0	57.3467	55.8679	53.1911	44.1023	45.3433	45.3763
104.0 : K	57.3884	56.9758	54.6395	43.7954	45.8177	46.2092
105.0	56.8333	57.7570	55.9673	43.2487	46.1488	46.9849
106.0	55.7463	58.2207	57.1720	42.4851	46.3406	47.7029

$r = 0.10$: $b = 0.10$: $K = 104.0$: $H = 102.0$.

Case 1:

$$V = Xe^{-rT}\phi(\varphi x_1 - \varphi\sigma\sqrt{T}) - Xe^{-rT}(H/S)^{2\mu}\phi(\eta y_1 - \eta\sigma\sqrt{T}) \qquad (16.3.17)$$

$$\eta = 1, \varphi = 1$$

Case 2:

$$V = Xe^{-rT}\phi(\varphi x_1 - \varphi\sigma\sqrt{T}) - Xe^{-rT}(H/S)^{2\mu}\phi(\eta y_2 - \eta\sigma\sqrt{T}) \qquad (16.3.18)$$

$$\eta = 1, \varphi = 1$$

Table 16.21 gives example valuations for a down-and-out cash-or-nothing call.

Up-and-out cash-or-nothing call has the following payoff:

$$X : \forall\tau \leq T, S(\tau) < H \vee S(T) > K; else, 0$$

Case 1: $K > H$;

$$V = 0 \qquad (16.3.19)$$

TABLE 16.21. Down-and-out cash-or-nothing call option valuations

Asset Price	Time = 0.25			Time = 0.5		
	$\sigma = 0.15$	$\sigma = 0.20$	$\sigma = 0.30$	$\sigma = 0.15$	$\sigma = 0.20$	$\sigma = 0.30$
106.0	7.8708	5.7992	3.6460	6.3608	4.4681	2.6542
105.0	6.2168	4.4863	2.7748	4.9927	3.4422	2.0157
104.0 : K	4.3556	3.0797	1.8752	3.4818	2.3558	1.3601
103.0	2.2834	1.5827	0.9494	1.8201	1.2084	0.6880
102.0 : H	0.0000	0.0000	0.0000	0.0000	0.0000	0.0000
101.0	0.0000	0.0000	0.0000	0.0000	0.0000	0.0000
100.0	0.0000	0.0000	0.0000	0.0000	0.0000	0.0000
99.0	0.0000	0.0000	0.0000	0.0000	0.0000	0.0000

$r = 0.10$, $b = 0.10$: $X = 17$: $H = 102.0$: $K = 104.0$

Case 2: $K < H$;

$$V = Xe^{-rT}\phi(\varphi x_1 - \varphi\sigma\sqrt{T}) - Xe^{-rT}\phi(\varphi x_2 - \varphi\sigma\sqrt{T}) + Xe^{-rT}(H/S)^{2\mu}\phi(\eta y_1 - \eta\sigma\sqrt{T}) - Xe^{-rT}(H/S)^{2\mu}\phi(\eta y_2 - \eta\sigma\sqrt{T}) \qquad (16.3.20)$$

$$\eta = -1, \varphi = 1$$

Table 16.22 provides example valuations for an up-and-out cash-or-nothing call options. If $K > H$, the value of this option is zero. The values for H and K are therefore reversed from our previous examples.

The down-and-out asset-or-nothing call option has a payoff given by:

$$S(T) : \forall\tau \le T, S(\tau) > H \vee S(T) > K; \textit{else}, 0$$

Case 1:

$$V = Se^{(b-r)T}\phi(\varphi x_1) - Se^{(b-r)T}(H/S)^{2(\mu+1)}\phi(\eta y_1) \qquad (16.3.21)$$

$$\eta = 1, \phi = 1$$

Case 2:

$$V = Se^{(b-r)T}\phi(\varphi x_2) - Se^{(b-r)T}(H/S)^{2(\mu+1)}\phi(\eta y_2) \qquad (16.3.22)$$

$$\eta = 1, \phi = 1$$

Example valuations are shown in Table 16.23.

TABLE 16.22. Up-and-out cash-or-nothing call option valuations

Asset Price	Time = 0.25			Time = 0.5		
	$\sigma = 0.15$	$\sigma = 0.20$	$\sigma = 0.30$	$\sigma = 0.15$	$\sigma = 0.20$	$\sigma = 0.30$
99.0	0.2557	0.1137	0.0350	0.0946	0.0410	0.0124
100.0	0.2115	0.0928	0.0282	0.0753	0.0327	0.0099
101.0	0.1612	0.0702	0.0213	0.0557	0.0243	0.0074
102.0 : K	0.1073	0.0468	0.0142	0.0363	0.0160	0.0049
103.0	0.0527	0.0231	0.0071	0.0176	0.0079	0.0024
104.0 : H	0.0000	0.0000	0.0000	0.0000	0.0000	0.0000
105.0	0.0000	0.0000	0.0000	0.0000	0.0000	0.0000
106.0	0.0000	0.0000	0.0000	0.0000	0.0000	0.0000

$r = 0.10$: $b = 0.10$: $K = 102.0$: $H = 104.0$: $X = 17.0$.

TABLE 16.23. Down-and-out asset-or-nothing call option valuations

Asset Price	Time = 0.25			Time = 0.5		
	$\sigma = 0.15$	$\sigma = 0.20$	$\sigma = 0.30$	$\sigma = 0.15$	$\sigma = 0.20$	$\sigma = 0.30$
106.0	52.8559	40.1038	26.7915	44.8701	32.8829	21.3053
105.0	41.6577	30.9765	20.3694	35.1597	25.3030	16.1680
104.0 : K	29.1402	21.2403	13.7558	24.4890	17.3018	10.9036
103.0	15.2620	10.9080	6.9617	12.7918	8.8702	5.5137
102.0 : H	0.0000	0.0000	0.0000	0.0000	0.0000	0.0000
101.0	0.0000	0.0000	0.0000	0.0000	0.0000	0.0000
100.0	0.0000	0.0000	0.0000	0.0000	0.0000	0.0000
99.0	0.0000	0.0000	0.0000	0.0000	0.0000	0.0000

$r = 0.10$: $b = 0.10$: $K = 104.0$: $H = 102.0$

The up-and-out asset-or-nothing call option has payoff given by:

$$S(T) = \forall \tau \leq T, S(\tau) < H \vee S(T) > K; \, else, 0$$

Case 1: $K > H$; $V = 0$.

Case 2: $K < H$;

$$\begin{aligned} V = & Se^{(b-r)T}\phi(\varphi x_1) - Se^{(b-r)T}\phi(\varphi x_2) + Se^{(b-r)T}(H/S)^{2(\mu+1)}\phi(\eta y_1) \\ & - Se^{(b-r)T}(H/S)^{2(\mu+1)}\phi(\eta y_2) \end{aligned} \tag{16.3.23}$$

$$\eta = -1, \varphi = 1$$

Table 16.24 gives example valuations for up-and-out asset-or-nothing call options.

TABLE 16.24. Up-and-out asset-or-nothing call option valuations

Asset Price	Time = 0.25			Time = 0.5		
	$\sigma = 0.15$	$\sigma = 0.20$	$\sigma = 0.30$	$\sigma = 0.15$	$\sigma = 0.20$	$\sigma = 0.30$
99.0	1.5444	0.6870	0.2113	0.5715	0.2477	0.0747
100.0	1.2777	0.5602	0.1705	0.4548	0.1976	0.0597
101.0	0.9735	0.4239	0.1284	0.3363	0.1470	0.0446
102.0 : K	0.6479	0.2824	0.0856	0.2192	0.0968	0.0296
103.0	0.3181	0.1398	0.0426	0.1063	0.0476	0.0147
104.0 : H	0.0000	0.0000	0.0000	0.0000	0.0000	0.0000
105.0	0.0000	0.0000	0.0000	0.0000	0.0000	0.0000
106.0	0.0000	0.0000	0.0000	0.0000	0.0000	0.0000

$r = 0.10$: $b = 0.10$: $K = 102.0$: $H = 104.0$.

16.3.5. Out Asset limited Valuations

The following four options have a positive payoff if the barrier is not hit and the asset finishes below the specified level.

For a down-and-out cash-or-nothing put option the payoff is:

$$X(T) : \forall \tau \leq T, S(\tau) > H \vee S(T) < K; \, else, 0$$

Case 1: $K > H$;

$$\begin{aligned} V = Xe^{-rT}\phi(\varphi x_1 - \varphi\sigma\sqrt{T}) - Xe^{-rT}\phi(\varphi x_2 - \varphi\sigma\sqrt{T}) \\ +Xe^{-rT}(H/S)^{2\mu}\phi(\eta y_1 - \eta\sigma\sqrt{T}) - Xe^{-rT}(H/S)^{2\mu}\phi(\eta y_2 - \eta\sigma\sqrt{T}) \quad (16.3.24) \\ \eta = 1, \varphi = -1 \end{aligned}$$

Case 2: $K < H$; $V = 0$

Table 16.25 shows example valuations for down-and-out cash-or-nothing put options.

The up-and-out cash-or-nothing put has a payoff of:

$$X(T) : \forall \tau \leq T, S(\tau) < H \vee S(T) < K; \, else, 0$$

Case 1: $K > H$;

$$V = Xe^{-rT}\phi(\varphi x_2 - \varphi\sigma\sqrt{T}) - Xe^{-rT}(H/S)^{2\mu}\phi(\eta y_2 - \eta\sigma\sqrt{T}) \quad (16.3.25)$$

$$\eta = -1, \varphi = -1$$

Case 2: $K < H$;

$$V = Xe^{-rT}\phi(\varphi x_1 - \varphi\sigma\sqrt{T}) - Xe^{-rT}(H/S)^{2\mu}\phi(\eta y_1 - \eta\sigma\sqrt{T}) \quad (16.3.26)$$

$$\eta = -1, \varphi = -1$$

TABLE 16.25. Down-and-out cash-or-nothing option valuations

Asset Price	Time = 0.25			Time = 0.5		
	$\sigma = 0.15$	$\sigma = 0.20$	$\sigma = 0.30$	$\sigma = 0.15$	$\sigma = 0.20$	$\sigma = 0.30$
106.0	0.1700	0.0822	0.0269	0.0604	0.0290	0.0094
105.0	0.1406	0.0652	0.0206	0.0486	0.0226	0.0072
104.0 : K	0.1019	0.0456	0.0141	0.0345	0.0156	0.0049
103.0	0.0545	0.0237	0.0071	0.0182	0.0081	0.0025
102.0 : H	0.0000	0.0000	0.0000	0.0000	0.0000	0.0000
101.0	0.0000	0.0000	0.0000	0.0000	0.0000	0.0000
100.0	0.0000	0.0000	0.0000	0.0000	0.0000	0.0000
99.0	0.0000	0.0000	0.0000	0.0000	0.0000	0.0000

$r = 0.10$: $b = 0.10$: $K = 104.0$: $H = 102.0$: $X = 17.0$.

TABLE 16.26. Up-and-out cash-or-nothing option valuations

Asset Price	Time = 0.25			Time = 0.5		
	$\sigma = 0.15$	$\sigma = 0.20$	$\sigma = 0.30$	$\sigma = 0.15$	$\sigma = 0.20$	$\sigma = 0.30$
99.0	3.8476	3.1667	2.3671	2.2520	1.9619	1.5583
100.0	2.5015	2.0799	1.5670	1.4513	1.2825	1.0295
101.0	1.2107	1.0206	0.7768	0.6987	0.6275	0.5098
102.0 : H	0.0000	0.0000	0.0000	0.0000	0.0000	0.0000
103.0	0.0000	0.0000	0.0000	0.0000	0.0000	0.0000
104.0 : K	0.0000	0.0000	0.0000	0.0000	0.0000	0.0000
105.0	0.0000	0.0000	0.0000	0.0000	0.0000	0.0000
106.0	0.0000	0.0000	0.0000	0.0000	0.0000	0.0000

$r = 0.10$: $b = 0.10$: $K = 104.0$: $H = 102.0$: $X = 17.0$.

Table 16.26 gives example valuations for up-and-out cash-or-nothing put options.

A down-and-out asset-or-nothing put has a payoff given by:

$$S(T) : \forall \tau \leq T, S(\tau) > H \vee S(T) < K; else, 0$$

Case 1: $K > H$;

$$V = Se^{(b-r)T}\phi(\varphi x_1) - Se^{(b-r)T}\phi(\varphi x_2)$$
$$+Se^{(b-r)T}(H/S)^{2(\mu+1)}\phi(\eta y_1) - Se^{(b-r)T}(H/S)^{2(\mu+1)}\phi(\eta y_2) \qquad (16.3.27)$$
$$\eta = 1, \varphi = -1$$

Case 2: $K < H$; 0

Table 16.27 shows examples of down-and-out asset-or-nothing option valuations.

The up-and-out asset-or-nothing put option has a payoff given by:

$$S(T) : \forall \tau \leq T, S(\tau) < H \vee S(T) < K; else, 0$$

TABLE 16.27. Down-and-out asset-or-nothing option valuations

Asset Price	Time = 0.25			Time = 0.5		
	$\sigma = 0.15$	$\sigma = 0.20$	$\sigma = 0.30$	$\sigma = 0.15$	$\sigma = 0.20$	$\sigma = 0.30$
106.0	1.0331	0.4999	0.1632	0.3668	0.1761	0.0571
105.0	0.8549	0.3963	0.1255	0.2952	0.1374	0.0436
104.0 : K	0.6194	0.2769	0.0854	0.2096	0.0949	0.0295
103.0	0.3314	0.1439	0.0434	0.1108	0.0490	0.0150
102.0 : H	0.0000	0.0000	0.0000	0.0000	0.0000	0.0000
101.0	0.0000	0.0000	0.0000	0.0000	0.0000	0.0000
100.0	0.0000	0.0000	0.0000	0.0000	0.0000	0.0000
99.0	0.0000	0.0000	0.0000	0.0000	0.0000	0.0000

$r = 0.10$: $b = 0.10$: $K = 104.0$: $H = 102.0$.

TABLE 16.28. Up-and-out asset-or-nothing valuations

Asset Price	Time = 0.25			Time = 0.5		
	$\sigma = 0.15$	$\sigma = 0.20$	$\sigma = 0.30$	$\sigma = 0.15$	$\sigma = 0.20$	$\sigma = 0.30$
99.0	34.6039	27.9076	19.7255	20.5636	17.0057	12.2748
100.0	27.3956	22.1883	15.7347	16.0452	13.4086	9.7549
101.0	20.1858	16.4712	11.7458	11.6893	9.8897	7.2610
102.0 : K	13.1251	10.8249	7.7801	7.5393	6.4699	4.7997
103.0	6.3547	5.3145	3.8583	3.6327	3.1678	2.3774
104.0 : H	0.0000	0.0000	0.0000	0.0000	0.0000	0.0000
105.0	0.0000	0.0000	0.0000	0.0000	0.0000	0.0000
106.0	0.0000	0.0000	0.0000	0.0000	0.0000	0.0000

$r = 0.10$: $b = 0.10$: $K = 102.0$: $H = 104.0$.

Case 1: $K > H$;

$$V = Se^{(b-r)T}\phi(\varphi x_2) - Se^{(b-r)T}(H/S)^{2(\mu+1)}\phi(\eta y_2) \tag{16.3.28}$$
$$\eta = -1, \varphi = -1$$

Case 2: $K < H$;

$$V = Se^{(b-r)T}\phi(\varphi x_1) - Se^{(b-r)T}(H/S)^{2(\mu+1)}\phi(\eta y_1) \tag{16.3.29}$$
$$\eta = -1, \varphi = -1$$

Table 16.28 gives example valuations for up-and-out asset-or-nothing put options.

Valuations for each of the binary barrier options are provided by the methods of class **Binarybars.** The class is shown in Listing 16.2. This class provides a set of public methods to implement valuations for the twenty eight binary barrier options. Each public method provides access to the calculating class **Digitalbarrier**. The constructor of **Binarybars** provides the interest rates, barrier level and the adjustment, if required, for discrete time periods.

```
package FinApps;
import static java.lang.Math.*;
public class Binarybars {

  /** Creates a new instance of Binarybars */
  public Binarybars(double rate, double yield, double barlevel,
                    int period) {
    r=rate;
    y=yield;
    barrier=barlevel;
    tau=period==1?(1.0/(24*365)):period==2?(1.0/(365.0)):period==3?
               (1.0/(52.0)):period==4?(1.0/(12.0)):0.0;
```

```
    }
  private double r;
  private double y;
  private double h;
  private double tau;
  private double barrier;

  public double downInch(double asset, double cash, double volatility,
                          double time) {
    h= barrier>asset? (barrier*exp(sqrt(tau)*volatility*0.5826)):
                     barrier;//Correction for discrete time (from
                     continuous)
    h= barrier<asset? (barrier*exp(-sqrt(tau)*volatility*0.5826)):
                     barrier;
    if(asset<=h)
      return cash;
    Digitalbarrier s=new Digitalbarrier(r,y,h);
    double barriervaldich= s.barops(asset,0.0,cash,volatility,time,1);
    return barriervaldich;
    }
  public double upInch(double asset ,double cash, double volatility,
                          double time) {
    h= barrier>asset? (barrier*exp(sqrt(tau)*volatility*0.5826)):
                     barrier;
    h= barrier<asset? (barrier*exp(-sqrt(tau)*volatility*0.5826)):
                     barrier;
    if(asset>=h)
      return cash;
    Digitalbarrier s=new Digitalbarrier(r,y,h);
    double barriervaluich= s.barops(asset,0.0,cash,volatility,time,2);
    return barriervaluich;
  }
  public double upInah(double asset , double volatility,double time) {
    h= barrier>asset? (barrier*exp(sqrt(tau)*volatility*0.5826)):
                     barrier;
    h= barrier<asset? (barrier*exp(-sqrt(tau)*volatility*0.5826)):
                     barrier;
    if(asset>=h)//No ned to continue
      return asset;//the asset value is the barrier at hit
    double cash=h;//X=H
    Digitalbarrier s=new Digitalbarrier(r,y,h);
    double barriervaluiah= s.barops(asset,0.0,cash,volatility,time,4);
    return barriervaluiah;
    }
  public double downInah(double asset , double volatility,double time) {
    h= barrier>asset? (barrier*exp(sqrt(tau)*volatility*0.5826)):
                     barrier;
    h= barrier<asset? (barrier*exp(-sqrt(tau)*volatility*0.5826)):
                     barrier;
    if(asset<=h)
      return asset;
    double cash=h;//X=H
    Digitalbarrier s=new Digitalbarrier(r,y,h);
    double barriervaldiah= s.barops(asset,0.0,cash,volatility,time,3);
```

```
  return barriervaldiah;
}
public double downIncx(double asset, double cash, double volatility,
                        double time) {
  h= barrier>asset? (barrier*exp(sqrt(tau)*volatility*0.5826)):
                       barrier;
  h= barrier<asset? (barrier*exp(-sqrt(tau)*volatility*0.5826)):
                       barrier;
  Digitalbarrier s=new Digitalbarrier(r,y,h);
  double barriervaldicx= s.barops(asset,0.0,cash,volatility,time,5);
  return barriervaldicx;
}
public double upIncx(double asset , double cash, double volatility,
                        double time) {
  h= barrier>asset? (barrier*exp(sqrt(tau)*volatility*0.5826)):
                       barrier;
  h= barrier<asset? (barrier*exp(-sqrt(tau)*volatility*0.5826)):
                       barrier;
  Digitalbarrier s=new Digitalbarrier(r,y,h);
  double barriervaluicx= s.barops(asset,0.0,cash,volatility,time,6);
  return barriervaluicx;
}
public double downInax(double asset , double volatility,double time) {
  h= barrier>asset? (barrier*exp(sqrt(tau)*volatility*0.5826)):
                       barrier;
  h= barrier<asset? (barrier*exp(-sqrt(tau)*volatility*0.5826)):
                       barrier;
  Digitalbarrier s=new Digitalbarrier(r,y,h);
  double barriervaldiax= s.barops(asset,0.0,0.0,volatility,time,7);
  return barriervaldiax;
}
public double upInax(double asset , double volatility,double time) {
  h= barrier>asset? (barrier*exp(sqrt(tau)*volatility*0.5826)):
                       barrier;
  h= barrier<asset? (barrier*exp(-sqrt(tau)*volatility*0.5826)):
                       barrier;
  Digitalbarrier s=new Digitalbarrier(r,y,h);
  double barriervaluiax= s.barops(asset,0.0,0.0,volatility,time,8);
  return barriervaluiax;
}
public double downOcon(double asset , double cash, double volatility,
                        double time) {
  h= barrier>asset? (barrier*exp(sqrt(tau)*volatility*0.5826)):
                       barrier;
  h= barrier<asset? (barrier*exp(-sqrt(tau)*volatility*0.5826)):
                       barrier;
  if(asset<=h)
    return 0.0;
  Digitalbarrier s=new Digitalbarrier(r,y,h);
  double barriervaldoc= s.barops(asset,0.0,cash,volatility,time,9);
  return barriervaldoc;
}
public double upOcon(double asset , double cash, double volatility,
                        double time) {
```

```
  h= barrier>asset? (barrier*exp(sqrt(tau)*volatility*0.5826)):
                    barrier;
  h= barrier<asset? (barrier*exp(-sqrt(tau)*volatility*0.5826)):
                    barrier;
  if(asset>=h)
    return 0.0;
  Digitalbarrier s=new Digitalbarrier(r,y,h);
  double barriervaluoc=s.barops(asset,0.0,cash,volatility,time,10);
  return barriervaluoc;
}
public double downOaon(double asset , double volatility,double time) {
    h= barrier>asset? (barrier*exp(sqrt(tau)*volatility*0.5826)):
                    barrier;
  h= barrier<asset? (barrier*exp(-sqrt(tau)*volatility*0.5826)):
                    barrier;
  if(asset<=h)
    return 0.0;
  Digitalbarrier s=new Digitalbarrier(r,y,h);
  double barriervaldoa=s.barops(asset,0.0,0.0,volatility,time,11);
  return barriervaldoa;
}
public double upOaon(double asset , double volatility,double time) {
  h= barrier>asset? (barrier*exp(sqrt(tau)*volatility*0.5826)):
                    barrier;
  h= barrier<asset? (barrier*exp(-sqrt(tau)*volatility*0.5826)):
                    barrier;
  if(asset>=h)
    return 0.0;
  Digitalbarrier s=new Digitalbarrier(r,y,h);
  double barriervaluoa=s.barops(asset,0.0,0.0,volatility,time,12);
  return barriervaluoa;
}
public double downIconc(double asset , double strike, double cash,
                        double volatility,double time) {
  h= barrier>asset? (barrier*exp(sqrt(tau)*volatility*0.5826)):
                    barrier;
  h= barrier<asset? (barrier*exp(-sqrt(tau)*volatility*0.5826)):
                    barrier;
  Digitalbarrier s=new Digitalbarrier(r,y,h);
  double barriervaldic=s.barops(asset,strike,cash,volatility,
                    time,13);
  return barriervaldic;
}
public double upIconc(double asset , double strike, double cash,
                      double volatility,double time) {
  h= barrier>asset? (barrier*exp(sqrt(tau)*volatility*0.5826)):
                    barrier;
  h= barrier<asset? (barrier*exp(-sqrt(tau)*volatility*0.5826)):
                    barrier;
  Digitalbarrier s=new Digitalbarrier(r,y,h);
  double barriervaluicc=s.barops(asset,strike,cash,volatility,
                    time,14);
  return barriervaluicc;
}
```

```
public double downIaonc(double asset , double strike, double cash,
                        double volatility,double time) {
  h= barrier>asset? (barrier*exp(sqrt(tau)*volatility*0.5826)):
                    barrier;
  h= barrier<asset? (barrier*exp(-sqrt(tau)*volatility*0.5826)):
                    barrier;
  Digitalbarrier s=new Digitalbarrier(r,y,h);
  double barriervaldiac=s.barops(asset,strike,cash,volatility,
                    time,15);
  return barriervaldiac;
}
public double upIaonc(double asset , double strike, double cash,
                        double volatility,double time) {
  h= barrier>asset? (barrier*exp(sqrt(tau)*volatility*0.5826)):
                    barrier;
  h= barrier<asset? (barrier*exp(-sqrt(tau)*volatility*0.5826)):
                    barrier;
  Digitalbarrier s=new Digitalbarrier(r,y,h);
  double barriervaluiac=s.barops(asset,strike,cash,volatility,
                    time,16);
  return barriervaluiac;
}
public double downIconp(double asset , double strike, double cash,
                        double volatility,double time) {
  h= barrier>asset? (barrier*exp(sqrt(tau)*volatility*0.5826)):
                    barrier;
  h= barrier<asset? (barrier*exp(-sqrt(tau)*volatility*0.5826)):
                    barrier;
  Digitalbarrier s=new Digitalbarrier(r,y,h);
  double barriervaldicp=s.barops(asset,strike,cash,volatility,
                    time,17);
  return barriervaldicp;
}
public double upIconp(double asset , double strike, double cash,
                        double volatility,double time) {
  h= barrier>asset? (barrier*exp(sqrt(tau)*volatility*0.5826)):
                    barrier;
  h= barrier<asset? (barrier*exp(-sqrt(tau)*volatility*0.5826)):
                    barrier;
  Digitalbarrier s=new Digitalbarrier(r,y,h);
  double barriervaluicp=s.barops(asset,strike,cash,volatility,
                    time,18);
  return barriervaluicp;
}
public double downIaonp(double asset , double strike, double cash,
                        double volatility,double time) {
  h= barrier>asset? (barrier*exp(sqrt(tau)*volatility*0.5826)):
                    barrier;
  h= barrier<asset? (barrier*exp(-sqrt(tau)*volatility*0.5826)):
                    barrier;
  Digitalbarrier s=new Digitalbarrier(r,y,h);
  double barriervaldiap= s.barops(asset,strike,cash,volatility,
                        time,19);
  return barriervaldiap;
```

```
}
public double upIaonp(double asset , double strike, double cash,
                      double volatility,double time) {
  h= barrier>asset? (barrier*exp(sqrt(tau)*volatility*0.5826)):
                    barrier;
  h= barrier<asset? (barrier*exp(-sqrt(tau)*volatility*0.5826)):
                    barrier;
  Digitalbarrier s=new Digitalbarrier(r,y,h);
  double barriervaluiap=s.barops(asset,strike,cash,volatility,
                    time,20);
  return barriervaluiap;
}
public double downOconc(double asset , double strike, double cash,
                      double volatility,double time) {
  h= barrier>asset? (barrier*exp(sqrt(tau)*volatility*0.5826)):
                    barrier;
  h= barrier<asset? (barrier*exp(-sqrt(tau)*volatility*0.5826)):
                        barrier;
  if(asset<=h)
    return 0.0;
  Digitalbarrier s=new Digitalbarrier(r,y,h);
  double barriervaldocc=s.barops(asset,strike,cash,volatility,
                    time,21);
  return barriervaldocc;
}
public double upOconc(double asset , double strike, double cash,
                      double volatility,double time) {
  h= barrier>asset? (barrier*exp(sqrt(tau)*volatility*0.5826)):
                    barrier;
  h= barrier<asset? (barrier*exp(-sqrt(tau)*volatility*0.5826)):
                    barrier;
  if(asset>=h)
    return 0.0;
  Digitalbarrier s=new Digitalbarrier(r,y,h);
  double barriervaluocc=s.barops(asset,strike,cash,volatility,
                    time,22);
  return barriervaluocc;
}
public double downOaonc(double asset , double strike, double cash,
                      double volatility,double time) {
  h= barrier>asset? (barrier*exp(sqrt(tau)*volatility*0.5826)):
                    barrier;
  h= barrier<asset? (barrier*exp(-sqrt(tau)*volatility*0.5826)):
                    barrier;
  if(asset<=h)
    return 0.0;
  Digitalbarrier s=new Digitalbarrier(r,y,h);
  double barriervaldoac=s.barops(asset,strike,cash,volatility,
                    time,23);
  return barriervaldoac;
}
public double upOaonc(double asset , double strike, double cash,
                      double volatility,double time) {
  h= barrier>asset? (barrier*exp(sqrt(tau)*volatility*0.5826)):
```

```
                    barrier;
  h= barrier<asset? (barrier*exp(-sqrt(tau)*volatility*0.5826)):
                    barrier;
  if(asset>=h)
    return 0.0;
  Digitalbarrier s=new Digitalbarrier(r,y,h);
  double barriervaluoac=s.barops(asset,strike,cash,volatility,
                    time,24);
  return barriervaluoac;
}
public double downOconp(double asset , double strike, double cash,
                      double volatility,double time) {
  h= barrier>asset? (barrier*exp(sqrt(tau)*volatility*0.5826)):
                    barrier;
  h= barrier<asset? (barrier*exp(-sqrt(tau)*volatility*0.5826)):
                    barrier;
  if(asset<=h)
    return 0.0;
  Digitalbarrier s=new Digitalbarrier(r,y,h);
  double barriervaldocp= s.barops(asset,strike,cash,volatility,
                    time,25);
  return barriervaldocp;
}
public double upOconp(double asset , double strike, double cash,
                      double volatility,double time) {
  h= barrier>asset? (barrier*exp(sqrt(tau)*volatility*0.5826)):
                    barrier;
  h= barrier<asset? (barrier*exp(-sqrt(tau)*volatility*0.5826)):
                    barrier;
  if(asset>=h)
    return 0.0;
  Digitalbarrier s=new Digitalbarrier(r,y,h);
  double barriervaluocp=s.barops(asset,strike,cash,volatility,
                    time,26);
  return barriervaluocp;
}
public double downOaonp(double asset , double strike, double cash,
                      double volatility,double time) {
  h= barrier>asset? (barrier*exp(sqrt(tau)*volatility*0.5826)):
                    barrier;
  h= barrier<asset? (barrier*exp(-sqrt(tau)*volatility*0.5826)):
                    barrier;
  if(asset<=h)
    return 0.0;
  Digitalbarrier s=new Digitalbarrier(r,y,h);
  double barriervaldoap=s.barops(asset,strike,cash,volatility,
                    time,27);
  return barriervaldoap;
}
public double upOaonp(double asset , double strike, double cash,
                      double volatility,double time) {
  h= barrier>asset? (barrier*exp(sqrt(tau)*volatility*0.5826)):
                    barrier;
  h= barrier<asset? (barrier*exp(-sqrt(tau)*volatility*0.5826)):
```

```
                    barrier;
  if(asset>=h)
    return 0.0;
  Digitalbarrier s=new Digitalbarrier(r,y,h);
  double barriervaluoap=s.barops(asset,strike,cash,volatility,
                    time,28);
  return barriervaluoap;
}
```

LISTING 16.2. Class **Binarybars**

References

Bermin, H. P. (1996). "Time and Path Dependent Options: The Case of Time Dependent Inside and Outside Barrier Options," Paper presented at the Third Nordic Symposium on Contingent Claims Analysis in Finance, Iceland, May.

Heynen, R. C. and H. M. Kat (1994): "Partial Barrier Options," *Journal of Financial Engineering*, 3, 253–274.

Reiner, E. and M. Rubinstein (1991). "Unscrambling the Binary Code" *Risk*, 4(9), 75–83.

17
Special Case Barrier Options

Special case barriers are variations on the standard or digital variety. There are two categories of variation; firstly the case where there is more than a single underlying.The second category is the case of partial time monitoring of the triggering event.

17.1. Partial Time Options

A partial time single asset barrier option has a restricted time period, during which the underlying asset is monitored for hits. The monitoring period can be at the beginning of the option life, or at the end section of the option's life. The valuation formulae for barrier options where the barrier is inactive for part of the option lifetime have been developed by

Heynen & Kat (1994). For partial time barrier options where the monitoring period is at the beginning and ends some time prior to expiry of the option lifetime (partial-barrier option). There are up-and-in and down-and-in variants together with up-and-out and down-and-out call variants.

For partial time barrier options where the monitoring period starts at some time into the option lifetime and ends at expiry (forward-starting barrier options) there are up and in, up and out and the corresponding down variants. Partial time end options have two types. Type B1 is where the event of a barrier being hit or crossed (in any direction) is sufficient to knock-out the option. Type B2 is the case where down-and-out options are knocked-out when the barrier is breached from above, or up-and-out options are knocked-out when the barrier is breached from below.

The partial-barrier out formula is:

$$C_{uo} = Se^{(b-r)T}\left[M(d_1, \eta f_3; \eta\rho) - \left(\frac{H}{S}\right)^{2(\mu+1)} M(f_1, \eta f_5; \eta\rho)\right]$$
$$- Xe^{-rT}\left[M(d_2, \eta f_4; \eta\rho) - \left(\frac{H}{S}\right)^{2\mu} M(f_2, \eta f_6; \eta\rho)\right] \quad (17.1.1)$$

Where $\eta = 1$ for a down and out call and $\eta = -1$ for an up and out call.

$M(x_1, x_2; \rho) \equiv \int_{-\infty}^{x_1} \int_{-\infty}^{x_2} \frac{1}{2\pi\sqrt{1-\rho^2}} e^{\left\{\frac{1}{2(1-\rho^2)}\left[z_1^2 - 2\rho z_1 z_2 + z_2^2\right]\right\}} dz_2 dz_1$, is the bivariate normal distribution function.

The remaining parameters are:

$$d_1 = \frac{\log(S/X) + (b + \sigma^2/2)T}{\sigma\sqrt{T}}, d_2 = d_1 - \sigma\sqrt{T}.$$

$$f_1 = \frac{log(S/X) + 2\log(H/S) + (b + \sigma^2/2)T}{\sigma\sqrt{T}}$$

$$f_2 = f_1 - \sigma\sqrt{T}$$

$$f_3 = \frac{\log(S/H) + (b + \sigma^2/2)t}{\sigma\sqrt{t}}, f_4 = f_3 - \sigma\sqrt{t}.$$

$$f_5 = f_3 + \frac{2\log(H/S)}{\sigma\sqrt{t}}, f_6 = f_5 - \sigma\sqrt{t}.$$

$$\mu = \frac{b - \sigma^2/2}{\sigma^2}, \rho = \sqrt{\frac{t}{T}}$$

The partial time in option can be found from the in out parity relationship, where the up and in call value is derived from a plain vanilla call minus the up and out call value so that:

$$C_{ui} = C_{bs} - C_{uo} \text{ And } C_{di} = C_{bs} - C_{do}.$$

Partial time option valuations with a range of termination times for monitoring are shown in Table 17.1. The table shows a range of different monitoring period end times. The valuations are for down-and-in (Din), up-and-out (Uout), down-and-out (Dout) and up-and-in (Uin) options.

For forward-starting barrier options there are two valuation categories defined as B1 and B2. The B1 category covers options that are knocked-out by simply hitting the barrier. The direction of hit is immaterial. B2 type options have a down-and-out and an up-and-out variant. For the former, the option is knocked-out once the barrier is hit from above. For the latter case, the option is knocked out if the barrier is hit from below.

There are two cases for type B1. The situation where the strike price is above the barrier level and the path dependency is relatively straightforward, this is shown in formula 17.1.2. The other situation is where the strike price is below the barrier level. The path dependency is more complex and takes into account the increased number of probabilistic routes for asset price movement in relation to the strike price.

For the case where X > H;

$$C_{oB1} = Se^{(b-r)T}\left[M(d_1, f_3; \rho) - \left(\tfrac{H}{S}\right)^{2(\mu+1)} M(f_1, -f_5; -\rho)\right]$$
$$-Xe^{-rT}\left[M(d_2, f_4, \rho) - \left(\tfrac{H}{s}\right)^{2\mu} M(f_2, -f_6, -\rho)\right] \qquad (17.1.2)$$

TABLE 17.1. Partial Time barrier option valuations

Asset Price	Strike Price	Call Type	Time 0.0	Time 0.20	Time 0.40	Time 0.60	Time 0.80	Time 1.00
		A	Valuation	Valuation	Valuation	Valuation	Valuation	Valuation
		Din	10.1811	10.1811	10.1811	10.1811	10.1811	10.1811
85.0	90.0	Uout	10.1811	6.7621	3.4386	1.6806	0.6898	0.1165
		Dout	0.000	0.000	0.000	0.000	0.000	0.000
		Uin	0.000	3.419	6.7425	8.5006	9.4913	10.0646
		Din	11.6574	11.6574	11.6574	11.6574	11.6574	11.6574
95.0	100.0	Uout	11.6574	1.9029	0.6863	0.2504	0.0638	0.000
		Dout	0.000	0.000	0.000	0.000	0.000	0.000
		Uin	0.000	9.7545	10.971	11.407	11.5935	11.6574
		Din	0.000	5.9882	6.628	6.8292	6.8964	6.9074
105.0	110.0	Uout	0.000	0.000	0.000	0.000	0.000	0.000
		Dout	13.1376	7.1495	6.5096	6.3084	6.2412	6.2303
		Uin	13.1376	13.1376	13.1376	13.1376	13.1376	13.1376
		Din	17.1612	17.1612	17.1612	17.1612	17.1612	17.1612
95.0	90.0	Uout	17.1612	3.1887	1.3313	0.6028	0.2376	0.0392
		Dout	0.000	0.000	0.000	0.000	0.000	0.000
		Uin	0.000	13.9725	15.8299	16.5583	16.9235	17.122
		Din	0.000	8.9739	10.0181	10.3911	10.5506	10.6051
105.0	100.0	Uout	0.000	0.000	0.000	0.000	0.000	0.000
		Dout	18.6416	9.6677	8.6234	8.2505	8.091	8.0365
		Uin	18.6416	18.6416	18.6416	18.6416	18.6416	18.6416
		Din	0.000	1.6056	2.6045	2.9957	3.1378	3.162
115.0	110.0	Uout	0.000	0.000	0.000	0.000	0.000	0.000
		Dout	20.1249	18.5193	17.5204	17.1292	16.9871	16.9629
		Uin	20.1249	20.1249	20.1249	20.1249	20.1249	20.1249

$H = 100.0$, $r = b = 0.10$, $\sigma = 0.25$

When $X < H$;

$$\begin{aligned}
Cout_{B1} = {} & Se^{(b-r)T}\left[M(-g_1, -f_3; \rho) - \left(\frac{H}{S}\right)^{2(\mu+1)} M(-g_1, f_5; -\rho)\right] \\
& -Xe^{-rt}\left[M(-g2, -f_4; \rho) - \left(\frac{H}{S}\right)^{2\mu} M(-g_4, f_6; -\rho)\right] \\
& -Se^{(b-r)T}\left[M(-d_1, -f_3; \rho) - \left(\frac{H}{S}\right)^{2(\mu+1)} M(-f_1, f_5; -\rho)\right] \\
& +Xe^{-rt}\left[M(-d_2, -f_4; \rho) - \left(\frac{H}{S}\right)^{2\mu} M(-f_2, f_6; -\rho)\right] \\
& +Se^{(b-r)T}\left[M(g_1, f_3; \rho) - \left(\frac{H}{S}\right)^{2(\mu+1)} M(g_3, -f_5; -\rho)\right] \\
& -Xe^{-rt}\left[M(g_2, f_4; \rho) - \left(\frac{H}{S}\right)^{2\mu} M(g_4, -f_6; -\rho)\right] \qquad (17.1.3)
\end{aligned}$$

Where,

$$g_1 = \frac{\log(S/H) = (b+\sigma^2/2)T}{\sigma\sqrt{T}}, \; g_2 = g_1 - \sigma\sqrt{T}.$$

$$g_3 = g_1 + \frac{2\log(H/S)}{\sigma\sqrt{T}}, \; g_4 = g_3 - \sigma\sqrt{T}.$$

Table 17.2 shows valuations for a range of B type options. The times refer to the start of the monitoring period from option valuation date. The assumption is that the termination date is 1.0.

The valuation of a down-and-out type B2 call is with $X < H$ given by:

$$C_{doB2} = Se^{(b-r)T}\left[M(g_1, f_3; \rho) - \left(\frac{H}{S}\right)^{2(\mu+1)} M(g_3, -f_5; -\rho)\right]$$
$$-xe^{-rT}\left[M(g_2, f_4; \rho) - \left(\frac{H}{S}\right)^{2\mu} M(g_4, -f_6; -\rho)\right] \quad (17.1.4)$$

TABLE 17.2. Forward-start option valuations

Asset Price	Strike Price	Call Type	Time 0.0	Time 0.20	Time 0.40	Time 0.60	Time 0.80	Time 1.00
		B1,B2	Valuation	Valuation	Valuation	Valuation	Valuation	Valuation
85.0	90.0	B1 Call	0.1165	1.0158	3.1288	5.2557	7.2793	10.1811
		B2 D	0.000	0.000	0.000	0.000	0.000	0.000
		B2 U	20.299	17.1764	13.188	9.9001	6.6323	0.7234
95.0	100.0	B1 Call	0.000	4.155	6.9477	8.9624	10.5002	11.6574
		B2 D	0.000	0.000	0.000	0.000	0.000	0.000
		B2 U	0.000	4.155	6.9477	8.9624	10.5002	11.6574
105.0	110.0	B1 Call	6.2303	9.1486	10.9537	12.1366	12.8809	13.1376
		B2 D	6.198	9.097	10.8824	12.0395	12.7359	12.4807
		B2 U	0.000	0.000	0.000	0.000	0.000	0.000
95.0	90.0	B1 Call	0.0392	5.2457	8.8749	11.6474	14.0027	17.1612
		B2 D	0.000	0.000	0.000	0.000	0.000	0.000
		B2 U	27.4532	12.8208	10.6464	8.7952	6.3141	0.7278
105.0	100.0	B1 Call	8.0365	11.8897	14.4092	16.2212	17.6027	18.6416
		B2 D	8.0365	11.8897	14.4092	16.2212	17.6027	18.6416
		B2 U	0.000	0.000	0.000	0.000	0.000	0.000
115.0	110.0	B1 Call	16.9629	17.5607	18.5678	19.3755	19.927	20.1249
		B2 D	16.8898	17.4833	18.4793	19.2692	19.7849	19.5813
		B2 U	0.000	0.000	0.000	0.000	0.000	0.000

H = 100.0, r = b = 0.10, $\sigma = 0.25$

For $X < H$ the up-and-out call is given by:

$$\begin{aligned}C_{uoB2} = {} & Se^{(b-r)T}\left[M(-g_1,-f_3;\rho)-\left(\frac{H}{S}\right)^{2(\mu+1)}M(-g_3,-f_5;-\rho)\right]\\ & -Xe^{-rT}\left[M(-g_2,-f_4;\rho)-\left(\frac{H}{S}\right)^{2\mu}M(-g_4,f_6;-\rho)\right]\\ & -Se^{(b-r)T}\left[M(-d_1,-f_3;\rho)-\left(\frac{H}{S}\right)^{2(\mu+1)}M(f_5,-f_1;-\rho)\right]\\ & +Xe^{-rT}\left[M(-d_2,-f_4;\rho)-\left(\frac{H}{S}\right)^{2\mu}M(f_6,-f_2;-\rho)\right] \qquad (17.1.5)\end{aligned}$$

The code for partial time single asset barriers is shown in Listing 17.1.

```
package FinApps;
import static java.lang.Math.*;
import BaseStats.Probnorm;
import BaseStats.Bivnorm;
public class Sbarrierptop {

public Sbarrierptop(double rate,double q,double volatility,
                   int period ) {
   crate=q;
   sigma=volatility;
   r=rate;
   b=crate==0.0?0.0:(b=crate!=r?(r-crate):r);
   tau=period==1?(1.0/(24*365)):period==2?(1.0/(365.0)):
       period==3?(1.0/(52.0)):
period==4?(1.0/(12.0)):0.0;

 }
 double tau;
 double crate;
 double d1;
 double d2;
 double sigma;
 double h;
 private static double b;
 private static double t1;
 private static double t2;
 private static double r;
 private static double f1;
 private static double f2;
 private static double f3;
 private static double f4;
 private static double f5;
 private static double f6;
 private static double mu;
 private static double rho;
 private static double g1;
```

```
private static double g2;
private static double g3;
private static double g4;
public double ptsoUocall(double s, double x, double barrier,
                          double time1, double time2)partial
                        {//partial time start out up and out callA
  barrier= x>s? (barrier*exp(sqrt(tau)*sigma*0.5826)):x<s?
               (barrier*exp(-sqrt(tau)
*sigma*0.5826)):barrier;// do any h adjustments now
  if(s>=barrier)
    return 0.0;
  params(s,x,barrier,time1,time2);
  double term1=(s*exp((b-r)*t2)*(M(d1,-f3,-rho)-pow((h/s),
               (2.0*(mu+1.0)))*M(f1,-f5,-rho)));
  double term2=(x*exp(-r*t2)*(M(d2,-f4,-rho)-pow((h/s),
               (2.0*mu))*M(f2,-f6,-rho)));

  return term1-term2;

}
public double ptsoDocall(double s, double x, double barrier,
                          double time1, double time2)partial
                          {// time start out down and out callA
  barrier= x>s? (barrier*exp(sqrt(tau)*sigma*0.5826)):x<s?
               (barrier*exp(-sqrt(tau)*sigma*0.5826)):barrier;
  if(s<=barrier)
    return 0.0;

  params(s,x,barrier,time1,time2);
  double term1=(s*exp((b-r)*t2)*(M(d1,f3,rho)-pow((h/s),
               (2.0*(mu+1.0)))*M(f1,f5,rho)));
  double term2=(x*exp(-r*t2)*(M(d2,f4,rho)-pow((h/s),
               (2.0*mu))*M(f2,f6,rho)));

  return term1-term2;

}
public double ptsoDicall(double s, double x, double barrier,
                          double time1, double time2)partial
                          {// time start out down and in callA
  barrier= x>s? (barrier*exp(sqrt(tau)*sigma*0.5826)):x<s?
               (barrier*exp(-sqrt(tau)*sigma*0.5826)):barrier;

  Blackscholecp b=new Blackscholecp(crate);
  double downout= ptsoDocall(s,x,barrier,time1,time2);
  b.bscholEprice(s,x,sigma,t2,r);
  re}turn (b.getCalle()-downout);
public double ptsoUicall(double s, double x,
                          double barrier,double time1, double time2)
                          {//partial time start out up and in callA
  barrier= x>s? (barrier*exp(sqrt(tau)*sigma*0.5826)):x<s?
               (barrier*exp(-sqrt(tau)*sigma*0.5826)):barrier;
  Blackscholecp b=new Blackscholecp(crate);
  double upout= ptsoUocall(s,x,barrier,time1,time2);
  b.bscholEprice(s,x,sigma,t2,r);
  return (b.getCalle()-upout);
```

```
}
public double pteoCall(double s, double x, double barrier,
                       double time1, double time2)
                       {//partial time end out call cob1
  params(s,x,barrier,time1,time2);
  double term1=(s*exp((b-r))*t2)*(M(d1,f3,rho)-(pow((h/s),
            (2.0*(mu+1.0))))*M(f1,-f5,-rho));
  double term2=(x*exp(-r*t2))*(M(d2,f4,rho)-(pow((h/s),
            (2.0*mu)))*M(f2,-f6,-rho));
  double term3=(s*exp((b-r)*t2)*(M(-g1,-f3,rho)-pow((h/s),
            (2.0*(mu+1.0)))*M(-g3,f5,-rho)));
  double term4=(x*exp(-r*t2)*(M(-g2,-f4,rho)-pow((h/s),
            (2.0*mu))*M(-g4,f6,-rho)));
  double term5=(s*exp((b-r)*t2)*(M(-d1,-f3,rho)-pow((h/s),
            (2.0*(mu+1.0)))*M(-f1,f5,-rho)));
  double term6=(x*exp(-r*t2)*(M(-d2,-f4,rho)-pow((h/s),
            (2.0*mu))*M(-f2,f6,-rho)));
  double term7=(s*exp((b-r)*t2)*(M(g1,f3,rho)-pow((h/s),
            (2.0*(mu+1.0)))*M(g3,-f5,-rho)));
  double term8=(x*exp(-r*t2)*(M(g2,f4,rho)-pow((h/s),
            (2.0*mu))*M(g4,-f6,-rho)));
  return x>h?(term1-term2):(term3-term4-term5+term6+term7-term8);
}
public double pEuocall(double s, double x, double barrier,
                        double time1, double time2)
                        {//partial end barrier x>h B2 up and out
  if(s>=barrier)
    return 0.0;
  if(x>=barrier)
    return pteoCall(s,x,barrier,time1,time2);
  barrier= x>s? (barrier*exp(sqrt(tau)*sigma*0.5826)):x<s?
            (barrier*exp(-sqrt(tau)*sigma*0.5826)):barrier;
  params(s,x,barrier,time1,time2);
  double term1=(s*exp((b-r)*t2)*(M(-g1,-f3,rho)-pow((h/s),
            (2.0*(mu+1.0)))*M(-g3,-f5,-rho)));
  double term2=(x*exp(-r*t2)*(M(-g2,-f4,rho)-pow((h/s),
            (2.0*mu))*M(-g4,f6,-rho)));
  double term3=(s*exp((b-r)*t2)*(M(-d1,-f3,rho)-pow((h/s),
            (2.0*(mu+1.0)))*M(f5,-f1,-rho)));
  double term4=(x*exp(-r*t2)*(M(-d2,-f4,rho)-pow((h/s),
            (2.0*mu))*M(f6,-f2,-rho)));
  return (term1-term2-term3+term4);
}
public double pEdocall(double s, double x, double barrier,
                       double time1, double time2)Assumes
                        {// x<h partial end barrierB2 down and out
  if(s<=barrier)
    return 0.0;
  if(x<=barrier)
    return pteoCall(s,x,barrier,time1,time2);
  barrier= x>s? (barrier*exp(sqrt(tau)*sigma*0.5826)):x<s?
            (barrier*exp(-sqrt(tau)*sigma*0.5826)):barrier;
```

```
    params(s,x,barrier,time1,time2);
    double term1=(s*exp((b-r)*t2)*(M(g1,f3,rho)-pow((h/s),
               (2.0*(mu+1.0)))*M(g3,-f5,-rho)));
    double term2=(x*exp(-r*t2)*(M(g2,f4,rho)-pow((h/s),
               (2.0*mu))*M(g4,-f6,-rho)));
    return (term1-term2);
  }
  private void params(double s,double x,double barrier,
                    double time1,double time2) {
    t1=time1;
    t2=time2;
    h=barrier;
    t1=t1==0.0?(1e-4):t1==1.0?(1.0-1e-12):t1;// restrict outer
                    extermes for approximations using M(a,b:p)
    double sigmas=((sigma*sigma)*0.5);
    d1=(log(s/x)+(b+sigmas)*t2)/(sigma*sqrt(t2));
    d2=d1-sigma*sqrt(t2);
    System.out.println("d1=="+d1+" d2=="+d2);
    f1=(log(s/x)+2.0*log(h/s)+(b+sigmas)*t2)/(sigma*sqrt(t2));
    f2=f1-sigma*sqrt(t2);
    f3=(log(s/h)+(b+sigmas)*t1)/(sigma*sqrt(t1));
    f4=f3-sigma*sqrt(t1);
    f5=f3+2.0*log(h/s)/(sigma*sqrt(t1));
    f6=f5-sigma*sqrt(t1);
    System.out.println("f1=="+f1+" f2=="+f2+" f3=="+f3+"
                       f4=="+f4+" f5=="+f5+" f6=="+f6);
    mu=(b-sigmas)/(sigma*sigma);
    rho=sqrt(t1/t2);
    g1=(log(s/h)+(b+sigmas)*t2)/(sigma*sqrt(t2));
    g2=g1-sigma*sqrt(t2);
    g3=g1+2.0*log(h/s)/(sigma*sqrt(t2));
    g4=g3-sigma*sqrt(t2);
  }
  public double putDouta(double s,double x,double barrier,
                       double time1,double time2) {
    if(s<=barrier)
      return 0.0;
    double callvalue=ptsoDocall(s,x,barrier,time1,time2);
    return (callvalue-puT.P_par.Argsval(s,x,h,1));
  }
  public double putUouta(double s, double x, double barrier,
                       double time1, double time2) {
    if(s>=barrier)
      return 0.0;
    double callvalue=ptsoUocall(s,x,barrier,time1,time2);
    return (callvalue-puT.P_par.Argsval(s,x,h,2));
  }
  public double putOutb1(double s, double x, double barrier,
                       double time1,double time2) {
    double callvalue=pteoCall(s,x,barrier,time1,time2);
    return (callvalue-puT.P_par.Argsval(s,x,h,3));
  }
  public double putDob2(double s, double x, double barrier,
```

```
                        double time1,double time2) {
  if(s<=barrier)
    return 0.0;
  double callvalue= pEdocall(s,x,barrier,time1,time2);
  return (callvalue-puT.P_par.Argsval(s,x,h,4));
}
public double putUob2(double s, double x, double barrier,
                        double time1,double time2) {
  if(s>=barrier)
    return 0.0;
  double callvalue=pEuocall(s,x,barrier,time1,time2);
  return (callvalue-puT.P_par.Argsval(s,x,h,5));
}
private static double M(double a, double b, double r) {
  return Bivnorm.bivar_params.evalArgs(a,b,r);
}
private static double N(double x) {
  Probnorm p=new Probnorm();
  double ret=x>(6.95)?1.0:x<(-6.95)?0.0:p.ncDisfnc(x);//restrict
                        the range of cdf values to stable values
  return ret;
}
private enum puT {
P_par{
  public double Argsval(double s, double x, double h,int tp) {
    int sw=tp;
    double puterm=0.0;
    double z1=N(f4)-pow((h/s),(2.0*mu))*N(f6);
    double z2=N(-f4)-pow((h/s),(2.0*mu))*N(-f6);
    double z3=M(g2,f4,rho)-pow((h/s),(2.0*mu))*M(g4,-f6,-rho);
    double z4=M(-g2,-f4,rho)-pow((h/s),(2.0*mu))*M(-g4,f6,-rho);
    double z5=N(f3)-pow((h/s),(2.0*(mu+1.0)))*N(f5);
    double z6=N(-f3)-pow((h/s),(2.0*(mu+1.0)))*N(-f5);
    double z7=M(g1,f3,rho)-pow((h/s),(2.0*(mu+1.0)))*M(g3,-f5,-rho);
    double z8=M(-g1,-f3,rho)-pow((h/s),(2.0*(mu+1)))*M(-g3,f5,-rho);
    switch(sw) {
       case 1: puterm=DoA.valuesz(z1,z5,s,x);//down out type A
       break;
       case 2: puterm=UoA.valuesz(z2,z6,s,x);//up out type A
       break;
       case 3: puterm=oB1.valuesz2(z3,z4,z7,z8,s,x);//out type B1
       break;
       case 4: puterm=Dob2.valuesz(z3,z7,s,x);//down and out type B2
       break;
       case 5: puterm=Uob2.valuesz(z4,z8,s,x);//up and out type B2
       break;
    }

    return puterm;
  }
},

DoA{
```

```
    public double valuesz(double za, double zb,double s,double x) {
      double val=s*exp((b-r)*t2)*zb+x*exp(-r*t2)*za;

      return val;
    }
  },
  UoA{
    public double valuesz(double za, double zb,double s,double x) {
      double val=s*exp((b-r)*t2)*zb+x*exp(-r*t2)*za;

      return val;
    }
  },
  oB1{
    public double valuesz2(double za, double zb,double zc, double zd,
                          double s,double x) {
      double val=s*exp((b-r)*t2)*zd+x*exp(-r*t2)
                *zb-s*exp((b-r)*t2)*zc+x*exp(-r*t2)*za;

      return val;
    }
  },
  Dob2{
    public double valuesz(double za, double zb,double s,double x) {
      double val=s*exp((b-r)*t2)*zb+x*exp(-r*t2)*za;

      return val;

    }
  },

  Uob2{
    public double valuesz(double za, double zb,double s,double x) {
      double val=s*exp((b-r)*t2)*zb+x*exp(-r*t2)*za;
      return val;

    }
  };
  puT() {

  }
  public double valuesz(double a, double b,double c, double d) {

    throw new UnsupportedOperationException("Not yet implemented");
  }
  public double valuesz2(double za, double zb,double zc, double zd,
                        double s,double x) {
    throw new UnsupportedOperationException("Not yet implemented");
  }
  public double Argsval(double a, double b, double p,int tp) {

    throw new UnsupportedOperationException("Not yet implemented");
  }
}
```

LISTING 17.1. Class Sbarrierptop

17.2. Two Asset Options

In the two asset option, one of the underlying determines the degree to which the option is in or out of the money. The other underlying is associated with hitting the barrier. The valuation formulae are due to Heynen & Kat 1994.

The general formula for valuing a knock-out two asset barrier has parameters:

$$d_1 = \frac{\log(S_1/X) + (\mu_1 + \sigma_1^2)T}{\sigma_1\sqrt{T}}, \; d_2 = d_1 - \sigma_1\sqrt{T}.$$

$$d_3 = d_1 + \frac{2\rho\log(H/S_2)}{\sigma_2\sqrt{T}}, \; d_4 = d_2 + \frac{2\rho\log(\frac{H}{S_2})}{\sigma_2\sqrt{T}}.$$

$$f_1 = \frac{\log(H/S_2) - (\mu_2 + \rho\sigma_1\sigma_2)T}{\sigma_2\sqrt{T}}, \; f_2 = f_1 + \rho\sigma_1\sqrt{T}.$$

$$f_3 = f_1 - \frac{2\log(H/S_2)}{\sigma_2\sqrt{T}}, \; f_4 = f_2 - \frac{2\log(H/S_2)}{\sigma_2\sqrt{T}}.$$

$\mu_1 = b_1 - \sigma_1^2, \mu_2 = b_2 - \sigma_2^2/s.$

The knock-out valuation is given by:

$$v = \eta S_1 e^{(b_1-r)T}\left(M(\eta d_1, \varphi f_1; -\eta\varphi\rho) - e\left[\frac{2(\mu_2 + \rho\sigma_1\sigma_2)\log(\frac{H}{S_2})}{\sigma_2^2}\right] M(\eta d_3, \varphi f_3; -\eta\varphi\rho)\right)$$
$$-\eta X e^{-rT}\left(M(\eta d_2, \varphi f_2; -\eta\varphi\rho) - e\left[\frac{2\mu_2\log(\frac{H}{S_2})}{\sigma_2^2}\right] M(\eta d_4, \varphi f_4; -\eta\varphi\rho)\right) \quad (17.2.1)$$

There are four two-asset out options.

Call down-and-out. Payoff is:

$$\max[S_1 - X; 0]\,\forall t \le T;\, S_2(t) > H: else, 0. \quad \eta = 1, \varphi = -1.$$

Call up-and-out. Payoff is:

$$\max[S_1 - X; 0]\,\forall t \le T;\, S_2(t) < H: else, 0. \quad \eta = 1, \varphi = 1.$$

Put down-and-out. Payoff is:

$$\max[X - S_1; 0]\,\forall t \le T;\, S_2(t) > H: else, 0. \quad \eta = -1, \varphi = -1.$$

Put up-and-out. Payoff is:

$$\max[X - S_1; 0]\,\forall t \le T;\, S_2(t) < H: else, 0. \quad \eta = -1, \varphi = 1.$$

TABLE 17.3. Call down-and-out; call up-and-out two-asset barrier option valuations

X	H	TYPE	ρ = −0.5	ρ = −0.25	ρ = 0.0	ρ = 0.25	ρ = 0.5
75.0	85.0	C_{do}	5.6883	6.5722	7.4555	8.3424	9.2373
	95.0	C_{uo}	5.7702	5.0044	4.2621	3.5414	2.8388
80.0	85.0	C_{do}	4.0964	4.9246	5.7511	6.5825	7.4253
	95.0	C_{uo}	4.7203	3.9874	3.2877	2.6174	1.972
85.0	85.0	C_{do}	2.7561	3.4898	4.2237	4.9638	5.716
	95.0	C_{uo}	3.7131	3.0425	2.4146	1.8245	1.2695
90.0	85.0	C_{do}	1.7232	2.3288	2.9395	3.5575	4.1843
	95.0	C_{uo}	2.7872	2.2102	1.6805	1.1947	0.7538
95.0	85.0	C_{do}	0.9992	1.4611	1.9344	2.4151	2.8986
	95.0	C_{uo}	1.9848	1.5229	1.1058	0.7341	0.4126

$S_1 = 90.0, S_2 = 90.0, \sigma_1 = 0.20, \sigma_2 = 0.20; r = 0.10, b_1 = 0.10, b_2 = 0.10 : T = 0.5$

Valuation examples for down-and-out and up-and-out call options are given in Table 17.4. The valuation examples in Table 17.5 are for down-and-out and up-and-out put options. The examples are for a range of correlation coefficients. Table 17.3 shows that for a more highly correlated asset pair the value of the down-and-out call option increases. The value of the up-and-out option however decreases.

Table 17.4 shows that the more highly correlated the asset pairs are, the down-and-out put value decreases. Whereas, the up-and-out value increases.

'In' barrier options are constructed from the portfolio relation:

$$C_{di} = \prod(C_{bs} - C_{do})$$

TABLE 17.4. Put down-and-out; put up-and-out two-asset barrier option valuations

X	H	TYPE	ρ = −0.5	ρ = −0.25	ρ = 0.0	ρ = 0.25	ρ = 0.5
75.0	85.0	P_{do}	0.1698	0.1336	0.0937	0.0552	0.0233
	95.0	P_{uo}	0.0072	0.0251	0.0536	0.0902	0.1309
80.0	85.0	P_{do}	0.4546	0.3626	0.2659	0.1719	0.0879
	95.0	P_{uo}	0.0301	0.0809	0.152	0.2389	0.3369
85.0	85.0	P_{do}	0.9909	0.8045	0.6152	0.4299	0.2552
	95.0	P_{uo}	0.0956	0.2089	0.3517	0.5189	0.7072
90.0	85.0	P_{do}	1.8346	1.5201	1.2076	0.9002	0.6001
	95.0	P_{uo}	0.2426	0.4494	0.6904	0.9619	1.2643
95.0	85.0	P_{do}	2.9873	2.529	2.0791	1.6344	1.1911
	95.0	P_{uo}	0.513	0.8349	1.1886	1.5741	1.9959

$S_1 = 90.0, S_2 = 90.0, \sigma_1 = 0.20, \sigma_2 = 0.20; r = 0.10, b_1 = 0.10, b_2 = 0.10 : T = 0.5$

TABLE 17.5. Call down-and-in; call up-and-in two-asset barrier option valuations

X	H	TYPE	ρ = −0.5	ρ = −0.25	ρ = 0.0	ρ = 0.25	ρ = 0.5
75.0	85.0	C_{di}	13.207	12.3231	11.4398	10.5529	9.658
	95.0	C_{ui}	13.1251	13.8909	14.6332	15.3539	16.0565
80.0	85.0	C_{di}	10.4792	9.6511	8.8245	7.9932	7.1503
	95.0	C_{ui}	9.8553	10.5882	11.2879	11.9582	12.6036
85.0	85.0	C_{di}	7.9485	7.2148	6.4809	5.7408	4.9886
	95.0	C_{ui}	6.9916	7.6621	8.290	8.8801	9.4351
90.0	85.0	C_{di}	5.7268	5.1212	4.5105	3.8925	3.2658
	95.0	C_{ui}	4.6629	5.2398	5.7696	6.2553	6.6962
95.0	85.0	C_{di}	3.9033	3.4414	2.9681	2.4874	2.0039
	95.0	C_{ui}	2.9178	3.3796	3.7967	4.1684	4.4899

$S_1 = 90.0, S_2 = 90.0, \sigma_1 = 0.20, \sigma_2 = 0.20; r = 0.10, b_1 = 0.10, b_2 = 0.10 : T = 0.5$

Thus for a portfolio containing an in and out option the payoff is guaranteed to be that of a standard vanilla call. If a down option hits the barrier at time t_n. The 'in' option immediately becomes a call option.

If the 'in' option fails to drop down to become 'in', the down-and-out call remains a vanilla call option. Table 17.5 shows valuation examples for call down-and-in and up-and-in options. For a down-and-in option, the higher correlation between assets, the lower is the value. For the up-and-in option, a higher correlation gives a higher value to the option.

The relationship between valuation and the relative movement of asset S_2 to the barrier level is shown in Figure 17.1. As the barrier is approached, values increase, with in-the-money and out-of-the-money values showing relative divergence. The correlation coefficient between assets is 0. The parameters are:

$$S_1 = 90.0, S_2 = \{80.0., 90.0\}, \sigma_1 = 0.20, \sigma_2 = 0.20; r = 0.10, b_1 = 0.10,$$
$$b_2 = 0.10 : T = 0.5.$$

There are four two-asset in options.

Call down-and-in. Payoff is:

$$\max[S_1 - X; 0]\,\forall t \leq T; S_2(t) > H : else, 0$$

$C_{di} = \prod(C_{bs} - C_{do})$.

Call up-and-in. Payoff is:

$$\max[S_1 - X; 0]\,\forall t \leq T; S_2(t) < H : else, 0$$

$C_{ui} = \prod(C_{bs} - C_{uo})$.

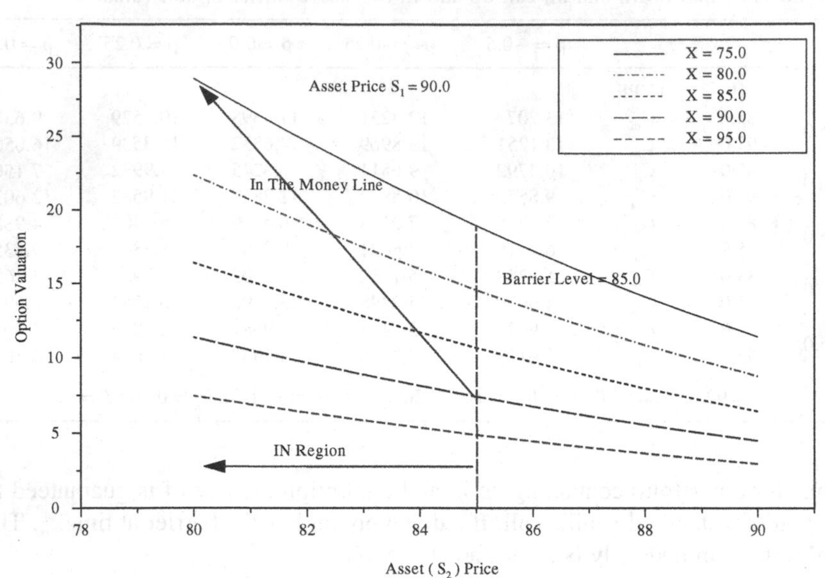

FIGURE 17.1. Call Dawn & in valuations.

Put down-and-in. Payoff is:

$$\max\left[X - S_1; 0\right] \forall t \leq T; S_2(t) > H : else, 0$$

$P_{di} = \prod(P_{bs} - P_{do})$.
Put up-and-in. Payoff is:

$$\max\left[X - S_1; 0\right] \forall t \leq T; S_2(t) < H : else, 0$$

$P_{ui} = \prod(P_{bs} - P_{uo})$.

Table 17.6 gives example valuations for put down-and-in & up-and-in options. In this case the value of down-and-in options increase with the positive correlation between both assets.

The graph of figure 17.2 shows the valuation curves for five different strike prices with $\rho = 0.0$ and parameters:

$$S_1 = 90.0, S_2 = \{80.0., 90.0\}, \sigma_1 = 0.20, \sigma_2 = 0.20; r = 0.10, b_1 = 0.10,$$
$$b_2 = 0.10 : T = 0.5.$$

Two asset barrier options are priced using code shown in Listing 17.2.

TABLE 17.6. Put down-and-in; put up-and-in two-asset barrier option valuations

X	H	TYPE	$\rho=-0.5$	$\rho=-0.25$	$\rho=0.0$	$\rho=0.25$	$\rho=0.5$
75.0	85.0	P_{di}	0.0677	0.1039	0.1438	0.1823	0.2142
	95.0	P_{ui}	0.2303	0.2125	0.1839	0.1474	0.1066
80.0	85.0	P_{di}	0.2194	0.3114	0.4081	0.5021	0.5861
	95.0	P_{ui}	0.6439	0.5931	0.522	0.435	0.3371
85.0	85.0	P_{di}	0.5682	0.7546	0.9439	1.1292	1.3039
	95.0	P_{ui}	1.4635	1.3502	1.2074	1.0402	0.852
90.0	85.0	P_{di}	1.2261	1.5406	1.853	2.1605	2.4606
	95.0	P_{ui}	2.8181	2.6113	2.3703	2.0988	1.7963
95.0	85.0	P_{di}	2.282	2.7403	3.1902	3.6349	4.0782
	95.0	P_{ui}	4.7563	4.4345	4.0807	3.6952	3.2734

$S_1=90.0, S_2=90.0, \sigma_1=0.20, \sigma_2=0.20; r=0.10, b_1=0.10, b_2=0.10; T=0.5$

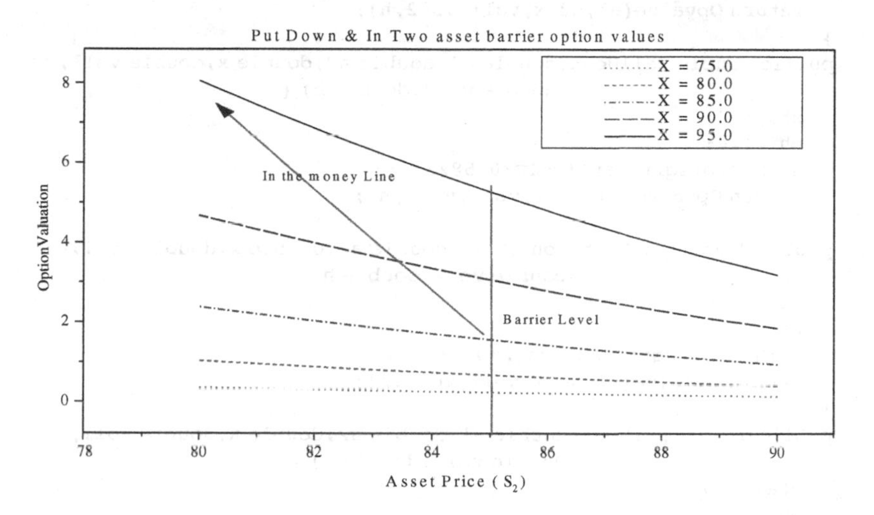

FIGURE 17.2. Put down & in valuations.

```
package FinApps;
import static java.lang.Math.*;
import BaseStats.Bivnorm;
public class Twobarrierop {
  public Twobarrierop(double q1,double q2, double rate, double time,
                       double rho,int period) {
    cratel=q1;
    crate2=q2;
    r=rate;
    t=time;
    tau=period==1?(1.0/(24*365)):period==2?(1.0/(365.0)):
          period==3?(1.0/(52.0)):period==4?(1.0/(12.0)):0.0;
    bl=cratel<0.0?0.0:(bl=cratel!=r?(r-cratel):r);
```

```
    b2=crate2<0.0?0.0:(b2=crate2!=r?(r-crate2):r);
    p=rho;
  }
  double b1;
  double b2;
  double tau;
  double crate1;
  double crate2;
  double t;
  double r;
  double p;
  double eta=0;
  double phi=0;
  public double callDout(double a1,double a2,double x,double vol1,
                          double vol2,double h) {
    eta=1.0;
    phi=-1.0;
    h=(h*exp(-sqrt(tau)*vol2*0.5826));
    return Opvalue(a1,a2,x,vol1,vol2,h);
  }
  public double callUout(double a1,double a2,double x,double vol1,
                          double vol2,double h) {
    eta=1.0;
    phi=1.0;
    h=(h*exp(sqrt(tau)*vol2*0.5826));
    return Opvalue(a1,a2,x,vol1,vol2,h);
  }
  public double putDout(double a1,double a2,double x,double vol1,
                          double vol2,double h) {
    eta=-1.0;
    phi=-1.0;
    h=(h*exp(-sqrt(tau)*vol2*0.5826));
    return Opvalue(a1,a2,x,vol1,vol2,h);
  }
  public double putUout(double a1,double a2,double x,double vol1,
                          double vol2,double h) {
    eta=-1.0;
    phi=1.0;
    h=(h*exp(sqrt(tau)*vol2*0.5826));
    return Opvalue(a1,a2,x,vol1,vol2,h);
  }
  public double callDin(double a1,double a2,double x,double vol1,
                          double vol2,double h) {
    eta=1.0;
    phi=-1.0;
    h=(h*exp(-sqrt(tau)*vol2*0.5826));
    Blackscholecp b=new Blackscholecp(b1);
    b.bscholEprice(a1,x,vol1,t,r);
    return (b.getCalle()-Opvalue(a1,a2,x,vol1,vol2,h));
  }
  public double callUin(double a1,double a2,double x,double vol1,
                          double vol2,double h) {
    eta=1.0;
    phi=1.0;
```

```
      h=(h*exp(sqrt(tau)*vol2*0.5826));
      Blackscholecp b=new Blackscholecp(b1);
      b.bscholEprice(a1,x,vol1,t,r);
      return (b.getCalle()- Opvalue(a1,a2,x,vol1,vol2,h));
   }
   public double putDin(double a1,double a2,double x,double vol1,
                          double vol2,double h) {
      eta=-1.0;
      phi=-1.0;
      h=(h*exp(-sqrt(tau)*vol2*0.5826));
      Blackscholecp b=new Blackscholecp(b1);
      b.bscholEprice(a1,x,vol1,t,r);
      return (b.getPute()- Opvalue(a1,a2,x,vol1,vol2,h));
   }
   public double putUin(double a1,double a2,double x,double vol1,
                          double vol2,double h) {
      eta=-1.0;
      phi=1.0;
      h=(h*exp(sqrt(tau)*vol2*0.5826));
      Blackscholecp b=new Blackscholecp(b1);
      b.bscholEprice(a1,x,vol1,t,r);
      return (b.getPute()- Opvalue(a1,a2,x,vol1,vol2,h));
   }


   private double M(double a, double b, double r) {

      return Bivnorm.bivar_params.evalArgs(a,b,r);
   }


   private double Opvalue(double s1,double s2,double x, double vol1,
                            double vol2,double h) {


      double sig1=(vol1*vol1);
      double sig2=(vol2*vol2);
      double mu1=(b1-(sig1*0.5));
      double mu2=(b2-(sig2*0.5));
      double d1=((log(s1/x)+(mu1+sig1)*t)/(vol1*sqrt(t)));
      double d2=(d1-vol1*sqrt(t));
      double d3=(d1+(2.0*p*log(h/s2))/(vol2*sqrt(t)));
      double d4=(d2+(2.0*p*log(h/s2))/(vol2*sqrt(t)));
      double f1=((log(h/s2)-(mu2+p*vol1*vol2)*t)/(vol2*sqrt(t)));
      double f2=(f1+(p*vol1*sqrt(t)));
      double f3=(f1-(2.0*log(h/s2)/(vol2*sqrt(t))));
      double f4=(f2-(2.0*log(h/s2)/(vol2*sqrt(t))));
      return (eta*s1*exp((b1-r)*t)*(M(eta*d1,phi*f1,-(eta*phi*p))-
exp(2.0*(mu2+p*vol1*vol2)*log(h/s2)/(sig2))
    *M(eta*d3,phi*f3,-(eta*phi*p)))
      -eta*x*exp(-r*t)*(M(eta*d2,phi*f2,-(eta*phi*p))
   -exp(2.0*mu2*log(h/s2)/(sig2))*M(eta*d4,phi*f4,-(eta*phi*p))));
   }
```

LISTING 17.2. Two Asset Barrier Option Valuation

17.3. Partial Time Two Asset Options

This option has a predefined monitoring period during which the asset (A_2) is observed for barrier hits.

The formula of Bermin (1996) can be used to value the option where monitoring takes place at some period in from the beginning of the option lifetime until the end. The option is knocked in or out if the asset price being monitored hits the barrier. If the barrier is hit prior to the monitoring period, the hit is ignored (option is 'protected'). The payoff from this option is dependent on the in-money/out-of-money status of a second asset (A_1) in relation to the option strike price.

The valuation of an out option is given by:

$$V = \eta A_1 e^{(b-r)T} \left\{ M\left(\eta d_1, \varphi_1; -\eta\varphi\rho\sqrt{t/T}\right) - e\left(\frac{2(\mu_2+\rho\sigma_1\sigma_2)\log(H/A_2)}{\sigma_2^2}\right) \right.$$
$$\left. \times M\left(\eta d_3, \varphi\varepsilon_3; -\eta\varphi\rho\sqrt{t/T}\right)\right\} - \eta X e^{-rT}\left\{ M\left(\eta d_2, \varphi\varepsilon_2; -\eta\varphi\rho\sqrt{t/T}\right) \right.$$
$$\left. -e\left(\frac{2\mu_2\log(H/A_2)}{\sigma_2^2}\right) M\left(\eta d_4, \varphi\varepsilon_4; -\eta\varphi\rho\sqrt{t/T}\right)\right\} \qquad (17.3.1)$$

Where,

$$d_1 = \frac{\log(A_1/X)+(\mu_1+\sigma_1^2)T}{\sigma_1\sqrt{T}}, \quad d_2 = d_2 - \sigma_1\sqrt{T}.$$

$$d_3 = d_1 + \frac{2\rho\log(H/A_2)}{\sigma_2\sqrt{T}}, \quad d_4 = d_2 + \frac{2\rho\log(H/A_2)}{\sigma_2\sqrt{T}},$$
$$\varepsilon_1 = \frac{\log(H/A_2)-(\mu_2+\rho\sigma_1\sigma_2)t}{\sigma_2\sqrt{t}}, \quad \varepsilon_2 = \varepsilon_1 + \rho\sigma\sqrt{t}$$

$$\varepsilon_3 = \varepsilon_1 = -\frac{2\log(H/A_2)}{\sigma_2\sqrt{t}}, \ \varepsilon_4 = \varepsilon_2 - \frac{2\log(H/A_2)}{\sigma_2\sqrt{t}}$$

$$\mu_1 = b_1 - \frac{\sigma_1^2}{2}, \ \mu_2 = b_2 - \frac{\sigma_2^2}{2}.$$

The binary variable η is 1 for a call and -1 for a put.; the binary variable φ is 1 for an up and out option and -1 for a down and out.

An in option valuation is given by:

1. For a call;

$$A_1 e^{(b-r)T} N(d_{a1}) - Xe^{-rT} N(d_{a2}) - \eta A_1 e^{(b-r)T} *$$
$$\left\{ M\left(\eta d_1, \varphi_1; -\eta\varphi\rho\sqrt{t/T}\right) - e\left(\frac{2(\mu_2 + \rho\sigma_1\sigma_2)\log(H/A_2)}{\sigma_2^2}\right)\right.$$
$$\left. \times M\left(\eta d_3, \varphi\varepsilon_3; -\eta\varphi\rho\sqrt{t/T}\right)\right\} - \eta Xe^{-rT}\left\{ M\left(\eta d_2, \varphi\varepsilon_2; -\eta\varphi\rho\sqrt{t/T}\right)\right.$$
$$\left. -e\left(\frac{2\mu_2\log(H/A_2)}{\sigma_2^2}\right) M\left(\eta d_4, \varphi\varepsilon_4; -\eta\varphi\rho\sqrt{t/T}\right)\right\} \tag{17.3.2}$$

2. For a put;

$$Xe^{-rt} N(-d_{a2}) - A_1 e^{(b-r)T} N(-d_{a1}) - \eta A_1 e^{(b-r)T} *$$
$$\left\{ M\left(\eta d_1, \varphi_1; -\eta\varphi\rho\sqrt{t/T}\right) - e\left(\frac{2(\mu_2 + \rho\sigma_1\sigma_2)\log(H/A_2)}{\sigma_2^2}\right)\right.$$
$$\left. \times M\left(\eta d_3, \varphi\varepsilon_3; -\eta\varphi\rho\sqrt{t/T}\right)\right\} - \eta Xe^{-rT}\left\{ M\left(\eta d_2, \varphi\varepsilon_2; -\eta\varphi\rho\sqrt{t/T}\right)\right.$$
$$\left. -e\left(\frac{2\mu_2\log(H/A_2)}{\sigma_2^2}\right) M\left(\eta d_4, \varphi\varepsilon_4; -\eta\varphi\rho\sqrt{t/T}\right)\right\} \tag{17.3.3}$$

Where,

$$d_{a1} = \frac{\log(A_1/X) + (b_1 + \sigma^2/2)T}{\sigma\sqrt{T}},$$
$$d_{a2} = \frac{\log(A_1/X) + (b - \sigma^2/2)T}{\sigma\sqrt{T}}, \equiv d_{a1} - \sigma\sqrt{T}.$$

$\eta = 1$ for a call.$\varphi = -1$ for a down and in. $\eta = 1$ for a down and in put and $\varphi = 1$ for an up and in put.

Formulae 17.3.2 and 17.3.3 are equivalent to taking a long Black-Scholes option position with a short partial time two asset barrier out option, with the same strike price and asset A_1. An in option can therefore be computed by a combination of a standard Black-Scholes call or put with a down or up, two asset partial barrier call or put.

Tables 17.7 and 17.8 show a range of values for partial time two asset barrier options. The tables show valuations for a range of barrier monitoring times. Table 17.7 gives valuations where the barrier is below the monitored asset price. Table 17.8 gives valuations where the barrier price is above the monitored asset price.

The implementation of partial-time two-asset barrier options is given in Listing 17.3.

```
package FinApps;
import static java.lang.Math.*;
import BaseStats.Probnorm;
import BaseStats.Bivnorm;

public class Ptwotbarrierop {

  public Ptwotbarrierop(double q1,double q2, double rate, double time,
                         double time2,double rho,int period) {

    cratel=q1;
    crate2=q2;
    r=rate;
    t=time;
    t2=time2;
    tau=period==1?(1.0/(24*365)):period==2?(1.0/(365.0)):period==3?
           (1.0/(52.0)):period==4?(1.0/(12.0)):0.0;
    b1=cratel==0.0?0.0:(b1=cratel!=r?(r-cratel):r);
    b2=crate2==0.0?0.0:(b2=crate2!=r?(r-crate2):r);
    p=rho;
  }
  double b1;
  double b2;
  double tau;
  double cratel;
  double crate2;
  double t;
  double t2;
  double r;
  double p;
  double eta=0;
  double phi=0;
  private double Opvalue(double a1,double a2,double x, double
sigma1,double sigma2,double h) {
    t=t==0.0?(1e-4):t==1.0?(1.0-1e-12):t;
           // restrict outer extermes for approximations using M(a,b:p)

    double sig1=(sigma1*sigma1);
    double sig2=(sigma2*sigma2);
    double mu1=(b1-(sig1*0.5));
    double mu2=(b2-(sig2*0.5));
    double d1=((log(a1/x)+(mu1+sig1)*t2)/(sigma1*sqrt(t2)));
    double d2=(d1-sigma1*sqrt(t2));
    double d3=(d1+(2.0*p*log(h/a2))/(sigma2*sqrt(t2)));
    double d4=(d2+(2.0*p*log(h/a2))/(sigma2*sqrt(t2)));
    double f1=((log(h/a2)-((mu2+p*sigma1*sigma2)*t))/
               (sigma2*sqrt(t)));
    double f2=(f1+(p*sigma1*sqrt(t)));
    double f3=(f1-(2.0*log(h/a2)/(sigma2*sqrt(t))));
    double f4=(f2-(2.0*log(h/a2)/(sigma2*sqrt(t))));
    return (eta*a1*exp((b1-r)*t2)*(M(eta*d1,phi*f1,-eta*phi*p
           *sqrt(t/t2))-exp(2.0*(mu2+p*sigma1*sigma2)*log(h/a2)/
           (sig2))*M(eta*d3,phi*f3,-eta*phi*p*sqrt(t/t2)))-eta*x
           *exp(-r*t2)*(M(eta*d2,phi*f2,-eta*phi*p*sqrt(t/t2))
           -exp(2.0*mu2*log(h/a2)/(sig2))*M(eta*d4,phi*f4,
           -eta*phi*p*sqrt(t/t2))));
```

```
}
private double M(double a, double b, double p) {
  return Bivnorm.bivar_params.evalArgs(a,b,p);
}

public double callDout(double a1,double a2,double x,double sigma1,
                         double sigma2,double h) {
  eta=1.0;
  phi=-1.0;
  h=(h*exp(-sqrt(tau)*sigma2*0.5826));
  return Opvalue(a1,a2,x,sigma1,sigma2,h);
}
public double callUout(double a1,double a2,double x,
                         double sigma1,double sigma2, double h) {
  eta=1.0;
  phi=1.0;
  h=(h*exp(sqrt(tau)*sigma2*0.5826));
  return Opvalue(a1,a2,x,sigma1,sigma2,h);
}
public double putDout(double a1,double a2,double x,double sigma1,
                        double sigma2, double h) {
  eta=-1.0;
  phi=-1.0;
  h=(h*exp(-sqrt(tau)*sigma2*0.5826));
  return Opvalue(a1,a2,x,sigma1,sigma2,h);
}
public double putUout(double a1,double a2,double x,double sigma1,
                        double sigma2, double h) {
  eta=-1.0;
  phi=1.0;
  h=(h*exp(sqrt(tau)*sigma2*0.5826));
  return Opvalue(a1,a2,x,sigma1,sigma2,h);
}
public double callDin(double a1,double a2,double x,double sigma1,
                        double sigma2, double h) {
  eta=1.0;
  phi=-1.0;
  h=(h*exp(-sqrt(tau)*sigma2*0.5826));
  Blackscholecp b=new Blackscholecp(crate1);
  b.bscholEprice(a1,x,sigma1,t,r);
  return (b.getCalle()-Opvalue(a1,a2,x,sigma1,sigma2,h));
}
  public double callUin(double a1,double a2,double x,double sigma1,
                         double sigma2, double h) {
  eta=1.0;
  phi=1.0;
  h=(h*exp(sqrt(tau)*sigma2*0.5826));
  Blackscholecp b=new Blackscholecp(b1);
  b.bscholEprice(a1,x,sigma1,t,r);
  return (b.getCalle()- Opvalue(a1,a2,x,sigma1,sigma2,h));
}
public double putDin(double a1,double a2,double x,double sigma1,
                       double sigma2, double h) {
  eta=-1.0;
```

```
    phi=-1.0;
    h=(h*exp(-sqrt(tau)*sigma2*0.5826));
    Blackscholecp b=new Blackscholecp(b1);
    b.bscholEprice(a1,x,sigma1,t,r);
    return (b.getPute()- Opvalue(a1,a2,x,sigma1,sigma2,h));
  }
  public double putUin(double a1,double a2,double x,double sigma1,
                      double sigma2, double h) {
    eta=-1.0;
    phi=1.0;
    h=(h*exp(sqrt(tau)*sigma2*0.5826));
    Blackscholecp b=new Blackscholecp(b1);
    b.bscholEprice(a1,x,sigma1,t,r);
    return (b.getPute()- Opvalue(a1,a2,x,sigma1,sigma2,h));
  }
```

LISTING 17.3. Partial-Time Two-Asset Barrier Option

17.4. Look Type Options

Look barrier options are monitored for hits of the barrier (B) during a part of the option lifetime. The option monitoring period begins at the start of the option lifetime (t) and ends at some time prior to expiry (T). A look barrier option has the characteristics of a partial time barrier option and a forward start fixed strike option. If the barrier is not hit during the monitoring period the option defaults to a fixed strike lookback.

TABLE 17.7. Partial-Time Two-Asset Barrier Options. Barrier < Strike

$\rho = -0.5$	T = 0.0	T = 1 month	T = 6 months	T = 1 year
Call down & out	17.0927	8.5751	3.2871	2.0861
Call up & in	23.5221	16.3336	16.6141	20.6072
Put up & in	1.625	1.164	2.1472	2.615
Put down & out	2.5764	1.6149	1.0771	0.990
$\rho = 0.0$				
Call down & out	17.0927	9.3944	4.6737	3.6397
Call up & in	20.2697	13.9004	14.7483	18.9084
Put up & in	2.3015	1.6416	2.451	2.8501
Put down & out	2.5764	1.416	0.7045	0.5486
$\rho = 0.5$				
Call down & out	17.0927	10.2018	6.1379	5.3482
Call up & in	17.3639	11.802	13.3067	17.7111
Put up & in	3.1666	2.3014	2.986	3.359
Put down & out	2.5764	1.2107	0.3743	0.1924

$A_1 = 105.0, A_2 = 100.0, X = 100.0, H = 95.0, \sigma_1 = 0.20, \sigma_2 = 0.25.T_2 = 1.0.$

TABLE 17.8. Partial-Time Two-Asset Barrier Options. Barrier > Strike

ρ = −0.5	T = 0.0	T = 1 month	T = 6 months	T = 1 year
Call up & out	13.2697	7.0019	3.4269	2.6283
Call down & in	11.9033	8.2868	10.2581	14.4513
Put up & out	3.7534	1.5117	0.3258	0.1171
Put down & in	5.5695	5.1506	5.4634	5.6073
ρ = 0.0				
Call up & out	13.2697	6.293	2.3531	1.4982
Call down & in	14.8536	10.5282	12.1345	16.2715
Put up & out	3.7534	1.780	0.6656	0.4238
Put down & in	4.258	4.1073	4.4917	4.6025
ρ = 0.5				
Call up & out	13.2697	5.5828	1.4138	0.6145
Call down & in	18.2087	13.1731	14.5503	18.743
Put up & out	3.7534	2.0475	1.0842	0.8759
Put down & in	3.1818	3.2986	3.834	3.9784

$A_1 = 100.0, A_2 = 100.0, X = 100.0, H = 105.0, \sigma_1 = 0.20, \sigma_2 = 0.25. T_2 = 1.0.$

The valuation formulae are due to Bermin (1996). The value is given by:

$$V = \eta \left\{ Se^{(b-r)T} \left(1 + \frac{\sigma^2}{2b}\right) \left[M\eta \left(\frac{m - \mu_2 t}{\sigma\sqrt{t}}, \frac{-\kappa + \mu_2 T}{\sigma\sqrt{T}}; -\rho \right) \right.\right.$$

$$\left. - e^{2\mu_2 H/\sigma^2} M\eta \left(\frac{m - 2H - \mu_2 t}{\sigma\sqrt{t}}, \frac{2H - \kappa + \mu_2 T}{\sigma\sqrt{T}}; -\rho \right) \right]$$

$$e^{-rT} X \left[M\eta \left(\frac{m - \mu_1 t}{\sigma\sqrt{t}}, \frac{-\kappa + \mu_1 T}{\sigma\sqrt{T}}; -\rho \right) \right.$$

$$\left. - e^{2\mu_1 H/\sigma^2} M\eta \left(\frac{m - 2H - \mu_1 t}{\sigma\sqrt{t}}, \frac{2H - \kappa + \mu_1 T}{\sigma\sqrt{T}}; -\rho \right) \right]$$

$$- e^{-rT} \left(\frac{\sigma^2}{2b} \right) \left[S \left(\frac{S}{X} \right)^{-2b/\sigma^2} M\eta \left(\frac{m + \mu_1 t}{\sigma\sqrt{t}}, \frac{-\kappa - \mu_1 T}{\sigma\sqrt{T}}; -\rho \right) \right.$$

$$\left. - B \left(\frac{B}{X} \right)^{-2b/\sigma^2} M\eta \left(\frac{m - 2H + \mu_1 t}{\sigma\sqrt{t}}, \frac{2H - \kappa - \mu_1 T}{\sigma\sqrt{T}}; -\rho \right) \right]$$

$$+ Se^{(b-r)T} \left[\left(1 + \frac{\sigma^2}{2b}\right) N\eta \left(\frac{\mu_2 (T - t)}{\sigma\sqrt{T - t}} \right) \right.$$

$$\left.\left. + e^{-b(T-t)} \left(1 - \frac{\sigma^2}{2b}\right) N\eta \left(\frac{-\mu_1 (T - t)}{\sigma\sqrt{T - t}} \right) \right] \gamma_1 - e^{-rT} X \gamma_2 \right\} \tag{17.4.1}$$

TABLE 17.9. Look-Barrier Up-and-Out Call Valuations

	$t = 0$			$t = 0.5$			$t = 1.0$		
$\sigma = 0.15$	$\sigma = 0.20$	$\sigma = 0.30$	$\sigma = 0.15$	$\sigma = 0.20$	$\sigma = 0.30$	$\sigma = 0.15$	$\sigma = 0.20$	$\sigma = 0.30$	Barrier
17.5212	21.5489	30.1874	1.4171	1.2239	1.0365	0.0177	0.008	0.0025	105.0
17.5212	21.5489	30.1874	4.2419	3.4686	2.7025	0.2388	0.112	0.0357	110.0
17.5212	21.5489	30.1874	7.7843	6.4569	4.9312	0.9397	0.4684	0.1576	115.0
17.5212	21.5489	30.1874	11.0593	9.6629	7.5509	2.1866	1.1788	0.4259	120.0
17.5212	21.5489	30.1874	13.4854	12.6042	10.3455	3.8041	2.2337	0.877	125.0

$S = 100.0$, $X = 100.0$, $T = 1.0$

TABLE 17.10. Look-Barrier Up-and-In Call Valuations

	t = 0			t = 0.5			t = 1.0		
$\sigma = 0.15$	$\sigma = 0.20$	$\sigma = 0.30$	$\sigma = 0.15$	$\sigma = 0.20$	$\sigma = 0.30$	$\sigma = 0.15$	$\sigma = 0.20$	$\sigma = 0.30$	Barrier
0.000	0.000	0.000	15.0683	18.6352	25.9924	11.6515	13.2616	16.7316	105.0
0.000	0.000	0.000	12.2435	16.3905	24.3264	11.4304	13.1577	16.6984	110.0
0.000	0.000	0.000	8.7011	13.4022	22.0976	10.7295	12.8013	16.5766	115.0
0.000	0.000	0.000	5.4261	10.1962	19.4779	9.4826	12.0909	16.3082	120.0
0.000	0.000	0.000	3.0001	7.2549	16.6833	7.865	11.036	15.8572	125.0

S = 100.0, X = 100.0, T = 1.0

$$\gamma_1 = \left\{\left[N\eta\left(\frac{\mathrm{H}-\mu_2 t}{\sigma\sqrt{t}}\right) - e^{2\mu_2\mathrm{H}/\sigma^2} N\eta\left(\frac{-\mathrm{H}-\mu_2 t}{\sigma\sqrt{t}}\right)\right]\right.$$

$$\left. - \left[N\eta\left(\frac{m-\mu_2 t}{\sigma\sqrt{t}}\right) - e^{2\mu_2\mathrm{H}/\sigma^2} N\eta\left(\frac{m-2\mathrm{H}-\mu_2 t}{\sigma\sqrt{t}}\right)\right]\right\} \tag{17.4.2}$$

$$\gamma_2 = \left\{\left[N\eta\left(\frac{\mathrm{H}-\mu_1 t}{\sigma\sqrt{t}}\right) - e^{2\mu_1\mathrm{H}/\sigma^2} N\eta\left(\frac{-\mathrm{H}-\mu_1 t}{\sigma\sqrt{t}}\right)\right]\right. \tag{17.4.3}$$

$$\left. - \left[N\eta\left(\frac{m-\mu_1 t}{\sigma\sqrt{t}}\right) - e^{2\mu_1\mathrm{H}/\sigma^2} N\eta\left(\frac{m-2\mathrm{H}-\mu_1 t}{\sigma\sqrt{t}}\right)\right]\right\} \tag{17.4.4}$$

$\eta = 1$ for up and out call; $\eta = -1$ for a down and out put.
$m = \min(\mathrm{H}, \kappa)$, for $\eta = 1$ and $m = \max(\mathrm{H}, \kappa)$, for $\eta = -1$.

$$\mathrm{H} = \log(B/S), \kappa = \log(X/S).$$

$$\mu_1 = b - \frac{\sigma^2}{2}, \mu_2 = b + \frac{\sigma^2}{2}, \rho = \sqrt{\frac{t}{T}}.$$

Tables 17.9 and 17.10 give a sample range of valuations for calls.

The implementation of look barrier options is done with code shown in Listing 17.4.

```
package FinApps;
import static java.lang.Math.*;
import BaseStats.Probnorm;
import BaseStats.Bivnorm;
public class Lookbarrier {

   public Lookbarrier(double rate,double q,double volatility,
                       int period) {

      crate=q;
      sigma=volatility;
      r=rate;
      b=crate==0.0?0.0:(b=crate!=r?(r-crate):r);
      tau=period==1?(1.0/(24*365)):period==2?(1.0/(365.0)):period==3?
(1.0/(52.0)):period==4?(1.0/(12.0)):0.0;

   }
   double tau;
   double crate;
   double sigma;
   double h;
```

```
double b;
double t1;
double t2;
double r;
double eta;
double heta;
double kappa;
double m;
double mu1;
double mu2;
double rho;
double gamma1;
double gamma2;
private void paRs(double s, double x, double barrier,
                  double time1, double time2) {

  time1=time1==0.0?(1e-6):time1==1.0?(1.0-1e-12):time1;
                                          //Avoid divide by zero
  t1=time1;
  t2=time2;
  double sig=(sigma*sigma);

  heta=log(barrier/s);
  kappa=log(x/s);
  m=eta==1.0?min(heta,kappa):max(heta,kappa);
  mu1=(b-(sig/2.0));
  mu2=(b+(sig/2.0));
  rho=sqrt(t1/t2);
  gamma1=gam(heta,mu2,m,t1,sigma);

  gamma2=gam(heta,mu1,m,t1,sigma);

public double callUout(double s, double x, double barrier,
                       double time1, double time2) {
  eta=1.0;

  paRs(s,x,barrier,time1,time2);

  double retval= lookbar(s,x,barrier);
  return retval;
}
public double putDout(double s, double x, double barrier,
                      double time1, double time2) {
  eta=-1.0;

  paRs(s,x,barrier,time1,time2);

  double retval= lookbar(s,x,barrier);
  return retval;
}
public double callUin(double s, double x, double barrier,
                      double time1, double time2) {
  eta=1.0;
```

```
    Parttfxlook p=new Parttfxlook(r,crate);
    double part=p.partFxCall(s,x,time1,time2,sigma);

    paRs(s,x,barrier,time1,time2);
    double retval= lookbar(s,x,barrier);

    return part-retval;
  }
  public double putDin(double s, double x, double barrier,
                       double time1, double time2) {
    eta=-1.0;
    Parttfxlook p=new Parttfxlook(r,crate);
    double part=p.partFxPut(s,x,time1,time2,sigma);
    paRs(s,x,barrier,time1,time2);

    double retval= lookbar(s,x,barrier);
    return part-retval;
  }
  private double gam(double h,double u,double m,double t,double vol) {
    double term1=N(eta*(h-u*t)/(vol*sqrt(t)))-exp(2.0*u*h/
                 (vol*vol))*N(eta* (-h-u*t)/(vol*sqrt(t)));
    double term2=N(eta*(m-u*t)/(vol*sqrt(t)))-exp(2.0*u*h/
               (vol*vol))*N(eta* (m-2.0*h-u*t)/(vol*sqrt(t)));

    return (term1-term2);
  }

  private double lookbar(double s, double x, double bar) {

    double sig=(sigma*sigma);
    double term1=s*exp((b-r)*t2)*(1.0+sig/(2.0*b))*(M(eta*
                   (m-mu2*t1)/(sigma*sqrt(t1)),
                   eta*(-kappa+mu2*t2)/(sigma*sqrt(t2)),-rho)
                   -exp(2.0*mu2*heta/sig)*M(eta*(m-2.0*heta
                   -mu2*t1)/(sigma*sqrt(t1)),eta*(2.0*heta-
                   kappa+mu2*t2)/(sigma *sqrt(t2)),-rho));
    double term2=-exp(-r*t2)*x*(M(eta*(m-mu1*t1)/(sigma*sqrt(t1)),
               eta*(-kappa+mu1*t2)/(sigma*sqrt(t2)),-rho)
               -exp(2.0*mu1*heta/sig)*M(eta*(m-2.0*heta-mu1*t1)/
               (sigma*sqrt(t1)),eta*(2.0*heta-kappa+mu1*t2)/
               (sigma*sqrt(t2)),-rho));
    double term3=-exp(-r*t2)*(sig/(2.0*b))*(s*pow((s/x),
               (-2.0*b/sig))*M(eta*(m+mu1*t1)/(sigma*sqrt(t1)),
               eta*(-kappa-mu1*t2)/(sigma*sqrt(t2)),-rho)
               -bar*pow((bar/x),(-2.0*b/sig))
               *M(eta*(m-2.0*heta+mu1*t1)/(sigma*sqrt(t1)),eta
               *(2.0*heta-kappa-mu1*t2)/(sigma*sqrt(t2)),-rho));
    double term4=s*exp((b-r)*t2)*((1.0+sig/(2.0*b))*N(eta*mu2
               *(t2-t1)/(sigma*sqrt(t2-t1)))+exp(-b*(t2-t1))
               *(1.0-sig/(2.0*b))*N(eta*(-mu1*(t2-t1))/
               (sigma*sqrt(t2-t1))))*gamma1-exp(-r*t2)*x*gamma2;
    double w=eta*(term1+term2+term3+term4);

    return w;
```

```
}
private static double M(double a, double b, double r) {

   return Bivnorm.bivar_params.evalArgs(a,b,r);
}
private static double N(double x) {
   Probnorm p=new Probnorm();

   double ret=x>(6.95)?1.0:x<(-6.95)?0.0:p.ncDisfnc(x);
                  //restrict the range of cdf values to stable values

   return ret;
}
```

LISTING 17.4. Look Barrier Options

References

Bermin (1996). "Combining Lookback Options and Barrier Options: The case of Look-Barrier Options," Paper, Department of Economics, Lund University.

Heynen, R. C. and H. M. Kat (1994). "Partial Barrier Options," *Journal of Financial Engineering*, 3, 253–274.

18
Other Exotics

This chapter deals with extensions to exotic type options. The exotics of previous chapters are further developed by introducing further complexity in adding an underlying condition or modifying a payoff condition. We also examine Asian options and currency options.

18.1. Two Asset Cash or Nothing

Options of this type are used to construct more complex derivative products. Heynen & Kat (1996) have developed valuation formulae to price several types of cash-or-nothing options. The valuations depend on the position of an asset S being above or below the strike price K. The fixed cash value X is paid out.

The two asset cash-or-nothing call is valued by:

$$V_c = Xe^{-rT}M(d_{1,1}, d_{2,2}; \rho) \tag{18.1.1}$$

This option has a payoff of:

$$X : iff, S_{1T} > K_1 \wedge S_{2T} > K_2; else, 0$$

The two asset cash-or-nothing put has a price given by:

$$V_p = Xe^{-rT}M(-d_{1,1}, -d_{2,2}; \rho) \tag{18.1.2}$$

This option has a payoff given as:

$$X : iff, S_{1T} < K_1 \wedge S_{2T} < K_2; else, 0$$

A two asset cash-or-nothing up-down is valued by:

$$V_D^U = Xe^{-rT}M(d_{1,1}, -d_{2,2}; -\rho) \tag{18.1.3}$$

The payoff is:

$$X : iff, S_{1T} < K_1 \wedge S_{2T} > K_2; else, 0$$

TABLE 18.1. Two Asset Cash-or-nothing valuations

	Two Asset Cash-Or-Nothing Options.								
TYPE	T = 1 Month			T = 6 Months			T = 1 Year		
	ρ = −0.5	ρ = 0.0	ρ = 0.5	ρ = −0.5	ρ = 0.0	ρ = 0.5	ρ = −0.5	ρ = 0.0	ρ = 0.5
1	0.3229	0.4272	0.4686	0.721	1.0381	1.3064	0.7704	1.1121	1.4262
2	0.3194	0.4238	0.4652	0.6463	0.9634	1.2317	0.6529	0.9946	1.3087
3	0.1488	0.0445	0.003	0.7369	0.4198	0.1515	0.9123	0.5707	0.2566
4	4.1757	4.0714	4.0299	2.6997	2.3826	2.1143	2.2799	1.9382	1.6241

$S_1 = S_2 = 50.0$; $X_1 = 55.0$, X2 = 45.0; CASH = 5.0; r = 0.08, q1 = 0.04, q2 = 0.03.

For a two asset cash-or-nothing down-up option the price is given by:

$$V_U^D = Xe^{-rT}M(-d_{1,1}, -d_{2,2}; -\rho) \tag{18.1.4}$$

This option has payoff of:

$$X : iff, S_{1T} > K_1 \wedge S_{2T} < K_2; else, 0.$$

Where,

$$d_{i,j} = \frac{\log(S_i/K_j) + (b_i - \sigma_i^2/2)T}{\sigma_i\sqrt{T}} \tag{18.1.5}$$

Table 18.1 gives a range of valuations for two asset cash-or-nothing options. For the cash-or-nothing call; as the correlation increases so does the option value. This option also increases with time to expiry. For the cash-or-nothing put option; the value increases with time and also increases with correlation.

In the case of an up-down option; the correlation increase produces a decrease in value, the option value increases with time to expiry. The down and up option reduces in value with increasing correlation. This option exhibits a decrease in value with an increased time to expiry. The different characteristics of each type of two asset-cash-or-nothing option allow the construction of a wide range of financial instruments to protect or exploit advantages in asset movements. Combinations of various strike prices with the same type of option allows for trade-off between asset price movements, between different strikes (insurance).

The code implementing two asset cash-or-nothing options is shown in Listing 18.1.

```
package FinApps;
import static java.lang.Math.*;
import BaseStats.Probnorm;
import BaseStats.Bivnorm;
```

```
public class Tassetcontop {

   public Tassetcontop(double asset1, double asset2, double r,double q1,
      double q2, double p) {
      crate1=q1;
      crate2=q2;
      s1=asset1;
      s2=asset2;
      rate=r;
      rho=p;
      b1=crate1==0.0?0.0:(b1=crate1!=r?(r-crate1):r);
      b2=crate2==0.0?0.0:(b2=crate2!=r?(r-crate2):r);
   }
    static double b1;
   static double b2;
   double crate1;
   double crate2;
   static double s1;
   static double s2;
   static double rate;
   static double rho;

   public double taconCall(double x1, double x2, double t, double sigma1,
      double sigma2, double cash) {
      return valU.T.Args(x1,x2,t,sigma1,sigma2,cash,1);
   }
   public double taconPut(double x1, double x2, double t, double sigma1,
      double sigma2, double cash) {
      return valU.T.Args(x1,x2,t,sigma1,sigma2,cash,2);
   }
   public double taconDup(double x1, double x2, double t, double sigma1,
      double sigma2, double cash) {
      return valU.T.Args(x1,x2,t,sigma1,sigma2,cash,4);
   }
   public double taconUdown(double x1, double x2, double t, double sigma1,
      double sigma2, double cash) {
      return valU.T.Args(x1,x2,t,sigma1,sigma2,cash,3);
   }
   private static double M(double a, double b, double r) {

      return Bivnorm.bivar_params.evalArgs(a,b,r);
   }
   private static double N(double x) {
      Probnorm p=new Probnorm();
      double ret=x>(6.95)?1.0:x<(-6.95)?0.0:p.ncDisfnc(x);//restrict
      the range of cdf values to stable values
      return ret;
   }
   private enum valU {
      T{
         public double Args(double x1, double x2, double t,
         double sigma1, double sigma2, double k, int tp) {
            int sw=tp;
```

```
        double puterm=0.0;

        switch(sw) {
          case 1: puterm=t1.m(x1,x2,sigma1,sigma2,t);// type 1
          break;
          case 2: puterm=t2.m(x1,x2,sigma1,sigma2,t);// type 2
          break;
          case 3: puterm=t3.m(x1,x2,sigma1,sigma2,t);// type 3
          break;
          case 4: puterm=t4.m(x1,x2,sigma1,sigma2,t);// type 4
          break;
        }

        return k*exp(-rate*t)*puterm;
      }
    },

    t1{
      public double m(double a, double b, double sigma1,
      double sigma2, double t) {
        double term1=paR_1.d1(a,sigma1,t);
        double term2=paR_2.d2(b,sigma2,t);
        return M(term1,term2,rho);
      }
    },

    t2{
      public double m(double a, double b,double sigma1,
      double sigma2, double t) {
        double term1=paR_1.d1(a,sigma1,t);
        double term2=paR_2.d2(b,sigma2,t);
        return M(-term1,-term2,rho);
      }
    },

    t3{
      public double m(double a, double b,double sigma1,
      double sigma2, double t) {
        double term1=paR_1.d1(a,sigma1,t);
        double term2=paR_2.d2(b,sigma2,t);
        return M(term1,-term2,-rho);
      }
    },

    t4{
      public double m(double a, double b,double sigma1,
      double sigma2, double t) {
        double term1=paR_1.d1(a,sigma1,t);
        double term2=paR_2.d2(b,sigma2,t);
        return M(-term1,term2,-rho);
      }
    },

    paR_1{
      public double d1(double x, double sigma, double t) {
```

```
            double sig=(sigma*sigma);
            double ans=(log(s1/x)+(b1-sig*0.5)*t)/(sigma*sqrt(t));
            return ans;
        }
    },

    paR_2{
        public double d2(double x, double sigma, double t) {
            double sig=(sigma*sigma);
            double ans=(log(s2/x)+(b2-sig*0.5)*t)/(sigma*sqrt(t));
            return ans;
        }
    };
        valU() {

        }
        public double m(double a, double b,double c, double d, double f) {

        throw new UnsupportedOperationException("Not yet implemented");
    }

        public double d2(double x, double sigma, double t) {
        throw new UnsupportedOperationException("Not yet implemented");
    }

        public double d1(double x, double sigma, double t) {
        throw new UnsupportedOperationException("Not yet implemented");
    }

        public double Args(double x1, double x2, double t, double sigma1,
        double sigma2, double k, int tp) {

        throw new UnsupportedOperationException("Not yet implemented");
        }
    }
```

Listing 18.1. Computation of Two Asset cash-or-nothing Options

18.2. Gap Option

A gap option has a payoff of zero if the asset price drops to or below a set price level (strike price 1). If the asset price is above the set price level, the payoff is a standard call or put value with a different strike price (strike 2).The payoff is given as:

$$\text{Call}: \quad 0; if, S_T \leq X_1 : else, S_T - X_1 \tag{18.2.1}$$

$$\text{Put}: \quad 0; if, S_T \geq X_1 : else, X_1 - S_T \tag{18.2.2}$$

The valuation method is a simple extension of the standard Black-Scholes valuation formulae. The parameter for determining the probability of the asset price (d_1) is the ratio of (S_T/X_1), this provides us with the probability value $N(d_1)$, which determines the expectation $Se^{rT}N(d_1)$; the expected value of the asset price when $S > X_1$. The standard Black-Scholes formula determines parameter d_2 in relation to the strike X_1, so that $X_1N(d_2)$ is the 'usual' term for the probability of the strike X_1 being paid times the strike (X_1). The Gap valuation substitutes the strike price used to develop the original probability defining parameters, with a different strike amount for paying out (X_2). Giving $X_2N(d_2)$ as the probability of the strike X_1 being paid, times an amount X_2. The Gap option allows a payoff amount to be dependent on the probability of an alternative value.

When the difference between strike prices is such that the option value is negligible, the option is referred to as a pay-later. It is possible to derive the payoff strike that will produce a pay later-option by approximating through an iterative procedure.

The formula for valuing a call is given by:

$$C = Se^{(b-r)T}N(d_1) - X_2e^{-rT}N(d2) \tag{18.2.3}$$

For a put option:

$$P = X_2e^{-rT}N(-d_2) - Se^{(b-r)T}N(-d_1) \tag{18.2.4}$$

Where the parameters are:

$$d_1 = \frac{\log(S/X_1) + (b + \sigma^2/2)T}{\sigma\sqrt{T}}, d_2 = d_1 - \sigma\sqrt{T}.$$

Example 18.0

An option is being considered as a pay-later. The risk free rate is 7%, the asset price is $77.580. The first strike is being set at $77.880. The underlying asset volatility is 25% per annum and the option will expire in 6 months. What would the second strike need to be for a pay-later put?

$$d_1 = \frac{\log(S/X_1) + (b + \sigma^2/2)T}{\sigma\sqrt{T}} = \frac{\log(77.580/77.880) + (0.07 + 0.25^2/2)*0.5}{0.25\sqrt{0.5}}$$

$$= 0.2645$$

$$d_2 = 0.2645 - 0.25^*\sqrt{0.5} = 0.0877.$$

$$N(-d_1) = N(-0.2645) = 0.3956$$
$$N(-d_2) = N(-0.0877) = 0.4650$$

$$P = X_2e^{-rT}N(-d_2) - Se^{(b-r)T}N(-d_1) =$$
$$X_2'e^{-0.07*0.5} *0.4650 - 77.580e^{0.0} *0.3956 = 0.0000 \tag{18.2.5}$$

TABLE 18.2. Gap Option Valuations with various strike prices

Strike X_2	T = 0.5		T = 1.0	
	Call	Put	Call	Put
53.0	8.1611	1.3382	10.6741	2.091
55.0	6.6809	1.7892	9.3259	2.6075
57.0	5.2007	2.2402	7.9776	3.1241
59.0	3.7204	2.6911	6.6294	3.6406
61.0	2.2402	3.1421	5.2811	4.1571
63.0	0.7599	3.5931	3.9329	4.6737
65.0	−0.7203	4.044	2.5846	5.1902

$S = 58.0, X_1 = 52.0, r = 0.07, b = 0.07, \sigma = 0.25.$

Where X_2' is iteratively calculated, so that 18.2.8 is zero, to the required degree of precision. The value of X_2' is calculated to be \$68.3617. If this value is placed into 18.2.7 for the value of X_2, the resulting put value is 4.3E-6, which is well within the accuracy required for zero value.

With strike prices having arbitrary values the possibility exists for negative valuations.

Table 18.2 shows valuations for Gap call and put options with various values of strike (X_2). The table 18.2 shows the effect of two expiry times. For six months to expiry the call price is always less than that with one year to expiry. Similarly for a put option, the longer time to expiry gives a higher value.

As the strike price increases, the value of a call decreases and the value of a put increases, as would be expected from a standard Black-Scholes option. When the strike is at 65.0, the call valuation is negative.

Gap options are implemented in code shown in Listing 18.2.

```
package FinApps;
import static java.lang.Math.*;
import BaseStats.Probnorm;
import CoreMath.NewtonRaphson;
public class Gapop extends NewtonRaphson{

  public Gapop(double rate, double yield, double time) {
    r=rate;
    crate=yield;
    t=time;
    b=crate==0.0?0.0:(b=crate!=r?(r-crate):r);

  }
  double b;
  double crate;
  double r;
  double t;
  double d1;
  double d2;
```

```
double asset;
double strike1;
double strike2;
double vol;
int typeflag=0;
public double gapCall(double s, double x1, double x2, double sigma)
{
    paRam(s,x1,sigma);
    return s*exp((b-r)*t)*N(d1)-x2*exp(-r*t)*N(d2);
}

public double paylateCall(double s,double st1,double st2,
    double sigma) {
    typeflag=-1;
    asset=s;
    strike1=st1;
    strike2=st2;
    vol=sigma;
    accuracy(1e-6,20);
    return newtraph(st2);
}

public double paylatePut(double s,double st1,double st2,
    double sigma) {
    asset=s;
    strike1=st1;
    strike2=st2;
    vol=sigma;
    accuracy(1e-6,20);
    return newtraph(st2);
}
public double gapPut(double s, double x1, double x2
    double sigma) {
    paRam(s,x1,sigma);
    return x2*exp(-r*t)*N(-d2)-s*exp((b-r)*t)*N(-d1);
}
private void paRam(double s, double x1, double sigma ) {
    double sig=(sigma*sigma);
    d1=(log(s/x1)+(b+sig*0.5)*t)/(sigma*sqrt(t));
    d2=d1-sigma*sqrt(t);
}
private double N(double x) {
    Probnorm p=new Probnorm();


    double ret=x>(6.95)?1.0:x<(-6.95)?0.0:p.ncDisfnc(x);
        //restrict the range of cdf values to stable values


    return ret;
}
public double newtonroot(double rootinput) {
```

```
    double solution;

    solution= typeflag==-1?(0.0-gapCall(asset,strike1,rootinput,
      vol)):(0.0-gapPut(asset,strike1,rootinput,vol));

    return solution;
}
```

LISTING 18.2. Gap Options

18.3. Soft Barrier Options

A soft barrier option has a range of upper and lower levels defining a single barrier. The knock-in knock-out values are reduced by a proportion of the asset price above the lower level (for down-and-out), having crossed the upper level. Or, the value is reduced by a proportion of the asset price below the upper level (for up-and-out), having crossed the lower level.

The soft barrier option is valued with the following formula derived from the work of Hart & Ross (1994):

$$V = \frac{1}{U-L}\Bigg\{\eta S e^{(b-r)T} S^{-2\mu} \frac{(SX)^{\mu+0.5}}{2(\mu+0.5)}\left[\left(\frac{U^2}{SX}\right)^{\mu+0.5} N(\eta d_1) - \lambda_1 N(\eta d_2) - \left(\frac{L^2}{SX}\right)^{\mu+0.5} N(\eta\varepsilon_1) + \lambda_1 N(\eta\varepsilon_2)\right]$$
$$- \eta X e^{-rT} S^{-2(\mu-1)} \frac{(SX)^{\mu-0.5}}{2(\mu-0.5)}\left[\left(\frac{U^2}{SX}\right)^{\mu-0.5} N(\eta d_3) - \lambda_2 N(\eta d_4) - \left(\frac{L^2}{SX}\right)^{\mu-0.5} N(\eta\varepsilon_3) + \lambda_2 N(\eta\varepsilon_4)\right]\Bigg\} \tag{18.3.1}$$

Where,

$$d_1 = \frac{\log(U^2/SX)}{\sigma\sqrt{T}} + \mu\sigma\sqrt{T}, \quad d_2 = d_1 - (\mu+0.5)\sigma\sqrt{T}$$

$$d_3 = \frac{\log(U^2/SX)}{\sigma\sqrt{T}} + (\mu-1)\sigma\sqrt{T}, \quad d_4 = d_3 - (\mu-0.5)\sigma\sqrt{T}$$

$$\varepsilon_1 = \frac{\log(L^2/SX)}{\sigma\sqrt{T}} + \mu\sigma\sqrt{T}, \quad \varepsilon_2 = \varepsilon_1 - (\mu+0.5)\sigma\sqrt{T}$$

$$\varepsilon_3 = \frac{\log(L^2/SX)}{\sigma\sqrt{T}} + (\mu-1)\sigma\sqrt{T}, \quad \varepsilon_4 = \varepsilon_3 - (\mu-0.5)\sigma\sqrt{T}$$

$$\lambda_1 = e^{-\frac{1}{2}[\sigma^2 T(\mu+0.5)(\mu-0.5)]}, \quad \lambda_2 = e^{-\frac{1}{2}[\sigma^2 T(\mu-0.5)(\mu-1.5)]}$$

$$\mu = \frac{b + \frac{1}{2}\sigma^2}{\sigma^2}.$$

The binary variable: $\eta = 1$ for a call; $\eta = -1$ for a put.

Table 18.3 gives valuations for down-and-out soft-barrier call options. Table 18.4 gives valuations for soft-barrier down-and-in call options.

TABLE 18.3. Soft-Barrier Down-and-out Call Option Valuations

	T = 1 Month			T = 6 Months			T =1 Year		
Lower	$\sigma = 0.15$	$\sigma = 0.25$	$\sigma = 0.35$	$\sigma = 0.15$	$\sigma = 0.25$	$\sigma = 0.35$	$\sigma = 0.15$	$\sigma = 0.25$	$\sigma = 0.25$
105.0	2.0959	2.9943	3.528	4.4907	4.7716	4.8566	5.6108	5.3822	5.2365
100.0	2.1201	3.2625	4.1493	5.2808	6.2215	6.6392	7.0315	7.3277	7.3807
95.0	2.1222	3.3196	4.3704	5.5723	7.107	7.9556	7.775	8.7532	9.155
90.0	2.1229	3.3394	4.458	5.6905	7.6215	8.8863	8.155	9.7565	10.5846
85.0	2.1232	3.3492	4.5022	5.7503	7.9227	9.5257	8.3619	10.4447	11.7085
80.0	2.1234	3.3552	4.5288	5.7862	8.1097	9.9628	8.4874	10.9152	12.5748
75.0	2.1235	3.3591	4.5464	5.8101	8.235	10.2675	8.5712	11.2433	13.2354
70.0	2.1236	3.3619	4.5591	5.8272	8.3246	10.4878	8.631	11.4806	13.739
65.0	2.1237	3.3641	4.5685	5.840	8.3917	10.6533	8.6758	11.659	14.1272
60.0	2.1238	3.3657	4.5759	5.850	8.4439	10.7821	8.7107	11.7977	14.4318

S = 110.0, X = 110.0, U = 105.0.

TABLE 18.4. Soft-Barrier Down-and-in Call Option Valuations

	T = 1 Month			T = 6 Months			T =1 Year		
Lower	$\sigma = 0.15$	$\sigma = 0.25$	$\sigma = 0.35$	$\sigma = 0.15$	$\sigma = 0.25$	$\sigma = 0.35$	$\sigma = 0.15$	$\sigma = 0.25$	$\sigma = 0.25$
105.0	17.267	11.7881	11.3104	19.5934	15.0104	16.2397	22.850	18.2896	20.3218
100.0	7.5336	6.5019	7.0886	10.4741	10.5316	12.6437	13.4153	13.8532	16.8014
95.0	3.8857	3.6815	4.4121	6.0301	7.3491	9.750	8.2844	10.4398	13.7966
90.0	2.5905	2.4602	2.9992	4.0438	5.2851	7.5458	5.6554	7.9476	11.3087
85.0	1.9429	1.8452	2.2507	3.033	4.0141	5.9487	4.2467	6.2073	9.3123
80.0	1.5543	1.4761	1.8005	2.4264	3.2152	4.8242	3.3974	5.0145	7.7539
75.0	1.2952	1.2301	1.5004	2.022	2.6795	4.0312	2.8312	4.1857	6.5594
70.0	1.1102	1.0544	1.2861	1.7331	2.2967	3.4566	2.4267	3.5884	5.6487
65.0	0.9714	0.9226	1.1253	1.5165	2.0096	3.0246	2.1234	3.1399	4.9483
60.0	0.8635	0.8201	1.0003	1.348	1.7863	2.6885	1.8875	2.791	4.3994

S = 100.0, X = 100.0, U = 105.0.

The code implementing a soft barrier option is shown in Listing 18.3

```
package FinApps;
import static java.lang.Math.*;
import BaseStats.Probnorm;
public class Softbarrier {

   public Softbarrier(double rate,double q,double volatility,
      int period) {

      crate=q;
      sigma=volatility;
      r=rate;
      b=crate==0.0?0.0:(b=crate!=r?(r-crate):r);
      tau=period==1?(1.0/(24*365)):period==2?(1.0/(365.0)):period==3?
         (1.0/(52.0)):period==4?(1.0/(12.0)):0.0;
   }
   double tau;
   double crate;
   double sigma;
   double h;
   double b;
   double r;
   double eta;

   public double downOcall(double s, double x, double up, double low,
   double t) {
      eta=1;
      if(up==low)// It is a standard barrier with one level
      {
         Sbarrierop sb=new Sbarrierop(r,crate,t,0);
         return sb.downOcall(s,x,sigma,low,0.0);
      }
      Blackscholecp bs=new Blackscholecp(crate);
      eta=1.0;
      if(s<low)// down and out.. stop here
         return 0.0;
      bs.bscholEprice(s,x,sigma,t,r);
      double barval=barrierVal(s,x,up,low,t);
      double adjusted =(bs.getCalle()-barval);
      return adjusted ;
   }
   public double upOput(double s, double x, double up, double low,
   double t) {
      eta=-1;
      if(up==low)// It is a standard barrier with one level
      {
         Sbarrierop sb=new Sbarrierop(r,crate,t,0);
         return sb.upOput(s,x,sigma,low,0.0);
      }
      if(s>up)
         return 0.0;
      Blackscholecp bs=new Blackscholecp(crate);
      eta=1.0;
      bs.bscholEprice(s,x,sigma,t,r);
```

```
    double barval=barrierVal(s,x,up,low,t);
    double adjusted =(bs.getPute()-barval);
    return adjusted ;
}
public double downIcall(double s, double x, double up, double low,
double t) {
    eta=1;
    if(up==low)// It is a standard barrier with one level
    {
        Sbarrierop sb=new Sbarrierop(r,crate,t,0);
        return sb.downIcall(s,x,sigma,low,0.0);
    }
    double barval=barrierVal(s,x,up,low,t);
    return barval;
}
public double upIput(double s, double x, double up, double low,
double t) {
    eta=-1;
    if(up==low)// It is a standard barrier with one level
    {
        Sbarrierop sb=new Sbarrierop(r,crate,t,0);
        return sb.upIput(s,x,sigma,low,0.0);
    }
    double barval=barrierVal(s,x,up,low,t);
    return barval;
}
private double barrierVal(double s, double x,double up,
double low, double t ) {// call down and in...put up and in
eta set to -1 for put

    double sig=(sigma*sigma);
    double mu=(b+sig/2)/sig;
    double d1=(log((up*up)/(s*x))/(sigma*sqrt(t)))
    +mu*sigma*sqrt(t);
    double d2=d1-((mu+0.5)*sigma*sqrt(t));
    double d3=(log(up*up/(s*x))/(sigma*sqrt(t)))+(mu-1.0)
    *sigma*sqrt(t);
    double d4=d3-((mu-0.5)*sigma*sqrt(t));
    double eps1=(log(low*low/(s*x))/(sigma*sqrt(t)))
    +mu*sigma*sqrt(t);
    double eps2=eps1-(mu+0.5)*sigma*sqrt(t);
    double eps3=(log(low*low/(s*x))/(sigma*sqrt(t)))
    +(mu-1.0)*sigma*sqrt(t);
    double eps4=eps3-(mu-0.5)*sigma*sqrt(t);
    double UPlambda1=exp(-0.5*(sig*t*(mu+0.5)*(mu-0.5)));
    double UPlambda2=exp(-0.5*(sig*t*(mu-0.5)*(mu-1.5)));
    double term1=s*exp((b-r)*t)*pow(s,(-2.0*mu))
        *(pow(s*x,(mu+0.5))/(2.0*(mu+0.5)));
    double term2=(pow(up*up/(s*x),(mu+0.5))*N(eta*d1)
    -UPlambda1*N(eta*d2)-pow((low*low/(s*x)),(mu+0.5))
    *N(eta*eps1)+UPlambda1*N(eta*eps2));
    double term3=-x*exp(-r*t)*pow(s,(-2.0*(mu-1.0)))
    *(pow(s*x,(mu-0.5))/(2.0*(mu-0.5)));
    double term4=(pow(up*up/(s*x),(mu-0.5))*N(eta*d3)-lambda2
```

```
    *N(eta*d4)-pow(low*low/(s*x),(mu-0.5))*N(eta*eps3)
    +lambda2*N(eta*eps4));

        return (eta*1.0/(up-low)*(term1*term2+term3*term4));

}
private static double N(double x) {
    Probnorm p=new Probnorm();
    double ret=x>(6.95)?1.0:x<(-6.95)?0.0:p.ncDisfnc(x);//restrict
    the range of cdf values to stable values
    return ret;
}
```

LISTING 18.3. Soft Barrier Option

18.4. Sequential Barrier Type Options

Sequential barrier options are an extension of the standard barrier options introduced by Merton (1973). These options have proven popular in the OTC market In a sequential barrier option the effect of hitting a barrier is to trigger either the creation or ending of a barrier option rather than the creation or ending of a put or call option.

The analysis of sequential barrier options can be made in terms of portfolios of standard and double barrier options. Pricing formulae and more extensive background are developed by Pfeffer (2001).

Sequential barrier options are analysed in terms of the eight standard barrier combinations (four call and four puts) and two double barriers (double call and double put). Sequential barrier options are classed into the four call options; Up-in down-and-out, down-in up-and-out, up-out down-and-in and the down-out up-and-in. There are four complementary put options. The up-in down-and-out call, is a down-and-out call option that comes into existence only after the asset hits the barrier from below, at which point the option is knocked-in as a down-and-out call. The down-in up-and -out sequential barrier is knocked in when the asset drops to the barrier from above, at which point it becomes an up-and-out call option. For these two options the hitting or crossing of the out barrier prior to hitting the in barrier, has no effect, since the option is not instantiated until the in barrier is crossed.

The payoff from a sequential barrier option is given by:

$$UIDOC = for, H_u > H_l;\ \tau_u < T : \max(S_T - X\ ,0) \forall S_{T-\tau_u} > H_l.$$

$$UIDIC = for, H_u > H_l;\ \tau_u < T : \max(S_T - X\ ,0),\ \text{if for some time, } (S_{T-\tau_u} \le H_l)$$

$$DIUIC = for, H_u > H_l;\ \tau_l < T : \max(S_T - X\ ,0),\ \text{if for some time, } (S_{T-\tau_l} \ge H_u)$$

$$DIUOC = for, H_u > H_l;\ \tau_l < T : \max(S_T - X\ ,0) \forall S_{T-\tau_l} < H_u$$

$$UODOC = for, H_u > H_l;\, \tau_u > T: \max(S_T - X\ ,0)\forall(S > H_l \wedge S < H_u)$$
$$UODIC = for, H_u > H_l;\, \tau_u > T: \max(S_T - X\ ,0),\ \text{if for some time, } (S \leq H_l)$$
$$DOUOC = for, H_u > H_l;\, \tau_l > T: \max(S_T - X\ ,0)\forall(S < H_u \wedge S > H_l)$$
$$DOUIC = for, H_u > H_l;\, \tau_l > T: \max(S_T - X\ ,0),\ \text{if for some time, } (S > H_u)$$

Where τ_u is the first time of crossing the upper barrier level and τ_1, is the first time passage for the lower barrier level. H_1, is the lower barrier value and H_u is the upper barrier value.

Up-and-in down-and-out call option (UIDOC)

The UIDOC pricing formula takes into account the probability associated with the asset firstly hitting the upper barrier, instantiating the option for a down-and-out call, then remaining above the lower barrier until expiry. An UIDOC can be expressed as a portfolio short a UIDIC and long a UIC, this can be rearranged as; UIDIC = UIC – UIDOC. This pays off when the up-and-in call hits the upper barrier and the UIDOC is extinguished by crossing the lower barrier. It can be shown that an UIDIC can be valued by:

$$UIDIC = d_1 S \left(\frac{H_l}{H_u}\right)^{2\lambda} N(\zeta) - dK \left(\frac{H_l}{H_u}\right)^{2\lambda-2} N(\zeta - v) \tag{18.4.1}$$

Where,

$$\zeta = \begin{cases} \zeta_1;\, for, K > H_l \\ \zeta_2;\, for, K < H_l, \end{cases} \quad \zeta_1 = \frac{1}{v}\log\left(\frac{H_l^2 S}{H^2 K}\right) + \lambda v, \quad \zeta_2 = \frac{1}{v} log\left(\frac{H_l S}{H^2}\right) + \lambda v.$$
$$d = e^{-rT}\ , d_1 = e^{-bT}\ , v = \sigma\sqrt{T}\ , \lambda = \frac{1}{2} + \left(\frac{(r-b)}{\sigma^2}\right).$$

Without considering the relative position of strike prices, the general valuation is:

$$\begin{aligned} UIC = {} & d_1 SN(x\ ) - d(K)N(x - v) + d_1 S \left(\frac{H_u}{S}\right)^{2\lambda} (N(y) - N(y_1)) \\ & - dK \left(\frac{H_u}{S}\right)^{2\lambda-2} (N(y - v) - N(y_1 - v)) \end{aligned} \tag{18.4.2}$$

Where,

$$x = \frac{1}{v}\log\left(\frac{S}{H_u}\right) + \lambda v\ , y = \frac{1}{v}\log\left(\frac{H_u^2}{SK}\right) + \lambda v\ , y_1 = \frac{1}{v}\log\left(\frac{H_u}{S}\right) + \lambda v.$$

Figure 18.1 shows the characteristic for an up-and-in down-and-out call, together with a down-and-out call. The parameters are:

$$r = 0.06, b = 0.0, t = 1.0, \sigma = 0.25, X = 60.0, U = 110.0, L = 75.0$$

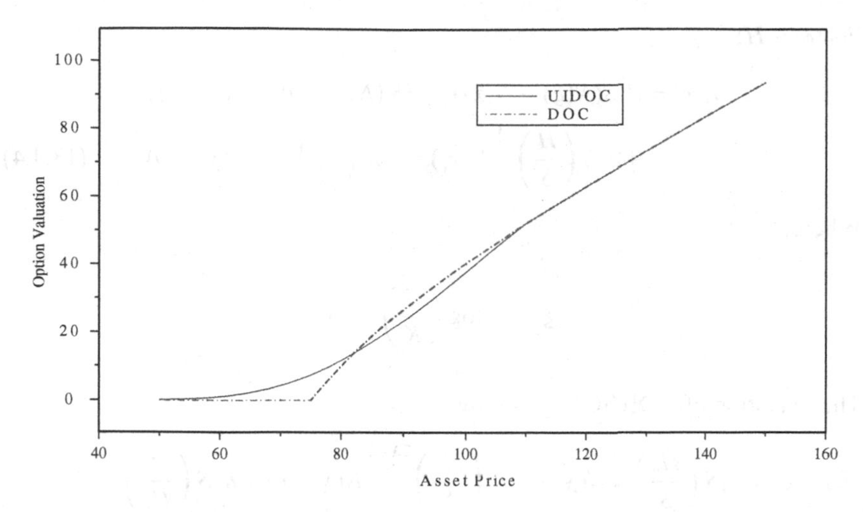

FIGURE 18.1. Combined UIDOC and DOC.

The value of a UIDOC is small at the lower asset prices, which are below the lower barrier level. At this point the down-and-out call option is out and has a value of zero. The sequential barrier has not yet been knocked in, so it retains value. Beyond an asset price of 75.0., the down-and-out call option takes on non-zero values and the UIDOC gradually increases. The value of a DOC is greater than the UIDOC on this gradient of the value curve, until we reach the point at which the asset price equals the upper barrier. When the point at which $S = H_u$ is reached the value of a UIDOC is identical to that of aDOC.

Down-and-in up-and-out call (DIUOC)

By hitting the lower barrier (H_l) from above, a DIUOC becomes a UOC option. The UOC option has a barrier level H_u. The DIUOC can be viewed as a portfolio of a long down-and-in call position with a short down-and-in up-and-in call position;

$$DIUOC = DIC - DIUIC.$$

Allowing for the relative position of strike price to barrier level the valuation of a DIC is given by:

For $K > H$;

$$DIC = d_1 S \left(\frac{H}{S}\right)^{2\lambda} N(Y) - dK \left(\frac{H}{S}\right)^{2\lambda-2} N(y - v) \tag{18.4.3}$$

For $K > H$;

$$DIC = d_1 S(N(\xi) - N(x)) - dK(N(\xi - v) - N(x - v)) + d_1 S\left(\frac{H}{S}\right)^{2\lambda} N(y_1) - dK\left(\frac{H}{S}\right)^{2\lambda-2} N(y_1 - v) \quad (18.4.4)$$

Where,

$$\xi = \frac{1}{v}\log\left(\frac{S}{K}\right) + \lambda v.$$

The valuation of a DIUIC is given by:

$$DIUIC = d_1 S\left(\frac{H_u}{S}\right)^{2\lambda} N(\chi) - dK\left(\frac{H_u}{S}\right)^{2\lambda-2} N(\chi - v) + d_1 S\left(\frac{H_l}{H_u}\right)^{2\lambda} *(N(\zeta_1) - N(\zeta_2)) - dK\left(\frac{H_1}{H_u}\right)^{2\lambda-2}(N(\zeta_1 - v) - N(\zeta_2 - v)) \quad (18.4.5)$$

Where,

$$\chi = \frac{1}{v}\log\left(\frac{H_u^2}{SH_l}\right) + \lambda v$$

Figure 18.2 shows the characteristic of a down-and-in up-and-out call with parameters:

$$r = 0.06, b = 0.0, t = 1.0, \sigma = 0.25, X = 60.0, U = 110.0, L = 75.0$$

The graph shows that values for the up-and-out call are the same as for the down-and-in-up and out call, until the asset price equals the lower barrier value. At this point, the DIUOC is knocked-in. The UOC retains greater value over the curve until the asset price approaches the upper barrier. At an asset price of 110.0, the UOC is knocked out. Note the sequential option retains value after the knock-out event.

Up-and-out-down and-in option (UODIC)

The order of barrier hits is not important in this sequential option. The knock-out triggering event will terminate the option if the upper barrier is hit prior to or subsequent to the knock-in event. With the UIDOC and DIUOC options, the option has to triggered prior to it being extinguished.

The UODIC valuation can be viewed as a portfolio of long an up-and-out call and short a double barrier call;

$$UODIC = UOC - DBC \quad (18.4.6)$$

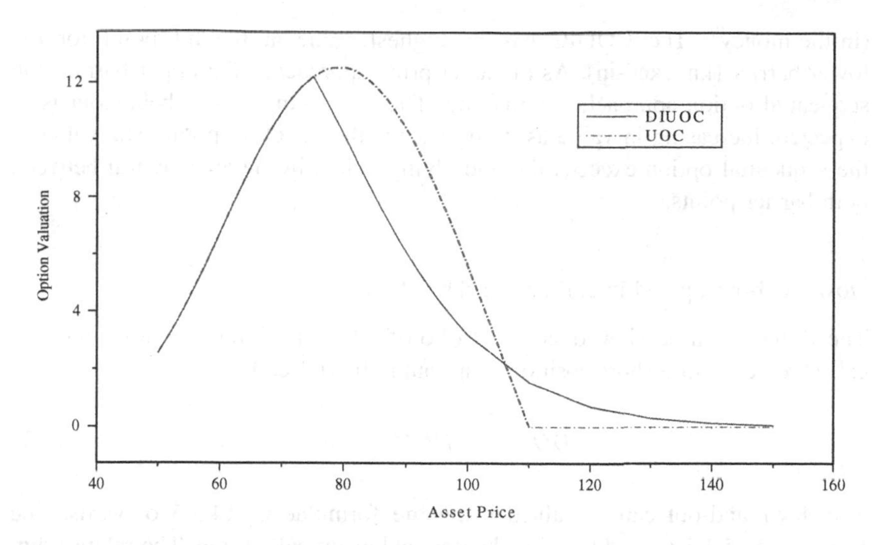

FIGURE 18.2. Characteristic of DIUOC.

An up-and-out call is valued using the formulae in 14.2.5 and 14.2.6. The double barrier option is valued using 15.1.1. Figure 18.3 shows the characteristic for a UODIC, together with the underlying (DIC). The parameters are:

$$r = 0.06, b = 0.0, t = 1.0, \sigma = 0.25, X = 60.0, U = 110.0, L = 75.0$$

Figure 18.3 plots the valuation of the underlying DIC. For asset price below the lower barrier, the underlying has a higher value than the sequential option

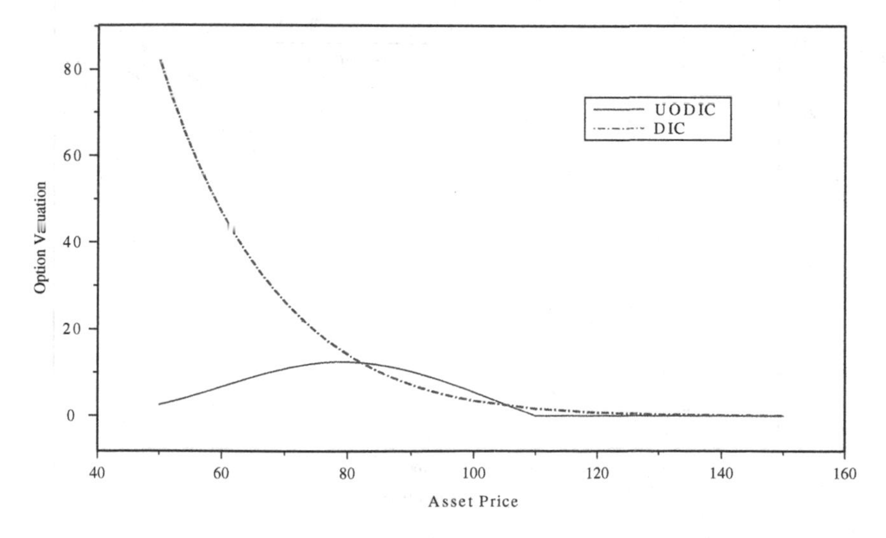

FIGURE 18.3. Characteristic of UODIC.

(in-the-money). The UODIC has its highest value at the hit point for the lower barrier (knocked-in). As the asset price approaches the upper barrier, the sequential option approaches extinction. The underlying (asset) behaviour is as expected, increasing in value as it approaches the knock-in point. The value of the sequential option exceeds the underlying value, in the area central between both barrier points.

Down-and-out-up and-in call option (DOUIC)

The DOUIC can be viewed as a portfolio of a long position in a down-and-out call together with a short position in a double barrier call;

$$DOUIC = DOC - DBC \tag{18.4.7}$$

The down-and-out call is valued using the formulae of 14.2.3 onwards. The formula of 15.1.1 is used to value the double barrier call option. The relationship shown in 18.3 can be argued in terms of the barrier hit sequence for each option. Consider a situation where the asset price hits the lower barrier first. The DOUIC, DOC and DBC are all extinguished and the relationship holds. Next, consider the situation where the asset price hits the upper barrier level first. The DOUIC is in as a call, the DOC is a call and the DBC is knocked out; the relationship call = call, is satisfied. Finally, consider the case of no barrier hits in the option lifetime; the DOUIC is extinguished, the DOC is a call and the DBC is a call. This satisfies the relationship call = call.

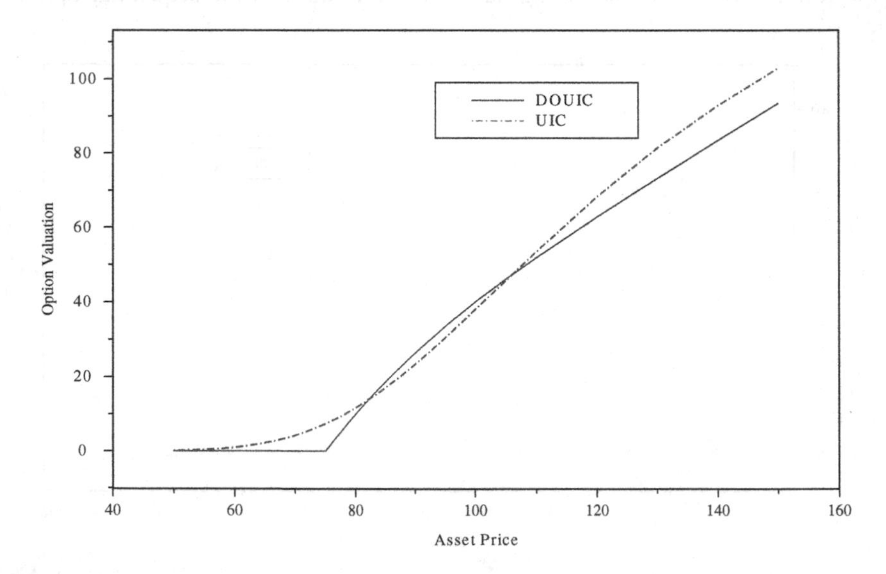

FIGURE 18.4. Characteristic of DOUIC.

The argument above is equally applicable to the relationship of 18.2. With the first case being an upper barrier as a first hit, second case the lower barrier as first hit. The final case being no barrier hits.

Figure 18.4 shows the characteristic for a DOUIC and its underlying UIC. The parameters are:

$$r = 0.06, b = 0.0, t = 1.0, \sigma = 0.25, X = 60.0, U = 110.0, L = 75.0$$

Figure 18.4 shows the DOUIC extinguished at the lower barrier. The UIC increases in-the-money after the upper barrier is hit (knocked-in) and the asset price increases. The DOUIC also increases at a lower rate (reflecting the probability of the option being knocked-out). At the point where the asset price equals the knock-in barrier the DOUIC is equivalent to the UIC option (as expected). At the point where the asset price hits the lower (knock-out) barrier, the DOUIC is extinguished and the UIC has a value which reflects the probability of the option being knocked-in; around the mid points between barrier levels, the DOUIC has a value greater than it's underlying.

The final two sequential options UODOC and DOUOC are both valued by the double barrier option formulae .

Up-and-in-down and-out put (UIDOP)

This option requires the asset price to hit the upper barrier to instantiate the down-and-out put option.

The valuation is given by:

$$UIDOP = UIP + d_1 S \left(\frac{H_u}{S}\right)^{2\lambda} N(-\chi) - dK \left(\frac{H_u}{S}\right)^{2\lambda-2} N(-\chi + v) -$$
$$d_1 S \left(\frac{H_1}{H_u}\right)^{2\lambda} (N(\zeta_1) - N(\zeta_2)) + dK \left(\frac{H_1}{H_u}\right)^{2\lambda-2} (N(\zeta_1 - v) - N(\zeta_2 - v)) \tag{18.4.8}$$

Figure 18.5 gives the characteristic for a UIDOP together with the underlying DOP. The parameters are:

$$r = 0.06, b = 0.0, t = 1.0, \sigma = 0.25, X = 90.0, U = 95.0, L = 65.0$$

The UIDOP is not knocked-in until the asset price hits 95.0. After this point the underlying becomes a DOP, which is shown by the dotted line following the DOP curve. The value of the UIDOP at the tails prior to the upper barrier level and beyond the lower barrier level is zero. The maximum value of the UIDOP prior to being knocked-in is a function of the residual probability of the asset price rising to the knock-in price. The drop to zero is an artefact of the relative strike to asset price for the put.

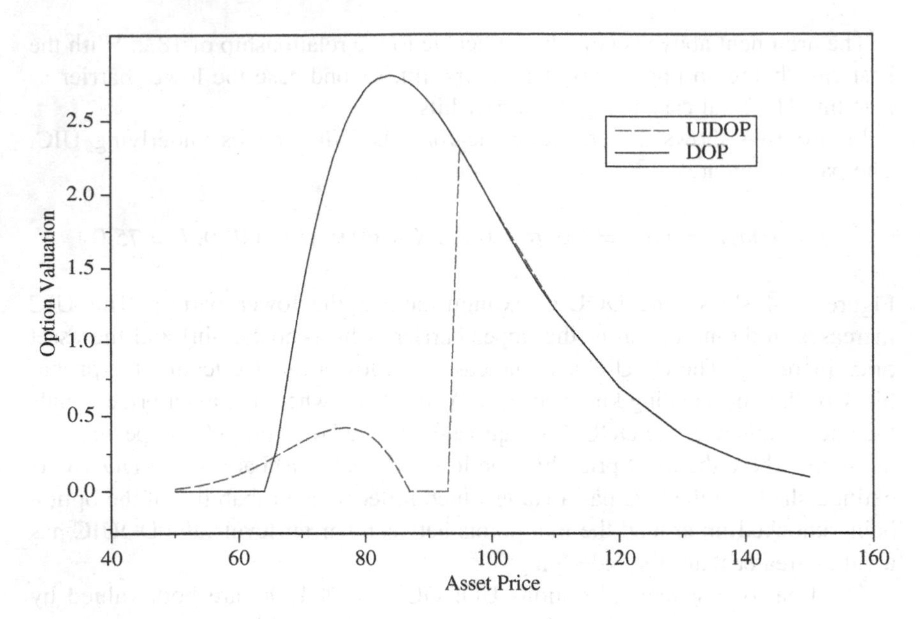

FIGURE 18.5. Characteristic of UIDOP.

Down-and-in up-and-out put (DIUOP)

The DIUOP is valued by:
For $(K > H_u)$

$$DIUOP = dKN(-\chi + v) - d_1 SN(-\chi) + dK\left(\frac{H_l}{S}\right)^{2\lambda-2}(N(y_1 - v) - N(\chi - v)) -$$

$$d_1 S\left(\frac{H_l}{S}\right)^{2\lambda}(N(y_1) - N(\chi)) + d_1 S\left(\frac{H_u}{H_l}\right)^{2\lambda} N(-\zeta_2) - dK\left(\frac{H_u}{H_l}\right)^{2\lambda-2} N(-\zeta_2 + v) \tag{18.4.9}$$

For $(K < H_u)$

$$DIUOP = dKN(-x + v) - d_1 SN(-x) + dK\left(\frac{H_l}{S}\right)^{2\lambda-2}(N(y_1 - v) - N(y - v)) -$$

$$d_1 S\left(\frac{H_l}{S}\right)^{2\lambda}(N(y_1) - N(y)) + d_1 S\left(\frac{H_u}{H_l}\right)^{2\lambda} N(-\zeta_1) - dK\left(\frac{H_u}{H_l}\right)^{2\lambda-2} N(-\zeta_1 + v) \tag{18.4.10}$$

Figure 18.6 gives the characteristic for a DIUOP with its underlying UOP. With parameters:

$$r = 0.06, b = 0.0, t = 1.0, \sigma = 0.25, X = 90.0, U = 95.0, L = 65.0$$

Figure 18.6 shows the point at which the DIUOP hits the lower barrier, giving the same profile as the underlying UOP. The value of the UOP is zero at the

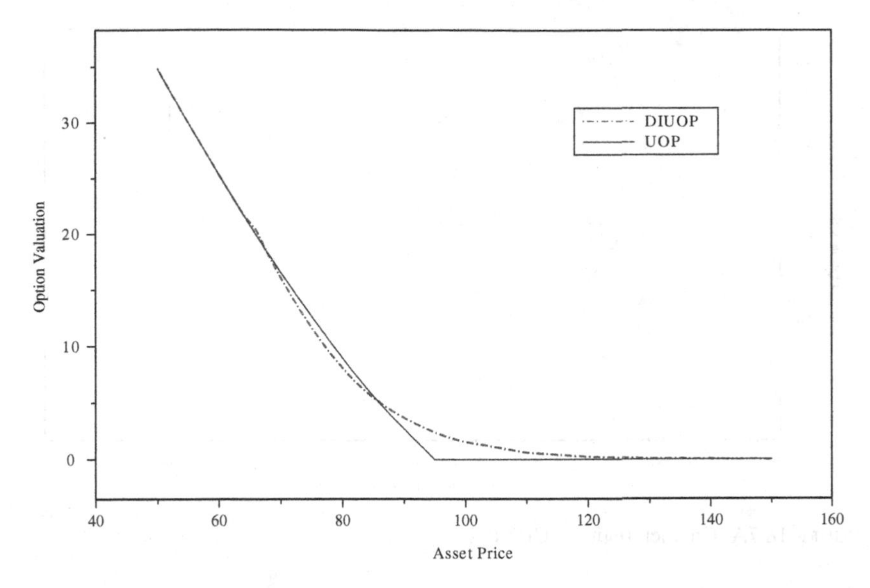

FIGURE 18.6. Characteristic of DIUOP.

upper barrier level, in line with expectation. The DIUOP however maintains residual value. This is a value which reflects the probability of a down hit.

Up-and-in down-and-in put (UIDIP)

This option is instantiated when the asset price hits the upper barrier level, it then becomes a DIP.

The valuation is straightforward as:

$$UIDIP = UIP - UIDOP$$

The UIDOP is valued by the formulae discussed above.

Figure 18.7 gives the characteristic for a UIDIP together with its underlying. The parameters are:

$$r = 0.06, b = 0.0, t = 1.0, \sigma = 0.25, X = 90.0, U = 95.0, L = 65.0$$

Figure 18.7A shows the point at the lower barrier level where the DIP is knocked-in. The UIDIP rises above the DIP expectation towards the upper barrier. At this point the value of a UIDIP rises above the DIP due to the probability of an upper barrier hit. The UIDIP shown in Figure 18.7B more clearly demonstrates the discontinuity at the upper barrier level.

Down-and-in up-and-in put (DIUIP)

The DIUIP option knocks-in a UIP option when the lower barrier is hit. The valuation is given by:

$$DIUIP = DIP - DIUOP$$

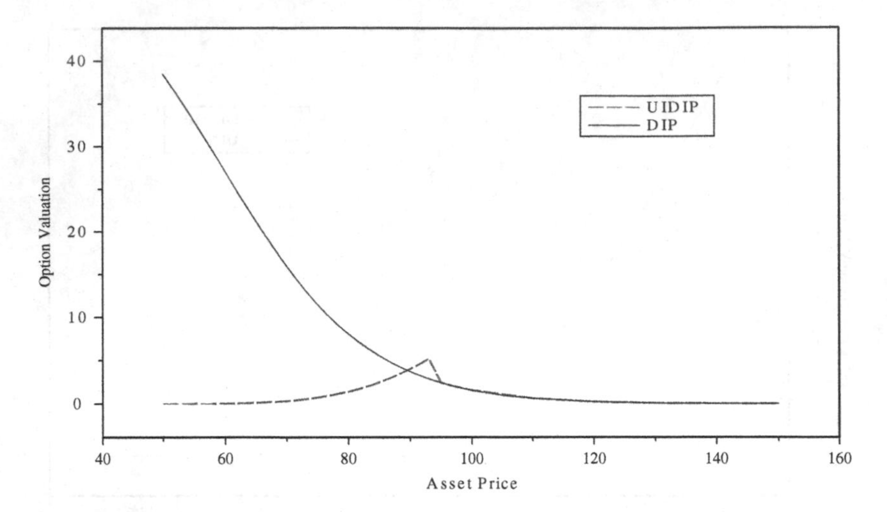

FIGURE 18.7A. Characteristic of UIDIP.

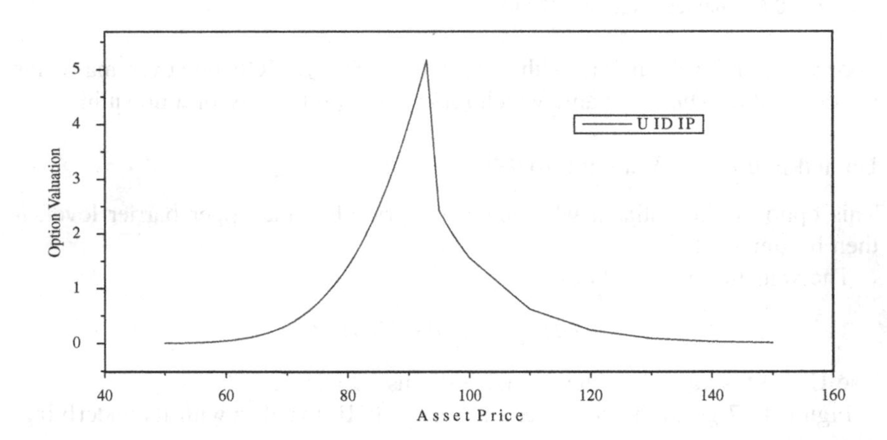

FIGURE 18.7B. UIDIP discontinuity.

The DIP is valued by formulae discussed in chapter 17. The DIUOP is valued by formulae discussed above. Figure 18.8A. shows the characteristic of a DIUIP with its underlying UIP. The parameters are:

$$r = 0.06, b = 0.0, t = 1.0, \sigma = 0.25, X = 90.0, U = 95.0, L = 65.0$$

Figure 18.8A shows the DIUIP being knocked-in at the lower barrier to create the UIP. There is a residual value around the barrier, this demonstrated in Figure 18.8B. below.

Sequential Barrier options are valued using the class in Listing 18.4.

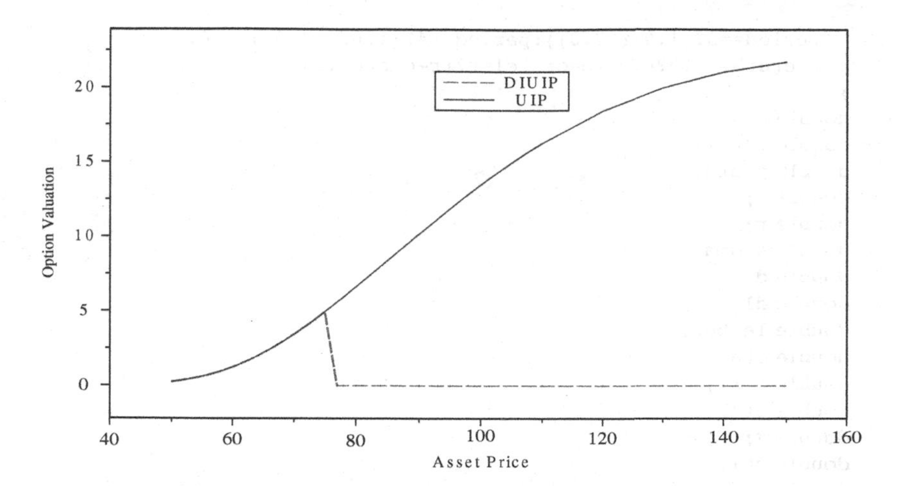

FIGURE 18.8A. Characteristic of DIUIP.

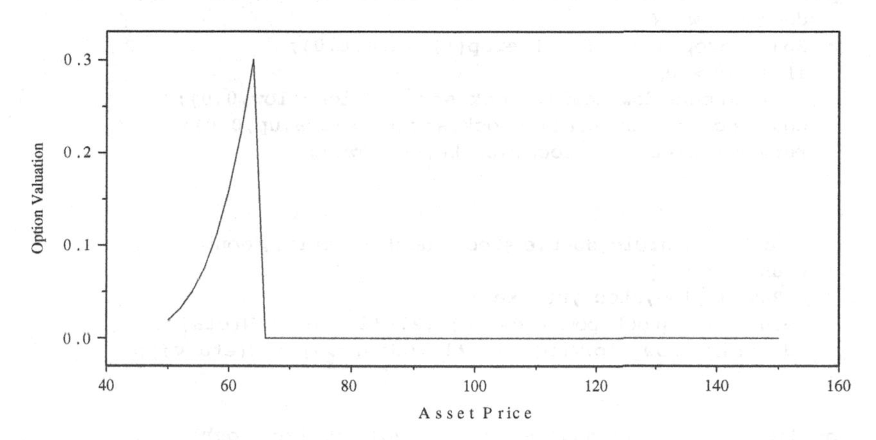

FIGURE 18.8B. DIUIP residual value.

```
package FinApps;
import static java.lang.Math.*;
import BaseStats.Probnorm;
public class Sequentbar {

   public Sequentbar(double rate,double q,double time,
      double volatility, int period ) {
      crate=q;
      r=rate;
      t=time;
      //0 continuos,1 hourly,2 daily,3 weekly, 4 monthly
      sigma=volatility;
      tau=period==1?(1.0/(24*365)):period==2?(1.0/(365.0)):
```

```
    period==3?(1.0/(52.0)):period==4?(1.0/(12.0)):0.0;
    b=crate==0.0?0.0:(b=crate!=r?(r-crate):r);
}
double b;
double tau;
double crate;
double t;
double r;
double sigma;
double d;
double d1;
double lambda;
double eta;
double eta1;
double eta2;
double v;
double chi;
double xo;
double gamma;
double gamma1;
public double uiDoc(double stock, double strike, double up,
    double low ) {
    Sbarrierop sb=new Sbarrierop(r,crate,t,0);
    if(stock>=up)
        return sb.downOcall(stock,strike,sigma,low,0.0);
    double cui=sb.upIncall(stock,strike,sigma,up,0.0);
    return (cui-uiDic(stock,strike,up,low));
}


public double uiDic(double stock, double strike, double up,
    double low ) {
    paRam(up,low,stock,strike);
    return (d1*stock*pow((low/up),(2.0*lambda))*N(eta)
    -d*strike*pow((low/up),(2.0*lambda-(2.0)))*N(eta-v));

}

public double diUic(double stock, double strike, double up,
    double low ) {
    paRam(up,low,stock,strike);
    double term1=d1*stock*pow((up/stock),(2.0*lambda))*N(chi)-d
    *strike*pow((up/stock),(2.0*lambda-(2.0)))*N(chi-v);

    double term2=d1*stock*pow((low/up),(2.0*lambda))
    *(N(eta1)-N(eta2))-d*strike*pow((low/up),(2.0*lambda-(2.0)))
    *(N(eta1-v)-N(eta2-v));
    return(term1+term2);
}

public double diUoc(double stock, double strike, double up,
    double low ) {
    Sbarrierop sb=new Sbarrierop(r,crate,t,0);
    double uoc=sb.upOcall(stock,strike,sigma,up,0.0);
    if(stock<=low)
        return uoc;
```

```
    double cdi=sb.downIcall(stock,strike,sigma,low,0.0);
    return (cdi-diUic(stock,strike,low,up));

}
public double uoDic(double stock, double strike, double up,
    double low ) {
    Sbarrierop sb=new Sbarrierop(r,crate,t,0);
    Dbarrierh sh=new Dbarrierh(r,crate,t,sigma,0);
    double dbc=sh.uoDoc(stock,strike,up,low);
    double uoc=sb.upOcall(stock,strike,sigma,up,0.0);
    return (uoc-dbc);
}

public double doUic(double stock, double strike, double up,
    double low ) {
    Sbarrierop sb=new Sbarrierop(r,crate,t,0);
    Dbarrierh sh=new Dbarrierh(r,crate,t,sigma,0);
    double dbc=sh.uoDoc(stock,strike,up,low);
    double doc=sb.downOcall(stock,strike,sigma,low,0.0);
    return (doc-dbc);
}

private void paRam(double up, double low, double s, double x) {
    d=exp(-r*t);
    d1=exp(-b*t);
    lambda=(((r-b)/(sigma*sigma))+0.50);
    v=(sigma*sqrt(t));
    eta1=((1.0/v)*log((low*low*s)/(up*up*x))+(lambda*v));
    eta2=((1.0/v)*log((low*s)/(up*up))+(lambda*v));
    eta=x>=low?eta1:eta2;
    chi=((1.0/v)*log((up*up)/(low*s))+(lambda*v));
    xo=((1.0/v)*log((s)/(up))+(lambda*v));
    gamma=((1.0/v)*log((up*up)/(s*x))+(lambda*v));
    gamma1=((1.0/v)*log((up)/(s))+(lambda*v));

    }

public double uoDip(double stock, double strike, double up,
    double low ) {
    Sbarrierop sb=new Sbarrierop(r,crate,t,0);
    Dbarrierh sh=new Dbarrierh(r,crate,t,sigma,0);
    double uop=sb.upOput(stock,strike,sigma,up,0.0);
    double dbp=sh.uoDop(stock,strike,up,low);
    return (uop-dbp);
}

public double doUip(double stock, double strike, double up,
    double low ) {
    Sbarrierop sb=new Sbarrierop(r,crate,t,0);
    Dbarrierh sh=new Dbarrierh(r,crate,t,sigma,0);
    double dop=sb.downOput(stock,strike,sigma,up,0.0);
    double dbp=sh.uoDop(stock,strike,up,low);
    return (dop-dbp);
}

public double uiDop(double stock, double strike, double up,
    double low ) {//up>low,strike>low,stock<up,for a
```

```
    positive payoff

    Sbarrierop sb=new Sbarrierop(r,crate,t,0);
    if(stock>=up)
      return sb.downOput(stock,strike,sigma,low,0.0);
    double uip=sb.upIput(stock,strike,sigma,up,0.0);
    paRam(up,low,stock,strike);
    double term= d1*stock*pow((up/stock),(2.0*lambda))
    *N(-chi)-d*strike*pow((up/stock),((2.0*lambda)-2.0))
    *N(-chi+v)-d1*pow((low/up),(2.0*lambda))
    *(N(eta1)-N(eta2));
    double term1=d*strike*pow((low/up),((2.0*lambda)-2.0))
    *(N(eta1-v)-N(eta2-v));
    return (uip+term+term1);
  }
  public double uiDip(double stock, double strike, double up,
    double low ) {
    Sbarrierop sb=new Sbarrierop(r,crate,t,0);
    if(stock>=up)
      return sb.downIput(stock,strike,sigma,low,0.0);
    double uip=sb.upIput(stock,strike,sigma,up,0.0);
    return(uip-uiDop(stock,strike,up,low));
  }
  public double diUop(double stock, double strike, double up,
    double low ) { //Assumes up<low,strike>up,stock>up
    Sbarrierop sb=new Sbarrierop(r,crate,t,0);
    if(stock<=low)
      return sb.upOput(stock,strike,sigma,up,0.0);

    double reverse=up;
    up=low;low=reverse;

    paRam(up,low,stock,strike);
    if(strike>=low) {
      double term1= d*strike*N(-xo+v)-d1*stock*N(-xo)+d*strike
    *pow((up/stock),((2.0*lambda)-2.0))*(N(gamma1-v)-N(chi-v));
      double term2=d1*stock*pow((up/stock),(2.0*lambda))
    *(N(gamma1)-N(chi))+d1*stock*pow((low/up),(2.0*lambda))
    *N(-eta2)-d*strike*pow((low/up),((2.0*lambda)-2.0))
    *N(-eta2+v);
      return (term1-term2);
    }
      double term1=d*strike*N(-xo+v)-d1*stock*N(-xo)
    +d*strike*pow
    ((up/stock),((2.0*lambda)-2.0))*(N(gamma1-v)-N(gamma-v));
      double term2=d1*stock*pow((up/stock),(2.0*lambda))
    *(N(gamma1)-N(gamma))+d1*stock*pow((low/up),(2.0*lambda))
    *N(-eta1)-d*strike*pow((low/up),((2.0*lambda)-2.0))
    *N(-eta1+v);
      return (term1-term2);

  }
  public double diUip(double stock, double strike, double up,
    double low ) {
    Sbarrierop sb=new Sbarrierop(r,crate,t,0);
```

```
    if(stock<=low)
      return sb.upIput(stock,strike,sigma,up,0.0);
    return(sb.downIput(stock,strike,sigma,low,0.0)-diUop
    (stock,strike,up,low));
}

private static double N(double x) {
    Probnorm p=new Probnorm();
    double ret=x>(6.95)?1.0:x<(-6.95)?0.0:p.ncDisfnc(x);//restrict
    the range of cdf values to stable values
    return ret;
}
```

LISTING 18.4. Sequential Barrier Option Valuation

18.5. Supershares

Supershares are based on the original work of Hakansson. The original idea was the creation of a superfund which consisted of a portfolio of assets that would have shares issued against their value. The notion being similar to a mutual fund. The Supershare offers the holder a right to receive a dollar based value as a proportion of the assets in a superfund, given that the asset values finish within an upper and lower boundary value at expiry. Otherwise the Supershare expires worthless.

The ASE listed Superunits in 1992. The CBOE listed securities called Supershares based on the same principle. The history of exchange traded funds has been mixed and to-date there is a proliferation in the so-called ETF's available. The basic valuation model is based on the Black-Scholes economy.

$$V = \frac{Se^{(b-r)T}}{X} N(d_1) - \frac{Se^{(b-r)T}}{X} N(d_2) \tag{18.5.1}$$

The first term can be viewed as the discounted value of the upside in owning the fund. The second term is the discounted cost of ownership.

The parameters are given by:

$$d_1 = \frac{\log(S/X_L) + (b + \sigma^2/2)T}{\sigma\sqrt{T}}, \quad d_2 = \frac{\log(S/X_U) + (b + \sigma^2/2)T}{\sigma\sqrt{T}}.$$

The payoff is given by:

$$\forall S > X_L \wedge < H_U : S/X_L; \, else, 0$$

Example 18.1

A Supershare option has three months to expiry. The price is $105.0, the underlying volatility is 25% and the risk-free rate is 6%. The lower boundary is $98.0 and the upper boundary is $112.0. What is the value?

$$r = 0.06, b = 0.0, t = 0.25, \sigma = 0.25, S = 105.0, X_L = 98.0, X_U = 112.0.$$

$$d_1 = \frac{\log(S/X_L) + (b + \sigma^2/2)T}{\sigma\sqrt{T}} = \frac{\log(105.0/98.0) + (0 + 0.25^2/2)*0.25}{0.25\sqrt{0.25}}$$
$$= 0.6144$$

$$d_2 = \frac{\log(S/X_U) + (b + \sigma^2/2)T}{\sigma\sqrt{T}} = \frac{\log(105.0/112.0) + (0 + 0.25^2/2)*0.25}{0.25\sqrt{0.25}}$$
$$= -0.4538$$

$$N(d_1) = N(0.6144) = 0.7305$$
$$N(d_2) = N(-0.4538) = 0.3249$$

$$V = \frac{Se^{(b-r)T}}{X} N(d_1) - \frac{Se^{(b-r)T}}{X} N(d_2) = \frac{105.0e^{-0.06*0.25}}{X} *0.7305$$
$$- \frac{105.0e^{-0.06*0.25}}{X} *0.3249 = 0.42805$$

The valuation is \$0.43.

The class **Supershare** contains the code implementing the Supershare option. This is shown in Listing 18.5.

```
package FinApps;
import static java.lang.Math.*;
import BaseStats.Probnorm;

public class Supershare {
    public Supershare(double rate,double q,double time,
    double volatility) {
        crate=q;
        r=rate;
        t=time;
        sigma=volatility;
        b=crate==0.0?0.0:(b=crate!=r?(r-crate):r);
    }

    double t;
    double r;
    double crate;
    double b;
    double sigma;

    public double sShare(double s, double xl, double xu) {
        double sig=(sigma*sigma);
        double d1=(log(s/xl)+(b+(sig*0.5))*t)/(sigma*sqrt(t));
        double d2=(log(s/xu)+(b+(sig*0.5))*t)/(sigma*sqrt(t));
```

```
        Probnorm prob=new Probnorm();
        return(s*exp((b-r)*t)/(xl))*prob.ncDisfnc(dl)
        -(s*exp((b-r)*t)/(xl))*prob.ncDisfnc(d2);
    }
```

LISTING 18.5. Supershare option Valuation

18.6. Asian Options

An Asian option is of the path dependent type in that the payoff is related to the past asset prices, being averaged. There are two types of averages; the arithmetic average and the geometric average. Arithmetic averaged options are difficult to value and no satisfactory closed-form solution exists. For the geometric average, fairly straightforward analytical methods exist. An average rate Asian is one where the asset price is the average asset value and the average strike is one where the strike price is the average strike.

Asian options are very useful in hedging thinly traded assets as the hedge is much cheaper than maintaining a portfolio of assets. Asian options are widely used in organisations exposed to outstanding foreign accounts. It is possible to hedge exchange rates by taking an average over the period to be hedged. The arithmetic option is most useful in these circumstances.

There have been several approaches to developing solutions to average rate options. The early work of Kemna and Vorst (1990) produced a closed-form solution for the geometric average rate Asian. The work of Turnbull and Wakeman (1991) used an analytical approximation by moment matching the arithmetic average with the lognormal property of asset price. The work of Levy (1992) used moment matching in an approach where approximations are made in discrete sampling of the average. The Levy approach is the basis of a popular and still widely used technique. The Levy methodology is the basis of a wide range of sophisticated (and some very simple but effective) Asian options. There are more complex methods that extend the analytical approach (see Zhang (2001) but require significant additional computation.

18.6.1. Geometric Average Rate Option

The valuation of this option is given by:

$$C = Se^{(\overline{b}-r)T}N(d_1) - Xe^{-rT}N(d_2) \tag{18.6.1}$$

$$P = Xe^{-rT}N(-d_2) - Se^{(\overline{b}-r)T}N(-d_1) \tag{18.6.2}$$

Where,

$$d_1 = \frac{\log(S\ /\ X) + (\overline{b} + \overline{\sigma}^2/2)T}{\overline{\sigma}\sqrt{T}}, \quad d_2 = d_1 - \overline{\sigma}\sqrt{T}$$

The model is based on a standard Black-Scholes formula, with the volatility and cost of carry averaged, based on the moments described by Kemna & Vorst:

$$\overline{\sigma} = \frac{\sigma}{\sqrt{3}} \tag{18.6.3}$$

$$\overline{b} = \frac{1}{2}\left(b - \frac{\sigma^2}{6}\right) \tag{18.6.4}$$

Example 18.2

A geometric average rate option has six month's to expiry. The underlying price is £100.0 with a volatility of 25%. The strike price is £105.0 and the risk free rate is 6% with the cost of carry 7%. What is the valuation of the option?

$$\overline{\sigma} = \frac{\sigma}{\sqrt{3}} = \frac{0.25}{\sqrt{3}} = 0.2886, \quad \overline{b} = \frac{1}{2}\left(0.07 - \frac{0.25^2}{6}\right) = 0.0141$$

$$d_1 = \frac{\log(S/X) + (\overline{b} + \overline{\sigma}^2/2)T}{\overline{\sigma}\sqrt{T}} = \frac{\log(100.0/105.0) + (0.0141 + 0.2886^2/2)*0.5}{0.2886\sqrt{0.5}}$$

$$= -0.10225$$

$$d_2 = d_1 - \overline{\sigma}\sqrt{T} = --0.10225 - 0.2886\sqrt{0.5} = -0.30638$$

$$N(-d_1) = N(--0.10225) = 0.5407$$

$$N(-d_2) = N(--0.30638) = 0.6203$$

$$P = Xe^{-rT}N(-d_2) - Se^{(\overline{b}-r)T}N(-d_1) = 105.0e^{-0.06*0.5}*0.6203$$
$$- 100.0e^{(0.0141-0.06)0.5}*0.5407 = 10.363$$

The put value is £10.36.

The class for valuing Kemna & Vorst geometric averages is shown in Listing 18.6.

```
package FinApps;
import static java.lang.Math.*;
import BaseStats.Probnorm;

public class Asiangeop {

   public Asiangeop(double rate,double yield,double volatility) {

      =rate;
```

```
        crate=yield;
        sigma=volatility;
        b=yield;
    }
    double r;
    double crate;
    double b;
    double sigma;
    double d1;
    double d2;
    double sigadj;
    double badj;
    private void paraM(double s, double x, double t) {
        double adj=sqrt(3);
        sigadj=(sigma/adj);
        double sig=(sigadj*sigadj);
        badj=(b-((sigma*sigma)/6.0))*0.5;
        d1=(log(s/x)+(badj+sig*0.5)*t)/(sigadj*sqrt(t));
        d2=(d1-sigadj*sqrt(t));
    }
    public double geoCall(double s, double x, double t) {
        paraM(s,x,t);
        return s*exp((badj-r)*t)*N(d1)-x*exp(-r*t)*N(d2);
    }
    public double geoPut(double s, double x, double t) {
        paraM(s,x,t);
        return x*exp(-r*t)*N(-d2)-s*exp((badj-r)*t)*N(-d1);
    }
    private static double N(double x) {
        Probnorm p=new Probnorm();
        double ret=x>(6.95)?1.0:x<(-6.95)?0.0:p.ncDisfnc(x);//restrict
        the range of cdf values to stable values
        return ret;
    }
```

LISTING 18.6. Valuation of Geometric Average Rates

18.6.2. Arithmetic Approximations

Turnbull & Wakeman method

This approximation method makes use of a standard Black-Scholes valuation with the adjusted cost of carry and variance to match the mean and variance of an arithmetic moment. The asset price is observed for a set time during which the averaging is done. At this time the asset price is $\overline{S}$. When the option is in the average period the strike is adjusted to $\hat{X}$ and the resulting valuation multiplied by the coefficient T_2/T.

$$C = Se^{(\overline{b}-r)T_2}N(d_1) - Xe^{-rT_2}N(d_2) \tag{18.6.5}$$

$$P = Xe^{-rT_2}N(d_2) - Se^{(\overline{b}-r)T_2}N(d_1) \tag{18.6.6}$$

Where

$$d_1 = \frac{\log(S/X) + (\overline{b} + \overline{\sigma}^2/2)T_2}{\overline{\sigma}\sqrt{T_2}}, \; d_2 = d_1 - \overline{\sigma}\sqrt{T_2}$$

$$\overline{\sigma} = \sqrt{\frac{\log(m_2)}{T} - 2\overline{b}} \qquad (18.6.7)$$

$$\overline{b} = \frac{\log(m_1)}{T} \qquad (18.6.8)$$

The arithmetic moments are:

$$m_1 = \frac{e^{bT} - e^{b\tau}}{b(T-\tau)} \qquad (18.6.9)$$

$$m_2 = \frac{2e^{(2b+\sigma^2)T}}{(b+\sigma^2)(2b+\sigma^2)(T-\tau)^2} + \frac{2e^{(2b+\sigma^2)\tau}}{b(T-\tau)^2}\left[\frac{1}{2b+\sigma^2} - \frac{e^{b(T-\tau)}}{b+\sigma^2}\right] \qquad (18.6.10)$$

$$\hat{X} = \left(\frac{T}{T_2}X - \frac{T_1}{T_2}\overline{S}\right) \qquad (18.6.11)$$

T is original time to maturity,
T_2 is the time remaining,
$T_1 = T - T_2$, is the observed average period.
τ is the time to the start of the averaging observation period.

Example 18.3

An approximate arithmetical option has an asset price of \$100.0 and a strike of \$102.0. The averaging period is to start at the option beginning and lasts for two months. The option lifetime is six months, the risk free rate is 6% and the underlying volatility is 20%. The cost of carry is 3%. During the average period the asset price is estimated at \$96.0. What is the value of a put option?

$S = 100.0$, $\overline{S} = 96.0$, $X = 102.0$, $T = 0.5$, $T_2 = 0.333$, $\tau = 0.0$, $r = 0.06$, $b = 0.03$, $\sigma = 0.20$

$$m_1 = \frac{e^{bT} - e^{b\tau}}{b(T-\tau)} = \frac{e^{0.03*0.5} - e^{0}}{0.03(0.5-0)} = 1.0075$$

$$m_2 = \frac{2e^{(2b+\sigma^2)T}}{(b+\sigma^2)(2b+\sigma^2)(T-\tau)^2} + \frac{2e^{(2b+\sigma^2)\tau}}{b(T-\tau)^2}\left[\frac{1}{2b+\sigma^2} - \frac{e^{b(T-r)}}{b+\sigma^2}\right]$$

$$= \frac{2e^{(2*0.03+0.20^2)0.5}}{(0.03+0.20^2)(2*0.03+0.20^2)(0.5-0.0)^2} + \frac{2e^{(2*0.03+0.20^2)*0.0}}{0.030(0.5-0.0)^2}$$

$$\left[\frac{1}{2^{*}0.03+0.20^{2}}-\frac{e^{0.030(0.5-0.0)}}{0.030+0.20^{2}}\right]=1.02$$

$$\hat{X}=\left(\frac{T}{T_2}X-\frac{T_1}{t_2}\overline{S}\right)=\left(\frac{0.5}{0.333}102.0-\frac{0.167}{0.333}96.0\right)=105.0$$

$$\overline{b}=\frac{\log(m_1)}{T}=\frac{\log(1.0075)}{0.5}=0.015$$

$$\overline{\sigma}=\sqrt{\frac{\log(m_2)}{T}-2\overline{b}}=\sqrt{\frac{\log(1.0219)}{0.5}-2^{*}0.015}=0.1157$$

$$\text{if } T_1>0.0 \text{ then } X=\hat{X} \text{ else } X=X$$

$$d_1=\frac{\log(S/X)+(\overline{b}+\overline{\sigma}^2/2)T_2}{\overline{\sigma}\sqrt{T_2}}=\frac{\log(100.0/105)+(0.015+0.1157^2/2)0.333}{0.1157\sqrt{0.333}}$$

$$=-0.1879$$

$$N(d_1)=0.267,\,N(d_2)=0.245$$

$$d_2=d_1-\overline{\sigma}\sqrt{T_2}=d_1-0.1157\sqrt{0.333}=-0.2547$$

$$P=Xe^{-rT_2}N(d_2)-Se^{(\overline{b}-r)T_2}N(d_1)=105.0e^{-0.06^{*}0.333}*0.245$$

$$-100^{*}e^{(0.015-0.06)0.333}*0.267=3.6306$$

The value of this approximation is \$3.63.

The Turnbull & Wakeman method is implemented in the class **Asianturwakeop** shown in Listing 18.7.

```
package FinApps;
import static java.lang.Math.*;
public class Asianturwakeop {
   public Asianturwakeop(double rate,double yield,double volatility) {
      r=rate;
      crate=yield;
      sigma=volatility;
      b=yield;
      b=crate==0.0?0.0:(b=crate!=rate?(rate-crate):rate);
   }
   double r;
   double crate;
   double b;
   double sigma;
   double d1;
   double d2;
   double sigadj;
```

```
double badj;
public double tWakeput(double s,double sa, double x, double t,
   double t2, double tau) {
   paraM(s,x,t,t2,tau);
   double t1=t-t2;
   double q=r-badj;
   Blackscholecp bs=new Blackscholecp(q);
   x=t1>0.0?(((t/t2)*x)-((t1/t2)*sa)):x;
   bs.bscholEprice(s,x,sigadj,t2,r);
   return t1>0.0?(bs.getPute()*t2/t):(bs.getPute());
}
public double tWakecall(double s,double sa, double x, double t,
   double t2, double tau) {
   paraM(s,x,t,t2,tau);
   double t1=t-t2;
   double q=r-badj;
   Blackscholecp bs=new Blackscholecp(q);
   x=t1>0.0?(((t/t2)*x)-((t1/t2)*sa)):x;
   bs.bscholEprice(s,x,sigadj,t2,r);
   return t1>0.0?(bs.getCalle()*t2/t):(bs.getCalle());
}
private void paraM(double s, double x, double t, double t2,
   double tau) {
   double sig=(sigma*sigma);
   double moment1=(exp(b*t)-exp(b*tau))/(b*(t-tau));
   double term1=2.0*exp((2.0*b+sig)*t)/((b+sig)*(2.0*b+sig)
   *pow((t-tau),(2.0)));
   double term1a=2.0*exp(2.0*(b+sig)*tau)/(b*pow((t-tau),(2.0)));
   double term2=(1.0/(2.0*b+sig)-exp(b*(t-tau))/(b+sig));
   double moment2=(term1+(term1a*term2));
   badj=log(moment1)/(t);
   sigadj=(sqrt(log(moment2)/(t)-2.0*badj));
   double sig2=(sigadj*sigadj);
   d1=(log(s/x)+(badj+sig2*0.5)*t2)/(sigadj*sqrt(t2));
   d2=(d1-sigadj*sqrt(t2));
}
```

LISTING 18.7. Arithmetic Average Approximation

18.6.3. Levy Method

This method uses the technique of moment matching, where the arithmetic average is approximated by a lognormal variable. The first two moments coincide with the true arithmetic distribution. The basic model is ubiquitous in Asian arithmetic models. Further work is described in Lord (2005).

This approximation is valued by:

$$C = \overline{S}N(d_1) - \hat{X}e^{-rT}N(d_2) \tag{18.6.12}$$

$$P = C - \overline{S} + \hat{X}e^{-rT_2} \tag{18.6.13}$$

Where

$$\bar{S} = \frac{S}{Tb}\left(e^{(b-r)T_2 - e^{-rT_2}}\right) \tag{18.6.14}$$

$$d_1 = \frac{1}{\sqrt{V}}\left[\frac{\log(D)}{2} - \log(\hat{X})\right] \tag{18.6.15}$$

$$d_2 = d_1 - \sqrt{V} \tag{18.6.16}$$

$$m = \frac{2S^2}{b+\sigma^2}\left[\frac{e^{(2b+\sigma^2)T_2 - 1}}{2b+\sigma^2} - \frac{e^{bT_2 - 1}}{b}\right] \tag{18.6.17}$$

$$V = \log(D) - 2\left[rT_2 + \log(\bar{S})\right] \tag{18.6.18}$$

$$\hat{X} = X - \frac{T - T_2}{T} S_{av} \tag{18.6.19}$$

$$D = m/T_2 \tag{18.6.20}$$

Example 18.4

Using the same relevant parameters as in Example 18.2. The approximate arithmetical option has an asset price of $100.0 and a strike of $102.0. The remaining time to maturity is four months. The option lifetime is six months, the risk free rate is 6% and the underlying volatility is 20%. The cost of carry is 3%. During the average period the asset price is estimated at $96.0. What is the value of a put option?

$$m = \frac{2S^2}{b+\sigma^2}\left[\frac{e^{(2b+\sigma^2)T_2-1}}{2b+\sigma^2} - \frac{e^{bT_2-1}}{b}\right] = \frac{2{*}100.0^2}{0.03+0.20^2}$$

$$\left[\frac{e^{(2{*}0.03+0.20^2)0.3333-1}}{2{*}0.3333+0.20^2} - \frac{e^{0.03{*}0.03-1}}{0.03}\right] = 1127.077$$

$$D = m/T_2 = 1127.077/0.3333 = 4508.3095$$

$$\overline{S} = \frac{S}{Tb}\left(e^{(b-r)T_2-e^{-rT_2}}\right) = \frac{100.0}{0.5{*}0.03}\left(e^{(0.03-0.06)0.3333-e^{-0.06{*}0.3333}}\right) = 65.6679$$

$$V = \log(D) - 2[rT_2 + \log(\overline{S})] = \log(4508.3095) - 2[0.06{*}0.3333 + \log(65.6679)] = 0.0044$$

$$\hat{X} = X - \frac{T - T_2}{T} S_{av} = 102.0 - \frac{0.5 - 0.3333}{0.5} * 96.0 = 69.993$$

$$d_1 = \frac{1}{\sqrt{V}} \left[\frac{\log(D)}{2} - \log(\hat{X}) \right] = \frac{1}{\sqrt{0.0044}} \left[\frac{\log(4508.3095)}{2} - \log(69.993) \right] = -0.6223$$

$$d_2 = d_1 - \sqrt{V} = -0.6223 - \sqrt{0.0044} = -0.6891$$

$$C = \overline{S}N(d_1) - \hat{X}e^{-rT}\ N(d_2) = 65.6679{*}0.2668 - 69.993e^{-0.068{*}0.5{*}}0.2453 = 0.6896$$

$$C - \overline{S} + \hat{X}e^{-rT_2} = 0.6896 - 65.6679 + 69.993e^{-0.06{*}0.3333} = 3.6294.$$

The put is worth \$3.63. Comparing the computed values of Levy's method and the Turnbull & Wakeman Method:

Levy; 3.294: Turnbull & Wakeman; 3.6306.

There is a small difference, which is in favour of the Levy method as the more accurate approximation.

The class which computes the Levy approximation is shown in Listing 18.8.

```
package FinApps;
import static java.lang.Math.*;
import BaseStats.Probnorm;
public class Asianlevyop {
   public Asianlevyop(double rate,double yield,double volatility) {
      r=rate;
      crate=yield;
      sigma=volatility;
      b=yield;
      b=crate==0.0?0.0:(b=crate!=rate?(rate-crate):rate);
   }
   double r;
   double crate;
   double b;
   double sigma;
   double d1;
   double d2;
   double se;
   double xadj;
   public double levyCall(double s,double sa, double x, double t,
   double t2) {
      paraM(s,sa,x,t,t2);
      return se*N(d1)-xadj*exp(-r*t2)*N(d2);
   }
   public double levyPut(double s,double sa, double x, double t,
```

```
double t2) {
   paraM(s,sa,x,t,t2);
   double val=-se+xadj*exp(-r*t2);
   return (se*N(d1)-xadj*exp(-r*t2)*N(d2))-se+xadj*exp(-r*t2);
}
private void paraM(double s,double sa, double x, double t, double t2) {
   double sig=(sigma*sigma);
   xadj=x-((t-t2)/(t))*sa;
   double m=(2.0*(s*s)/(b+sig))*((exp((2*b+sig)*t2)-1.0)
   /(2.0*b+sig)-(exp(b*t2)-1.0)/(b));
   se=s/(t*b)*(exp((b-r)*t2)-exp(-r*t2));
    double d=(m/(t*t));
   double v=log(d)-2.0*(r*t2+log(se));
   d1=1/sqrt(v)*(log(d)/(2.0)-log(xadj));
   d2=d1-sqrt(v);
}
private static double N(double x) {
   Probnorm p=new Probnorm();
    double ret=x>(6.95)?1.0:x<(-6.95)?0.0:p.ncDisfnc(x);//restrict
   the range of cdf values to stable values
   return ret;
}
```

LISTING 18.8. Levy Arithmetic average approximation

18.7. Quantos

Quanto options were originally due to Derman et al. (1990) and later refinement by Reiner (1992). A quanto (cross country derivative) involves two currencies. The payoff is dependent on the behaviour of the underlying denominated in one of the currencies and the payoff is made in the other currency. An example is the market variable being the S&P500 (denominated in $) and the payoff being in sterling (£), quantos are traded OTC and on the exchanges.

18.7.1. Fixed Exchange Valuation

The valuation for a fixed exchange rate option can be divided into two subsets; the option has value denominated in the domestic currency, or the value is denominated in the foreign currency. For the former case:

$$C = S_{EP}(Ae^{(r_f-r-y-\rho_A\sigma_{EP})T}N(d_1) - Xe^{-rt}N(d_2)) \tag{18.7.1}$$

$$P = S_{EP}\left(Xe^{-rT}N(-d_2) - Ae^{(r_f-r-y-\rho_A\sigma_{EP})T}N(-d_1)\right) \tag{18.7.2}$$

For the latter case:

$$C = AS_{EP}\left(Ae^{(r_f-r-y-\rho_A\sigma_{EP})T}N(d_1) - Xe^{-rT}N(-d_2)\right) \tag{18.7.3}$$

$$P = AS_{EP}\left(Xe^{-rT}N(-d_2) - Ae^{(r_f-r-y-\rho_A\sigma_{EP})T}N(-d_1)\right) \tag{18.7.4}$$

Where,

$$d_1 = \frac{\log(A/X) + (r_f - y - \rho_A\sigma_{EP} + \sigma_A^2/2)T}{\sigma_A\sqrt{T}} \tag{18.7.5}$$

$$d_2 = d_1 - \sigma_E\sqrt{T} \tag{18.7.6}$$

A = underlying asset price (foreign denominated)
X = delivered price (foreign denominated)
S_{EP} = exchange rate (predetermined units of domestic per unit of foreign)
S_E = spot exchange rate (units of foreign per unit of domestic)
σ_{EP} = volatility of domestic exchange rate
σ_A = volatility of underlying asset
r = domestic rate
r_f = foreign rate
y = proportional dividend yield of the underlying asset
ρ = correlation coefficient between the domestic rate and the underlying asset.

Example 18.5

A quanto call has three months to expiry. The index is value at 106.0 and the delivery price is fixed at 110.0. The exchange rate is 1.75 units of domestic currency to 1 unit of foreign currency and the volatility is 9%. The underlying yield is 3% per annum and the underlying volatility is 15% per annum. The domestic interest rate is 5.6% and the foreign rate is 4%. The correlation between the underlying and the exchange rate is 0.35. What is the valuation given in the domestic currency?

The parameters are:

$$A = 106.0, X = 110.0, S_{EP} = 1.75, \sigma_{EP} = 0.09, y = 0.03, \sigma_A = 0.15, r = 0.056,$$

$$r_f = 0.04, \rho = 0.35$$

$$d_1 = \frac{\log(A/X) + (r_f - y - \rho_A\sigma_E\rho + \sigma_A^2/2)T}{\sigma_A\sqrt{T}} =$$

$$\frac{\log(106.0/110.0) + (0.04 - 0.03 - 0.35*0.09 + \sigma_A^2/2)0.25}{0.15\sqrt{0.25}} = -0.4388$$

$$d_2 = d_1 - \sigma_A\sqrt{T}d_2 = -0.4388 - 0.15\sqrt{0.25} = -0.5138$$

$$N(d_1) = N(-0.4388) = 0.3304, \quad N(d_2) = N(-0.5138) = 0.3036$$

$$C = S_{EP}\left(Ae^{(r_f - r - y - \rho_A \sigma_{EP})T} N(d_1) - Xe^{-rT} N(d_2)\right) =$$

$$1.75\left(106.0e^{(0.04-0.056-0.03-0.35*0.09)0.25} * 0.3304 - 110.0e^{-0.056*0.25} * 0.3036\right) = 2.868$$

The value is 2.87.

The class implementing quantos is shown in Listing 18.9.

```
package FinApps;
import static java.lang.Math.*;
import BaseStats.Probnorm;
public class Quantop {
   public Quantop(double rate,double yield,double ratef,double p,
   double time ) {
      r=rate;
      rf=ratef;
      crate=yield;
      rho=p;
      t=time;
   }
   double r;
   double rf;
   double crate;
   double b;
   double rho;
   double t;
   double d1;
   double d2;
   public double callDom(double s, double x,double e, double sigd,
   double siga) {
      paraM(s,x,sigd,siga);
      return e*(s*exp((rf-r-crate-rho*siga*sigd)*t)*N(d1)-x*exp(-r*t)
   *N(d2));
   }
   public double callFrn(double s, double x,double e,double ef,
   double sigd, double siga) {
      paraM(s,x,sigd,siga);
      return ef*e*(s*exp((rf-r-crate-rho*siga*sigd)*t)*N(d1)
   -x*exp(-r*t) *N(d2));
}
   public double putFrn(double s, double x,double e,double ef,
   double sigd, double siga) {
      paraM(s,x,sigd,siga);
      return e*ef*(x*exp(-r*t)*N(-d2)-s*exp((rf-r-crate-rho*siga
```

```
    *sigd)*t) *N(-d1));
  }
  public double putDom(double s, double x,double e, double sigd,
  double siga) {
    paraM(s,x,sigd,siga);
    return e*(x*exp(-r*t)*N(-d2)-s*exp((rf-r-crate-rho*siga*sigd)
  *t) *N(-d1));
  }
  private void paraM(double s, double x, double sigd, double siga) {
    d1=(log(s/x)+(rf-crate-rho*siga*sigd+(siga*siga*0.5))*t)
  /(siga*sqrt(t));
    d2=d1-siga*sqrt(t);
    System.out.println("d1=="+d1+" d2=="+d2);
  }
  private static double N(double x) {
    Probnorm p=new Probnorm();
    double ret=x>(6.95)?1.0:x<(-6.95)?0.0:p.ncDisfnc(x);//restrict
  the range of cdf values to stable values
    return ret;
  }
  }
```

LISTING 18.9. Class for valuation of Quantos

18.7.2. Foreign Exchange Option

The equity linked foreign exchange option is used to link the value of the forward price of a stock or stock index to the quantity of the underlying. This option is developed by Reiner 1992, the valuation is given by:

Option in domestic currency

$$C = S_E A e^{-yT} N(d_1) - XAE^{(r_f - r - y - \rho_A \sigma_E)T} N(d_2) \tag{18.7.7}$$

$$P = AXe^{(r_f - r - y - \rho_A \sigma_E)T} N(-d_2) - AS_E e^{-yT} N(-d_1) \tag{18.7.8}$$

Option in foreign currency

$$C = Ae^{-yT} N(d_1) - XS_{fd} Ae^{(r_f - r - y - \rho_A \sigma_E)T} N(d_2) \tag{18.7.9}$$

$$P = AXS_{fd} e^{(r_f - r - y - \rho_A \sigma_E)T} N(-d_2) - Ae^{-yT} N(-d_1) \tag{18.7.10}$$

Where,

$$d_1 = \frac{\log(S_E/X) + (r - r_f + \rho_A \sigma_E + \sigma_E^2/2)T}{\sigma_E \sqrt{T}} \tag{18.7.11}$$

$$d_2 = d_1 - \sigma_E \sqrt{T} \tag{18.7.12}$$

A = underlying asset price (foreign denominated)
X = currency strike price (domestic)
S_{fd} = exchange rate (units of foreign currency per unit of domestic currency)
S_E = spot exchange rate (units of domestic currency per unit of foreign currency)
σ_E = volatility of domestic exchange rate
σ_A = volatility of underlying asset
r = domestic rate
r_f = foreign rate
y = proportional dividend yield of the underlying asset
ρ = correlation coefficient between the domestic rate and the underlying asset.

Example 18.6

An FX put option with six months to expiry has the index at 110.0, the exchange rate at 1.6 and the strike at 1.7. The index has a volatility of 17%, the currency has a volatility of 11%. The domestic rate is 5.5%, the foreign rate is 3% and the index yield is 2%. If the correlation between currency and index is at -0.35, what is the value in domestic currency?

$$d_1 = \frac{\log(S_E/X) + (r - r_f + \rho_A \sigma_E + \sigma_E^2/2)T}{\sigma_E \sqrt{T}} =$$

$$\frac{\log(1.6/1.7) + (0.055 - 0.03 + -0.35*0.11 + 0.11_E^2/2)0.5}{0.11\sqrt{0.5}} = -0.6218$$

$$d_2 = d_1 - \sigma_E \sqrt{T} = -0.6218 - 0.11\sqrt{0.5} = -0.6996$$

$$N(-d_1) = N(--0.6218) = 0.7329\ ,\ N(-d_2) = N(--0.6996) = 0.7579$$

$$P = AXe^{r_f - r - y - \rho_A \sigma_E)T} N(-d_2) - AS_E e^{-yT} N(-d_1) =$$

$$110.0*1.7e^{(0.03-0.055-0.02--0.35*0.11)0.5}*0.7579 - 110.0*1.6e^{-0.02*0.5}*0.7329 = 11.311$$

The value in domestic currency is 11.31.

The class implementing FX options is shown in Listing 18.10

```
package FinApps;
import static java.lang.Math.*;
import BaseStats.Probnorm;
public class Foreignxop {
   public Foreignxop( double rate,double yield,double ratef,double p,
   double time ) {
      r=rate;
      rf=ratef;
      crate=yield;
      rho=p;
      t=time;
   }
```

```
double r;
double crate;
double rf;
double rho;
double t;
double d1;
double d2;
public double callDom(double s, double x, double e, double sigd,
double siga) {
   paraM(s,x,e,sigd,siga);
   return e*s*exp(-crate*t)*N(d1)-x*s*exp((rf-r-crate-rho*siga
   *sigd) *t)*N(d2);
}
public double callFrn(double s, double x, double e, double sigd,
double siga) {
   paraM(s,x,e,sigd,siga);
   return s*exp(-crate*t)*N(d1)-e*x*s*exp((rf-r-crate-rho*siga
   *sigd) *t)*N(d2);
}
public double putFrn(double s, double x, double e, double sigd,
double siga) {
   paraM(s,x,e,sigd,siga);
   return e*x*s*exp((rf-r-crate-rho*siga*sigd)*t)*N(-d2)-s*exp
(-crate*t)*N(-d1);
}
public double putDom(double s, double x, double e, double sigd,
double siga) {
   paraM(s,x,e,sigd,siga);
   return x*s*exp((rf-r-crate-rho*siga*sigd)*t)*N(-d2)-e*s*exp
(-crate*t)*N(-d1);
}
private void paraM(double s, double x,double e, double sigf,
double siga) {
   d1=(log(e/x)+(r-rf+(rho*sigf*siga+(sigf*sigf*0.5)))*t)
/(sigf*sqrt(t));
   d2=d1-sigf*sqrt(t);
}
private static double N(double x) {
   Probnorm p=new Probnorm();
   double ret=x>(6.95)?1.0:x<(-6.95)?0.0:p.ncDisfnc(x);//restrict
the range of cdf values to stable values
   return ret;
}
```

LISTING 18.10. Foreign Exchange Options

References

Derman, E., P. Karasinski and J. S. Wecker (1990). "Understanding Guaranteed Exchange-Rate Contracts in Foreign Stock Investments," *International Equity Strategies*. Goldman Sachs, June.

Hart, I. and M. Ross (1994). "Striking Continuity," *Risk Magazine*, 7(6).

Heynen, R. C. and H. M. Kat (1996). "Discrete Partial Barrier Options with a moving Barrier," *Journal of Financial Engineering*, 5(3), 199–210.

Kemna, A. and A. Vorst (1990). "A Pricing Method for Options Based on Average Asset Values," *Journal of Banking and Finance*, 14, 113–129.

Levy, E. (1992). "Pricing European Average Rate Currency Options," *Journal of International Money and Finance*, 11, 474–491.

Lord, C. (2005). "A Motivation for Conditional Moment Matching," *Modelling and Research* (UC-R-355), Rabobank International, Utrecht. The Netherlands.

Merton, R. (1973). "Theory of Rational Option Pricing," *Bell Journal of Economics and Management Science*, 4, 141–183.

Reiner, E. (1992). "Quanto Mechanics," *Risk Magazine*, 5, 59–63.

Turnbull, S.M. and L. M. Wakeman (1991). "A Quick Algorithm for Pricing European Average Options," *Journal of Financial and Quantitative Analysis*, 26, 377–389.

Pfeffer, D. (2001). "Sequential Barrier Options," *Algo Research Quarterly*, 4(3).

Zhang, J. E. (2001). "A Semi-analytical Method for Pricing and Hedging Continuously Sampled Arithmetic Average Rate Options," *Journal of Computational Finance*, 5(1).

Appendix 1

Listings for CoreMath classes.

```
package CoreMath;
import static java.lang.Math.*;
public abstract class ContFract
{         public double prec=1E-30;
          private double nume;
          private double denom;
          private double interim;
          private double lentzval;
          private double oldans;
          private int n=1;
      double[] vars=new double[2];
          abstract void computeFract(int n);
          public void setInitial(double numerator, double denominator)
          {

                    nume=numerator;
                    denom=denominator;

          }
          public void setInt(int n)
          {
                    this.n=n;
          }
          private int getInt()
          {
                    return n;
          }

          public void setFrac(double initial )
          {
                    //initial value of Fn //
                    lentzval=initial;

          }
          public double floorvalue(double x)
          {
                    return abs(x)<Csmallnumber.getSmallnumber()?
                           Csmallnumber.getSmallnumber():x;
          }
```

```
public double getFrac()
{
        return lentzval;
}
public void evalFract()// lentzs method...................//
{
        int i=getInt();

while(abs(oldans-1.0)>prec)// terminating criteria //
{

        computeFract( ++i);//
        set up the a, b, x and initial values for //
        denom=floorvalue((vars[1]+ vars[0]*denom));//
        array contains numerator and denominator //
        denom=(1.0/denom);
        nume=floorvalue((vars[1]+ (vars[0]/nume)));
        oldans=nume*denom;
        lentzval*=(nume*denom);// Cn * Dn //


}
setFrac(lentzval);
            }
            }
```

LISTING A1. Continued Fraction// Uses the Lentz algorithm to compute the continued fraction

```
package CoreMath;
public class Csmallnumber
{
        static private double negmacprec=0;
        static private int radix=0;
        static private double smallnumber=0;
        private static void cRad()
        {
                double a=1.0d;
                double tmp1,tmp2;
                do{a+=a;
                        tmp1=a+1.0d;
                        tmp2=tmp1-a;
                }
                while(tmp2-1.0d!=0.0d);
                double b=1.0d;
                while(radix==0)
                {
                        b+=b;
                        tmp1=a+b;
                        radix=(int)(tmp1-a);
        }
        }
        private static void snum()
        {
                double floatradix=getRadix();
```

```
            double inverseradix=1.0d/floatradix;
            double fullmantissa=1.0d-floatradix
            *getnegmachineprec();
            while(fullmantissa!=0.0d)
            {
                    smallnumber=fullmantissa;
                    fullmantissa*=inverseradix;
            }

    }
    public static int getRadix()
    {
            if(radix==0)
            cRad();
            return radix;
    }
    private static void compnegprec()
    {
            double frad=getRadix();
            double invrad=1.0d/frad;
            negmacprec=1.0d;
            double tmp=1.0d-negmacprec;
            while(tmp-1.0d!=0.0d)
            {
                    negmacprec*=invrad;
                    tmp=1.0d-negmacprec;
            }
    }
    public static double getnegmachineprec()
    {
            if(negmacprec==0)
            compnegprec();
            return negmacprec;
    }
    public static double getSmallnumber()
    {
            if(smallnumber==0)
            snum();
            return smallnumber;
    }
}
```

LISTING A2. Class to compute basic machine parameters. Returns smallest number available (After Press et al and extensions by Besset, DH)

```
package CoreMath;
import java.util.*;
import java.text.*;
public abstract class Derivative//
{
   public abstract double deriveFunction(double fx); //
   returns a double...... the function//
   public double h=1e-6;//
   DEFAULT degree of accuracy in the calculation//
```

```
   public double derivation(double InputFunc) {
      double value=0.0;
      double X2=0.0;
      double X1=0.0;
      X2=deriveFunction(InputFunc-h);
      X1=deriveFunction(InputFunc+h);
      value=((X1-X2)/(2*h));
      return value;
   }
   public double secondderiv(double InputFunc) {
      double value=0.0;
      double X2=0.0;
      double X1=0.0;
      double basefunction=0.0;
      X2=deriveFunction(InputFunc-h);
      X1=deriveFunction(InputFunc+h);
      basefunction=deriveFunction(InputFunc);
      value=((X1+X2-(2*basefunction))/(h*h));
      return value;
   }
}
```

LISTING A3. Abstract Class that computes first & second derivative

```
package CoreMath;
import static java.lang.Math.*;
public class Errf extends PolyEval// After Abramovitz et al //
{
        private static double p=0.3275911;
        private double[] polycofs={0.0,0.254829592, -0.284496736,
        1.421413741, -1.453152027, 1.061405429};//
        to use Horner's rule add constant//
        private double[] answers=new double[2];
        private double derivative;
        private double compliment;
        private void setDeriv(double der)
        {
                derivative=der;
        }
        public double getDerivative()
        {
                return derivative;
        }
        private void setComp(double errorf,double sign)
        {// adjusted for -ve x values//
                compliment= sign<0?(2-(1-errorf)):1-errorf;

        }
        public double getErfc()
        {
                return compliment;
        }

        public void polycoeffs(double[] input)
```

```
        {
                setcoefficients(input);
        }
        public double erf(double x)
        {               //THIS USE'S HORNERS RULE//
                double sign=x;
                x=abs(x);
                double norm=exp(-pow(x,2));
                double norm2=norm*2/sqrt(PI);//
                derivative of the error function as norm2//
          setDeriv(norm2);
          if(x==0)
        {
        setComp(0,0);
        return 0;
        }
        else
        {
                double t=1.0/(1.0+p*x);
        polycoeffs(polycofs);
                answers=evalDerv(t);
                double v0=(1.0-answers[0]*norm);
                setComp(v0,sign);// compliment is set prior
                                //to adjusting for sign
                return sign<0?-v0:v0;
                                }
}
}
```

LISTING A4. Class to compute the error function

```
package CoreMath;
import java.util.*;
import static java.lang.Math.*;
public final class Function {
  static double factaray[]=new double[200];
  static double vals[]={76.18009172947146,-86.50532032941677,
  24.01409824083091,-1.231739572450155,0.1208650973866179e-2,
  -0.5395239384953e-5};// Lanczos coefficient values
  protected static double lgamma(double x)//
  returns the log of the values.. for large values
  {
     return x>1? log(x+5.5)*(x+0.5)-(x+5.5)+log(coeffs(x)
     *sqrt(2*PI)/x):(x<=1.0&&x>0?lgamma(x+1)-log(x):
     (x<0?log(reflect(x)):Double.NaN));
  }
  public static double gamma(double x) // implements the algorithm
  {
     double g=0;
     if(x>1) {
        g=(exp(log(x+5.5)*(x+0.5)-(x+5.5))*coeffs(x)*sqrt(2*PI)/x);
```

```
    } else
      if(x<=1.0&&x>0) {
      g=(gamma(x+1)/x);
      } else
        if(x<0) {
      g=(PI)/(gamma(1-x)*(sin(PI*x)));
        }
    return g;
  }
  public static double reflect(double x)//
  negative non-integer values only
  {return
      (PI)/(gamma(1-x)*(sin(PI*x)));
  }
  private static double coeffs(double x) {
    double ans=1.000000000190015;
    double term=x;
    for(int i=0;i<6;i++) {
      term+=1;
      ans+=vals[i]/term;
    }
    return ans;
  }

  public static double factorial( int x) {
    //maximum to prevent overflow  170 //
    return x <2?1:(x>40?exp(lgamma(x+1.0)):x*factorial(x-1));
  }
  public static double binom(int n, int k) {
    // max 100 n 17 k Binomial coefficient no of ways that
    k things can be selected from n items//
    return floor( exp(lgamma(n+1.0)-lgamma(n-k+1.0)-lgamma
                    (k+1.0)));
  }
  public static double beta( double x, double y) {
    return exp(lgamma(x)+lgamma(y)-lgamma(x+y));
  }
  public static double betainv( double x, double y) {
    // returns same as 1/beta //
    return exp(lgamma(x+y)-lgamma(x)-lgamma(y));
  }
  public static double lbeta(double x, double y) {
    return (lgamma(x+y)-lgamma(x)-lgamma(y));
  }
  public static double logbeta(double x, double y) {
    return (lgamma(x)+lgamma(y)-lgamma(x+y));
  }
}
```

LISTING A5. Class dealing with basic functions

```
package CoreMath;
import static java.lang.Math.sqrt;
   public Genwiener() {
   }
   private double constdrift;
   private double wienervalue;
   private void setDrift(double driftval) {
      constdrift=driftval;
   }
   public double getDrift() {
      return constdrift;
   }
   private void setWiener(double wienval) {
      wienervalue=wienval;
   }
   public double getwienerVal() {
      return wienervalue;
   }
   public double genWienerproc(double drift, double t, double sd) {
      Wiener w=new Wiener();
      double deltaz;
      double driftvalue;
      double deltax;
      deltaz=w.wienerProc(t);
      setWiener(deltaz);
      driftvalue=drift*t;
      setDrift(driftvalue);
      deltax=(driftvalue+(sd*deltaz));
      return deltax;
   }
}
```

LISTING A6. Class to compute the Wiener process

```
package CoreMath;
import static java.lang.Math.*;

public final class Ibeta extends ContFract {
   private static double a=0;
   private static double b=0;
   private static double x=0;
   private double ans;
private void setFraction(double aval,double bval,double xval) {
   a=aval;
   b=bval;
   x=xval;
}

private void setAns(double val) {
   ans=val;
}
public double getAns() {
   return ans;
```

```
}

public double betai(double a1, double a2, double xval) {
   double d;
   double result;
   double interim;
   double lead;
   lead=eval(a1,a2,xval);// get the leading factor//
   if((xval<0.0)||(xval>1.0))
      return 0;//an error to be handled//
   if(xval==0)
      return 0;
   if(xval==1.0)
      return 1;
   if(xval<(a1+1.0)/(a1+a2+2.0)) {
      setFraction(a1,a2,xval);

      d=floorvalue(1.0-(a+b)*x/(a+1.0));//initial value //
      d=1.0/d;
      setInitial(1.0,d);
      setFrac(d);
      evalFract();//get the CF //
      result=lead*getFrac()/a;

      setAns(result);

   }

   else {
      setFraction(a2,a1,(1.0-xval));
      System.out.println("x<a+1/a+b+2 is NOT TRUE");

      d=floorvalue(1.0-(a+b)*x/(a+1.0));//initial value //
      d=1.0/d;
      setInitial(1.0,d);
      setFrac(d);
      evalFract();//get the CF //
      result=1.0-lead*getFrac()/a;
      //result=1.0-lead*betacf(a,b,x)/a;  nice test //
      setAns(result);
   }
   return getAns();
}
private double eval(double a1, double a2, double x)//
produces the leading factor to multiply the continued fraction value ,
this is independent of it being inverse or not//
{
   double leadinfactor;
   if(x==0)
      return 0;
   if(x==1)
      return 1;
   leadinfactor=exp(Function.lbeta(a1,a2)+a1*log(x)+a2*log(1-x));
   return leadinfactor;
}
protected void computeFract(int n) //
```

```
implements the data required to compute the continued fraction //
{
   // this is sensitive to the inverse so variables are changeable //
   vars[1]=1.0;// for ibeta the numerator is fixed at 1 ..The b's//
   int i=n/2;
      int j=2*i;

      vars[0]=j==n?x*i*(b-i)/((a-1.0+j)*(a+j)):0.0-(a+i)*(a+b+i)
      *x/((a+j)*(a+1.0+j));
      return;// the latest update to be evaluated at the super class //
   }
private double betacf(double a, double b, double x ) //
After Press et al does the continued fraction only..needs lead for
       ibeta//
{
   int n, n2;
   double retval, aa, del;
   double[] lentzfactors= new double[2];
   lentzfactors[0]=1.0;
   lentzfactors[1]= floorvalue(1.0 - (a+b)* x / (a+1.0));//1 //
   lentzfactors[1]=1.0/lentzfactors[1];
   retval=lentzfactors[1];
   n=1;
   del=0;
   while(abs(del-1.0)>prec) {
      n2 = n + n;
      aa = (b-n) * n * x / ( (a-1.0+n2) * (a+n2) );
      lentzfactors[1]=floorvalue( 1.0 + aa * lentzfactors[1]);
      lentzfactors[0]=floorvalue( 1.0 + aa / lentzfactors[0]);
      lentzfactors[1]=1.0/lentzfactors[1];
      retval*=lentzfactors[0]*lentzfactors[1];
      aa = 0.0 - (a+n) * ((a+b)+n) * x / ( (a+n2) * (a+1.0+n2) );
      lentzfactors[1]=floorvalue( 1.0 + aa * lentzfactors[1]);
      lentzfactors[0]=floorvalue( 1.0 + aa / lentzfactors[0]);
      lentzfactors[1]=1.0/lentzfactors[1];
      del=lentzfactors[0]*lentzfactors[1];
      retval *= del;
      n++;
   }

   return (retval);
 }

}
```

Listing A7. Class to compute Incomplete Beta

```
package CoreMath;
import static java.lang.Math.*;
public class Igamma extends ContFract
{
private double a;
private double b;
```

```
private double x;
private double igamma;
private double igammaq;
private double upperigam;
private double tricomi;
private double gamcomp;
private double qval;
private double pval;
private double chisquare;
private double chisquarec;
public void setB(double bval)
{
b=bval;
}
private double getB()
{
return b;
}

private void setUpgam(double value)
{
upperigam=value;
}
private double getUpgam()
{
return upperigam;
}
private void setTricom(double value)
{
tricomi=value;
}
private double getTricom()
{
return tricomi;
}

private void setCumQ(double value)
{
qval=value;
}
private double getCumQ()
{
return qval;
}
private void setCumP(double value)
{
pval=value;
}
private double getCumP()
{
return pval;
}

private void setChis(double chival)
{
```

```
chisquare=chival;
}
private double getChis()
{
return chisquare;
}
private void setChisc(double chival)
{
chisquarec=chival;
}
private double getChisc()
{
return chisquarec;
}
private void setGam(double value)
{
igamma=value;
}
public double getGam()
{
return igamma;
}
private void setGamq(double value)
{
igammaq=value;
}
public double getGamq()
{
return igammaq;
}
private void setComp(double compliment)
{
gamcomp=compliment;
}
public double getComp()
{
return gamcomp;// returns the compliment of igamma//
}
public double errf(double xerr)
{//error function for all x values positive and negative//
double aconst;
double xvar;
double sign=xerr;
if(xerr==0)
return 0;
aconst=0.5;// constant for igamma //
xvar=xerr*xerr;
double ans=igamrp(aconst,xvar);
return sign<0? -ans:ans;
}
public double errfc(double xerr)
{// complimentary error func
```

```
double aconst;
double xvar;
if( xerr==0)
return 1;
aconst=0.5;
double sign=xerr;
xvar=xerr*xerr;
double erf=igamrq(aconst,xvar);
return sign<0?(2-erf):erf;
}
private double rgami(double a, double b)
{
return (b>a+1.0)?igamrq(a,b):igamrp(a,b);
}
public double tricomi(double aval, double b)//
little gamma if divided by gama a gives P(ax) lower gamma ratio//
{
a=aval;
x=b;
double serdev=computeSeries();
double lead=leadhilo();
double value=lead*serdev;
setTricom(value);
return getTricom();
}
public double igamup(double aval,double bval)//
R(a,x)if divided by gamma a becomes Q(a,b) //
{
setInt(0);
double fn;// R(a,x) //
double leadfactor;
a=aval;
x=bval;
double minval=floorvalue(0.0);
double val=x-a+1.0;
double den=1.0/val;
double num=1.0/minval;
setInitial(num,den);
setB(val);
setFrac(den);// CONTINUED FRACTION //
evalFract();
fn=getFrac();
//leadfactor=leadhilo();
leadfactor=leadhilo();
double value=leadfactor*fn;
setUpgam(value);
return getUpgam();
}
public double igamrp(double aval, double valx)//
computelower incomplete gamma function ratio//
{
a=aval;
```

```
x=valx;
double serdev=computeSeries();
//double lead=leadhilo();
double lead=leadvalue();//tricomi/rrrrra//
double value=lead*serdev;
setGam(value);// P(a,x) //
return getGam();
}
public double igamrq(double aval, double valx)//
computeupper incomplete gamma function ratio Q//
{

setInt(0);
double fn;// R(a,x) //
double leadfactor;
a=aval;
x=valx;
double minval=floorvalue(0.0);
double val=floorvalue(x-a+1.0);
double den=1.0/val;
double num=1.0/minval;
setInitial(num,den);
setB(val);
setFrac(den);// CONTINUED FRACTION //
evalFract();
fn=getFrac();
leadfactor=leadvalue();
double value=leadfactor*fn;
setGamq(value);
return getGamq();

}
protected void computeFract(int n)
{

b+=2;// auxillary bn=x-a+2n for n=0.......| x+2,x+4,x+6 etc'//
vars[0]=n*(a-n);
vars[1]=b;
return;
}

private double computeSeries()
{
double sum;
double terms;
double summation;
double termprevious=0;
terms=1.0/a; // initialize the series //
summation=a;
sum=1.0/a;
while(abs(sum-termprevious)>prec)
{
termprevious=sum;
summation+=1.0;
terms*=x/summation;
```

```
sum+=terms;
}
return sum;
}

private double leadvalue()// lead value for igamma ratio //
{
double value;
value=exp(log(x)*a-x-Function.lgamma(a));
return value;
}
private double leadhilo()
{
double value;
value=(exp(-x)*pow(x,a));
return value;//
returns lower or upper incomplete gamma lead values.....not the ratio//
}
public final double gcf(double a, double x)
{
this.a=a;
this.x=x;
int i=1;
double retval=0;
double fn,an, b, c, d, den, h;
double minval=floorvalue(0.0);
b = x + 1.0 - a;
c = 1.0 / minval;
d = 1.0 / b;
fn=d;
den=d;
while (abs(den - 1.0) > 1e-30)
{
an = i * (a - i);
b += 2.0;
d = floorvalue(an * d + b);
c =floorvalue( b + an / c);
d = 1.0 / d;
den = floorvalue(d * c);
fn*=den;

i++;
// if (abs(del - 1.0) < 2e-15) break;
}
double lead=leadhilo();
double ret=lead*fn;
return (ret);
}

public final double poisscumQ(int k,double xval)//
x is expected mean, also returns the prob that
p(x)is between 0 and k-1 NOT at most k//
{
if(xval<0)
return 0;
```

```
if(k<1)
return exp(-k);
this.a=k;
this.x=xval;
double value=rgami(a,x);
return x>a+1.0?getGamq():1-getGam();

}
public final double poisscumP(int k, double xval)
{
if(x<0)
return 0;
if(k<1)
return exp(-k);
this.a=k;
this.x=xval;
double value=rgami(a,x);
return x>a+1.0?1-getGamq():getGam();
}
public final double chisqr(double chival,int fval)//
lower tail//
{// CDF //
double v=fval/2.0;
double kisqr=chival/2.0;
double divisor=Function.gamma(v);
double tric=tricomi(v,kisqr);
double value=tric/divisor;
setChis(value);
return chival==0? 0 :getChis();

}
public final double chisqrc(double chival,int fval)//
Upper tail for x2<35 v<=30//
{

double value=chisqr(chival,fval);
setChisc(1.0-value);
return chival==0?1:getChisc();
}
public final double chisqrpdf(double x, int fval)
{
double num= pow(x,((0.5*fval)-1))*exp(-0.5*x);
double den=Function.gamma(0.5*fval)*pow(2,0.5*fval);
return num/den;
}
public final double chisqrc2(double chival,int fval)//
Upper tail for x2<35 v<=30//
{
double v=fval/2.0;
double kisqr=chival/2.0;
double divisor=Function.gamma(v);
double upgam=igamup(v,kisqr);
//igamratio(v,kisqr);
double value=upgam/divisor;
```

```
//setChisc(value);
setChisc(upgam);
return chival==0? 1 :getChisc();
}

}
```

LISTING A8. Class to compute the Incomplete Gamma function. & dependent distributions

```
package CoreMath;
import java.util.*;
import java.lang.*;
import static java.lang.Math.*;
public class Interpolate {
   public double lagrange(double[][] valpairs,double xval) {
      int n=valpairs.length;
      double pn=0.0;
      for(int i=0;i<n;i++) {
         double px=1;
         for(int j=0;j<n;j++) {
            if(j!=i)
               px*=((xval-valpairs[j][0])/
(valpairs[i][0]-valpairs[j][0]));
         }
         pn+=px*valpairs[i][1];
      }

      return pn;

   }

   public double neville(double[][] valpairs, double xval)//
interpolates for the given value xval Neville type interpolation//
   {
      // for positive values only //
      double prec=1e-2;
      int k=0;
      int ky=0;
      int n=valpairs.length;
      double[][] kpvals= new double[n][2];
      double x;
      double v;
      double nume;
      double denom;
      double newp=0;
      double compareval;
      int counter=0;
      int m=0;
      int indx= 1;
         ArrayList<Double> pvalues=new ArrayList<Double>();
      TreeMap<Double, Double> h= new TreeMap<Double, Double>();//
      sorted//
```

```
        for(int i=0;i<n;i++) {
          if(xval>valpairs[i][0] ) {
            h.put(new Double(xval-(valpairs[i][0])),
            new Double(valpairs[i][1]));

          } else {
            h.put(new Double(abs(xval-valpairs[i][0])),
            new Double(-valpairs[i][1]));// mark the negative terms//
          }
        }
        Iterator<Double> kee=h.keySet().iterator();
        Iterator<Double> val=h.values().iterator();
        while(val.hasNext()) {
        x=val.next().doubleValue();
        v= (x > 0.0) ? kee.next().doubleValue()-xval:xval+kee.next().
        doubleValue();// Reconstituting the values //


        x=abs(x);
        v=abs(v);
        kpvals[k][0]=v;
        kpvals[k][1]=x;
        if(counter>=1) {
          pvalues.add(ky,new Double(((((xval-kpvals[ky][0])
          *kpvals[ky+1][1] )+((kpvals[ky+1][0]-xval)
          *kpvals[ky][1]))/(kpvals[ky+1][0]-kpvals[ky][0]))));
          double res= pvalues.get(ky).doubleValue();
          ky++;
        }
        counter++;
        k++;
      }
      while(!pvalues.isEmpty()) {
        indx++;
        compareval=pvalues.get(0).doubleValue();

        for(m=0;m<pvalues.size()-1;m++) {
          nume= (((xval-kpvals[m][0])
          *(pvalues.get(m+1).doubleValue()))+(kpvals[indx+m][0]-xval)
          *(pvalues.get(m).doubleValue()));
          denom=(kpvals[(indx+m)][0]-kpvals[m][0]);
          newp=nume/denom;
          pvalues.set(m,new Double(newp));
        }
        if((abs(compareval-newp))<prec) {
            return compareval;
        }
        pvalues.remove(m);
        }
        return newp;
      }
    }
```

Listing A9. Class to compute Interpolation

```
package CoreMath;
import static java.lang.Math.*;
import java.util.*;
* Computes the generalised Wiener process where the parameters
are functions of the underlying variable
public class Itoprocess {

   public Itoprocess() {
   }
   private double sdchange;
   private double meanvalue;
   private double changebase;
   private void setChange(double changevalue) {
      changebase=changevalue;
   }
   public double getBaseval() {
      return changebase;
   }
   private void setSd(double sd) {
      sdchange=sd;
   }
   public double getSd() {
      return sdchange;
   }
   private void setMean(double drift) {
      meanvalue=drift;
   }
   public double getMean() {
      return meanvalue;
   }

   /**
   *
   * @param mu mean value
   * @param sigma The variance
   * @param timedelta time periods for each step
   * @param basevalue the starting value
   * @return The change in the base value
   */
   public double itoValue(double mu, double sigma, double timedelta,
   double basevalue) {

      setSd(basevalue*(sigma*sqrt(timedelta)));
      Genwiener g=new Genwiener();
      mu=mu*basevalue;
      sigma=sigma*basevalue;
      double change=( g.genWienerproc(mu, timedelta, sigma));
      setChange(change);
      setMean(g.getDrift());
      return change;
   }
}
```

LISTING A10. Class to compute Ito process

```
package CoreMath;
import static java.lang.Math.*;
public class Partialdiff {// takes the form: y^2x/yx^2..fractional input

  /** Creates a new instance of Partialdiff */
  public Partialdiff() {
  }
  private double diffone;
  private double difftwo;
  private void setOne(double fdiff) {
    diffone=fdiff;
  }

  private void setTwo(double sdiff) {
    difftwo=sdiff;
  }
  public double getFdiff() {
    return diffone;
  }

  public double getSdiff() {
    return difftwo;
  }

  public void diffValues(double[] coefficients, double x)//
  equation terms from 1 to n, x the the value of the variable (numerator)
  {
    double dp2=0.0;
    double dp=0.0;
    int cnt=0;
    int n=-1;
    double[] firstdiff=new double[coefficients.length];
    for(double d:coefficients) {
      firstdiff[cnt]=(d*n*pow(x,n-1));
      dp2+=((d*n*(n-1))*pow(x,n-2));
      dp+=firstdiff[cnt];
      cnt++;
      n—;
    }
    setOne(dp);
    setTwo(dp2);
  }

}
```

LISTING A11. Class to compute first terms in differential

Appendix 2

Package BaseStats

```
package BaseStats;
public enum Bivnorm {

     bivar_params{
         public double evalArgs(double a, double b, double p)
                   double paramval =(a*b*p);
                    int caseval=0;
                   int casea=0;
                   int caseb=0;
                   double val1=0.0;
                   double val2=0.0;
                   double densityf=0.0;
                   if(paramval<=0) {

caseval=(a<=0&b<=0&p<=0)?1:(a<=0&b>=0&p>=0)?
2:(a>=0&b<=0&p>=0)?3:(a>=0&b>=0&p<=0)?4:caseval;
                   } else {
                      int signa;
                      int signb;
                      signa=(signum(a)==(0|1))?1:-1;
                      signb=(signum(b)==(0|1))?1:-1;
                      double p1=((((p*a)-b)*signa)/
                      (sqrt((a*a)-2*p*a*b+(b*b))));
                      double p2=((((p*b)-a)*signb)/
                      (sqrt((a*a)-2*p*a*b+(b*b))));
                      double delta=((1-signa*signb)/4.0);
                      double ba=0.0;

casea=(a<=0&ba<=0&p1<=0)?1:(a<=0&ba>=0&p1>=0)?
2:(a>=0&ba<=0&p1>=0)?3:(a>=0&ba>=0&p1<=0)?4:caseval;
                      double aa=b;

caseb=(aa<=0&ba<=0&p2<=0)?1:(aa<=0&ba>=0&p2>=0)?
2:(aa>=0&ba<=0&p2>=0)?3:(aa>=0&ba>=0&p2<=0)?4:ca seval;

val1=casea==1?(allvars_less.valuesz(a,0.0,p1)):casea==2?
(limita_less.valuesz(a,0.0,p1)):casea==3?
(limitb_less.valuesz(a,0.0,p1)):(varrho_less.valuesz(a,0.0,p1));

val2=caseb==1?(allvars_less.valuesz(b,0.0,p2)):caseb==2?
(limita_less.valuesz(b,0.0,p2)):caseb==3?
(limitb_less.valuesz(b,0.0,p2)):
(varrho_less.valuesz(b,0.0,p2));
```

```
                    densityf=((val1+val2)-delta);
                    return densityf;
                }
                switch(caseval) {
                    case 1:densityf=allvars_less.valuesz(a,b,p);
                    break;
                    case 2:densityf= limita_less.valuesz(a,b,p);
                    break;
                    case 3: densityf=limitb_less.valuesz(a,b,p);
                    break;
                    case 4: densityf=varrho_less.valuesz(a,b,p);
                    break;

                }

                return densityf; }
    },
    limita_less{
        double valuesz(double a,double b,double p) {
                Probnorm pn=new Probnorm();

                return ((pn.ncDisfnc(a))-(pn.cumBiv(a,-b,-p)));
        }


    },

    allvars_less {
        double valuesz(double a,double b,double p) {
                Probnorm pn=new Probnorm();
                return pn.cumBiv(a,b,p);

        }
    },
    limitb_less{
        double valuesz(double a,double b,double p) {
                Probnorm pn=new Probnorm();
                return (pn.ncDisfnc(b)-pn.cumBiv(-a,b,-p));

        }

    },
    varrho_less{
        double valuesz(double a,double b,double p) {
                Probnorm pn=new Probnorm();
                return (((pn.ncDisfnc(a)+pn.ncDisfnc(b))-1)+pn.
                    cumBiv(-a,-b,p));
        }

    };
    Bivnorm() {
    }
    double xvalue;
    double valuesz(double a, double b, double p) {
        throw new UnsupportedOperationException
                ("Not yet implemented");
```

```
      }
      public double evalArgs(double a, double b, double p) {
          throw new UnsupportedOperationException
          ("Not yet implemented");
          }
      }
```

LISTING B12. Class to compute Bivariate normal distribution

```
import CoreMath.*;
import static java.lang.Math.*;
import BaseStats.Bivnorm;

public class Probnorm {
private double bivareprob=0.0;

//Based on the normalised data in general, z=x-u/s

        public double ncDisfnc(double zval)//
        THE normal cumulative distribution
        {
   zval= zval>35.0?35.0:zval<-35.0?-35.0:zval;//
   Reasonable limits for very large values
     lgamma ig=new lgamma();
                  return 0.5*(1+ig.errf(zval/sqrt(2)));
        }


public double npdfDisfnc(double zval)// normal pdf
{
        return (exp(-pow(zval,2)/2)/sqrt(2*PI));
}
public double ndfP(double zval)//
normal (cumulative) distribution function (normalised form)..usual
{    lgamma ig=new lgamma();
        return 0.5*(ig.errf(zval/sqrt(2)));
}

public double ncumRange(double x1, double x2)//
Prob of a normal variable having a value in the range x1..x2
{ lgamma ig=new lgamma();
    return (0.5*(ig.errf(x1/sqrt(2.0))-ig.errf(x2/sqrt(2.0))));
}
public double nhazard(double zval)//
normal hazard function
{
        return npdfDisfnc(zval)/ncDisfnc(-zval);
}
public double cnhazard(double zval)//
normal cumulative hazard function
{
        return -log(1-ncDisfnc(zval));
}

public double nsurvfnc(double zval)//
```

```
normal survival function
{
        return 1-ncDisfnc(zval);
}
public double probnsd(double n)//
The probability that a measurement falls within n sd's
{
  lgamma ig=new lgamma();
  return ig.errf(n/sqrt(2.0));
}
public double cBiv(double a, double b, double p)
{

  return 1.0;
}

public double cumBiv(double a, double b, double p)//
Cumulative Bivariate distribution parameters
{// ONLY WHERE a*b*p IS POSITIVE for direct use
  double[][] coeffs={
    {0.24840615,0.10024215},
    {0.39233107,0.48281397},
    {0.21141819,1.0609498},
    {0.033246660,1.7797294},
    {0.00082485334,2.6697604}
  };
  double fa=(a/(sqrt(2*(1-p*p))));
  double fb=(b/(sqrt(2*(1-p*p))));
  double lead=(sqrt(1.0-p*p)/PI);
  double func=0.0;
    for(int i=0;i<5;i++)
    {
    for(int j=0;j<5;j++)
    {
        func+=((coeffs[i][0]*coeffs[j][0])
        *exp(fa*(2*coeffs[i][1]-fa)+fb*(2*coeffs[j][1]-fb)+2
        *p*(coeffs[i][1]-fa)*(coeffs[j][1]-fb)));

    }
    }
  return(lead*func);
  }
}
```

LISTING B13. Class to compute some probability distributions

```
package BaseStats;
import java.io.*;
public class inputmod {
          public static String readString()
          {
                    BufferedReader buffin= new BufferedReader
                    (new InputStreamReader(System.in), 1);
                    String string=" ";
```

```
try
{
        string=buffin.readLine();
}
catch (IOException ex)
{
        System.out.println(ex);
}return string;

        }
        public static int readInt()
        {
                return Integer.parseInt
                (readString());
        }
        public static double readDouble()
        {
                return Double.parseDouble
                (readString());
        }
        public static byte readByte()
        {
                return Byte.parseByte
                (readString());
        }
        public static long readLong()
        {
                return Long.parseLong
                (readString());
        }
        public static float readFloat()
        {
                return Float.parseFloat
                (readString());
        }
        public static short readShort()
        {
                return Short.parseShort
                (readString());
        }
}
.
```

LISTING B14. Class to handle console input/output

References

Besset, D.H. (2001). *Object-Oriented Implementation of Numerical Methods. An introduction with Java and smalltalk.* Morgan Kaufmann.

Press et al. (1987). *Numerical Recipes in C.* Cambridge University Press.

Index